Biomathematics

Volume 17

Mathematical Ecology

An Introduction

Edited by
Thomas G. Hallam and Simon A. Levin

With 84 Figures

Springer-Verlag
Berlin Heidelberg New York
London Paris Tokyo

Thomas G. Hallam
Department of Mathematics and Program in Ecology
University of Tennessee
Knoxville, TN 37996-1300, USA

Simon A. Levin
Section of Ecology & Systematics
and Ecosystems Research Center
Cornell University
Ithaca, NY 14853-0239, USA

Mathematics Subject Classification (1980): 92A15, 92A17, 92A10

ISBN-13:978-3-642-69890-3   e-ISBN-13:978-3-642-69888-0
DOI: 10.1007/978-3-642-69888-0

Library of Congress Cataloging in Publication Data
Mathematical ecology. (Biomathematics; v. 17)
Papers presented during the first weeks of the Autumn Course on Mathematical
Ecology, held at the International Centre for Theoretical Physics, Miramare-Trieste,
Italy, November-December 1982, under the sponsorship of the UNESCO and the
International Atomic Energy Agency.
Bibliography: p. Includes index.
1. Ecology – Mathematical models – Congresses. 2. Ecology – Mathematics – Congresses.
I. Hallam, T.G. (Thomas G.) II. Levin, Simon A. III. Unesco. IV. International
Atomic Energy Agency. V. Autumn Course on Mathematical Ecology
(1982: Miramare (Italy)) VI. Series.
QH541.15.M3M365 1986   574.5'0724   85-27905
ISBN-13:978-3-642-69890-3 (U.S.)

© Springer-Verlag Berlin Heidelberg 1986
Softcover reprint of the hardcover 1st edition 1986

2141/3140-543210

Dedication

*In November and December, 1982, the Autumn Course in
Mathematical Ecology was held at the International Centre for
Theoretical Physics in Trieste, Italy, under the sponsorship of
UNESCO and the International Atomic Energy Agency.
The course participants, who came primarily from developing
countries, were bright, energetic, and highly motivated. The impact
these participants had on the course lectures and, consequently,
upon the development of these lecture notes, was substantial.
This book is respectfully dedicated to each of these participants.*

Acknowledgments

Our gratitude is expressed to the many individuals who influenced the structure and functioning of the 1982 Autumn Course on Mathematical Ecology and the eventual publication of these lecture notes.

Professor Giovanni Vidossich, Director of Mathematics Courses at the International Centre for Theoretical Physics (ICTP), suggested that we organize a course on mathematical ecology and fielded the many requests of the course directors with much personal interest and professional competence. Ms. Sandra Holmes, the ICTP course administrative assistant, was helpful to the participants and lecturers alike, with her assistance often extending beyond expectation. Professors Abdus Salam and Luciano Bertocchi graciously extended many courtesies and the use of the facilities of ICTP to the course. The assistance of all at ICTP was greatly appreciated.

Efforts of many have gone into the production of this book-authors, reviewers, technical staff, and moral supporters. We thank all of these for their contributions to this volume. Special mention is given to the 1983–84 Mathematical Ecology Seminar participants (E. Brandstetter, E. Clark, B. Cochran, J. de Luna, S. Ellner, L. Gross, C. Hom, J. Li, S. Pimm, and G. Sugihara) at the University of Tennessee who spent much time reading and commenting upon manuscripts. Their suggestions have improved the substance and presentation of these notes. Ms. Cindi Blair and Mrs. Evelyn Cook provided editorial help and handled technical typing with flair. Carole Hom contributed considerably to the utility of this book through composition of a comprehensive index.

Finally, Professor Lou Gross deserves special recognition for his invaluable assistance in many facets of the course. Dr. Gross presented many interesting lectures on several diverse areas, spent much time interacting with course participants, organized field trips, and handled local logistics. When one of the course directors (T.G.H.) was relegated to the local hospital by a ruptured appendix, Lou adroitly handled the additional obligations of a course director. Thank you, Lou, for a job well done, and thanks to all who contributed to the course and to these notes.

Thomas G. Hallam and Simon A. Levin

Preface

There is probably no more appropriate location to hold a course on mathematical ecology than Italy, the country of Vito Volterra, a founding father of the subject. The Trieste 1982 Autumn Course on Mathematical Ecology consisted of four weeks of very concentrated scholasticism and aestheticism. The first weeks were devoted to fundamentals and principles of mathematical ecology. A nucleus of the material from the lectures presented during this period constitutes this book.

The final week and a half of the Course was apportioned to the Trieste Research Conference on Mathematical Ecology whose proceedings have been published as Volume 54, Lecture Notes in Biomathematics, Springer-Verlag.

The objectives of the first portion of the course were ambitious and, probably, unattainable. Basic principles of the areas of physiological, population, community, and ecosystem ecology that have solid ecological and mathematical foundations were to be presented. Classical terminology was to be introduced, important fundamental topics were to be developed, some past and some current problems of interest were to be presented, and directions for possible research were to be provided. Due to time constraints, the coverage could not be encyclopedic; many areas covered already have merited treatises of book length. Consequently, preliminary foundation material was covered in some detail, but subject overviews and area syntheses were presented when research frontiers were being discussed. These lecture notes reflect this course philosophy.

Another goal of the course was to maintain a biological perspective with a balance between theoretical aspects and applications. Resource management, infectious diseases and epidemiology, foraging theory, crop yield models, and ecotoxicology were topics discussed in the course as areas where theoretical developments have been applied with vigor. Introductions and surveys of the first two of these topics comprised the final and important unit of this book.

The 1982 Autumn Course on Mathematical Ecology was an intoxicating experience for the lecturers and the participants. We hope that these lecture notes will ferment additional interest in the area of mathematical ecology, and look forward with anticipation to the second course in mathematical ecology, to be held in Trieste November 10–December 12, 1986.

Thomas G. Hallam and Simon A. Levin

Foreword

The wide range of topics covered by mathematical ecology is a reflection of the diversity that one can find throughout the field of ecology.

As discussed in Levin (1987), the study of ecology has its roots in the basic investigations of naturalists, who seek to understand the ecological and evolutionary relationships among species and their relationships to their environment. These studies usually have been retrospective, aimed at understanding how the universe we observe came to be. To explain why we see what we see, we must imbed our studies in a broader context, encompassing both what is and what is not. We must abstract and imagine, and construct a feasible world much bigger than reality; only then can we explain why evolution has taken the course it has. Such explanation must involve some combination of chance and necessity, of determinism and historical accident (Jacob 1982; Levin 1980, 1981).

Studies of this sort have been the mainstays of theoretical ecology, and occupy a major portion of this book. There is, however, a second major branch of ecology, encompassing such topics as the management of renewable and nonrenewable resources, the control of pest species, the protection of the environment against anthropogenic stresses, and the epidemiology of disease. In these applied investigations, the focus is on the future rather than the past, and the objectives are management and control rather than understanding and explanation.

In recent years, more and more of the efforts of theoreticians have been devoted to applied problems – fisheries, forestry, agriculture, epidemiology, and ecotoxicology. As the need to address environmental problems becomes more and more urgent, both in developed and in developing nations, the need for an integration of the basic and applied becomes more and more pressing. The course which gave rise to this book was motivated by a recognition of these needs; and in assembling this text, we have attempted to reflect these needs and the intellectual diversity of the subject.

Simon A. Levin and Thomas G. Hallam

Bibliography

Jacob, F. (1982). The Possible and the Actual. Pantheon Books, New York. 70 + viii pp.
Levin, S.A. (1980). Mathematics, ecology, and ornithology. The Auk 97 (2): 422–25
Levin, S.A. (1981). The role of theoretical ecology in the description and understanding of populations in heterogeneous environments. American Biologist (21): 865–875
Levin, S.A. (1987). Mathematical ecology. McGraw-Hill Encyclopedia of Science and Technology. In press

Table of Contents

Part V. Applied Mathematical Ecology

List of Authors

Jon M. Conrad, Department of Agricultural Economics, Cornell University, Ithaca, NY 14853, USA

James C. Frauenthal, Bell Laboratories, Crawford's Corner Road, Holmdel, NJ 07733, USA

Louis J. Gross, Department of Mathematics and Graduate Program in Ecology, The University of Tennessee, Knoxville, TN 37996-1300, USA

William S.C. Gurney, Department of Applied Physics, University of Strathclyde, Glasgow G4 ON6, Scotland

Thomas G. Hallam, Department of Mathematics and Graduate Program in Ecology, University of Tennessee, Knoxville, TN 37996-1300, USA

Alan Hastings, Department of Mathematics, University of California, Davis, CA 95616, USA

Ray R. Lassiter, Athens Environmental Research Laboratory, U.S. Environmental Protection Agency, College Station Road, Athens, GA 30613, USA

Simon A. Levin, Ecology and Systematics, E 347 Corson Hall, Cornell University, Ithaca, NY 14853-0239, USA

Robert M. May, Biology Department, Princeton University, Princeton, NJ 08544, USA

Roger M. Nisbet, Department of Applied Physics, University of Strathclyde, Glasgow G4 ON6, Scotland

Luigi M. Ricciardi, Dipartimento di Matematica e Applicazioni, Università degli Studi di Napoli, Via Mezzocannone 8-Cap. 80134 Napoli, Italy

Michael Turelli, Department of Genetics, University of California, Davis, CA 95616, USA

Part I. Introduction

Ecology: An Idiosyncratic Overview

Louis J. Gross

The word "ecology" conjures up disparate images. It is part of the rhetoric of political parties, cries for salvation of dwindling species, attempts to increase utilization of renewable resources, determined efforts of scientists to understand our natural world, and blatant commercialization of products from laundry detergents to oil rigs. Aside from its use as an adjective to modify virtually every scientific research field, the jargon "ecologically sound" is used to justify a plethora of so-called development schemes. In daily usage, ecology is regularly confused with environmentalism, and to those in the business community an ecologist may well be considered automatically an obstructionist. Despite all this lack of agreement about definitions, ecology has had a major impact during the past two decades on changing the attitudes of humanity towards our world. Photographs of our glowingly beautiful orb as viewed from the moon notwithstanding, our conscious acceptance of the finiteness and interdependence of processes on this planet is only slowly developing.

Some might well argue that the reduction of the startling complexity of our natural world to a system of scientific thought is a debasement of the amazing intricacies that surround us. Yet anyone only slightly aware of the diversity of life that exists in even the most depauperate environments must have some feeling of awe towards this creation that we often take for granted. One of the major implications of modern ecological research is the immensity of the task before us, if we really wish to comprehend not just the structure of our natural world, but also the mechanisms that have caused that structure to evolve. Our ignorance is manifest. Acceptance of this ignorance, however, should not imply unwillingness to reduce its scope. In many ways, a scientific approach to studying the natural systems in which we live not only increases our knowledge, and thus our ability to avoid mishandling the systems, but also develops our appreciation for its beauty.

What Is It All About?

When discussing the science of ecology, it is quite important to limit properly the scope of the discussion. Indeed, essentially every field of biology could be considered a part of ecology. One of the major impacts of ecology on scientific thought over the last several decades has indeed been the introduction of an ecological perspective to quite disparate fields. Medical science often deals with the environment internal to the human body and its effects on microorganisms;

Biomathematics, Vol. 17, Mathematical Ecology
Edited by T. G. Hallam and S. A. Levin
© Springer-Verlag Berlin Heidelberg 1986

agricultural science, with the influence of environment on managed ecosystems; physiology and neurobiology, with internal chemical environments and cellular response to them; while fields such as botany, zoology, and psychology all include ecological components. The importance of biotic interactions along with abiotic ones has been appreciated only gradually.

Given that ecology touches upon virtually every field of biology, and additionally requires knowledge of many non-biological sciences, how do we go about actually defining the field? Unfortunately, there are as many different definitions as there are textbook writers, a situation which is bound to discourage those with a mathematically-inclined penchant for exactitude. Historically, the term goes back to Henry Thoreau, though Ernst Haeckel first defined ecology as "Haushaltslehre de Nature" – the study of the economics of nature. [See Krebs (1978) for a history of the term.] Some examples of modern definitions of ecology are:

(i) the scientific study of the distribution and abundance of organisms (Krebs, 1978);
(ii) the study of the natural environment, particularly the interrelationships between organisms and their surroundings (Ricklefs, 1980);
(iii) environmental biology (Odum, 1971).

These differences in definitions are also backed up by the very different viewpoints authors use in describing the subject. Krebs (1978) emphasizes distribution and abundance, Remmert (1980) puts more emphasis on physiological aspects, Odum (1971) pursues a systems-theory approach, while Emlen (1973) attempts a synthesis from an evolutionary perspective. Perhaps because of the highly diverse subject matter of the field and the non-agreement on a definition, it has become common for ecology to be split up into a large number of subdisciplines. Examples include physiological ecology, behavioral ecology, population ecology, human ecology, evolutionary ecology, systems ecology, and many others. Despite all this specializing, there is general agreement on the basic subject matter included in the field. For our purposes, I shall group this into four areas – physiological, population, community, and ecosystem. Below, I give a brief overview of the questions addressed by each of these divisions. It should be kept in mind that these areas are in no sense independent – ecological research is highly interdisciplinary and often requires perspectives from different levels of the natural system under consideration.

It is worthwhile throughout this discussion to be aware that at this point in its development, the science of ecology has only one underlying paradigm – that of the theory of evolution. An evolutionary perspective, although often not explicitly stated, guides much of the current thinking on population and community structuring, organism adaptations to the environment, and ecosystem functioning. Although there are divergent opinions as to the mechanisms by which evolution acts [see for example, D. S. Wilson (1978) for one such alternative to the neo-Darwinian mechanisms of natural selection], this in no way reduces the importance of evolutionary theory in formulating and evaluating hypotheses about ecological phenomena. The laws of chemistry and physics are also fundamental to an understanding of many ecological processes, especially at the physiological level; however, they do little to provide any overall structure to the field. In many respects, ecology still lacks the capability to extend the results of a particular ecological study to similar systems in other regions of the world, or to species or conditions other than those investigated. Potentially, ecological theory may give us some indication as to how representative a particular observation or experiment is of the world in general. At present, except under fairly restricted circumstances, our theories simply cannot handle the complexities of the real world. The trade-offs between realism, precision, and generality (Levins, 1968) are particularly evident in ecology. There is certainly much room for improvement of the theories, but real advances in basic understanding will be made only through coupling theory with careful observation and experiment.

Physiological Ecology

Generally, physiological ecology refers to the study of the direct effects of the physical environment on individual organisms. Its emphasis is on how factors such as temperature, water availability, radiant energy, and wind affect the distribution of organisms, how the organisms adapt to variations in these factors within their lifespan, and how these factors have produced selective forces which bring about evolutionary change. Although abiotic factors are often the ones under study, organisms have the capability to modify their environment and thus biotic interactions also come into play. The structuring of the overstory in a forest canopy affects the environment of the understory and an animal may through movement modify its environment, an example being the shuttling behavior of lizards for the purpose of thermoregulation. It is this feedback between organism and environment, operating on time scales from fractions of seconds to years depending upon the organism and the particular process, which serves as the focus for much research in this area.

The maintenance of homeostasis, by which we mean that certain metabolic processes are regulated to stay within a range that the individual can tolerate, is a central problem that is faced by all lifeforms. Examples are the maintenance of water saturated conditions within a leaf, proper osmotic potential in salt-water fish, and the relatively fixed body temperatures in homeotherms. One goal of research is to determine what physiological mechanisms limit the tolerance of individuals for environmental extremes and how these tolerance limits change due to genetic and developmental influences. Because of this interest in tolerance limits, a very common research tool is to study organisms in very harsh environments, in which the physiological constraints on the organism and the adaptations to cope with environmental extremes are quite evident. Studies at timberline (Tranquillini, 1979), or in hot, dry deserts (Osmond et al., 1980) are examples of this approach. Homeostatic mechanisms ultimately have their origin in biochemical processes (Hochachka and Somero, 1973), and although this is recognized by ecologists, it is relatively rare for studies to be carried down to the biochemical level. The level of integration considered is the organ or whole individual.

To a great extent physiological ecology deals with biophysics. The reductionist approach to this area, in which all environmental influences acting on an organism are analyzed according to established laws of physics and chemistry, has been quite fruitful. [See Campbell (1977) or Gates (1980) for a detailed analysis.] Here, for example, the heat and energy loads on an organism are analyzed by taking into account conduction, convection, radiation, and evaporation. This leads to predictions based upon physical principles about which leaf sizes and shapes, how much fur or feathers, and what type of behavior are to be expected in particular environments. Another process which is heavily researched in this area is photosynthesis, with emphasis on how light, temperature, humidity, and atmospheric concentrations of carbon dioxide affect the capacity of a plant to transform solar energy into organic carbon compounds. A combination of laboratory and field observations are usually undertaken. Perhaps the most difficult aspect of biophysical work is dealing with the multiplicity of factors which often affect any particular physiological process. Establishing organism response to changes in any single environmental factor may be easily accomplished, but coupling this with responses to other factors requires assumptions about the additive or multiplicative effects of various inputs. Since real environments are quite dynamic on time scales within which physiological processes respond, the step of moving from experimentally controlled environments to those which organisms actually encounter is one in which mathematical modelling can be quite useful. [See Hesketh and Jones (1980) for a detailed study of this approach to photosynthesis.]

Inherent in the reductionist approach is that whole organism response can be deduced from analysis of the processes which make it up. In this view the whole is indeed the sum of the parts. An alternative approach is the holistic one, which implies that there are properties of natural systems – the jargon is "emergent properties" – which arise due to the structure of the system and could not be predicted from knowledge of the subcomponents alone. Although usually associated with ecological work at the community or ecosystem level, this concept also comes into play at the physiological level, especially for evolutionary questions. It may be virtually impossible to consider all the environmental factors acting on an individual, and so concentration occurs on the one or two influences, called "limiting factors," which are found to have the most impact on the organism. The reductionist approach thus has its limitations, and an alternative is to consider phenotypic strategies for adaptation to environment. Here, within constraints set by the known biophysics, evolution is viewed as selecting individuals with phenotypes which are best suited for particular environments. A criterion for what is meant by "best" must be defined; for example, net photosynthetic gains in a leaf, or rate of food intake by foraging animals. The optimum phenotype is then chosen according to this criterion, subject to physiological constraints, and then comparison is made with actual observations. The assumption is that there is sufficient genetic variation to enable evolution to bring about this optimum, and that the proper criterion has been chosen. This is subject to controversy (Lewontin, 1978), but the approach has led to a body of theory that some consider very useful (Pyke et al., 1977). Certainly, the strategic approach serves as a way to tie together what we know about biophysics with the logical evolutionary consequences of that knowledge.

The results of physiological ecology find application to crop and forest growth analysis (de Wit and Goudriaan, 1978). Knowledge of plant response to environmental factors plays a major role in current research on such problems as assimilate partitioning, the effects of fertilization and irrigation, the energy potential of biomass, and the capacity of crop varieties to adapt to changing environmental conditions. Knowledge from the physiological level is critical to the construction of valid systems models of not only single-species crops, but also natural communities. Since physiological experiments are generally more readily carried out than those at the community level, in some senses we have a much better understanding of processes at the level of the individual than at that of the community. However, we are still far from a complete understanding of even the most basic physiological processes.

Population Ecology

At this level, the science investigates the dynamics and structure of populations of a given species. The definition of population depends upon the scale of interest. A field ecologist may study all the fish of a species within a particular lake, a system of lakes and rivers, or the whole world. In theoretical analyses, this variety of scales is often ignored, but they do lead to quite different assumptions about the mechanisms controlling population growth and decay. The chief questions of interest in population ecology concern how a population is structured in terms of age, size, and genotype; how this structure changes both temporally and spatially; and what factors external and internal to the population regulate this structure. [For an excellent overview of the field, see Hutchinson (1978).]

From the point of view of a field biologist who may be interested in establishing the size and structure of a population, a standard practice is to estimate the death and birth rates for individuals in different classes, the most usual being age classes. This allows the construction of a life table, consisting of a number of variables as functions of age, including survivorship, mortality rate, and expectation of further life, which describe the mortality schedule of the population. When this is combined with a fecundity schedule, it is possible to make predictions about future age structures, given an initial age distribution. The applicable techniques are the same as those in human demography (Keyfitz, 1968), the major difficulty being the statistical aspects (Poole, 1978). For plant populations, fecundity and mortality often depend more upon size than age, and although there are complications introduced due to non-linear relationships between these, size structured methods have been applied (Caswell and Werner, 1978). The development rate of an individual may often be environmentally dependent, and thus transformations to such variables as degree days are common in studies of agricultural dynamics, along with other physiological time scales for insect development (Curry et al., 1978).

Population biology is perhaps the most mathematically developed area of ecology, with a long history of interest by mathematicians in the problems associated with the dynamics of populations. Early studies of the population fluctuations of small mammals and of a variety of organisms in the laboratory lent themselves well to a mathematical formulation. Far and away the greatest emphasis has been on animal populations, leading to the development of fairly sophisticated models in both discrete and continuous time (Freedman, 1980), with and without delays due to maturation times (Cushing, 1974), along with stochastic models (Ludwig, 1974). A great deal of recent research deals with the effects of spatially non-uniform environments (Levin, 1976) and the diffusive spread of organisms (Okubo, 1980). Despite all the theory which has been developed, it is unclear how predictive the models are for situations other than the laboratory (Nisbet and Gurney, 1982). Part of the reason for this is the temporal heterogeneity associated with any natural environment. There is a long-standing controversy within population biology regarding whether populations are mostly regulated by density-independent factors (often considered abiotic) or density-dependent ones (often due to biotic interaction). The theories go a long way towards analyzing the effects of density-dependent factors, but provide much less information on abiotic factors, due to their complexity and variability. It is at this point that input from the physiological studies is needed to construct realistic theories of population structure. This combination of physiological ecology with demography serves as the focus of the rapidly growing study of plant population biology (Harper, 1977) a field which has developed relatively slowly in comparison to animal ecology at this level.

Another area of great current interest is the study of life histories, by which is meant the reproduction, growth, and senescence patterns exhibited by a population, which determine its long-run behavior. In a similar manner to the strategic approach to studying physiological adaptations mentioned earlier, the idea here is to consider, as variables under genetic control, such factors as time of first reproduction, number of offspring per clutch, number of reproductive phases per lifespan, and energy expended per offspring. It is then assumed that evolution acts to maximize some measure of population growth, for example the mean population fitness. Predictions may then be made about the circumstances in which certain life histories are to be expected. [See Stearns (1976) for a review.] Again, this approach can be subject to the criticism levelled at optimization in evolutionary theory, but it can be tied into a population genetics framework (Roughgarden, 1979) and has proven very useful in deriving hypotheses about ecological phenomena which are testable.

Behavioral ecology is another field which has become intensively studied in recent years, the goals being to analyze specific behavioral traits of animals in light of ecological constraints. The evolution of co-operation within social groups, territoriality, parental care, and dominance hierarchies may be viewed from a game theory perspective, leading to the theory of evolutionarily stable strategies (Maynard Smith, 1982). Conflicts of interest (as measured in terms of an organism's fitness) are settled in this theory through an evolutionary dynamics towards a Nash equilibrium solution of a game, meaning a solution which is stable under perturbations of the strategies played. Again, controversy has arisen regarding how appropriate this solution is to evolutionary biology (Lewontin, 1978), but often it can be demonstrated analytically that this solution is consistent with calculations from population genetics which track the full genetic trajectory of the situation (Maynard Smith, 1982).

The approaches of population biology have great applicability to wildlife management practices, especially regarding the problem of setting harvesting levels so as to attempt to maintain optimum yield (Clark, 1976). Models of population dynamics are central to the understanding and application of optimal control techniques for containing pest outbreaks (Vincent, 1981). Large portions of the sciences of forestry, fishery biology, and wildlife management rest upon the theories of population structuring, so that improvements in our theories will lead hopefully to better resource utilization.

Community Ecology

When all the populations of species within a prescribed region are considered as an interacting unit, we call this a community. Although intuitively it is relatively easy to perceive different communities, such as prairie grassland, mixed-deciduous forest, or ephemeral pond, the actual classification of community types historically involved vast effort on the part of many ecologists, often with quite a bit of acrimony. Part of the contention involved whether a community should be perceived as a "superorganism" with its own emergent properties not derivable from those of the populations which make it up. The debate is hardly settled, but it is probably fair to say that most ecologists view the superorganism concept as generally untenable, though there are indeed properties of a community, such as succession, which are inherent in the interactions between populations, and not in the populations themselves. For a discussion of the history of classification, see Whittaker (1962), or Golley (1977). A wide variety of methods are regularly used by ecologists, particularly analysis of species changes along an environmental gradient and numerous multivariate statistical techniques (Gauch, 1982), in attempts to elucidate correlations between compositional changes and environmental variations within the community.

The central problems of modern community ecology are the studies of diversity and stability, trophic structure, and community dynamics. In all of these, it is common to consider just isolated parts of a particular community. Thus in a predator-prey system such as the lynx-hare, effects of the hare on grass or of lynx droppings on fly populations are ignored. This reduction of a real world community to an isolated portion is utilized in field, laboratory, and modelling studies. Its justification is based upon the assumption that the major impacts on the populations you consider occur completely within the sub-system chosen. This very much depends upon the scale of the problem under investigation, for if other biotic influences change on a slow time scale relative to the populations you are considering, then they may be viewed as fixed. For example, in models of spruce-budworm outbreaks, the response of the trees is much slower than the insects, and thus the dimensionality of the problem is reduced (Ludwig et al., 1978).

A large theory of community dynamics has developed through the analysis of systems of ordinary differential equations to describe population density changes caused by competition, commensalism, and predator-prey types of interactions. The theory for two-species systems is quite well developed, though higher order systems involve considerably greater complications and it isn't clear how results from two-species systems carry over to multi-species situations. Although full dynamic solutions can be derived for many two-species cases, often in higher dimensions one is limited to an analysis of the equilibria. Thus there has been great emphasis placed on studying the local asymptotic stability of these equilibria. The biological relevance of this definition of stability is quite debatable (Lewontin, 1969); although due to ease of analysis, it is often the only form of stability investigated. A number of other measures of stability have been suggested and investigated, including the persistence of the system (no species approach extinction) and resilience (time constant for return to equilibrium following a perturbation) (see Maynard Smith, 1974). Associated with the investigation of stability is the question of whether stability is enhanced by higher diversity, meaning the number of species in a community. A typical example of this would be the decrease in diversity as one moves away from the tropics. Results from certain multi-species models indicate that stability may be decreased by increased diversity (May, 1973), although the area is still quite murky, partially due to the quite different definitions of stability used by various investigators (Robinson and Valentine, 1979).

The manner in which multi-species assemblages are structured as a hierarchy of energy flows within the system revolves around its trophic structure. This is based upon who eats whom, and the relative proportions of the populations which are plants, herbivores, carnivores, omnivores, and detritivores. Within any community the trophic interactions form a complex pattern, often more of a web than the classical food chain. Analysis of the structure of these food webs (Cohen, 1978; Pimm, 1982) has led to some generalizations about them, such as the ability to represent them as an interval graph. The analysis of such webs is closely related to theoretical studies of niche partitioning, meaning the manner in which limiting resources are partitioned among populations, each of which has its own functional roles, or niches, within the community. See Whittaker and Levin (1975) for an exposition of the niche concept.

Results from community ecology can play a major role in settling such issues as the proper site and pattern of wildlife refuges in order to preserve endangered species (Terborgh, 1975). In this view, the results of island biogeography, which attempts to relate species diversity to such factors as island or refuge area, and inter-island distances (MacArthur and Wilson, 1967), provides a framework for the analysis of this problem. Although arguments can be made about the limitations of our current ability to apply such theories (Simberloff and Abele, 1976), the theory serves a focal point for further analysis. Community theory may be ultimately quite useful also in providing predictive estimates of the potential for pest outbreaks (Conway and Murdie, 1972), an area of great importance in agriculture.

Ecosystems

An ecosystem may be defined as the collection of communities within some region, taken together with the environmental influences upon them. Thus the communities contained within a certain watershed, bay, or old field could all be considered as ecosystems, particularly if the emphasis of their analysis concerns biotic-environment interactions and feedbacks throughout the system. Major questions of interest in ecosystem studies concern the flow of nutrients, energy, and biomass through the system. Coupled with this are the patterns of successional change in the system through time, and how these affect such factors as primary and secondary production by the system. Here primary production refers to the energy content of materials produced by plants from solar energy inputs, while secondary production refers to energy obtained by consumers from plants used for growth and reproduction. See Whittaker (1970) for estimates of the productivity of a variety of ecosystems.

It is perhaps worthwhile noting at this point that the division between ecosystem and community ecology is not a sharp one and some authors would include most of the topics mentioned in the above paragraph as being community metabolism (Krebs, 1978). Despite this, I choose to make this a separate category of ecological research because the approach to answering the above questions is mostly that of systems analysis. By this I mean that the system is broken down, sometimes quite arbitrarily, into a set of compartments; for example primary producers, consumers, and detritivores in a simple ecosystem model. The flows of the currency of interest, such as energy, between compartments are then estimated, and a model constructed. The model is analyzed, tested against an independent set of data, and then utilized to answer questions of interest. This approach served as the basis of the International Biological Program, which attempted to collect immense quantities of data on particular ecosystems around the world, utilize computer-based systems models to collate all this information, and then use the model to analyze the system. Although the IBP definitely aided our understanding of many natural systems, the original hopes of simulating realistically a system via a computer were generally unrealized. Even the simplest natural systems are far too complex for the interactions to be suitably well understood to meet the demands of model construction. Even if adequate data were available to specify the functional forms for the interactions in the models, the enormous number of parameters necessary are quite difficult to specify even approximately with the available data. There is thus the danger that virtually any result desired can be obtained from the models, just by choosing appropriate parameter values. Despite this, IBP resulted in a collection of very fascinating studies, for example on the convergence of form and structure in mediterranean-type ecosystems around the globe (Miller, 1981). Systems approaches have been quite useful tools in forming an underlying structure for the study of a wide variety of ecosystems problems. See the series of books by Patten (1971).

Despite the difficulty imposed on ecosystem studies due to the great complexity of the interactions involved, it is at this level that the great number of public policy questions concerning human activities and their ecological impacts must be answered. The effects of environmental toxicants, construction of power plants which use rivers and lakes for cooling, mining in wilderness areas, and hosts of other actions of society pose challenging questions for ecologists. Unfortunately, the answers are required now and there is rarely time for adequate background research to be undertaken. There is probably some doubt that we shall ever be able to understand completely even a single natural system, but it is clear that with adequate time for study we shall be able to give, if not perfect, then at least rough estimates as to the effects of particular human activities on natural systems. One example is that of clear cutting a forest, the effects of which only become apparent after many years of study (Bormann and Likens, 1979). No matter how complex a computer model is constructed, it is worthless without the understanding of the underlying processes that can only be determined from field and laboratory studies.

Conclusions and the Role of Mathematics

The above review is idiosyncratic in that I have purposely emphasized some areas and left others virtually unmentioned. I have definitely slanted my remarks towards theoretical ecology, and have not discussed field techniques and statistical methods. There is certainly no lack of fascinating statistical problems which crop up in ecological studies; but I consider them to be a part of statistics, not ecology. I have not tried to give anything like a complete overview of the subject, but wished only to mention briefly those areas that I consider essential and about which a theoretician should be informed. My own biases have certainly crept in, and one should keep in mind that ecology is a highly contentious field. Others will undoubtedly argue with my choice of topics and my comments.

I view mathematical modelling as having three potential uses, not necessarily independent, in any field. A model may be descriptive, in the sense that it synthesizes the available information on a process with no real attempt to explain the underlying mechanism. An example would be a regression fit to data, a model in the statistical sense. A model could also be explanatory in that it makes certain underlying assumptions about the process under study, and derives the logical implications of those assumptions. An example would be the effect of heat loading on a leaf in which the physics of conduction, convection, and evaporation heat loss are applied to explain leaf temperatures. Thirdly, a model may be constructed for the purpose of predicting the response of the system to factors which haven't been observed. An example would be determining the effects of the rise in temperature in a lake, caused by a power plant, on fish populations. All of these uses come into play in ecological applications, though I think it is fair to say that it is the explanatory and predictive aspects that are potentially the most important. To date, most models are explanatory; and often through their construction, they point out to us areas which need further study. This tendency for the process of model construction to display clearly our ignorance of a certain aspect of the system under study is very useful. For example, the most complex physiologically-based crop growth models still cannot compete with relatively simple regression models in terms of their ability to forecast yields. However, the mechanistic models have clearly aided our understanding of crop systems merely by focussing our attention on certain processes – respiration is one – which require more study.

In my view, there are two different possible attitudes towards applying mathematics to biological problems. In one of these, a model already constructed is either analyzed further or extensions made to it, such as by considering time-dependent solutions or adding stochastic variation. The emphasis is mathematical with little or no attempt to tie together the mathematical results with biological observation. It is essentially a mathematical exercise, with a bit of biological justification. I refer to this as biomathematics, though I realize there are many biomathematics departments in which this is not the approach. In comparison, consider the situation in which a biological problem is investigated with mathematical tools, but the mathematics is considered of purely secondary interest, as a means to an end. Here the mathematical techniques which are utilized are not specified beforehand by the expertise of the researcher, but are chosen as appropriate for the particular biological problem. The objective is to derive biological conclusions which are testable, not to develop elegant mathematics, though that may indeed occur. This I refer to as mathematical biology. A great deal of mathematical work in ecology has been biomathematics; and, although the results can be fun as mathematics, too often they are either irrelevant to biological problems, or completely outside the realm of real-world testability. Very few ecological models can stand up to intense scrutiny of their assumptions, simply because of our lack of knowledge. If the models are constructed with a firm basis in biological fact, however, there is the likelihood that their analysis will lead to results which are testable. Good applied mathematics is also good science.

Given that one is mathematically trained, and the objective is to do mathematical biology rather than biomathematics, how does one proceed? In my experience, there are two routes: either learn the necessary biology, or find a biologist who is willing to collaborate. Each path has its advantages, and to be truly successful, a bit of both is undoubtedly best. Becoming an expert in a certain biological discipline can be extremely time-consuming, but it is probably the best way to ensure that any modelling efforts undertaken are firmly rooted in observation. When one immerses oneself in a field, the major open questions become apparent, and the biological intuition developed aids the rather iterative process of model construction. Collaboration has the benefit of not requiring as much time spent learning the biology; but to be effective, a certain minimum effort is necessary on the mathematician's part to learn at least the basics. As there are few who have the ability or desire to become biologists (in my definition a biologist is someone who can construct a real-world counterexample to any statement a theoretician makes), collaborative work is essential, even for those with biological expertise in one discipline. Happily I have found a very positive attitude growing among biologists, especially ecologists, towards the utility of mathematics in aiding our understanding of ecological systems. This can certainly be enhanced by further efforts on the part of theoreticians to learn the relevant biology and view modelling as a path to a biological end, not a mathematical one.

References

Bormann, F.H., Likens, G.E. (1979). Pattern and Process in a Forested Ecosystem. Springer-Verlag, N.Y.

Campbell, G.S. (1977). An Introduction to Environmental Biophysics. Springer-Verlag, N.Y.

Caswell, H., Werner, P.A. (1978). Transient behavior and life history analysis of teasel (Dipsacus sylvestrus Huds.). Ecology 59: 53–66

Clark, C. (1976). Mathematical Bioeconomics. Wiley-Interscience, N.Y.

Cohen, J.E. (1978). Food Webs and Niche Space. Princeton Univ. Press, Princeton, N.J.

Conway, G.R., Murdie, G. (1972). Population models as a basis for pest control. Pages 195–213 in J.N.R. Jeffers (ed.), Mathematical Models in Ecology. Blackwell, Oxford

Curry, G.L., Feldman, R.M., Sharpe, P.J.H. (1978). Foundations of stochastic development. J. Theor. Biol. 74: 397–410

Cushing, J.M. (1974). Integrodifferential Equations and Delay Models in Population Dynamics. Springer-Verlag, Heidelberg

de Wit, C.J., Goudriaan, J. (1978). Simulation of ecological processes. Centre for Agric. Publ. & Docum., Wageningen

Emlen, J.M. (1973). Ecology: An Evolutionary Approach. Addison-Wesley, Reading, Mass.

Freedman, H.I. (1980). Deterministic Mathematical Models in Population Biology. Marcel-Dekker, N.Y.

Gates, D.M. (1980). Biophysical Ecology. Springer-Verlag, N.Y.

Gauch, H.G., Jr. (1982). Multivariate Analysis in Community Ecology. Cambridge Univ. Press, Cambridge

Golley, F.B. (ed.) (1977). Ecological Succession. Dowden, Hutchinson & Ross, Stroudsburg, P.A.

Harper, J.L. (1977). Population Biology of Plants. Academic Press, London

Hesketh, J.D., Jones, J.W. (eds.) (1980). Predicting Photosynthesis for Ecosystem Models. Vol. I & II. CRC Press, Boca Raton, Florida

Hochachka, P.W., Somero, G.N. (1973). Strategies of Biochemical Adaptation. Saunders, Philadelphia, PA.

Hutchinson, G.E. (1978). An Introduction to Population Ecology. Yale Univ. Press, Provincetown, RI.

Keyfitz, N. (1968). Introduction to the Mathematics of Population. Addison-Wesley, Reading, Mass.

Krebs, C.J. (1972). Ecology: The Experimental Analysis of Distribution and Abundance. Harper & Row, N.Y.

Levin, S.A. (1976). Population dynamic models in heterogeneous environments. Ann. Rev. Ecol. System. 7: 287–310

Levins, R. (1968). Evolution in Changing Environments. Princeton Univ. Press, Princeton, N.J.

Lewontin, R.C. (1969). The Meaning of Stability. Brookhaven Symp. Biol. 22: 13–24

Lewontin, R.C. (1978). Fitness, survival, and optimality. Pages 3–21 in D.H. Horn, R. Mitchell, & G.R. Stairs (eds.), Analysis of Ecological Systems. Ohio St. Univ. Press, Columbus, OH.

Ludwig, D. (1974). Stochastic Population Theories. Springer-Verlag, Heidelberg

Ludwig, D., Jones, D.D., Holling, C.S. (1978). Qualitative analysis of insect outbreak systems: the spruce budworm and forest. J. Anim. Ecol. 47: 315–332

MacArthur, R.H., Wilson, E.O. (1967). The Theory of Island Biogeography. Princeton, Univ. Press, Princeton, N.J.

May, R.M. (1974). Stability and Complexity in Model Ecosystems. Princeton Univ. Press, Princeton, N.J.

Maynard Smith, J. (1974). Models in Ecology. Cambridge Univ. Press, Cambridge

Maynard Smith, J. (1982). Evolution and the Theory of Games. Cambridge Univ. Press, Cambridge

Miller, P.C. (ed.) (1981). Resource Use by Chapporal and Mattoral. A Comparison of Vegetation Function in Two Mediterranean Type Ecosystems. Springer-Verlag, N.Y.

Nisbet, R.M., Gurney, W.S.C. (1982). Modelling Fluctuating Populations. Wiley, Chichester

Odum, E.P. (1971). Fundamentals of Ecology. Saunders, Philadelphia, PA.

Okubo, A. (1980). Diffusion and Ecological Problems: Mathematical Models. Springer-Verlag, Heidelberg

Osmond, C.B., Bjorkman, O., Anderson, D.J. (1980). Physiological Processes in Plant Ecology – Toward a Synthesis with *Atriplex*. Springer-Verlag, N.Y.

Patten, B.C. (1971–76). Systems Analysis and Simulation in Ecology. Vol. I–IV. Academic Press, N.Y.

Pimm, S.L. (1982). Food Webs. Chapman & Hall, N.Y.

Poole, R.W. (1978). The statistical prediction of population fluctuations. Ann. Rev. Ecol. System. 9: 427–448

Pyke, G.H., Pulliam, H.R., Charnov, E.L. (1977). Optimal foraging: a selective review of theory and tests. Quart. Rev. Biol. 52: 137–154

Remmert, H. (1980). Ecology. Springer-Verlag, Berlin

Ricklefs, R.E. (1980). Ecology. Nelson and Sons, Middlesex

Robinson, J.V., Valentine, W.D. (1979). The concepts of elasticity, invulnerability and invadeability. J. theor. Biol. 81: 91–104

Roughgarden, J. (1979). Theory of population genetics and evolutionary ecology. MacMillan, N.Y.

Simberloff, D.S., Abele, L.G. (1976). Island biogeography theory and conservation practice. Science 191: 285–286

Stearns, S.C. (1976). Life-history tactics: a review of the ideas. Quart. Rev. Biol. 51: 3–46

Terborgh, J. (1975). Faunal equilibrium and the design of wildlife preserves. P. 369–380 in F.B. Golley & E. Medina (eds.), Tropical ecological systems. Springer-Verlag, N.Y.

Tranquillini, W. (1979). Physiological ecology of the alpine timberline. Springer-Verlag, Berlin

Vincent, T.L. (1981). Proc. of a workshop on control theory applied to renewable resource management and ecology. Springer-Verlag, Heidelberg

Whittaker, R.H. (1962). Classification of natural communities. Bot. Rev. 28: 1–239

Whittaker, R.H. (1970). Communities and ecosystems. MacMillan, N.Y.

Whittaker, R.H., Levin, S.A. (eds.) (1975). Niche: theory and application. Dowden, Hutchinson & Ross, Stroudsberg, PA

Wilson, D.S. (1978). The natural selection of populations and communities. Benjamin/Cummings, Menlo Park, Calif.

Part II. Physiological and Behavioral Ecology

Biophysical Ecology:
An Introduction to Organism Response to Environment

Louis J. Gross

Methodology

The direct effects of environmental factors on the physiology and behavior of individuals comprise the building blocks upon which ecology is based. In the reductionist view (Gates, 1980), all ecological interactions may be carried down to the level of the cell and individual through careful analysis of physical and chemical processes. This approach, involving the detailed analysis of the underlying processes in a natural system, is in practice limited by the complexity of biological systems. Interactions may be viewed at levels from the molecular, through tissue, organ, and individual, and beyond to population, community and ecosystem. The viewpoint of biophysical ecology is that general ecological perspectives may be obtained by considering the effects of environments on individuals and that knowledge of processes at this level is critical to understanding the structure of populations and communities. The known results from scientific fields as varied as chemical kinetics, fluid flow, thermodynamics, atmospheric physics, and soil mechanics are applied to structure a cohesive framework for analysing organism response to environmental factors. A great advantage of this framework is that it applies independent of the type of organism, animal or plant, terrestrial, aquatic or arboreal. The same physical principles underly all of nature. In practice, of course, this advantage is ameliorated by the great range of environments which organisms inhabit, each providing its own unique challenges for organism survival. This is not meant to imply that approaches at higher levels of integration than the organism may not also be widely applicable across organism lines. However, there is little agreement about the universality of higher level theories, whereas physical and chemical laws are accepted (and often envied) by ecologists.

Three alternative approaches to ecological investigation-descriptive, functional, and evolutionary – have been mentioned by Krebs (1978). Descriptive methodology is natural history in the sense of merely observing characteristics of organisms in the field without any explicit attempt to explain them. A functional approach attempts to determine the proximate causes of responses of organisms, for example through correlations between environments and species distributions. The evolutionary approach attempts to study the ultimate causes of organism response through the dynamics of the evolutionary process. It is only within the past few decades that the emphasis of ecological studies has shifted from the descriptive to the functional and evolutionary. For the study of physiological

Biomathematics, Vol. 17, Mathematical Ecology
Edited by T. G. Hallam and S. A. Levin
© Springer-Verlag Berlin Heidelberg 1986

adaptations, Calow and Townsend (1981) describe two methodologies which are intermediary between the above. The first of these, a posteriori, takes as given that certain organism characters are observed to be correlated with particular environments, and then asks what it is about these characters which aid the individual in surviving and reproducing in its environment. For example, it may be observed that leaf sizes in hot desert conditions are generally much smaller than in more moist environments. Given this, one may attempt to explain it in terms of heat loading on the leaf as it affects leaf photosynthetic rate and water status. The second methodology, a priori, assumes a range of potential characteristics are accessible to the organism and asks which of these is the most appropriate for certain ecological circumstances. Thus, in the leaf example, the approach would be to assume that a range of leaf sizes is possible and then determine, according to some criterion of fitness, which sizes would be most adaptive in dry versus moist habitats.

The a posteriori approach, consisting of a combination of the descriptive and functional views, inherently has the potential to view all characteristics of organisms as adaptive. By an adaptation is meant any character of an individual which gives that individual a selective advantage relative to individuals without that character. As has been pointed out (Maynard Smith, 1978; Gould and Lewontin, 1979), however, there is no reason to believe that all traits are adaptive. Evolution may lag behind and be out of phase with a changing environment, and certain traits may be merely a side-effect of other traits which are adaptive. Care must therefore be taken in assigning adaptive significance to traits observed to be correlated with certain environments. Another difficulty surrounds the multiplicity of selective forces operating in any particular environment. It may be very difficult to separate out any one factor to use as a criterion for selection. For example, although one might expect the distribution of fat in vertebrates to be determined by the needs for thermoregulation, a critical analysis shows such factors as buoyancy, sexual and social signalling, and locomotory mechanics to be much more significant (Pond, 1978).

The proper choice of fitness criteria is also a difficulty in the a priori approach, which is an amalgam of the functional and evolutionary methods. Evolutionary mechanisms at the level of population genetics are not considered, but rather it is assumed that phenotypic traits can be closely linked to fitness (in the genetics sense). Also the phenotypes which are optimal for the chosen selection criterion are assumed feasible in that they are on an evolutionary trajectory which could be reached from an ancestor population. This is accomplished by building into the approach constraints which limit the allowable set of phenotypes. Determining what the appropriate constraints should be is very difficult – usually one resorts to the use of observations to specify constraints, thus intermingling the a posteriori method with the a priori. These assumptions are all open to criticism (Lewontin, 1978; but when care is taken, this approach is often preferable to the a posteriori one (Calow and Townsend, 1981).

My procedure in what follows is to first give a brief review of the environmental factors most crucial to organism behavior. I will then discuss several cases of applications of biophysical techniques to investigations of adaptation. Hopefully these will give some indication of the diversity of problems and the highly interdisciplinary efforts needed to understand them. The approach will be a priori as discussed above. I draw heavily from the books of Campbell (1977), Gates (1980), and Townsend and Calow (1981). It will be noted that I do not particularly emphasize the mathematics involved in these analyses. Often the analysis is either straight-forward calculus, or is beyond the tractability of analytic methods. Quite frequently in my view, it is the simplest models which give us the greatest insight, though the more complex ones may enable us to synthesize and analyze the effects of multiple interacting factors. As I emphasized in the introductory chapter in this volume, it is important not to lose sight of the biological goals of the research while tramping joyfully through the mathematical woods which surround them.

Environmental Factors

I. Radiation

Essentially all the energy utilized by life forms on this planet is derived from solar radiation. The energy incident upon the earth's upper atmosphere, the so-called solar constant, is approximately $1360\,\mathrm{Wm}^{-2}$, although this is subject to variation. The energy actually incident upon the earth's surface may be determined from a very complicated combination of factors such as time of year, location, cloud cover, elevation, and time of day. Instantaneous values of global radiation (the sum of direct beam solar radiation and diffuse skylight) on a horizontal surface may vary from 50 to $1250\,\mathrm{Wm}^{-2}$, depending mainly upon sky conditions (Gates, 1980, p. 127). Much of the loss of radiation is due to refraction and diffraction in the high atmosphere, reflection from clouds, and absorption by the atmosphere. Not only does this large variation in radiation affect the light conditions at a site, but temperature, windiness, and moisture are also all affected. Total global radiation at a fixed site is highly variable with a dynamic pattern due to cloud and atmospheric turbidity changes induced on top of diurnal patterns. These dynamic patterns can be quite significant determinants of organism behavior. See for example Christian et al. (1983), which uses biophysical models to examine seasonal changes in body temperature and habitat use for an iguana.

   Global radiation has direct effects on primary productivity at a site. The efficiency of gross primary productivity, measured as the world average percent of incident radiation converted to biomass, is approximately 0.2%. Considered on a net basis, when respiratory losses are subtracted, this is reduced to 0.09%. The maximum potential gross primary productivity of a managed crop is estimated to be 8%, while the average annual value for U.S. crops is about 1.7%. The conversion efficiency of natural ecosystems varies from about 4.4% for a tropical forest, to 0.04% for a Nevada desert, all measured on a gross basis (Gates, 1980, p. 135). As can be seen, a very small fraction of incident energy is actually productively captured by plants, partially due to the inefficiencies of the photosynthetic process, and in some systems the incomplete capture of utilizable energy.

For many biophysical questions not only is the total radiation at a site important, but also its spectral composition. Although the spectrum of energy incident on the atmosphere is essentially that of a black body and follows Planck's law, the spectrum is markedly changed by the absorption and scattering characteristics of the atmosphere, particularly in the infra-red. The minor gases of water vapor, CO_2, and ozone absorb strongly in the infra-red, and this is one reason why there is concern about the potential increases in atmospheric CO_2. Since oxygen and ozone concentrations control the amount of ultra-violet radiation below 290 nm wavelength which reaches the lower atmosphere, the potential degradation effects of fluorocarbons from aerosol spray cans on these gases is also of concern.

The radiation spectrum at any given location in the earth's surface can change quite markedly through a day, particularly due to cloud cover and solar angle shifts. The spectrum also shifts with depth in a plant canopy since leaves transmit more in the green, from 500 to 580 nm, than in the red or blue, and have a very low absorptance in the near infra-red from 700 to 1500 nm. These properties of leaves are critical to their ability to photosynthesize and avoid overheating. Similarly, there are changes in radiation spectra with depth in bodies of water with rapid absorption of infra-red and higher attenuation in the red than in the blue-green. This can be expected to have significant effects on the visual processes of fish. It is worth keeping in mind that other organisms may be quite sensitive to light qualities outside the range of human perception. For example, reflectance properties of some flowers in the ultra-violet play an important role in pollination by insects (Silberglied, 1979), and the polarization of skylight may be significant during the migration of birds (Able, 1982).

II. Temperature

The thermal regime of a habitat is of critical importance to the survival, reproduction, and types of organisms which live there. At any particular location, temperatures are affected by an immense collection of factors in part because most terrestrial objects emit a broad spectrum of infra-red radiation, above 2.5 nm, and environmental temperatures are strongly coupled to thermal radiation from the surroundings. Because they have relatively low absorptances to short-wave radiation, below 2.5 nm, organism temperatures are mainly affected by long-wave radiation. Relationships between organism temperatures and environmental ones can be quite complex, as is discussed later under heat loading. It is also true that although environmental temperatures are one of the easiest measurements to make, and enormous amounts of data have been taken, much of this is of relatively little ecological significance (Remmert, 1980, p. 24). Organism temperatures may bear little relationship to the meteorological temperatures usually measured, even for poikilotherms. Behavioral actions such as sunbathing by grasshoppers on cool mornings to raise body temperature and solar tracking by arctic flowers, are just two of the many thermoregulatory mechanisms available. Due to the great variability of environmental temperatures, it is quite difficult to establish relationships with species distributions. However, energy budget models which take this variability into account have been successful in predicting the spatial and temporal distributions of some organisms, at least when other temperature dependent factors such as water loss aren't overriding (Waldschmidt and Tracy, 1983).

Temperatures may affect organisms quite differently at different stages of their life cycles, with especially strong effects on development and reproduction. Such processes as seed germination, flowering times, and larval development rates are strongly coupled with temperature. Again, it often isn't clear what aspect of temperature is most critical – maximum, minimum, averages, or ranges. Which measurement is meaningful is often specific to a given process. This is complicated by the ability of organisms to adapt, both genetically at the population level, and physiologically for individuals, to temperature ranges. Thus, although most temperate-zone plants are very tolerant of midwinter cold, early frosts may damage them due to inadequate time to adapt physiologically (Weiser, 1970). Although much laboratory work has been done to determine the temperature tolerance limits of organisms along with optimal temperature ranges for various physiological processes, applying these results to natural environments is complicated by the necessary transformation from environmental measurements to physiologically significant ones (Campbell, 1981).

III. Moisture

Water is critical to life, not only because it is the major constituent of biomass and is necessary for most metabolic processes, but also because of its importance in evaporative cooling. Aquatic environments present unique osmothic challenges to fresh-water organisms which have a higher osmotic pressure than their environment and thus are faced with the problem of keeping water out, whereas ocean-dwelling organisms have the reverse problem of keeping water in. Many terrestrial organisms must maintain themselves internally at water-saturated conditions, while the moisture content of their environment is often quite low. It is thus expected that organisms will have developed quite complex mechanisms to buffer themselves from changes in external moisture. For example most deep-sea dwelling organisms can tolerate only very small fluctuations in salinity of the surrounding water, whereas species in tidal zones and estuaries have developed active mechanisms to alter internal osmotic pressure to maintain a proper ionic balance. The stomata of leaves are critically important for the maintenance of internal water level in plants, and respond dynamically to a complex of factors including light, humidity, and CO_2 (Raschke, 1979).

As in the case of temperature, it is often quite difficult to obtain measurements of environmental moisture which are indeed physiologically significant in terms of water actually available to the individual. For terrestrial organisms, humidity, soil moisture (for plants), and drinking water (for animals) may all be important in setting tolerance limits for individuals. These moisture measurements must be tied together with the evaporative and transpirational needs of the organism for heat balance, and thus are highly dependent upon other environmental factors. Biophysically important measurements are usually made in terms of the water

potential, or chemical potential, of water in various components of the system. See
Nobel (1974) for an overview of water relations in plants, and Tracy (1975) for the
application of a mass balance approach to water budgets of animals.

IV. Nutrients

Organism nutritional requirements, in which I include utilization of gases such as
O_2 and CO_2, vary qualitatively and quantitatively among taxa. For more on
animal nutrition, see the chapter on foraging theory. Soil concentrations of
nutrients which are critical to plant growth, such as nitrogen and potassium, tend
to be extremely low, requiring active mechanisms within terrestrial plant roots in
order to obtain adequate supplies. Further complications arise due to the
importance of micro-organisms to nutrient supply, for example in legumes which
have symbiotic relationships with nitrogen-fixing bacteria in their root systems.
The difficulties involved in studying roots in vivo also contributes to the general
lack of theoretical development on the subject of plant nutrition. Generally,
experimental studies are unable to ascertain minimum and optimal nutrient
concentrations necessary for growth, partially due to the great plasticity of growth
response. Nutrient-stressed plants can change root-to-shoot ratios, photosynthesis, and root absorption capacity, and there are consistently different patterns
of nutrient response in plants from different soil fertility growth regimes (Chapin,
1980). All these complications combine to make mineral nutrition one of the least
developed areas of terrestrial plant growth analysis although there have been
models both of root growth (Plant, 1982), and of nutrient transport after
partitioning (Thornley, 1975). For an overview of mineral nutrition in terrestrial
plants, see Hewitt and Savage (1974). As a bit of contrast to the above, the
importance of nutrient limitations, particularly phosphorus, in controlling
phytoplankton community structure and biomass has led to the development of a
resource-based theory to explain their distribution (Tilman et al., 1982). For
aquatic macrophytes, however, it is still unclear how nutrients mobilized from
sediments through roots interrelate to those obtained from the water (Barko and
Smart, 1981).
    In contrast to the situation with soil nutrients, a great deal is known about the
CO_2 and O_2 requirements of plants and animals. CO_2 is the building block of all
organic matter, and critically affects photosynthesis. Its low concentration in the
atmosphere, approximately 340 parts per million, limits the growth of plants,
particularly crop plants under good growing conditions. I discuss its relationship
to photosynthesis later, and at this point just mention that its concentration in the
atmosphere is increasing due to fossil fuel burning. The implications of this
increase are complex and debatable, though most agronomists feel that it will lead
to increased crop yields. For land organisms, the availability of O_2 for respiratory
processes is generally non-limiting, however, aquatic species have evolved
adaptations to utilize the low levels of dissolved O_2 in water bodies. The
production of O_2 in water by aquatic plants is critical to fish life. There can be
complex interactions with nutrients and bacteria, leading to reduction in dissolved

O_2 and the death of fish. In this situation, the fish production in a nutrient-poor lake may well be much higher than in a nutrient-rich lake. See Schmidt-Nielsen (1975) for an overview of physiological oxygen requirements of animals.

V. Other Factors

A variety of other environmental factors have important influences on organisms in particular environments. Wind plays a critical role in convective heat exchange and evaporation water loss, as is discussed later. This has implications for such characters as leaf shape, body size, and fur thickness. Wind is a major factor limiting trees at timberline (Tranquillini, 1979) and greatly affects the morphology of plants. See Nobel (1981) for a review of the ecological effects of wind. Similarly, water currents can affect the distribution of fish within streams, and are critical to the larval stages of many marine invertebrates. For a development of fluid mechanics as applied to the problem of locomotion in fish, birds, and insects, see Childress (1981). The physical laws which govern flight have also been applied to investigate the evolution of vertebrate flight from an earlier jumping form (Caple et al., 1983). Fire, caused by lightning or spontaneous combustion, is a regular feature of many natural ecosystems. In a complex manner, it brings about regular changes in community structure and affects the nutrient availability at a site. Physiologically, some species of plants have developed adaptations to fire, including in some cases the production of seeds which only germinate after a fire. See Rundel (1981) for an overview of fire as an ecological factor.

All of the above mentioned environmental factors may interact in ways that often make predictions of organism responses quite difficult and process-specific. Some examples of interactions are given in the case studies below. In some cases, factors do indeed act independently, at least over certain ranges of the variables. For example, many plant species' distributional limits can be determined independently by soil conditions and climate. Generally, however, it is preferable to view an individual as an (often non-linear) integrator of the environmental influences upon it. Further difficulties arise due to the temporal and spatial variation of environmental factors, leading to possibilities for temporal acclimation to changing environments. Biophysical methods often provide the framework for analysis of response dynamics, but require enormous quantities of temporal data, which limits their utility. Thus, steady-state approaches are prevalent. It should also be kept in mind that biotic factors, such as local densities of individuals, along with their growth forms, play an important role in specifying organism response. For example, relative size structures of many plant parts depend greatly upon planting densities (Harper, 1977, Chap. 7). In what follows, however, I limit the discussion to cases of response to physical factors.

Case Studies

I. Heat Loading

One of the most useful of biophysical analyses is the study of the energy budgets of organisms. The basic physical principles underlying heat exchange are the same in plants and animals, and at steady-state may be expressed as (Gates, 1980, p. 20):

$$M + Q_a = R + C + \lambda E + G + X,\tag{1}$$

where M = metabolic rate,

$\quad Q_a$ = radiation absorbed,

$\quad R$ = radiation emitted,

$\quad C$ = energy transmitted by convection,

$\quad \lambda E$ = energy exchanged by evaporation or condensation,

$\quad G$ = energy exchanged by conduction,

$\quad X$ = energy put into or obtained from storage.

All these are expressed as rates per unit time per unit surface area, with units of Wm^{-2}. We may consider each of these components in turn, and then combine their effects. Note that this separation still averages the effects of each form of energy exchange over the whole organism, which introduces errors since there may be quite different interactions between the exchange factors on different parts of the organism. For example radiant absorption and convective cooling may be quite different for upper versus lower leaf surfaces, and using average values for each of these over the whole organism is incorrect. However, for purposes of investigating large-scale trends in comparative ecology, the approach is a good approximation (Bakken and Gates, 1975). The main variable will be organism surface temperature, T_s, which will be assumed uniform over the individual.

By the law of radiation,

$$R = \varepsilon\sigma(T_s + 273)^4,\tag{2}$$

where σ = Stefan-Boltzman constant,

$\quad \varepsilon$ = emissivity of the organism surface.

If the organism's environment is at a uniform temperature T_0, with the same emissivity as the organism, then the radiative energy exchange rate for an organism of surface area A is

$$\frac{1}{A}\frac{dE}{dt} = \varepsilon\sigma[(T_s + 273)^4 - (T_0 + 273)^4].\tag{3}$$

If T_s and T_0 are close, we may approximate (3) by linearizing the right-hand side to get

$$\frac{1}{A}\frac{dE}{dt} = K(T_s - T_0),\tag{4}$$

where $K = 4\varepsilon\sigma(T_0 + 273)^3$. Equation (4) is often called Newton's law of cooling though this has led to confusion in the literature as to whether convection and conduction should be included (Bakken and Gates, 1974). It is probably best to omit the use of the term completely.

Convective heat exchange with the surrounding air or water in the case of a uniform environment of temperature T_a is

$$C = h_c(T_s - T_a),\tag{5}$$

where h_c is the convection coefficient. The convection coefficient here is really an average over the organism surface, which may have a very complex boundary layer. The relationship between h_c and organism shape can be quite difficult to establish, but fairly accurate results are obtained by taking

$$h_c \propto D^m V^n,$$

where D is a characteristic dimension of the organism in the direction of air flow, V is wind speed, and m and n are constants. Estimated values for leaves are $m \approx -0.3$ and $n \approx 0.5$ (Gates, 1980, p. 298), indicating the effects that increasing wind speed and decreasing leaf size have on reducing the boundary layer thickness (proportional to h_c^{-1}) around a leaf. Quite similar values are used for animal models, though the use of animal-shaped metal castings in wind tunnels has provided the best estimates of h_c (Bakken and Gates, 1975). For applications to leaves, the conduction term in (1) is ignored, while for animals it is proportional to the surface area in contact with the ground and the temperature difference between ground and organism surface. The proportionality coefficient (of conduction) will depend greatly upon the fur, feathers, and structure of the organism (Gates, 1980, pp. 406–410). Estimates of the coefficient have been made for a variety of fur thicknesses (Skuldt et al., 1975).

The general form for evaporative heat loss is

$$\lambda E = \lambda(T_s)E(T_s, T_a, h, r),$$

where λ is the latent heat of vaporization, which is a function of organism surface temperature, and E is the rate of water loss per unit surface area, which depends on relative humidity h, and a resistance to water loss, r, along with the temperatures. For leaves, Gates (1980, p. 27) suggests

$$E = \frac{d_\ell(T_s) - h d_a(T_a)}{r_\ell + r_a},\tag{6}$$

where $d_\ell(\cdot)$ and $d_a(\cdot)$ are the saturation densities of water vapor in leaf intercellular air spaces and air, respectively, at the respective temperatures, r_ℓ is internal leaf resistance and r_a is boundary layer resistance. A similar form holds for water loss from animals either by sweating or respiration, in which case there is no r_ℓ term if the animal skin is wet from sweat. In all cases the boundary layer resistance r_a is approximately proportional to $V^{-0.5}$.

For a plant leaf, the above is sufficient to specify the thermal budget since metabolic rate and photosynthesis have no measurable effect on leaf temperatures. Thus in the simplest application of (1), we take $M=G=X=0$, and can solve numerically for the steady-state temperature at any level Q_a of absorbed radiation, using (2), (3), and (6). Through this analysis, we can determine the effects of transpiration, radiation, leaf dimension, and wind speed on leaf temperatures. There is good evidence that leaves do indeed undergo thermal regulation, with a variety of strategies taken to adjust leaf energy exchange (Gates, 1980, pp. 38–46). The relative advantages of smaller leaves in dryer, hotter environments may be explored with the above type of analysis. Since plant growth rates are correlated with photosynthesis, and photosynthetic rates depend upon leaf temperature, the above types of energy budget calculations are also useful in specifying potential plant growth rates under a variety of environmental conditions.

The application to animals is a bit more complex. The net heat production $M-\lambda E$ is the net heat produced by metabolic processes less the heat lost through breathing and perspiration. This may be taken as

$$M-\lambda E=\frac{T_b-T_s}{I}. \tag{7}$$

Here T_b is internal body temperature (not surface) and I is a measure of insulation, which depends upon fur, fat, and feather thicknesses. Since it is internal body temperature T_b that is critical to metabolic processes, one solves (7) for T_s and substitutes into (1) to obtain

$$M-\lambda E+Q_a-\varepsilon\sigma(T_b+273-I(M-\lambda E))^4$$
$$-k\left(\frac{V}{D}\right)^{0.5}[T_b-T_a-I(M-\lambda E)]-G=0. \tag{8}$$

Values of Q_a are one of the most difficult to estimate, though an approach by breaking the animal's surface into increments with different absorptances and summing does give some estimates (Gates, 1980, pp. 416–418).

By considering each segment of the energy budget in turn, it is possible to derive from (8) information about how such factors as fur thickness and coloration, body size, wind speed, and fat content affect the range of body temperatures a poikilotherm may experience, and the range of environmental conditions over which a homeotherm may maintain a fixed temperature. These techniques have been applied to a variety of organisms including hummingbirds (Southwick and Gates, 1975), rabbits (Kluger, 1975), deer (Moen and Jacobsen, 1975), and amphibians (Tracy, 1975). The approach is also useful in making comparisons across taxonomic lines of organisms under similar ecological conditions; for example in explaining the thermal conductance and basal metabolic rate differences between subterranean rodents and insectivores, and in explaining why, for burrowing mammals, males tend to be larger than females (Nevo, 1979).

The entire approach to energy budget is often viewed analogously to simple electrical circuits, where heat flows are analogous to current. For example, see

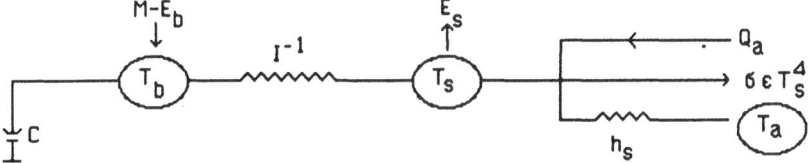

Fig. 1. E_b is heat loss in respiratory evaporation, E_s is heat loss from surface evaporation, C is the heat capacity of the body of the animal, and all other variables are as given in the text. This represents an animal with negligible conduction to the ground. See text for further explanation. (Redrawn from D. Gates: Biophysical Ecology, p. 385. New York-Berlin-Heidelberg: Springer 1980)

Fig. 1 for the case of negligible conductance to the ground. Arrows represent thermal currents not directly related to temperature, the nodes represent parts of the body/environment system at which temperature measurements may be made, and resistors correspond to thermal conductances between nodes. A fully dynamic analysis may be performed rather than just the steady-states as done above (Bakken and Gates, 1975), to examine the effects of such behaviors as shuttling from one environment to another, a very important behavioral mechanism of thermoregulation for many cold-blooded organisms (Crawford et al., 1983). In general, of course, time dependent solutions require further data on both the environment and potential organism responses to it. Many more complicated circuit analogs may be set up, for example to include conduction through the ground, and metabolic or respiratory rates which are body temperature dependent.

Although biophysical techniques to derive predictive models for the thermoregulatory behavior are now well developed, combining these with life history approaches to determine evolutionary patterns of adaptation has been rare. Kingsolver (1983) combines biophysical approaches with strategic modeling to determine how well adapted with respect to thermoregulation a butterfly species is for maximizing flight activity, and how environmental heterogeneity constrains thermoregulatory strategy. Ellner and Beuchat (1984) discuss a case in which there is no single physiologically optimum temperature, in particular for reptiles that bear live young. Instead, they derive, from an evolutionary perspective, the range of thermoregulatory behaviors that might be expected. Further work which introduces life history theory into biophysical modeling will add a new dimension to the problems attacked thus far.

II. Photosynthesis

As the most basic process for energy transformation on this planet, photosynthesis has been intensely studied, with the result that there is a wealth of data, but as yet no complete understanding of it at any level. The environmental inputs to photosynthesis are light, CO_2, water potential in the soil, humidity, wind, temperature, and soil mineral status. Photosynthetic rates for leaves depend upon many characters including biochemical and photochemical capacities (there are at least three different major types of photosynthesis), the nature and density of stomata (pores in leaf surfaces through which gases diffuse), the ability of the plant to withstand water stress, the form and size of leaves, and the extent of the root system. Securing the appropriate conditions for photosynthesis presents distinct adaptive problems to plants. Only about 2% of incoming solar radiation can be utilized, requiring the dissipation of energy to avoid overheating. Since CO_2 levels in the atmosphere are very low, and CO_2 is absorbed through passive diffusion, for every molecule of CO_2 absorbed, 300 to 1000 molecules of H_2O are lost. This trade-off between CO_2 uptake and water loss, due to the fact that they use the same diffusive pathway, is of major importance to questions of leaf form and size. Many attempts have been made to model various aspects of photosynthesis, but because of its complexity, the models have limited scope. See Hesketh and Jones (1980) for the most complete compendium of models to date.

The great majority of photosynthetic models deal with the leaf level and I here would like to give some indication of the most common approach. This uses an electrical circuit analog, as mentioned above for heat loading, only the fluxes are now fluxes of CO_2 rather than heat. The models compartmentalize a leaf and consider the resistances to flow of CO_2 between the compartments. An example is given in Fig. 2. This is a steady-state type of model which is analyzed using Kirchoff's law, meaning in this case mass balance at every junction in the circuit.

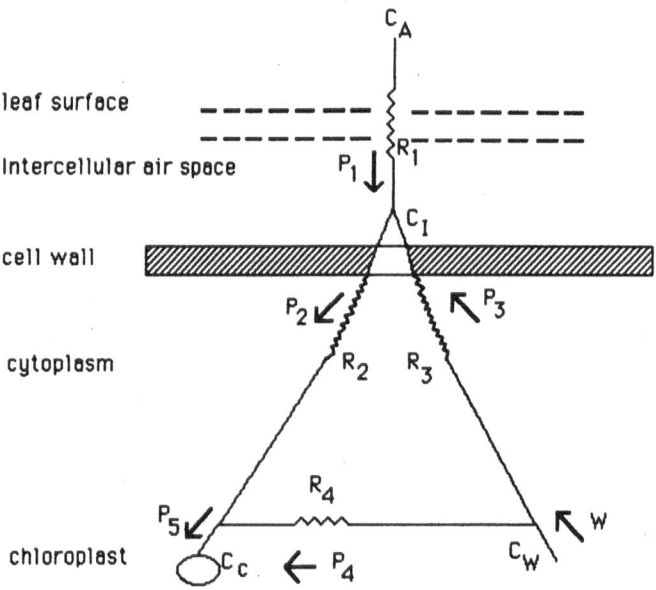

Fig. 2. Resistors represent resistances, R_i, to diffusive flow between various leaf compartments, P_i's are CO_2 fluxes between compartments, C_i's are CO_2 concentrations (air, intercellular air space, chloroplast, and respiratory), and W is the flux of respiratory CO_2. See text for analysis. (Redrawn from D. Gates: Biophysical Ecology, p. 496. New York-Berlin-Heidelberg: Springer 1980)

The underlying physical principle is Fick's law of diffusion

$$J = D \frac{\partial \varrho}{\partial x}, \tag{9}$$

where J is flux of CO_2, D is diffusivity of CO_2, ϱ is CO_2 concentration, and x is distance into the leaf. A discrete analog of (9) is used:

$$J = \frac{\varrho_1 - \varrho_2}{r}, \tag{10}$$

where ϱ_1 and ϱ_2 are CO_2 concentrations at two points in the circuit, and $r = \Delta x/D$ is the resistance across a distance Δx of some medium (such as inter-cellular air space or cytoplasm). Note that D is temperature and pressure dependent and the Δx's aren't easily specified, but various techniques have been employed to estimate the resistances R_i for the diffusive pathways in a leaf (Gates, 1980, pp. 328–329).

Analyzing the fluxes of CO_2 in Fig. 2 produces for example from (10)

$$P_1 = \frac{C_A - C_I}{R_1} \tag{11}$$

and similar forms for P_2, P_3, and P_4. From mass balance,

$$P_2 = P_1 + P_3, \quad W = P_3 + P_4, \quad P_5 = P_2 + P_4. \tag{12}$$

The flux P_5 must equal the rate of biochemical uptake of CO_2 at the chloroplast, which depends upon light, CO_2, and O_2 inputs. A Michaelis-Menten form is often used, such as

$$P_5 = \frac{P_M}{1 + \dfrac{K\beta}{C_c}}, \quad \beta = 1 + \frac{[O_2]}{K_0}, \tag{13}$$

where P_M is the CO_2 saturated rate of photosynthesis, K and K_0 are constants and $[O_2]$ is the O_2 concentration at the site of carbon fixation. The introduction of O_2 levels is needed because it inhibits the activity of the major enzyme of photosynthesis, ribulose 1–5 biphosphate carboxylase. P_M in turn depends upon light level L, and a variety of forms have been utilized such as

$$P_M = \frac{\alpha L}{\left(1 + \dfrac{\alpha^2 L^2}{P_{ML}^2}\right)^{1/2}}, \tag{14}$$

where P_{ML} is the photosynthetic rate at light and CO_2 saturation, and α is the maximum efficiency to light

$$\alpha = \lim_{L \to 0^+} \frac{dP_M}{dL}.$$

The measured quantities in the above are P_1, C_A, and L, while values of $R_1, ..., R_4$, W can be estimated. Using (11)–(14) and a bit of algebraic manipulation, it is possible to solve for the unknown fluxes $P_2, ..., P_4$, back substitute and arrive at

$$P_1 + W = \frac{P_M}{1 + \dfrac{K\beta}{C_A - P_1 S_1 - W S_2}}, \tag{15}$$

where S_1 and S_2 are functions of the R_i's. This gives a quadratic in P_1 which can be solved. The model can then be analyzed by varying the environmental inputs and making comparisons to observations to test its validity. Generally, this model gives qualitatively similar results to those observed in laboratory experiments. See Gates (1980, pp. 514–522) for a complete analysis.

The above model is, of course, quite limited in applicability, for it is inherently steady-state, ignores the spatial structuring of leaves, and the biochemistry is handled in a very simplistic manner. Yet, this approach has been most useful for combining the many factors which affect photosynthesis, and has stimulated much research to specify more accurately the parameters and functional forms used in the model. Models have been formulated which emphasize the biochemical dynamics of photosynthesis (Milstein and Bremermann, 1979) along with the photochemical energy absorption involved (Crill, 1977). Relatively few attempts have been made to analyze photosynthetic dynamics at the leaf level. Parkhurst (1977) analyzes the steady-state behavior of a continuous model of internal leaf CO_2 uptake

$$\frac{\partial C}{\partial t} = \frac{\partial}{\partial x}\left(p_1 D \frac{\partial C}{\partial x}\right) + \frac{\partial}{\partial y}\left(D p_2 \frac{\partial C}{\partial y}\right) + \frac{\partial}{\partial z}\left(D p_3 \frac{\partial C}{\partial z}\right) - U \tag{16}$$

for the purpose of investigating the effect of 3-dimensional structure on CO_2 uptake. Here C is CO_2 concentration, U is photosynthetic rate, p_i's are leaf porosities, and D is diffusivity. Gross (1981) analyzes the dynamic behavior of (16) to compare with data on photosynthetic responses to light changes and concludes that the dynamics of diffusion do not account for the observed dynamics. The dynamics of stomatal response to variations in CO_2, water, and light have been investigated by Cowan and Farquhar (1977), while Rand et al. (1981) have investigated the potential for Hopf bifurcations to occur in models of stomatal mechanics. At the leaf level, photosynthetic models for the dynamic response to external CO_2 changes (Kaitala et al., 1982) and environmental light variations (Gross, 1982) have been constructed, but as yet there are no models which combine these factors, or include others such as water availability, in a dynamic manner.

III. Leaf Size

It is possible to use the above results concerning heat loading and photosynthesis to analyze the question of what leaf sizes are expected to be most adaptive in a variety of environments. The central problem revolves around the fact that most of the radiant energy incident upon a leaf cannot be utilized by photosynthesis, so that much absorbed energy must be dissipated as heat through evaporation. Photosynthesis increases with leaf temperature over certain ranges, and leaf temperatures rise as leaf size increases, due to an increase in boundary layer thickness. Transpiration rates also increase with leaf size, but not in the same manner as photosynthesis. There are trade-offs involved in how high a leaf temperature can be supported by the root system's capacity to supply water and also by the reduction in photosynthesis at high temperatures. Thus a number of studies have attempted to establish the relationship between leaf size and benefit to the plant, and I here review some of these.

Because of the trade-offs between carbon gain and water use, it is unclear what might be the proper criterion to use for determining optimal leaf sizes. Parkhurst and Loucks (1972) use for their criterion water use efficiency, meaning the ratio of photosynthesis to transpiration. For photosynthetic rate, they use a simpler resistance analog than investigated above, letting

$$P = \frac{C_A - C_c}{R_A + R_M + R_S} f(T_s) g(L), \tag{17}$$

where R_A, R_M, and R_S are respectively air, mesophyll, and stomatal resistance and $f(\cdot)$ describes the temperature response of photosynthesis and $g(\cdot)$ the light response. Then using (6) for evaporation, and the simpler form of (1)

$$Q_a = \sigma\varepsilon(T_s + 273)^4 + h_c(T_s - T_A) + \lambda E \tag{18}$$

they choose leaf size so as to maximize water use efficiency, $P/\lambda E$, using a numerical technique. There are seven independent variables they investigate: convection coefficient h_c, air temperature T_A, relative humidity h, absorbed radiation Q_a, stomatal resistance R_S, mesophyll resistance R_M, and stomatal distribution. At first the dependence of photosynthesis and transpiration on leaf temperature is ignored, but these may be included later. This analysis produces, among others, the following predictions:
1. When radiation absorbed by the leaf is high, smaller leaf sizes have a higher water use efficiency.
2. When absorbed radiation and air temperature are low, smaller leaves should again prevail.
3. When absorbed radiation is low, but temperatures high, larger leaves should be favored.

These are generally born out by the available data, but it should be noted that the use of efficiency as an optimization criterion allows for the optimal leaf to be either infinitely large or infinitesimally small (Solbrig, 1981).

An alternative approach, taken by Givnish (1979), is to look at transpiration as having a photosynthetic "cost" in terms of maintaining root and shoot tissue to support a given level of transpiration. This view, similar to a cost-benefit analysis, uses net carbon assimilation as the optimization criterion, given by

$$N = (P - R - bE)A, \tag{19}$$

where N is net carbon gain, R, P, and E are daily rates per unit leaf area of nightly respiration, net photosynthesis, and transpiration respectively, and A is leaf area. The parameter b is a constant giving the metabolic costs of supplying a unit of transpirational water. Using a graphical approach, Givnish has investigated how the maximum value of N changes with leaf area depending upon solar radiation inputs and water supply. Qualitative predictions can be made such as:

1. Small leaves are favored over large leaves in very sunny and very shady environments, though for different reasons.
2. In intermediate radiation regimes, transpiration is rather independent of leaf size, so leaf size is at the point where photosynthesis alone is maximized.

These types of results have been tested against observations of vine leaves (Givnish and Vermeij, 1976), with generally good agreement. This approach has advantages in that it is possible to include in it a whole variety of other factors which could affect optimal leaf sizes including leaf thickness, costs of support structures such as petioles, and the role of nutrients.

Although the above has gone a long way in resolving questions such as under what circumstances plants should be energy maximizers (e.g. maximize photosynthesis) or cost minimizers (e.g. minimize water use), the criteria used are still very crude. The approaches consider only steady-state environments, and introduction of dynamics would complicate the picture extremely. Although the models give good qualitative predictions of general trends, they are generally at a loss to explain the great variability in leaf shapes and sizes which may be observed at any single location. The whole question of plant interactions is basically unexplored in the context and it may well be, as suggested by Givnish (1979), that for these situations criteria of maximal carbon gain need to be replaced with those based upon evolutionarily stable strategies.

Acknowledgements

I thank the Biomathematics Division of the Grassland Research Institute, Hurley, Great Britain, and especially Dr. J. H. M. Thornley for their hospitality while much of this was written. The support from a Faculty Research Participation award from Oak Ridge Associated Universities and the U.S. Department of Energy is greatly appreciated. Editorial comments from the mathematical ecology group at the University of Tennessee, Knoxville, were very helpful.

References

1. Able, K.P. (1982). Skylight polarization patterns at dusk influence migratory orientation in birds. Nature *299*: 550–551
2. Bakken, G.S., Gates, D.M. (1974). Notes on "Heat loss from a Newtonian animal." J. Theor. Biol. *45*: 283–292
3. Bakken, G.S., Gates, D.M. (1975). Heat transfer analysis of animals: some implications for field ecology, physiology, and evolution. Pages 255–290 in D. Gates and R. Schmerl (eds.), *ibid.*

4. Barko, J.W., Smart, R.M. (1981). Sediment-based nutrition of submersed macrophytes. Aquatic Botany *10*: 339–352
5. Calow, P., Townsend, C.R. (1981). Energetics, ecology, and evolution. Pages 3–19 in C.R. Townsend and P. Calow (eds.), *ibid.*
6. Campbell, G.S. (1977). An Introduction to Environmental Biophysics. Springer-Verlag, N.Y.
7. Campbell, G.S. (1981). Fundamentals of radiation and temperature relations. Pages 11–40 in O.L. Lange et al. (eds.), *ibid.*
8. Caple, G., Balda, R.P., Willis, W.R. (1983). The physics of leaping animals and the evolution of preflight. American Naturalist *12*: 455–467
9. Chapin, F.S. III (1980). The mineral nutrition of wild plants. Ann. Rev. Ecol. System. *11*: 233–260
10. Childress, S. (1981). Mechanics of Swimming and Flying. Cambridge University Press, New York
11. Christian, K., Tracy, C.R., Porter, W.P. (1983). Seasonal shifts in body temperature and use of microhabitats by Galapagos land iguanas *(Conolophus pallidus)*. Ecology *64*: 463–468
12. Cowan, I.R., Farquhar, G.D. (1977). Stomatal function in relation to leaf metabolism and environment. Symp. Soc. Exp. Biol. *31*: 471–505
13. Crawford, K.M., Spotila, J.R., Standora, E.A. (1983). Operative environmental temperatures and basking behavior of the turtle *Pseudemys scripta*. Ecology *64*: 989–999
14. Crill, P.A. (1977). The photosynthesis-light curve: a simple analog model. J. Theor. Biol. *64*: 503–516
15. Ellner, S., Beuchat, C.A. (1984). A model of optimal thermoregulation during gestation by *Sceloporus jarrovi*, a live-bearing lizard. Pages 15–28 in S. Levin and T.G. Hallam (eds.). Mathematical ecology, proceedings. Trieste 1982. Lecture Notes in Biomathematics, Vol. 54. Springer-Verlag, Berlin
16. Gates, D.M. (1980). Biophysical ecology. Springer-Verlag, N.Y.
17. Gates, D.M., Schmerl, R.B. (eds.) (1975). Perspectives of Biophysical Ecology. Springer-Verlag, N.Y.
18. Givnish, T. (1979). On the adaptive significance of leaf form. Pages 375–407 in O.T. Solbrig, S. Jain, G.B. Johnson, and P.H. Raven (eds.). Topics in Plant Population Biology. Columbia Univ. Press, N.Y.
19. Givnish, T., Vermeij, G.J. (1976). Sizes and shapes of liane leaves. Amer. Nat. *110*: 743–776
20. Gould, S.J., Lewontin, R.C. (1979). The spandrels of San Marco and the Panglossian paradigm: a critique of the adaptionist program. Proc. of the Royal Soc. of London B Biol. Sci. *205*: 581–598
21. Gross, L.J. (1981). On the dynamics of internal leaf carbon dioxide uptake. J. Math. Biol. *11*: 181–191
22. Gross, L.J. (1982). Photosynthetic dynamics in varying light environments: a model and its application to whole leaf carbon gain. Ecology *63*: 84–93
23. Harper, J.L. (1977). Population Biology of Plants. Academic, London
24. Hesketh, J.D., Jones, J.W. (1980). Predicting Photosynthesis for Ecosystems Models. CRC Press, Boca Raton, Fla.
25. Hewitt, E.J., Smith, T.A. (1974). Plant Mineral Nutrition. English Univ. Press, London
26. Kaitala, V., Hari, P., Vapaavuori, E., Salminen, R. (1982). A dynamic model for photosynthesis. Ann. Bot. *50*: 385–396
27. Kingsolver, J.G. (1983). Thermoregulation and flight in Colias butterflies: elevational patterns and mechanistic limitations. Ecology *64*: 534–545
28. Kluger, M.J. (1975). Energy balance in the resting and exercising rabbit. Pages 497–507 in D.M. Gates and R.B. Schmerl (eds.), *ibid.*
29. Krebs, C.J. (1978). Ecology. Harper and Row, N.Y.
30. Lange, O.L., Nobel, P.S., Osmond, C.B., Ziegler, H. (eds.) (1981). Physiological plant ecology I. Encyl. Pl. Physiol. Vol. 12A, Springer-Verlag, Berlin
31. Lewontin, R.C. (1978). Fitness, survival, and optimality. Pages 3–21 in D.H. Horn, R. Mitchell, and G.R. Stairs (eds.). Analysis of Ecological Systems. Ohio State Univ. Press, Columbus, Oh.
32. Maynard Smith, J. (1978). Optimization theory in evolution. Ann. Rev. Ecol. System. *9*: 31–56
33. Milstein, J., Bremermann, H.J. (1979). Parameter identification of the Calvin photosynthesis cycle. J. Math. Biol. *7*: 99–116

34. Moen, A.N., Jacobsen, N.K. (1975). Thermal exchange, physiology, and behavior of white-tailed deer. Pages 509–524 in D.M. Gates and R.B. Schmerl (eds.), *ibid.*
35. Nevo, E. (1979). Adaptive convergence and divergence of subterranean mammals. Ann. Rev. Ecol. System. *10*: 269–308
36. Nobel, P.S. (1974). Introduction to Biophysical Plant Physiology. Freeman, San Francisco
37. Nobel, P.S. (1981). Wind as an ecological factor. Pages 475–500 in O.L. Lange et al. (eds.), *ibid.*
38. Parkhurst, D.F. (1977). A 3-dimensional model for CO_2 uptake by continuously distributed mesophyll in leaves. J. Theor. Biol. *67*: 471–488
39. Parkhurst, D.F., Loucks, O.L. (1972). Optimal leaf size in relation to environment. J. Ecol. *60*: 505–537
40. Plant, R.E. (1982). A continuum model for root growth. J. Theor. Biol. *98*: 45–59
41. Pond, C.M. (1978). Morphological aspects and the ecological significance of fat distribution in wild vertebrates. Ann. Rev. Ecol. System. *9*: 519–570
42. Rand, R.H., Upadhyaya, S.K., Cooke, J.R., Stori, D.W. (1981). Hopf bifurcation in a stomatal oscillator. J. Math. Biol. *12*: 1–11
43. Raschke, K. (1979). Movements of stomata. Pages 383–441 in W. Haupt and M.E. Feinleib (eds.). Physiology of movements. Encycl. Pl. Phys. Vol. 7, Springer-Verlag, Berlin
44. Remmert, H. (1980). Ecology. Springer-Verlag, Berlin
45. Rundel, P.W. (1981). Fire as an ecological factor. Pages 501–538 in O.L. Lange et al. (eds.), *ibid.*
46. Schmidt-Nielsen, K. (1975). Animal Physiology. Adaptation and Environment. Cambridge Univ. Press, Cambridge
47. Silberglied, R.E. (1979). Communication in the ultra-violet. Ann. Rev. Ecol. System. *10*: 373–398
48. Skuldt, D.J., Beckman, W.A., Mitchell, J.W., Porter, W.P. (1975). Conduction and radiation in artificial fur. Pages 549–558 in D.M. Gates and R.B. Schmerl (eds.), *ibid.*
49. Solbrig, O.T. (1981). Energy, information, and plant evolution. Pages 274–299 in C.R. Townsend and P. Calow (eds.), *ibid.*
50. Southwick, E.E., Gates, D.M. (1975). Energetics of occupied hummingbird nests. Pages 417–442 in D.M. Gates and R.B. Schmerl (eds.), *ibid.*
51. Thornley, J.H.M. (1976). Mathematical Models in Plant Physiology. Academic Press, N.Y.
52. Tilman, D., Kilham, S.S., Kilham, P. (1982). Phytoplankton community ecology: the role of limiting nutrients. Ann. Rev. Ecol. System. *13*: 349–372
53. Townsend, C.R., Calow, P. (1981). Physiological Ecology: An Evolutionary Approach to Resource Use. Sinauer, Sunderland, Mass.
54. Tracy, C.R. (1975). Water and energy relations of terrestrial amphibians: insights from mechanistic modeling. Pages 325–346 in D.M. Gates and R.B. Schmerl (eds.), *ibid.*
55. Tranquillini, W. (1979). Physiological Ecology of the Alpine Timberline. Springer-Verlag, Berlin
56. Waldschmidt, S., Tracy, C.R. (1983). Interactions between a lizard and its thermal environment: implications for sprint performance and space utilization in the lizard *Uta stansburiana*. Ecology *64*: 476–484

An Overview of Foraging Theory

Louis J. Gross

One of the most active areas of ecological research concerns predation and herbivory. Of interest is how an individual allocates its time and energy in search of food. Animal nutritional requirements vary greatly, not only in quantity, but also in quality, and it is not my aim here to review the variety of feeding mechanisms employed, nor the physiology of alimentary systems and digestion. See Church (1975) and Davenport (1978) for a review of these. Rather, the question of ecological interest is: given certain physiological needs and constraints on feeding, what complex of behavioral traits are employed to meet these needs? My discussion will concern only animals, but analogous questions for plants were mentioned in the chapter on biophysical ecology concerning leaf sizes which maximize photosynthetic gains. Without the behavioral repertoire available to animals, plants respond to the problem of "foraging" for light, water, and nutrients through phenotypically plastic growth forms (Bradshaw, 1965).

The approach I describe for analyzing adaptive feeding strategies is the a priori methodology mentioned in the biophysics lecture. Several reviews of foraging theory have appeared (Schoener, 1971; Pyke et al., 1977; Townsend and Hughes, 1981; Krebs and Davies, 1984), and my purpose here is not to give a complete review, but rather an introduction to the various theoretical questions which have been attacked. I would like to emphasize that this area is one in which models have proliferated at a very rapid rate, with relatively few experimental tests which can readily discriminate between models. I will mention a few of the tests which have been done, but will not attempt by any means to be comprehensive. Pyke et al. (1977) give a good review of model testing, and Krebs and McCleery (1984) provide an updating of this, but there is tremendous activity in the field which rapidly causes any review, including this one, to become dated. Readers should consult volumes of the American Naturalist, Ecology, and the Journal of Animal Ecology for more recent work.

The approach taken in foraging theory is essentially an optimization one (Maynard Smith, 1978). The technique is threefold. First a set of possible behavioral responses of the animal must be delineated, this being the phenotype set of Maynard Smith. The phenotype set specifies the range of responses which it is assumed are available to a population through the action of evolution. Secondly, the state equations of the system must be described. These constraints describe how organism response is coupled to the environment and include responses which are not subject to change via evolution. For example, in most foraging models it is assumed that the perceptual abilities of individuals are inflexible and thus the distances over which a prey item can be located are fixed by the state

Biomathematics, Vol. 17, Mathematical Ecology
Edited by T. G. Hallam and S. A. Levin
© Springer-Verlag Berlin Heidelberg 1986

equations. Lastly, a fitness criterion must be defined, giving a function over the phenotype set which assigns a fitness to each phenotype. It is assumed that evolution acts to maximize this fitness, and the predicted phenotype from this maximization may then be compared to those actually observed in the field.

As mentioned in the biophysics chapter, this approach has its weaknesses. By dealing only with phenotypes, the underlying genetics of the evolutionary process is ignored and thus the approach is subject to many of the criticisms of optimality thinking in ecology (Oster and Wilson, 1978). Setting up the phenotype set may lead to circularity since this set is often delimited by what is known to be feasible from biological experience, and the model is then tested against that experience. The view of natural selection as an optimizing process ignores the history-dependence of evolution, which can be argued to be more of a "tinkering" which jury-rigs adaptations using material available at the time (Jacob, 1977). Thus the genetic variation necessary to reach an overall optimal phenotype may not be present, causing selection to move a population towards a local peak in Wright's adaptive topography. Since evolution doesn't act on individual characters of an organism in isolation of all others, the description of an appropriate fitness criterion may be extremely difficult, if not impossible. Similarly, individuals do not live in isolation and thus there may be frequency dependent selective forces operating, implying the necessity to investigate evolutionary stable behaviors, rather than optimal ones derived by ignoring interactions within the population. Due to all these problems, it is best when using this approach to carefully describe the particular assumption made, so that when tests are done it will be clear which hypotheses of the model are being tested. Aronson and Givnish (1983) point out the importance of hypothesis testing in this context. Unfortunately, for foraging applications often many more assumptions need to be included than are amenable to observational testing, but the models still provide one of the few means available to conceptualize the outcomes of different resource utilization patterns by animals (Townsend and Hughes, 1981).

The procedure is thus to first construct a model for how an organism feeds in terms of the habitat (such as patchy or not), the abundance of potential prey items and their relative values to the predator, and particular behavior of the predator such as flight ability. Included within the model are variables which are assumed to be under evolutionary control, such as fraction of each prey type attacked, or time in a day spent actively foraging. These control variables specify the foraging strategy. The model choice may depend upon the currency used to describe the system; often some measure of energy gain to the forager is used. The cost and benefits of the range of potential foraging strategies are computed in terms of this common currency. An optimal strategy is then determined based upon a given optimization criterion such as maximizing net energy gain per unit time, or total energy gain in a day. This approach has been applied to problems such as determining what food types to eat, which type of patch to feed in, how much time to spend in different patches, what patterns and speed of movement to use (Pyke et al., 1977), whether to forage in groups (Pulliam and Caraco, 1984), and what size territory to maintain (Davies and Houston, 1984). Applications to anthropology have also been quite extensive (Smith, 1983).

One of the most confusing aspects of the literature on foraging theory is the number of different optimization criteria that have been used. Those most frequently applied include the rate of net energy gain (Pulliam, 1974; Charnov, 1976; Oaten, 1977; De Benedictis et al., 1978; Waddington and Holden, 1979; Werner et al., 1983a), optimal use of time (Katz, 1974; De Benedictis et al., 1978; Craig et al., 1979; Pyke, 1979; Hixon, 1980; Hoffman, 1983; Abrams, 1984), and total food uptake or reward (Oster, 1976; Craig et al., 1979; Pyke, 1979; Hixon, 1980). Other criteria suggested include energy gain efficiency (De Benedictis et al., 1978; Pyke, 1979), survivorship (Caraco, 1979), utility (Caraco, 1980), specific nutrient maximization (Belovsky, 1978, 1984), mean time to reach satiation (Richards, 1983), opportunity cost (Winterhalder, 1983), and value for non-food items (McGinley, 1984). Though many of these studies cite data to indicate they have made a proper choice of criteria, arguments have arisen over the matter (De Benedictis et al., 1978; Pyke, 1977; Sih, 1979; Hixon et al., 1983). Also subject to debate is what constraints should be included when carrying out the optimization procedure (Oaten, 1977; Pyke, 1979; Martindale, 1983; Belovsky, 1984). Below I give examples of how some of these criteria are applied.

A few studies have attempted to compare model results using a number of different optimization criteria with experimental data, though the results are somewhat conflicting. De Benedictis et al. (1978), compare four different criteria for meal size in hummingbirds and find that the rate of net energy gain and energy gain efficiency are optimized and not the use of time or total net energy gain. Pyke (1979), in a comparison of four criteria in a model of sunbird foraging, determines that daily energy cost is minimized while net energy gain and energy gain efficiency are very poor criteria to use. Hixon et al. (1983) investigate whether migrant hummingbirds are time minimizers or energy maximizers, and conclude that the birds adjust territory area and time budgeting in a manner consistent with maximizing net energy gain, but constrained by a required sitting time for crop emptying. Munger (1984) compares predictions based on net energy gain with those on use of time for horned lizards feeding on patches of ants and generally the results uphold the energy gain predictions. Belovsky (1984) uses several field data sets on herbivores to compare a net energy gain per unit time model, a model based on food abundance weighted by net energy content, and a linear programming approach for maximization of critical nutrients. The conclusion is that nutrient maximization best predicted diet composition, with the other criteria providing highly inaccurate results. Clearly, this is one area in which further experimentation is needed.

My procedure in what follows is first to develop the simplest deterministic foraging models for the cases of optimal diet and patch choice, following Townsend and Hughes (1981). I will then cover some detailed models to give an indication how the simpler models can be easily elaborated, with sometimes contradictory results. Finally, I will mention some problems which I feel have been relatively unexplored theoretically.

Diet Choice

The basic diet choice model makes the following assumptions:

(a) The forager encounters food items individually, so the prey are not bunched in time or space.

(b) The food items are handled one at a time by the forager, and while handling the item the forager cannot pursue other activities.

(c) The environment is stationary, so that encounter rates with food items do not vary in time or space. The abundance and distribution of food items in the environment is assumed fixed.

(d) The food items are not dangerous to the forager in terms of inducing injury.

(e) The choice of food items taken by the forager doesn't affect future encounters with food, so that the food abundance is not reduced by the actions of the single forager.

(f) Encounters with food items are independent of each other, so that the history of encounters up to a certain time do not affect the probability that the next item encountered will be of a particular type.

(g) There is no explicit learning on the part of the foragers.

(h) The time between successive prey encounters is split into two periods: a search time which includes the time to search for and to recognize the next food item, and a handling time which includes the time to pursue, attack, and eat the food item.

(i) The food value, which can be measured by the forager in a single currency, and handling times are fixed for each food type.

(j) The forager is omniscient in that it knows what food items are available, their abundances and values without having to sample the environment. In what follows, I use the terms "prey item" and "food item" interchangeably.

Under these assumptions, each prey item, i, has a fixed energy content c_i, which is utilizable by the predator, and also has a fixed handling time h_i which represents the amount of time necessary for the predator to actually consume the prey item once it has been located. The density of prey type i is λ_i. The control parameters which specify the foraging strategy are the fractions, p_i, of prey type i which are actually attacked, given that a type i prey is encountered. For example, if there are only two prey types with $p_1 = 1$ and $p_2 = 0$, then the predator is a "specialist" on prey type 1 and ignores prey type 2. If $p_1 > 0$ and $p_2 > 0$, the predator is a "generalist" in the sense that it will attack both prey types. If $p_1 > p_2$, the predator is said to "prefer" prey type 1 over type 2. The optimization criterion considered is the mean rate of energy gain per unit time, and the optimal strategy will be that set of p_i's which maximizes this.

The time required for a predator to locate a prey item is assumed exponential,

$$P[T_s \leq t] = 1 - e^{-\lambda t}, \tag{1}$$

where T_s is the random variable for time since start of the search when the prey is located, and $\lambda = (\lambda_1 + \lambda_2 + \dots + \lambda_n)s$ is the product of the density of prey $(\lambda_1 + \dots + \lambda_n)$ and the area searched per unit time by the predator (s). This will hold for the situation when prey items are distributed according to a Poisson distribution in 2-space, and the predator moves at constant speed, searching a constant area per unit time, with no overlap with area already searched. In the above the phenotype set is the set of attack probabilities, $0 \leq p_i \leq 1$. The state constraints are that the predator travels at a constant speed, has a fixed set of potential prey items, $i = 1, 2, ..., n$ and has a fixed efficiency for handling each prey type, since the h_i's are fixed. For a non-mobile forager which sits in one location and captures prey as they arrive, the above is modified by taking $s = 1$ and λ_i is then the number of arrivals of type i prey per unit time at the forager's location. The model analysis follows Pulliam (1974).

Consider a single prey arrival. The mean time for one prey arrival to occur is

$$E[T_s] = \frac{1}{\lambda}, \tag{2}$$

where $E[\ ]$ is expectation. The total time from the start of the search for the first prey item to the start of the search for the second prey item is $T = T_s + T_p$, where T_p is the random variable for pursuit or handling time. By the law of total probability

$$E[T_p] = \sum_{i=1}^{n} h_i P \begin{bmatrix} 1^{st} \text{ prey arrival} \\ \text{is attacked} \end{bmatrix} \begin{matrix} 1^{st} \text{ prey arrival} \\ \text{is type } i \end{matrix} P \begin{bmatrix} 1^{st} \text{ prey arrival} \\ \text{is type } i \end{bmatrix}$$

$$= \sum_{i=1}^{n} h_i p_i \frac{\lambda_i}{\lambda} \tag{3}$$

so

$$E[T] = \frac{1}{\lambda} \left(1 + \sum_{i=1}^{n} h_i p_i \lambda_i \right). \tag{4}$$

Similarly, the caloric gain from the first prey arrival C, has mean

$$E[C] = \frac{1}{\lambda} \sum_{i=1}^{n} c_i p_i \lambda_i \tag{5}$$

and the fitness criterion is

$$F = \frac{E[C]}{E[T]} = \frac{\sum\limits_{i=1}^{n} c_i p_i \lambda_i}{1 + \sum\limits_{i=1}^{n} h_i p_i \lambda_i}. \tag{6}$$

The problem is then to choose $0 \leq p_i^* \leq 1$, $i = 1, ..., n$, to maximize (6).

As an example, consider the case of two prey types, and assume without loss of generality that $\frac{c_1}{h_1} > \frac{c_2}{h_2}$. This says that the per unit handling time caloric gain from type 1 prey is higher than that from type 2 prey. By differentiating (6), it is easy to show that $\dfrac{\partial F}{\partial p_1} > 0$, independent of the value of p_2. Thus the type 1 prey should always be included in the diet and $p_1^* = 1$. Also

$$\frac{\partial F}{\partial p_2} > 0 \quad \text{if} \quad \frac{c_2}{h_2} > \lambda_1 \left(c_1 - \frac{c_2}{h_2} h_1 \right) \tag{7}$$

and $\dfrac{\partial F}{\partial p_2} < 0$ if the reverse inequality holds. This implies that $p_2^* = 1$ if the condition (7) holds, while $p_2^* = 0$ otherwise. Thus, an optimal predator would specialize on type 1 prey if the encounter rate with these, λ_1, is high enough relative to the per unit time gains from type 2 prey. If λ_1 is not high enough, then at some value of λ_1 the optimal diet is a generalist one which includes both prey types whenever they arrive. Note that there is no partial preference for either prey type; it is always optimal to take either all or none of each prey type as they arrive. Also, the optimal choice is independent of the density, λ_2, of type 2 prey.

These results carry over to the case of several prey types. If the types are ranked so that $\dfrac{c_1}{h_1} > \dfrac{c_2}{h_2} > \dots > \dfrac{c_n}{h_n}$, then the analogous condition to (7) is

$$\frac{c_i}{h_i} > \sum_{j=1}^{i-1} \lambda_j \left(c_j - \frac{c_i}{h_i} h_j \right). \tag{8}$$

The prediction is that the most valuable prey type (as defined by the ranking) should always be accepted as encountered, but as the encounter rate with it falls, the next prey item in the ranking should be included. This continues for all prey types, with the less valuable prey items being sequentially included as the encounter rates with more valuable types drop. Again, either all or none of particular prey types are included in the optimal diet, e.g., $p_i^* = 0$ or $p_i^* = 1$ for each i.

The above approach has a large number of limitations. Biologically, it assumes that predators can rank prey types according to their food values, and remember these rankings. The food value of a prey item is assumed to be recognized by the predator immediately upon being perceived, which ignores any prey type dependent recognition time necessary. The optimization criterion ignores any limitations there may be on total time available for foraging, and is taken over a short time period, relative to organism lifespan. The time period chosen for this model includes only that portion of the day when the organism is actively foraging. Many of the assumptions listed above have been weakened in a variety of models, some of which are discussed below. A review of experimental tests of this model is given in Townsend and Hughes (1981). Some tests, such as Erichsen et al. (1980), show relatively good agreement with model predictions for the situation when birds are offered different sized food items which are disguised to all appear the same. Although the diet observed was close to the optimal one, the birds portrayed a partial preference for the poorer food type, which is not predicted by the above diet model.

One can ask how the above results might change with a similar model, but different optimization criteria. If the criterion was to maximize net energy gain in a fixed amount of foraging time, T, then the above holds exactly when $T \to \infty$. When T is large relative to mean time between encounters, then we still expect the above conclusions to hold approximately. If T is very small relative to time between encounters, then the optimum solution is to accept whatever food item is encountered. For intermediate values of T, this criterion appears not to have been investigated. Also the criterion of minimizing the total time spent handling prey, T_p^{tot}, in some fixed total foraging time T, subject to the constraint that total food intake be above some fixed level, has not been investigated. Such a criterion would be reasonable in an environment with abundant prey, but with a risk to the forager of being attacked by it's own predator while foraging. Experimental tests which investigate how foragers balance the conflicting demands of food requirements and predation risk (Cerri and Fraser, 1983; Werner et al., 1983b) indicate that behaviors are considerably modified from those based just on optimal diet choice.

Other criteria concerning such things as maximizing the probability of a certain minimum food intake or minimizing the variance of intake cannot be formulated in the above model without taking into account the underlying probabilistic structure of the model. The handling times and energy values above were all assumed fixed, ignoring the variability in these within any one prey type. This might only be proper if the within-type variability is small in comparison to between-type variability in these factors. The above also does not analyze the full stochastic nature of the arrival process of prey, which should be looked at as a locking-counter type of renewal process with cumulative energy gains. Only the means of the random variables are considered in the above and the optimization criterion is $\dfrac{E[C]}{E[T]}$, which will not in general be the same as $E\left[\dfrac{C}{T}\right]$. Equating these two has been called the fallacy of the averages and it has been pointed out as a potential flaw of the theory by Templeton and Lawlor (1981). However other authors (Gilliam et al., 1982; Turelli et al., 1982) argue that the above model is reasonable when either long foraging times are considered or large numbers of prey are consumed. When within-prey type variation in handling times is small relative to mean search time, the fallacy of the averages doesn't apply even when the number of prey attacked is as low as 20. Note also that in the above the criterion was computed at the end of a foraging bout, which is a random time, rather than an arbitrary fixed time. The conclusions of the above diet model do depend in part on the assumption of Poisson arrivals of individual prey types and the independence of types of successively encountered prey. A more general stochastic model is considered later, and there has been some work on the case of clumped prey distribution (Pulliam, 1974), but the analysis becomes considerably more complicated when a full stochastic formulation of the model is considered.

The above model explicitly considers the case when food items come in discrete packages which the forager encounters at distinct times and there is a cost to the organism for time spent foraging. A quite different situation arises in the case of filter feeders and deposit feeders in aquatic habitats which continuously sift food particles out of suspension or the sediment. In this case, energy value is determined by particle size and digestibility, handling times don't enter into the problem, and somewhat different models have been constructed to evaluate foraging behavior (Townsend and Hughes, 1981; Phillips, 1984). Another situation for which the above model assumptions are inappropriate is the case of grazers for which there is a continuum of "prey" types (e.g. leaf ages, sizes, heights on plants, etc.) which the herbivore ingests continuously. Models of herbivore foraging are mostly oriented towards maximizing certain nutrient intakes and avoiding overingestion of plant toxins (Owen-Smith and Novelli, 1982). Linear programming techniques have been applied to take into account a variety of constraints on such foragers (Belovsky, 1978, 1984).

Patch Use

In this situation, a predator is visualized as foraging on a number of different patches which vary in the quantity and quality of prey available. The patches could be discrete units such as a flower cluster foraged by bees for nectar, a tree trunk searched by woodpeckers for insects, or a slow moving pool in a stream, or the patches could just be the heterogeneity in an overall uniform environment, such as a relatively high density of flowers in an old field. Once entering a patch, the forager may pursue prey according to an optimal-diet-type model, except that the foraging activity reduces the prey density in the patch. Compare this with the diet model, in which the prey densities are constant in time. The question of interest concerns how a predator apportions its time between the various patches so as to maximize a fitness measure. Again, the currency is assumed to be caloric gain, and the fitness criterion is taken as the net rate of energy gain.

The model assumptions are:

(a) Patches can be grouped into types $1, 2, ..., n$. These patch types are associated with particular prey types, particular prey densities or sizes within the patch, or particular sizes of patch.

(b) Patches of type i are of constant density, so that $p_i =$ proportion of patches of type i, is constant. Thus, although foragers may deplete a patch, at any one time relatively few patches are depleted. This implies that the patches recover rapidly from foraging relative to the time between arrivals of foragers, or else that patches exist in a statistical equilibrium of types, which would imply that there is a constant density of foragers. In this sense, the model is non-dynamic.

(c) The forager is omniscient in that it knows when entering a patch what type it is and what the energy gain rate will be. It also knows this for the environment as a whole, thus the p_i's are known.

(d) There is a constant locomotory energy cost per unit travel time, E_r, for movement between patches and this is independent of patch type most recently visited.

(e) The travel time between patches is independent of the history of patch visits – it's mean value is T_r.

(f) For each patch type, there is an energy gain function $E_i(t)$ which gives the net gain from foraging in patch type i for time t. This function includes the subtraction of any metabolic or handling costs in the patch.

(g) The forager has control over t_i, the time spent foraging in a patch of type i.

Under these assumptions, the average time between leaving one patch and leaving the next is

$$T = T_r + \sum_{i=1}^{n} p_i t_i \qquad (9)$$

and the average net energy gain during this time is

$$E = \sum_{i=1}^{n} p_i E_i(t_i) - E_r T_r . \qquad (10)$$

The fitness criterion is

$$F = \frac{E}{T} = \frac{\sum_{i=1}^{n} p_i E_i(t_i) - E_r T_r}{T_r + \sum_{i=1}^{n} p_i t_i} \qquad (11)$$

and the objective is to choose t_i^*, $i = 1, 2, ..., n$ so as to maximize (11).

It is easily seen by differentiating (11) that if $\dfrac{dE_i}{dt}$ is a decreasing function, then the optimal t_i^*'s are such that

$$\frac{dE_i(t_i^*)}{dt} = \frac{E}{T} \quad \text{for all} \quad i = 1, ..., n, \qquad (12)$$

where E and T are calculated at the optimum $t^* = (t_1^*, ..., t_n^*)$. This says that the optimal time to leave a patch is when the instantaneous rate of net energy gain in that patch has reached the value of the overall rate of energy gain from all patches. This result, derived in Charnov (1976), is called the marginal value theorem. It implies that there will be a concentration of foraging effort to progressively fewer patches, until all patches are depleted to the average profitability of the habitat (Townsend and Hughes, 1981). If a comparison is made between habitats with different average profitabilities (E/T values), then the result implies that predators should remain a longer time in a patch with a given number of prey in the less profitable habitat than in a similar patch in the more profitable habitat. Note that the assumption that $\dfrac{dE_i}{dt}$ is decreasing implies that the forager depletes the patch.

Alternative forms, such as assuming that $\dfrac{dE_i}{dt}$ is increasing for small times to take into account learning effects, seem not to have been introduced in this setting, though other learning models have been described (McNair, 1980).

Some difficulties with the mathematical formulation of the above will be pointed out later. A variety of other models have been constructed for patch use, including the use of a fixed "giving up time". In this model a predator leaves a patch if it hasn't located a prey item there by the giving up time (Hassell and May, 1974). In the above model the giving up time isn't fixed, but depends upon the average availability of prey over all patches. Thus a fixed giving up time is a more reasonable assumption if the predator cannot assess the overall profitability of a site, so that the time spent in a patch depends upon some innate response to each patch by the predator. Another example of this is the gut-filling type of model in which the time to leave a patch depends upon how full the fut of the animal is (Cook and Cockrell, 1978). Unfortunately, the experimental tests to date do not exclude most of the above models.

Hassell (1980) investigates a host-parasite situation, concluding that the data agree with the conclusions of three different foraging models, but are unable to distinguish between their assumptions. A study of heteropterans feeding on mosquito larvae (Giller, 1980) also was unable to distinguish between the gut-filling and marginal value theorem-types of models. The implication is that the models are much too simplified to be able to include the effect of complex behaviors such as variable searching efficiency and extraction efficiencies which varied with position of prey in the catch sequence. In a study on patch use by horned lizards, Munger (1984) obtained results generally consistent with the marginal value theorem and showed the behavior did not follow the predictions of a particular giving up time model.

Difficulties

Deterministic models are the most common approach to optimal foraging theory (Schoener, 1971; Katz, 1974; De Benedictis et al., 1978; Caraco, 1979; Craig et al., 1979; Pyke, 1979; Hixon, 1980; Sih, 1980). These models assume a fixed rate of prey arrival, fixed food value of each prey type considered, and fixed times to handle and eat the prey. Oaten (1977) has given several arguments as to why the inherent stochasticity of foraging should not be ignored. It is important not only that prey encounters and prey quality (in terms of food content) are sampled from a distribution, but also that the predator must use this random information to determine its choice of future behavior. Other difficulties revolve around the free use of means of random variables in deterministic models, as is done in the patch use case above. Oaten concludes that the optimal behaviors deduced from deterministic models can be very poor choices when considered in a more realistic stochastic framework. In a study on bees, the model predictions were found to be completely inadequate to describe the observed foraging behavior since the effects of the variance of the bee's behavior patterns were not included (Waddington and Holden, 1979). In a diet choice model, Lucas (1983) shows that partial preferences for prey types can arise when the forager can monitor the variance in the number of high quality prey missed while handling a lower quality prey. Thus it appears a stochastic approach is essential to increase the realism of many foraging models.

Although a number of studies have included probabilistic formulations (Charnov, 1976; Sih, 1979; Taylor, 1979; Waddington and Holden, 1979), their approach is essentially a deterministic one in that they use only expected values of the relevant random variables in their models. This in effect ignores the underlying random nature of the systems. A few papers do carry out a more complete stochastic analysis (Paloheimo, 1971; Pulliam, 1974; Oster, 1976; Oaten, 1977; Caraco, 1980; Chesson, 1983; Clark and Mangel, 1984; Green, 1984), though these papers mostly use expected values of the optimization criteria in determining the optimal behavior.

Oster (1976) constructs a time and energy budget model for bumblebees as a semi-Markov process. The states of the process are the behavioral roles of the predator (i.e. hive activities, feeding, loafing, etc.), the holding time distribution in each state is assumed to depend only on the most recent transition, and the optimization criterion is the expected total reward from foraging. This approach has proved useful in analyzing the structure of a bee colony (Oster and Wilson, 1978). One of its limitations however, is that risk-taking behavior cannot be considered since it is expected reward which is optimized. Later I will describe the attempt of Caraco (1980) to model risk-taking.

A More Elaborate Patch Model

As mentioned above, there are problems with the deterministic approach to patch utilization models. For example, in the above patch model, the times spent in a patch of type i, t_i, were taken fixed for any given strategy of patch use, whereas in reality they are variable. This implies that the energy gains $E_i(t_i)$ are average gains for a mean time of t_i in patch type i. The difficulty is that for different strategies with the same mean time t_i in a patch, there could be quite different expected energy gains. For example, consider a patch of unit area with exactly one prey in it, and two different foragers, each of which search the patch systematically at the same speed of 1 unit area per unit time. Forager A searches half the patch regardless of what happens, while B always searches until it captures the prey, then leaves. Then A and B both search an average time $t = 1/2$ but for A the average gain is $E(t) = 1/2$ while for B, $E(t) = 1$. Thus to specify the model, which in this case means establishing $E_i(t_i)$, the strategy for choosing the t_i's must be known, which introduces a circularity in the model. This difficulty is pointed out by Oaten (1977), and I now proceed to describe his alternative formulation of the problem.

There are basically two stochastic elements introduced. First, prey captures are random events, with interarrival times specified by a stochastic process. Secondly, a predator does not know upon arriving at a patch how many prey are there, but has a prior distribution, based on past experience, for the probability the patch holds a certain number of prey. Thus a fraction q_i of patches have i prey items, where $\sum\limits_{i=0}^{\infty} q_i = 1$. The manner in which an individual might estimate the prior distribution $\{q_i\}$ and adjust it with experience involves statistical decision theory. This aspect of the problem is not considered here, but see McNamara and Houston (1980) for the application to foraging.

The assumptions are as follows (Oaten, 1977):

    i) The predator knows only q_i = probability there are i prey in any given patch, $i = 0, 1, 2, \ldots$, and not the actual number in a patch. There is only one type of prey, and there are an effectively infinite number of patches.

ii) Given that there are k prey in a patch and the interarrival times between prey captures are $T_1, T_2, ..., T_j, j \leq k$, then $(T_1, ..., T_j)$ have joint probability density $f(u_1, u_2, ..., u_j|k) = f(\tilde{u}_j|k)$, which is known to the predator.

iii) Prey are not replaced as they are eaten, so that $T_{k+1} = \infty$ if there were initially k prey in the patch.

iv) The time to travel between patches is τ, which is here considered fixed, but could be taken random.

v) The decision to leave a patch is based on τ, $\{q_i\}$, the functions $f(\cdot)$, and the experience in the patch, meaning the actual interarrival times $u_1, ..., u_j$ and the time since last capture v_{j+1}. The predator strategy is a sequence of giving up times $t_1, t_2(u_1), t_3(u_1, u_2), ...$ so that if prey have been caught at times $u_1, u_1 + u_2, u_1 + u_2 + u_3, ..., u_1 + ... + u_j$, then the predator will leave the patch if another prey has not been caught by time $t_{j+1}(u_1, ..., u_j)$ after the last prey capture.

vi) A strategy of giving up times is chosen to maximize the expected number of prey eaten per unit time defined by

$$R = \frac{E[G]}{\tau + \eta E[S]E[G]},\tag{13}$$

where G is the number of prey eaten in a patch, S is the time spent searching a patch, and η is the time spent handling each prey, here assumed constant. The value of η does not affect the optimal strategy, so set $\eta = 1$.

With these assumptions:

$$E[G] = \sum_{i=1}^{\infty} P[G \geq i]$$

$$= \sum_{i=1}^{\infty} \sum_{k=0}^{\infty} q_k P[G \geq i | k \text{ prey present initially}]$$

$$= \sum_{i=1}^{\infty} \sum_{k=i}^{\infty} q_k \int_0^{t_1} \int_0^{t_2} ... \int_0^{t_i} f(\tilde{u}_i|k) d\tilde{u}_i.\tag{14}$$

The expected search time is

$$E[S] = \sum_{i=1}^{\infty} E[(T_i \wedge t_i) \prod_{j=0}^{i-1} 1_{\{T_j \leq t_j\}}]$$

$$= \sum_{i=1}^{\infty} E[X_i],\tag{15}$$

where 1_A is the indicator random variable of the event A, i.e.,

$$1_A(w) = \begin{cases} 1 & \text{if } w \in A \\ 0 & \text{if } w \notin A \end{cases} \quad \text{for } w \text{ in the sample space},$$

$T_i \wedge t_i = \text{minimum}(T_i, t_i)$, and $t_0 = \infty$. Then

$$E[S] = \sum_{i=1}^{\infty} \sum_{k=0}^{\infty} q_k E[X_i | k \text{ prey present initially}]$$

$$= \sum_{i=1}^{\infty} \left\{ q_{i-1} E[X_i | i-1 \text{ prey present initially}] \right.$$

$$\left. + \sum_{k=i}^{\infty} q_k E[X_i | k \text{ prey present initially}] \right\}$$ (16)

and letting

$$b_{ik} = q_k E[X_i | k \text{ prey present initially}]$$

$$= q_k \int_0^{t_1} \cdots \int_0^{t_{i-1}} \int_0^{\infty} (t_i \wedge u_i) f(\tilde{u}_i | k) d\tilde{u}_i \quad \text{for} \quad k \geq i$$

one gets

$$E[S] = q_0 t_1 + \sum_{k=1}^{\infty} b_{1k}$$

$$+ \sum_{i=2}^{\infty} \left\{ q_{i-1} \int_0^{t_1} \cdots \int_0^{t_{i-1}} t_i f(\tilde{u}_{i-1} | i-1) d\tilde{u}_{i-1} + \sum_{k=i}^{\infty} b_{ik} \right\}.$$ (17)

The criterion (13) is now completely specified using (14) and (17).

Oaten (1977) describes a calculus of variations procedure to specify the t_i's which maximize (13) and derives a necessary condition for a solution. For the situation in which there are only a finite maximum number of prey possible in a patch, the optimum t_i's may be computed; however, there are analytical difficulties introduced by the possible non-uniqueness of the solution. The main result may be stated as: the optimal time to leave a patch, t_m^*, after the $(m-1)^{\text{st}}$ prey arrival, satisfies

$$R(t_m^* | \tilde{u}_{m-1}) + Z = R^*,$$ (18)

where $R(t_m | \tilde{u}_{m-1})$ is the conditional capture rate at time t_m given capture interarrival times $\tilde{u}_{m-1}$, R^* is the optimal overall prey capture rate, and Z is a function of $(\{q_i\}, m, \tilde{u}_{m-1})$ representing the rate of future success in the patch for an optimally acting predator. The marginal value theorem of the earlier patch use model essentially ignores the Z term, and chooses t_i's to satisfy (18) with $Z=0$. Oaten (1977) gives examples of situations in which a predator following the marginal value theorem would do arbitrarily poorly in comparison to a predator following the optimal strategy derived from (18). This arises due to the benefit a predator has in not leaving a patch immediately upon the time its rate of capture reaches the level of the habitat average, because remaining there may provide information that the yield is much richer than it appeared by prior experience.

The above formulation is so complex that it is difficult to see how simple behavioral rules that a forager might follow would lead to an optimal solution. To analyze this, Green (1984) considers a special case of the above under the assumption that each patch contains a fixed number of locations to search, with a binomially distributed number of these locations containing prey, where the parameter of the binomial distribution varies between patches according to a beta distribution. All this is "known" to the prey and it is then possible to derive and compare three rules as to when to leave a patch: a fixed-time rule, a giving-up time rule, and an assessment rule. The assessment rule here is the optimal stopping rule based on the present assessment of patch quality, and is determined from a dynamic programming approach. Green also argues that this assessment rule is robust to small changes in it's form, and is simple enough that foragers might be expected to use it. A somewhat different model (Clark and Mangel, 1984), making use of Bayesian techniques to analyze the advantages of flocking, concludes that the marginal value theorem is misleading due to the importance of probing in an uncertain environment and further indicates that giving-up-time ideas may be far too restrictive in cases of uncertainty.

The results of these further elaborations clearly point out the potential inadequacies of using a deterministic formulation when the underlying nature of the problem is stochastic. However, the above formulation is still not complete. The results can be readily extended to several prey types, but with the inclusion of random handling times, and finite numbers of patches so that it would be necessary to include the sampling effects on patches already visited, the analysis would rapidly become intractible. There has been essentially no work done on optimization criteria other than mean energy gain per unit time, and it is unclear whether other realistic criteria based on the probabilistic nature of foraging would produce very different predictions. Some other potential extensions are mentioned later.

A Utility Theory Criterion

One of the criticisms that may be aimed at virtually all the foraging models developed to date is their ignorance of the variability and unpredictability of habitats. As seen in the above models, the optimization criteria are couched in terms of expected values of energy gains. This ignores the potential importance of variance in these gains. By these criteria, the same strategy would be adaptive in an environment with a constant reward value of 1 say, as in an environment which had 0.5 probability of reward 0 and 0.5 probability of reward 2. In order to include certain aspects of environmental variance, Caraco (1980) develops an approach based on utility theory, which I here briefly describe.

Suppose that B, the net energetic benefit acquired during one foraging period, is a random variable with probability density function $f(b)$. We suppose there is a real-valued utility function, $U(B)$, which specifies the utility or preference of the predator for the energy gains given by the random variable B, relative to other potential gain distributions. Here $E[U(B)] < \infty$, and suppose U, U', and U'' are continuous and $U(\cdot)$ is monotone increasing. If fitness is a concave function of net energy, then it is reasonable that U should be concave, i.e., $U'>0$, $U''<0$. Then Jensen's inequality implies

$$E[U(B)] \leqq U(E[B])$$

and the forager prefers, in terms of utility, $E[B]$ with certainty over sampling from the distribution of B, $f(b)$. A fixed energy gain $\hat{B}$, which would give the same utility as the expected utility sampled from $f(b)$, i.e.

$$U(\hat{B}) = E[U(B)]$$

is called the *certainty equivalent*. We thus have, since U is monotone, $\hat{B}<E[B]$. This is called *risk-averse* behavior since the animal prefers to avoid the risk of doing poorly due to environmental variation in energy gains.

Alternatively, if U is convex, so that $U'>0$, $U''>0$, then $\hat{B}>E[B]$ and $E[U(B)] \geq U(E[B])$ and the forager is *risk-prone*. In this case, the forager prefers the risk of feeding in a stochastic environment over receiving the mean energy gain with certainty. The previous deterministic models are *risk-neutral* in that they do not differentiate between two alternative resources with the same expected rewards but different variances.

To apply utility theory, consider an environment with m food types, each with an energy gain probability density function $f_i(b)$, $i=1,2,...,m$. A resource utilization strategy would have the form

$$f(b) = \sum_{i=1}^{m} p_i f_i(b)$$

for the density associated with spending a fraction p_i of the day foraging on food type i. Here $\sum_{i=1}^{m} p_i = 1$ and the p_i's are control variables. The optimal strategy is that which maximizes the expected utility

$$E[U] = \int_{-\infty}^{\infty} U(b) f(b) db$$

subject to the constraints on the p_i's. Caraco (1980) has investigated this problem for a number of different utility functions. This allows comparisons to be made between conservative foragers that are risk-averse, and those that are risk-prone. Specific formulae can be derived to indicate how environmental variance affects the optimal strategy. A key assumption throughout this however, is that the forager knows the probability distribution for rewards for each resource. As mentioned before, if this is not assumed, statistical decision theory comes into play (McNamara and Houston, 1980). With reference to the statistical estimation of preference, see Chesson (1983).

There have been just a couple of experimental studies of the effects of risk on foraging behavior. Werner et al. (1983b) show that bluegill sunfish, which are vulnerable to largemouth bass, will modify their behavior and not forage in an optimal manner when bass are present. The fish balance the conflicting demands of foraging and predation risk, but this is done at the cost of significant loss in growth rate. In a different study, Cerri and Fraser (1983) hypothesize that foragers will take greater risk when potential benefits are high. Testing this with minnows, they determined that the fish did not take proportionately greater risks when the benefits were higher, and this sense do not seem to balance the conflicting demands. They present an alternative that is based on the time in a patch spent watching for predators, but in essence it seems that though both the species investigated in these two studies are risk-averse, the form of the utility functions involved are probably quite different.

Extensions

There are a variety of other problems which fit within the area of foraging theory which I have not discussed above. For a review of theory and experiment on size of territories, see Davies and Houston (1984). Schoener (1983) reviews a large group of territoriality models and concludes that optimal territory size changes in a very complex manner with food density and depends strongly upon how a variety of factors covary. Janetos (1982) compares sit-and-wait foraging to an actively searching behavior. Ford (1983), through an extensive simulation model, obtains results on the size and shape of optimal home range along with proportion of home range used per foraging bout, when the environment consists of a mixture of patches which vary in quantity and density of food. The spatial aspects of foraging have been particularly evaluated in determining the optimal search patterns of bees on flowers. See Pyke (1978) for derivation of rules for moving between patches of inflorescences and Best and Bierzychudek (1982) for predictions of movement patterns along a flower stalk which maximize the rate of net energy gain. Searching behaviors of predators can give rise to a consistent aggregative response to prey density, so that aggregation is not just correlated with the physical environment (Hildrew and Townsend, 1982). This aggregation can lead to stabilizing influences on predator prey systems (Hassell, 1978). Aggregation in groups as it relates to foraging, predation risk and reproductive success is reviewed in Pulliam and Caraco (1984). Clark and Mangel (1984) show that flocking may be evolutionarily stable due to sharing of information between group members. Giraldeau (1984) argues that foraging groups benefit from a pool of skills within the flock and this can generate frequency-dependent effects which promote individual specializations.

Foraging theories have gone a long way towards examining the potential benefits of alternative feeding strategies in a range of different ecological circumstances. It is clear that the theories have generally progressed beyond the capacity of the experimental work to date to differentiate between them. That the need for further experimental work has been recognized is observable from the large number of attempts to test the theories which are appearing in the current literature. However, there are still a wide range of theoretical questions which have not been adquately attacked. I feel that work on these would lead not only to potentially interesting theories, but also perhaps better specify the experiments necessary to adequately test the models. Below I give a list, by no means exhaustive, of the issues which I feel require further theoretical investigation.

1. Very little work has been done on situations in which foraging strategies change with the age, size, or sex of individuals. That these types of within-population feeding differences can be quite important is indicated by some experiments (Clark, 1980; Hoffman, 1983; Peters and Grubb, 1983).

2. The question of specific requirements for a variety of different nutrients (or food types) has been addressed rather infrequently (Westoby, 1974; Owen-Smith and Novellie, 1982; Belovsky, 1984). In some cases a single nutrient can be of overriding importance, as for example lack of nitrogen causes most deaths among young insects (White, 1978) and sodium limitations seem to govern foraging by moose (Belovsky, 1978). More often, a host of different nutrients are important, along with constraints induced by toxins in the forage (Freeland and Janzen, 1974). A few attempts to utilize linear programming techniques to incorporate various toxin constraints and minimal nutrient requirements have been made (Belovsky, 1984), and the importance of the dual has been noted (Altmann, 1984), but clearly much remains to be done concerning such constraints on foraging.

3. Some recent work has undertaken to tie together foraging theory with models of predator-prey population dynamics. Comins and Hassell (1979) deal with a host-parasitoid situation, Holt (1983) investigates how optimal foraging of a predator on two prey may constrain isocline shape, Sih (1984a) discusses spatial patterns which arise due to reciprocal behavioral responses by predator and prey, and Sih (1984b) examines how several optimization criteria of the predator affect density-dependent predation. The effects of optimal foraging on food web models is relatively unexplored, but see (Abrams, 1984) for the effects of foraging-time optimization on predator functional response in a three-layer web. Further work in this area should lead to a better understanding of coevolution.

4. Some authors have derived models for foragers which include the capacity to learn about the prey while foraging and adjust their behavior accordingly (McNair, 1980). A laboratory experiment on rats demonstrated that only a preference for sodium was innate, all other preferences were learned (Rogers, 1967). From field observations, it is clear that for some organisms the ability to determine prey profitability is learned, not inherited (Jaeger and Rubin, 1982), and further work in this quite difficult area is indicated.

5. How a predator deals with a changing environment is still a very open problem. Rather than the daily time period used by most foraging models, the question of long-term optimization criteria, say over an organism's lifespan, is somewhat neglected (Katz, 1974; Craig et al., 1979). Baranga (1983) points out that dietary variation of a monkey changes seasonally as chemical composition, moisture, and energy values of the food plants change. Reichman and Fay (1983) compared a caching and a non-caching rodent to analyze foraging behavior differences re long-term or short-term optimization. De Angelis et al. (1984) consider the changing environment within a day as it affects the food uptake of largemouth bass. These studies point out the need to consider nonstationary environments in many foraging situations.

6. Aside from the use of utility theory mentioned above, there is little work on the effects of the probabilistic nature of rewards on the optimization criteria. Such problems as minimizing the probability of being preyed upon by a higher-trophic level predator, while foraging, are basically not investigated (Heller, 1980). Here a full analysis of the stochastic nature of foraging is essential, extending the work of Paloheimo (1971). Criteria with constraints such as maximizing survival probability subject to obtaining a certain minimum daily intake of food are quite open problems, although Williams and Nichols (1984) do point out a dynamic programming approach which can be used to minimize variance around a pre-established, time-varying target reward level. For related work concerning optimal harvesting rules for minimizing variance subject to constraints on the mean, see Gleit (1979).

7. Attempts to tie together foraging theory with biophysical energy budget requirements have been few (Powell, 1979; Martindale, 1983), but are a logical next step once adequate knowledge is available on a particular organism's energetics.

8. Few studies have explicitly taken into account the sensory limitations foragers face in detecting food sources. Rice (1983) points out that different sensory abilities in foragers will lead to quite different foraging behaviors, Gendron and Staddon (1983) investigate search speeds for predators to balance the conflicting demands of food intake and detection of cryptic prey, while Staddon and Gendron (1983) apply signal detection theory to the case of searching for cryptic prey. These sensory limitations lead to interesting coevolutionary questions (Pyke, 1978b), an example being plants which deceive pollinators (Dafni, 1983).

Acknowledgements

I thank the Grassland Research Institute, Hurley, Great Britain for their hospitality while the initial version of this review was written. Comments and discussions with participants in the Ecology 6100 seminar on foraging theory at the University of Tennessee, Knoxville were extremely helpful and pointed out to me a number of studies of which I was previously unaware. I greatly appreciate the support of a Faculty Research Participation award, administered by Oak Ridge Associated Universities for the U.S. Department of Energy, held at the Atmospheric Turbulence and Diffusion Division, Air Resources Laboratory, National Oceanic and Atmospheric Administration.

References

Abrams, P.A. (1984). Foraging time optimization and interactions in food webs. Amer. Natur. *124*: 80–96

Altman, S.A. (1984). What is the dual of the energy-maximization problem? Amer. Natur. *123*: 433–441

Aronson, R.B., Givnish, T.J. (1983). Optimal central place foragers: a comparison with null hypotheses. Ecology *64*: 395–399

Baranga, D. (1983). Changes in chemical composition of food parts in the diet of Colobus monkeys. Ecology *64*: 558–673

Belovsky, G.E. (1978). Diet optimization in a generalist herbivore: the moose. Th. Pop. Biol. *14*: 105–134

Belovsky, G.E. (1974). Herbivore optimal foraging: a comparative test of three models. Amer. Natur. *124*: 97–115

Best, L.S., Bierzychudek, P. (1982). Pollinator foraging on foxglove *(Digitalis purpurea)*: a test of a new model. Evolution *36*: 70–79

Bradshaw, A.D. (1965). Evolutionary significance of phenotypic plasticity in plants. Adv. Genetics *13*: 115–155

Caraco, T. (1979). Time budgeting and group size: a theory. Ecology *60*: 611–617

Caraco, T. (1980). On foraging time allocation in a stochastic environment, Ecology *61*: 119–128

Cerri, R.D., Fraser, D.F. (1983). Predation and risk in foraging minnows: Balancing conflicting demands. Amer. Natur. *121*: 552–561

Charnov, E.L. (1976). Optimal foraging: attack strategy of a mantid. Amer. Natur. *110*: 141–151

Chesson, J. (1983). The estimation and analysis of preference and its relationship to foraging models. Ecology *64*: 1297–1304

Church, D.C. (1975). Digestive physiology and nutrition of ruminants. 2nd ed. O & B Books, Corvallis, Oregon

Clark, D.A. (1980). Age- and sex-dependent foraging strategies of a small mammalian omnivore. J. Anim. Ecol. *49*: 549–563

Clark, C.W., Mangel, M. (1984). Foraging and flocking strategies: information in an uncertain environment. Amer. Natur. *123*: 626–641

Comins, H.M., Hassell, M.P. (1979). The dynamics of optimally foraging predators and parasitoids. J. Anim. Ecol. *48*: 335–351

Cook, R.M., Cockrell, B.J. (1978). Predator ingestion rate and its bearing on feeding time and the theory of optimal diets. J. Anim. Ecol. *47*: 529–549

Craig, R.B., De Angelis, D.L., Dixon, K.R. (1979). Long- and short-term dynamic optimization models with application to the feeding strategy of the loggerhead shrike. Amer. Natur. *113*: 31–51

Dafni, A. (1983). Pollination of *Orchis caspia* – a nectarless plant which deceives the pollinators of nectivorous species from other plant families. J. Ecology *71*: 467–474

Davenport, H.W. (1978). A digest of digestion. 2nd ed. Year Book Medical Publisher, Chicago

Davies, N.B., Houston, A.I. (1984). Territory economics. Pages 148–169 in J.R. Krebs and N.B. Davies (eds.) ibid

De Angelis, D.L., Adams, S.M., Breck, J.E., Gross, L.J. (1984). A stochastic predation model: application to largemouth bass observations. Ecol. Model. *24*: 25–41

De Benedictis, P.A., Gill, F.B., Hainsworth, F.R., Pyke, G.F., Wolf, L.L. (1978). Optimal meal size in hummingbirds. Amer. Natur. *112*: 301–316

Erichsen, J.T., Krebs, J.R., Houston, A.I. (1980). Optimal foraging and cryptic prey. J. Anim. Ecol. *49*: 271–276

Freeland, W.J., Janzen, D.H. (1974). Strategies in herbivory by mammals: the role of plant secondary compounds. Amer. Natur. *108*: 269–289

Ford, R.G. (1983). Home range in a patchy environment: optimal foraging predictions. Amer. Zool. *23*: 315–326

Gendron, R.P., Staddon, J.E.R. (1983). Searching for cryptic prey: the effect of search rate. Amer. Natur. *121*: 172–186

Giller, P.S. (1980). The control of handling time and its effects on the foraging strategy of a heteropteran predator, *Notonecta*. J. Anim. Ecol. *49*: 699–712

Gilliam, J.F., Green, R.F., Pearson, N.E. (1982). The fallacy of the traffic policeman: a response to Templeton and Lawlor. Amer. Natur. *119*: 875–878

Giraldeau, L.A. (1984). Group foraging: the skill pool effect and frequency dependent learning. Amer. Natur. *124*: 72–79

Gleit, A. (1979). On optimal mean variance harvesting rules. Math. Bio. Sci. *45*: 179–200

Green, R.F. (1984). Stopping rules for optimal foragers. Amer. Natur. *123*: 30–40

Hassell, M.P. (1978). The dynamics of arthropod predator-prey systems. Princeton Univ. Press, Princeton, N.J.

Hassell, M.P. (1980). Foraging strategies, population models and biological control: a case study. J. Anim. Ecol. *49*: 603–628

Hassell, M.P., May, R.M. (1974). Aggregation of predators and insect parasites and its effect on stability. J. Anim. Ecol. *43*: 567–587

Heller, R. (1980). Foraging on potentially harmful prey. J. Theor. Biol. *85*: 807–813

Hildrew, A.G., Townsend, C.R. (1982). Predators and prey in a patchy environment: a freshwater study. J. Anim. Ecol. *51*: 707–815

Hixon, M.A. (1980). Food production and competitor density as the determinants of feeding territory size. Amer. Natur. *115*: 510–530

Hixon, M.A., Carpenter, F.L., Paton, D.C. (1983). Territory area, flower density and time budgeting in hummingbirds: an experimental and theoretical analysis. Amer. Natur. *122*: 366–391

Hoffman, S.G. (1983). Sex-related foraging behavior in sequentially hermaphroditic hogfishes (*Bodianus* spp.) Ecology *64*: 798–808

Holt, R.D. (1983). Optimal foraging and the form of the predator isocline. Amer. Natur. *122*: 521–541

Jacob, F. (1977). Evolution and tinkering. Science *196*: 1161–1166

Jaeger, R.G., Rubin, A.M. (1982). Foraging tactics of a terrestrial salamander: judging prey profitability. J. Anim. Ecol. *51*: 167–176

Janetos, A.C. (1982). Active foragers versus sit and wait predators: a simple model. J. theor. Biol. *95*: 381–385

Katz, P.L. (1974). A long-term approach to foraging optimization. Amer. Natur. *108*: 758–782

Krebs, J.R., Davies, N.B. (1984). An introduction to behavioural ecology. 2nd ed. Sinauer. Sunderland, Mass.

Krebs, J.R., McCleery, R.M. (1984). Optimization in behavioural ecology. Pages 91–121 in J. R. Krebs and N. B. Davies, (eds.) ibid.

Lucas, J.R. (1983). The role of foraging time constraints and variable prey encounter in optimal diet choice. Amer. Natur. *122*: 191–209

Martindale, S. (1983). Foraging patterns of nesting Gila woodpeckers. Ecology *64*: 888–898

Maynard Smith, J. (1978). Optimization theory in evolution. Ann. Rev. Ecol. System. *9*: 31–56

McGinley, M.A. (1984). Central place foraging for non-food items: determination of the stick size-value relationship of house building materials collected by eastern woodrats. Amer. Natur. *123*: 841–853

McNair, J.M. (1980). A stochastic foraging model with predator training effects: I. Functional response, switching, and run lengths. Th. Popul. Biol. *17*: 141–166

McNamara, J., Houston, A. (1980). The application of statistical decision theory to animal behavior. J. Theor. Biol. *85*: 673–690

Munger, J.C. (1984). Optimal foraging? Patch use by horned lizards (Iguanidae: *Phrynosoma*) Amer. Natur. *123*: 654–680

Oaten, A. (1977). Optimal foraging in patches: a case for stochasticity. Th. Popul. Biol. *12*: 263–285

Oster, G. (1976). Modeling social insect populations. I. Ergonomics of foraging and population growth in bumblebees. Amer. Natur. *110*: 215–245

Oster, G., Wilson, E.O. (1978). Caste and ecology in the social insects. Princeton Univ. Press, Princeton, N.J.

Owen-Smith, N., Novellie, P. (1982). What should a clever ungulate eat? Amer. Natur. *119*: 151–178

Peters, W.D., Grubb, T.C. Jr. (1983). An experimental analysis of sex specific foraging in the Downy Woodpecker. *Picoides pubescens*. Ecology *64*: 1437–1443

Phillips, N.W. (1984). Compensatory intake can be consistent with an optimal foraging model. Amer. Natur. *123*: 867–872

Powell, R.A. (1979). Ecological energetics and foraging strategies of the fisher *(Martes pennanti)*. J. Anim. Ecol. *48*: 195–212

Pulliam, H.R. (1984). On the theory of optimal diets. Amer. Natur. *108*: 59–74

Pulliam, H.R., Caraco, T. (1984). Living in groups: is there an optimal group size? Pages 122–147 in J. R. Krebs and N. B. Davies (eds.) ibid.

Pyke, G.H. (1978a). Optimal foraging: movement patterns of bumblebees between inflorescences. Theor. Popul. Biol. *13*: 72–98

Pyke, G.H. (1978b). Optimal foraging in bumblebees and coevolution with plants. Oecologia *36*: 281–293

Pyke, G.H. (1979). The economics of territory size and time budget in the golden-winged sunbird. Amer. Natur. *114*: 131–145

Pyke, G.H., Pulliam, H.R., Charnov, E.L. (1977). Optimal foraging: a selective review of theory and tests. Quart. Rev. Biol. *52*: 137–154

Reichman, O.J., Fay, P. (1983). Comparison of the diets of a caching and a noncaching rodent. Amer. Natur. *122*: 576–581

Rice, W.R. (1983). Sensory modality: an example of its effect on optimal foraging behavior. Ecology *64*: 403–406

Richards, L.J. (1983). Hunger and the optimal diet. Amer. Natur. *122*: 326–334

Rogers, W.L. (1967). Specificity of specific hungers. J. Comp. Physiol. Psych. *64*: 49–58

Schoener, T.W. (1971). Theory of feeding strategies. Ann. Rev. Ecol. System. *2*: 369–404

Schoener, T.W. (1983). Simple models of optimal feeding – territory size: a reconciliation. Amer. Natur. *121*: 608–629

Sih, A. (1979). Optimal diet: the relative importance of the parameters. Amer. Natur. *113*: 360–463

Sih, A. (1980). Optimal foraging: partial consumption of prey. Amer. Natur. *116*: 281–290

Sih, A. (1984a). The behavioral response race between predator and prey. Amer. Natur. *123*: 143–150

Sih, A. (1984b). Optimal behavior and density dependent predation. Amer. Natur. *123*: 314–326

Smith, E.A. (1983). Anthropological applications of optimal foraging theory: a critical review. Current Anthropology *24*: 625–651

Staddon, J.E.R., Gendron, R.P. (1983). Optimal detection of cryptic prey may lead to predator switching. Amer. Natur. *122*: 843–848

Taylor, R.J. (1979). The value of clumping to prey when detectability increases with group size. Amer. Natur. *113*: 299–301

Templeton, A.R., Lawlor, L.R. (1981). The fallacy of the averages in ecological optimization theory. Amer. Natur. *117*: 390–393

Townsend, C.R., Hughes, R.N. (1981). Maximizing net energy returns from foraging. Pages 86–108 in C. R. Townsend and P. Calow (eds.). Physiological ecology. An evolutionary approach to resource use. Sinauer, Sunderland, Mass.

Turelli, M., Gillespie, J.H., Schoener, T.W. (1982). The fallacy of the fallacy of the averages in ecological optimization theory. Amer. Natur. *119*: 879–884

Waddington, K.D., Holden, L.R. (1979). Optimal foraging: on flower selection by bees. Amer. Natur. *114*: 179–196

Werner, E.E., Mittelbach, G.G., Hall, D.J., Gilliam, J.F. (1983a). Experimental tests of optimal habitat use in fish: the role of relative habitat profitability. Ecology *64*: 1525–1539

Werner, E.E., Gilliam, J.F., Hall, D.J., Mittelbach, G.G. (1983b). An experimental test of the effects of predation risk on habitat use in fish. Ecology *64*: 1540–1548

Westoby, M. (1974). An analysis of diet selection by large generalist herbivores. Amer. Natur. *108*: 290–304

White, T.C.R. (1978). The importance of a relative shortage of food in animal ecology. Oecologia *33*: 71–86

Williams, B.K., Nichols, J.D. (1984). Optimal timing in biological processes. Amer. Natur. *123*: 1–19

Winterhalder, B. (1983). Opportunity-cost foraging models for stationary and mobile predators. Amer. Natur. *122*: 73–84

Part III. Population Ecology

Population Dynamics in a Homogeneous Environment

Thomas G. Hallam

Outline

1. Introduction to Population Theory

1.0 Introduction

Ecology, freely translated from the Greek expression, means "the study of the household of nature." A *population* is a collection of organisms, usually of the same species, that occupy a prescribed region and function together as an ecological entity. While most populations consist of a single species, the definition of a

Biomathematics, Vol. 17, Mathematical Ecology
Edited by T. G. Hallam and S. A. Levin
© Springer-Verlag Berlin Heidelberg 1986

population is intended to be sufficiently broad to include assemblages of species such as those that can interbreed to produce viable hybrids. Another example might be lichen populations where the algae and fungi are so closely associated that they function as a single species. Population ecology, in the sense meant here, refers to the structure and function of a collection of organisms as an autecological unit and, as such, addresses the more purely biological aspects of ecology.

The relationship between a population and its biogeochemical environment is often an intricate one. Populations are seldom dominated by their physical environment; indeed, they are often able to modify and even regulate, within constraints, their environment. The biology of the population, the biogeochemical characteristics of the environment and the feedback mechanisms between the collection of organisms and the environment are important aspects of population ecology.

The purpose of this chapter is to introduce some classical deterministic population models and some elementary analysis techniques for those models. It will not be possible to cover this field completely; however, the references provided should indicate the directions of other developments.

1.1 Characteristics of a Population

Before embarking on a study of population models, I introduce some terminology that is employed to describe a population.

There are many traditional, physical characteristics of populations that even a casual observer can delineate. These include the size and distribution of the population. The *density* of the population is the number of individual organisms per unit of space. The density of the population can be estimated from a population *census*; it is, of course, a nontrivial task to accurately census most natural populations. The manner in which organisms are distributed in space is referred to as the *dispersion* of the population. In the subsequent models, it will be assumed that the population is uniformly dispersed throughout the region. This situation is termed *spatially homogeneous*; hence, the origin of the title of this chapter.

There are ways that population size can fluctuate. Natality can occur; new organisms can arise from seeds, spores, or eggs. It is customary to refer to the rate of addition of new individuals to a population by reproduction as the *birth rate*. Mortality decreases population size; the rate at which organisms are lost from the population by death is called the *death rate*. Another change in population size can be effected through immigration and emigration. The rate at which organisms immigrate to or emigrate from the region used to define the population is called the *dispersal rate* of the population.

There are many other attributes of a population that help to characterize it. These include the sex, age, and organism size distributions as well as genetic characteristics. The proportion of organisms of different ages (or sizes) is often a fundamental characteristic of populations and this feature subsequently will be explored in detail in Drs. Nisbet, Gurney, and Frauenthal's notes (Chaps. 2 and 3). Other important traits that should be included in discussions of a population involve the evolutionary aspects of population ecology. Evolutionary phenomena are generally considered to be most important on time scales longer than the demographic time scales which these notes cover.

1.2 Modelling Aspects of Populations

The use of the language of mathematics to model or represent a phenomenon has several advantages. This interdisciplinary approach allows the use of tools from two disciplines – those of the scientific area plus those of mathematics. A traditional view of the use of mathematical models is to "predict" the behavior of a system. In biology, while prediction is highly desirable, this objective is probably one of the least fruitful avenues of research. Among the uses of mathematical modelling, many of which will be illustrated in these notes, are
1. Clarification of definitions of compartments, flows, and pathways,
2. Generation of hypotheses about the system,
3. Suggestion of experiments to validate these hypotheses,
4. Assistance in research planning,
5. Identification of poorly understood system mechanisms.

In any modelling process the *objectives* of the effort must be delineated clearly. When the objectives are specified, the *representation* of the scientific phenomenon in the language of mathematics results in a model. This procedure requires basic laws which, in ecology, are not plentiful. The contrast between existence of fundamental laws in the physical sciences and the lack of corresponding laws in biological sciences is striking. However, this is one reason that modelling in the biological sciences is fascinating and is evolving at a high rate. After the model is formulated, it must be *solved*. A specific problem can sometimes be solved analytically but often it is necessary to study the model numerically. In the modelling process, the next step is to *interpret* the solution in terms of the ecology of the system and quite possibly to then start the process over by redefining the objectives.

When the model is suitable according to the objectives, an important aspect of modelling is the *control* of the biological system. It is frequently necessary to modify or optimize system behavior (as in agricultural production) so the development of the theory of control system is fundamental.

Some steps (and perhaps, all of these) in the modelling process will be evident in the presentations of these notes.

1.3 Continuous and Discrete Models

If the domain of a deterministic dynamical system is a continuous set (usually $\mathbb{R}_+ = [0, \infty)$) then the model is called, briefly, a *continuous* model. In these notes, a continuous model will refer to an ordinary differential equation. If the domain of the model is a discrete subset of the integers, the model is said to be *discrete*. The discrete models discussed here will be difference equations. Examples of quantities measured continuously include inflation and temperature, while the census of a population is a discrete measurement. It is traditional to model certain discrete phenomena in a continuous manner. Notable physical examples include gas and hydrology measurements. As we will see, it is often convenient to model populations by a continuous model. If a population model has as a state variable the number of individual organisms, this would seem to force the model to be discrete. However, if the population is very large, the addition of another individual is a small change relative to the population size. In this situation, it might be reasonable to model the population in a continuous manner (often not only continuity but differentiability is assumed for such variables). Another justifiable way to view population state variables as continuous is to use biomass as the measured quantity.

1.4 Quantitative Behavior of Populations

To motivate the subsequent development, the dynamic structure of some laboratory and natural populations is presented.

How do Populations Behave at Densities that Allow Substantial Growth?

Figures 1.1, 1.2, and 1.3 indicate that some populations grow rapidly from low initial population sizes. From this realized growth, it can be inferred that the population is not at a level where some resource factor (such as a limiting nutrient or a restricted space requirement) is drastically hindering the growth of the population. A reasonable but tautological definition of a population density which promulgates at a high growth rate is that there are no constraints to limit severely expansion of the population. There exist certain constraints on densities of this type; in particular, they need not be very low densities (compare Figs. 1.3a and 1.3b). The *Allee principle* states that undercrowding can be an important factor limiting population growth. Limitations at low population densities might occur if

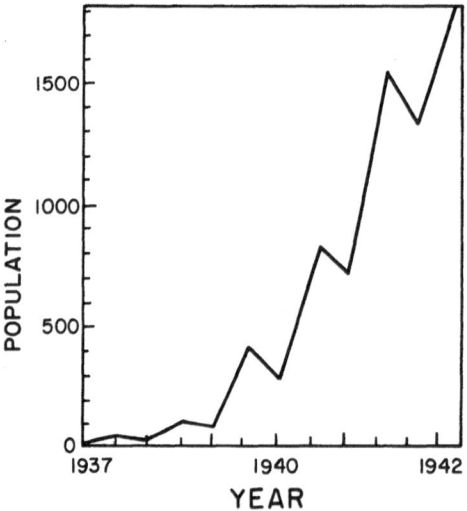

Fig. 1.1. Population of pheasant *(Phasianus colchicus torquatus)* introduced onto Protection Island and followed during the early part of its development (data of Einarsen; Hutchinson, 1978)

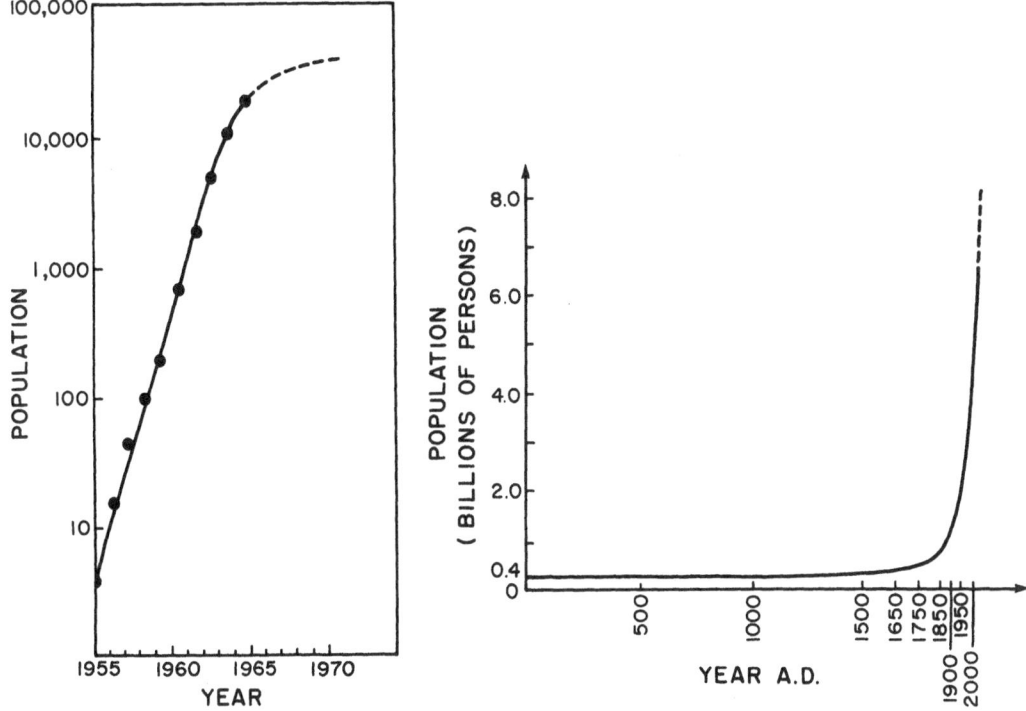

Fig. 1.2. Estimates of population of the *collar turtledove (Streptopelia decalocto)* in Great Britain since 1955. Logarithm scale (Hutchinson, 1978)

Fig. 1.3a. World population in the first two millennia of the Christian Era

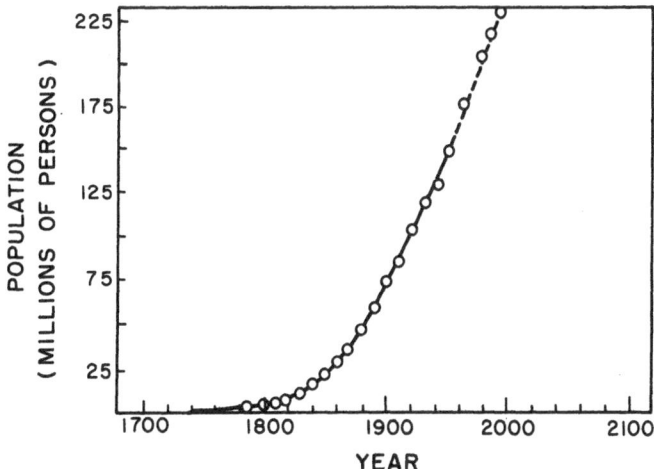

Fig. 1.3b. The human population of the U.S.

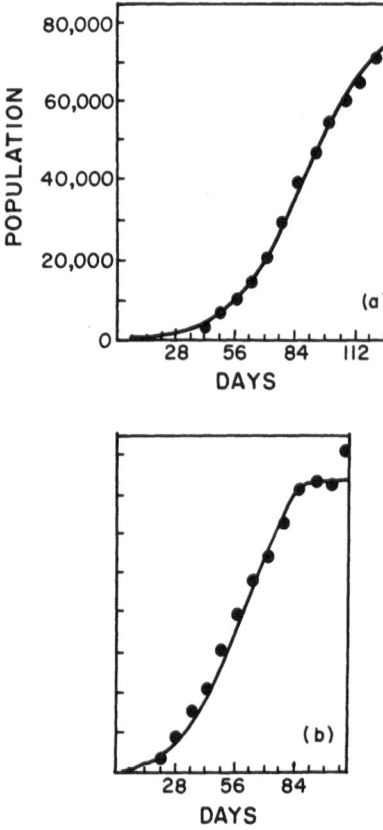

Fig. 1.4a, b. Logistic growth of colonies, (a) of Cyprian and (b) of Italian cultivars of the honeybee *(Apis mellifera)* (Bodenheimer, 1937a, b)

Fig. 1.5. Growth of *Saccharomyces cerevisiae* (Carlson, 1913)

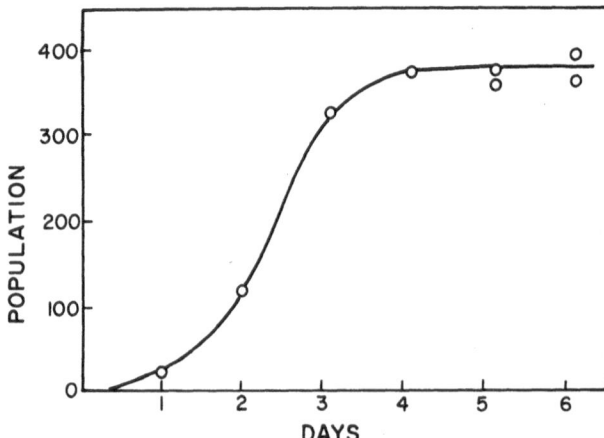

Fig. 1.6. Growth of a population of *Paramecium caudatum* (Gause, 1934)

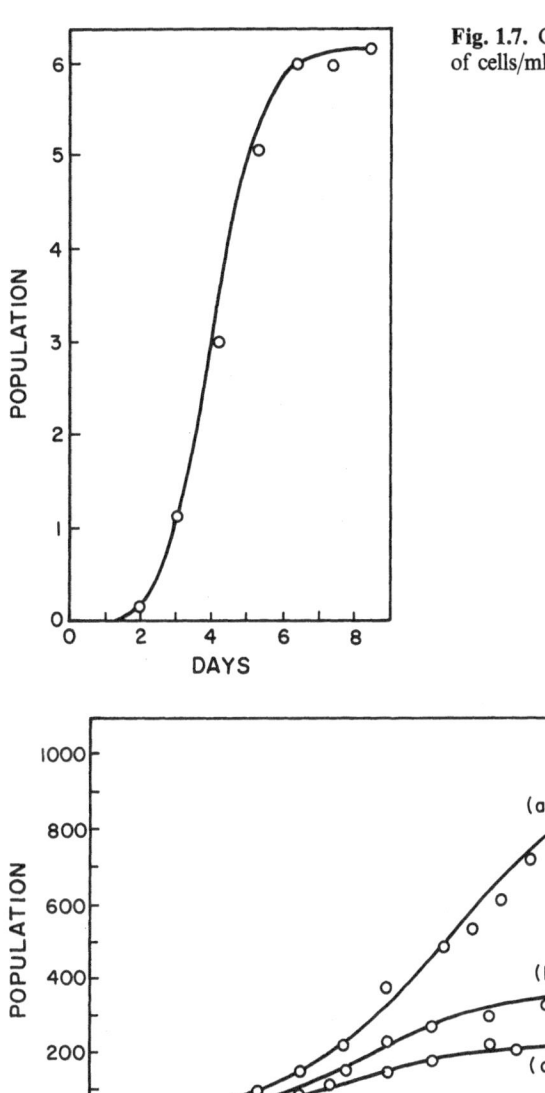

Fig. 1.7. Growth of *Escherichia coli* (in millions of cells/ml) (McKendrick and Kesava Pai, 1911)

Fig. 1.8. Growth of populations of *Drosophila melanogaster*, (a) wild type in a pint bottle; (b) stock homozygous or hemizygous for five recessives including vestigial wing, in same-sized bottle; (c) wild type in a half-pint bottle (Pearl, 1932; after Hutchinson, 1978)

the population is so widely dispersed that reproductive contacts are restricted and infrequent.

One of the simplest ways to model rapid growth employing a minimal number of parameters is to fit it exponentially. Two parameters are required here: one to indicate the initial size of the population and the second to represent the rate of increase.

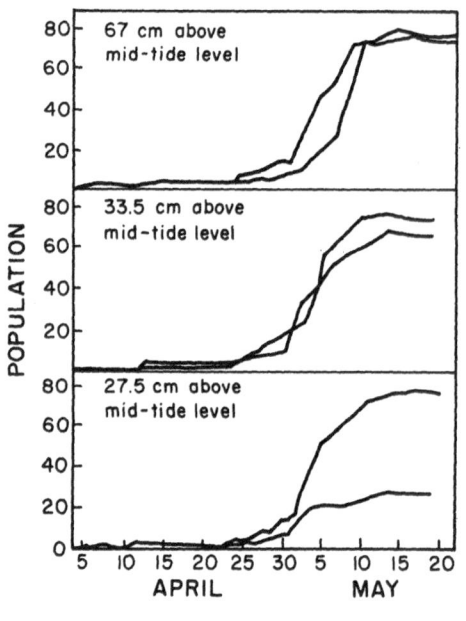

Fig. 1.9. Populations of the barnacle *Balanus balaniodes* (Connell, 1961)

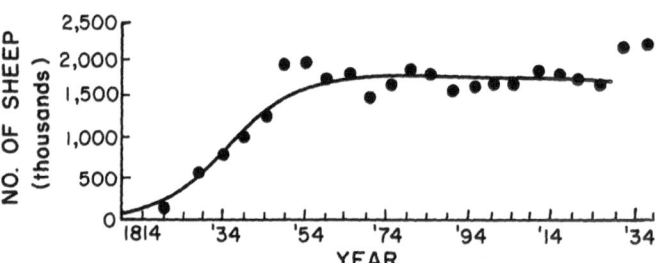

Fig. 10. Population growth of sheep introduced into Tasmania. The dots represent average numbers over five-year periods (Davidson, 1938)

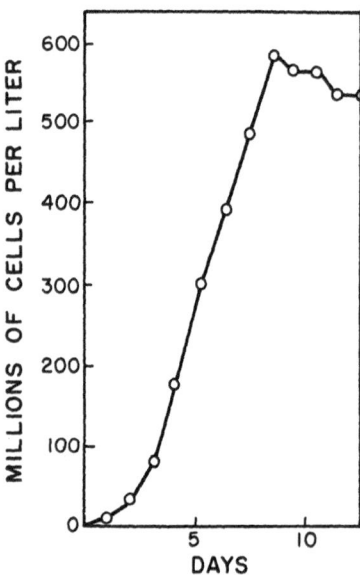

Fig. 1.11. Growth of the marine diatom *(Nitzchia clostidium)* in culture (Riley, 1943)

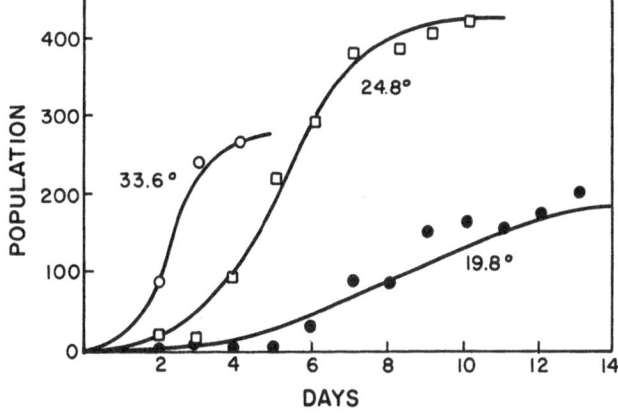

Fig. 1.12. Growth of populations of the water flea *Moina macrocopa* at three temperatures (Terao and Tanaka, 1928)

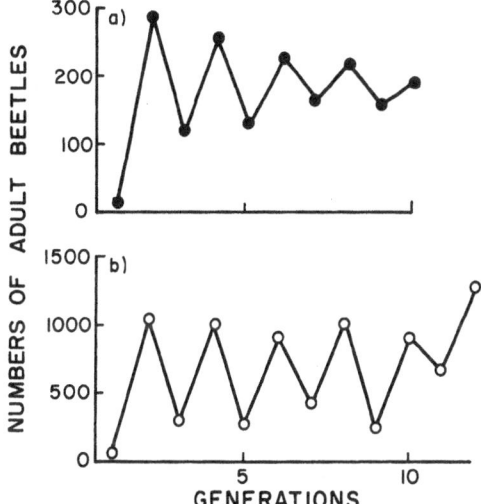

Fig. 1.13a, b. The number of adult beetles in successive generations: (a) *Callosobruchus maculatus* (Utida, 1967); (b) *Callosobruchus maculatus* (Fujii, 1968)

How do High Density Populations Grow?

As in the case of a growing population, my terminology describes situation relative to another. A high density population refers to a population that is limited by the availability of some resource. The degree of limitation should affect the growth of the population until there is a balance between available limiting resource and population utilization of that resource. This should lead to a steady state population, assuming that no other limiting factor becomes critical or that no ecological change occurs.

Figures 1.4, 1.5, 1.6, 1.7, 1.8, 1.9, 1.10, 1.11, 1.12, and 1.13 all reflect exponent-type growth for low initial densities and saturation effects at high densities.

Variation in Populations

Many influences, including environmental and genetic fluctuations, can change the behavior of a population. Figures 1.10, 1.11, 1.12, and 1.13 represent

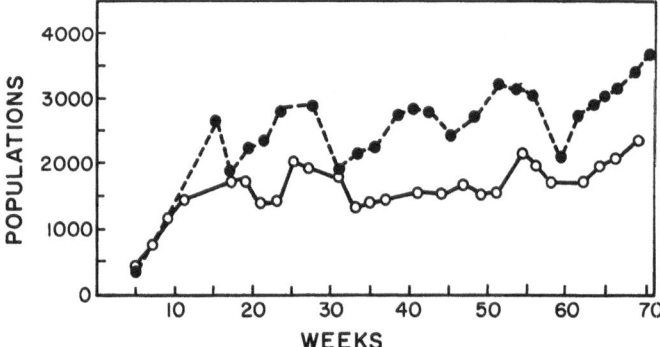

Fig. 1.14. Populations of *Drosophila birchi*, reared at 19 °C: one (○) from Popondetta, New Guinea; and one (×) a strain derived from mass hybridization of the Popondetta strain and of another from 200 km north of Sydney, Australia (Ayala, 1968)

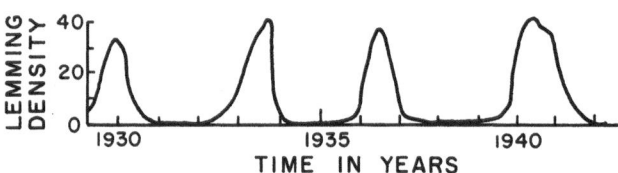

Fig. 1.15. The lemming population in the Churchill area in Canada (Shelford, 1943)

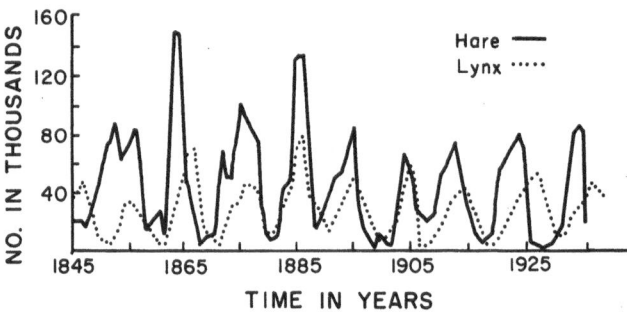

Fig. 1.16. Changes in the abundance of lynx and snowshoe hare (after D. A. MacLulich, 1937)

populations that exhibit some oscillatory behavior. Figure 1.14 represents a *Drosophila* population with some genetic regulation present. Figures 1.15, 1.16, 1.17, and 1.18 demonstrates that oscillations are a feasible mode of behavior for some populations.

Why Don't All Populations Behave Nicely?

The sample populations presented in this section have behaviors that range from nice (Figs. 1.4–1.12) to wild oscillations (Figs. 1.16 and 1.17). There are many factors, biotic and abiotic, which influence the behavior of a population: the physiological characteristics of a species, the relationship of a population to its environment and to other populations, the energetics, behavior and evolution of

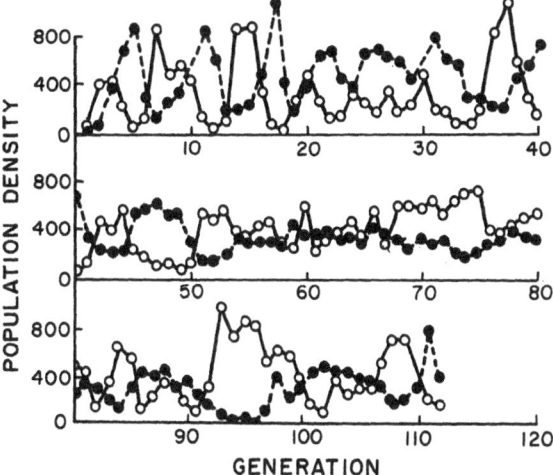

Fig. 1.17. Fluctuation in population density of interacting populations of a host, the bean weevil *Callosobruchus chinensis* (solid line) and a parasite, the abracenid wasp *Heterospilus prosopidis* (dashed line) (Utida, 1957)

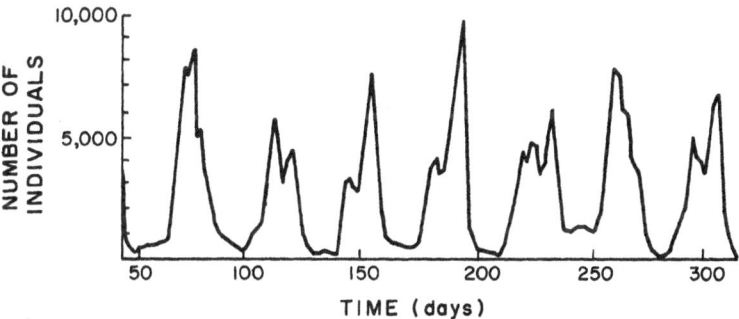

Fig. 1.18. Fluctuations in a *Lucilia Cuprina* blowfly population (Nicholson, 1954)

the population. All affect the dynamical development of the population. At various times, different factors may be dominant in the control of the population dynamics. For example, nutrient availability may limit population growth at one time while predation may be a limiting stress at another time. Indeed, the totality of these exogenous and endogenous forces acting upon a population determine the population behavior. With so many existing system influences, it is surprising that any population behaves nicely.

Effects of some limiting factors on population growth have been delineated; however, with the current state of knowledge of population ecology, there is much to be done before the governing principles of population dynamics can be ascertained.

Extinction

The above illustrations of population dynamics are concerned with population growth and survival. The problem of extinction has not been addressed. Populations go to extinction and they do so with a high frequency. Levins (1970) estimated that there have been 10^8–10^9 species that have disappeared since the advent of the Cambrian geological era. Why do populations crash? Some theoretical ideas will be presented later in these notes.

2. Discrete Models of Populations

2.0 Introduction to Exponential Growth

A population in which generations do not overlap or population size is not large is best modelled in a discrete manner. Suppose that the time variable is scaled so that the generation time is one, and let y_n represent the number of individual organisms in the population at time n. Then, the population size at time $n+1$ is given by a conservation law

$$y_{n+1} = y_n + by_n - dy_n + e_n, \qquad (2.1)$$

where the terms on the right in (2.1) are, respectively: the population size of the previous generation, the number of individuals born in the nth time interval, the number of individuals that died (or were removed for example, by predators) in the nth time interval, and the net number of individuals that emigrate from and immigrate to the region defining the population.

Let $r = 1 + b - d$; r may not be constant over an extended period of time. It varies with the environment, food supply, predation pressure, and many other factors that can affect the population. Let's explore the modelling effects of various hypotheses about the population as they are reflected through the net growth rate parameter r.

If r is a function of the population size, the model (2.1) is said to have a *density dependent* growth rate. For biologically realistic models of growth processes, certain restrictions on r are necessary. For example, suppose that the net exogenous input, e_n, is zero for all n, and that $r(y_n) \geq \alpha > 1$ or $r(y_n) \leq \delta < 1$ for all n (or for all n sufficiently large). These hypotheses have important implications for the growth of the population. If $r(y_n) \geq \alpha > 1$, then $y_{n+1} = r(y_n) y_n \geq \alpha y_n \geq \alpha^n y_0$, $n = 1, 2, \dots$. Thus, $\lim_{n \to \infty} y_n = \infty$ and the population grows; indeed, it grows beyond bounds imposed by physical constraints. The alternative hypothesis, $r(y_n) \leq \delta < 1$, results in extinction of the population; few populations can operate for long under these conditions.

The hypothesis above that led to exponential growth for the model yields conclusions that are, at best, ridiculous. MacArthur and Connell (1966) demonstrate this by considering a single organism, such as a bacterium, that reproduces by dividing into daughter organisms every twenty minutes. Under exponential growth, a bacterium and its progeny could produce a population that would be one foot deep over the entire earth in a 36 hour period. This population would, in a few thousand years, weigh as much as our universe and be expanding outward at the speed of light.

Braun (1975) has performed some interesting calculations about exponential growth and the world's population. Although he did so for the differential equation analog of the difference equation we are considering, he found, by assuming that people are able to live as members of aquatic communities as well as terrestrial ones, that: in approximately 500 years, each of us would have only 9 feet of surface area: in 600 years, each would have only one square foot per person; and in another 35 years, each person would have someone standing on his or her shoulders.

Clearly, with these kinds of numerics, there is something that is wrong with the model. Is there anything that can be saved? There is certainly evidence that populations can explode, and even some evidence that this explosion can be exponential for at least a finite time period. Swarms of grasshoppers, massive outbreaks of agriculture pests, multitudes of tent caterpillars, gypsy moths, and spruce budworms provide examples of the occurrence of exponential growth on a limited time scale. Many of the populations illustrated in Figs. 1.1–1.13 exhibit an exponential growth for low densities of the populations. Minimally, we observe that at small population sizes, the population growth can be approximated by an exponential function.

2.1 Density Dependence and Delays in Population Models

Returning to the model (2.1) and noting the above restrictions, what form might a generic density dependent growth rate take? If population size is small, the organisms could be sparsely distributed in their habitat, mating encounters could be infrequent, survival and development might be difficult. This could be reflected in $r(y_n)$ by requiring that $r(y_n) < 1$ if y_n is small. (This often called an Allee effect.) On the other hand, if y_n is large, overcrowding is likely since intraspecific competition will be strong and the assumption $r(y_n) < 1$ again seems reasonable. As we noted above there should be a population size where $r(y_n) > 1$; otherwise, extinction is automatic. Figure 2.1 illustrates a generic density dependent growth rate.

The points E_u and E_s of Fig. 2.1 where $r(E_u) = r(E_s) = 1$ are important for the analysis of

$$y_{n+1} = r(y_n)y_n; \tag{2.2}$$

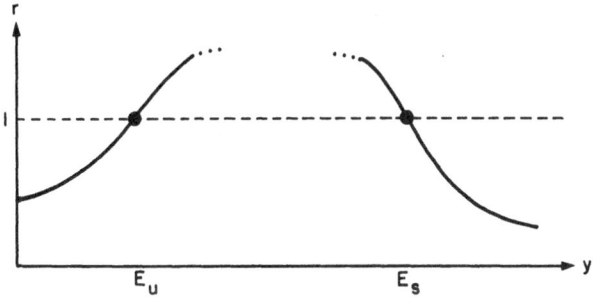

Fig. 2.1. A generic per capita growth rate reflecting effects of density dependence

namely, they are equilibrium values of (2.2) where the population size remains constant for all n ($r(y_n) = 1$), $n = 1, 2, \ldots$. The fact that a state is at equilibrium tells very little about the dynamical system. An egg standing on its end at an equinox is in equilibrium but it won't stay there long because the conditions which determine the equilibrium would change. Certain stability properties of the equilibrium would be desirable information.

A standard technique used in the stability analysis of an equilibrium is that of linearization. This process now is briefly described. Let E represent an equilibrium. Translate the equilibrium to the origin: $y_n - E = Y_n$. Using Taylor's theorem, linearize the system ignoring all higher order terms. Study the linearized system for stability since most nonlinear systems behave locally about an equilibrium like an associated linear system (Coddington and Levinson, 1955; Leunberger, 1979).

Employing this procedure leads to

$$r(y_n) \cong r(E) + r'(E)(y_n - E) = 1 - Er'(E) + r'(E)y_n;$$

hence, the resulting linearized equation is

$$Y_{n+1} = \lambda Y_n, \quad n = 1, 2, \ldots, \tag{2.3}$$

where $\lambda = 1 + r'(E)E$. The asymptotic behavior of the first order equation (2.3) with constant coefficients is indicated in Fig. 2.2; the solution of (2.3) is $Y_n = Y_0 \lambda^n$, $n = 1, 2, \ldots$.

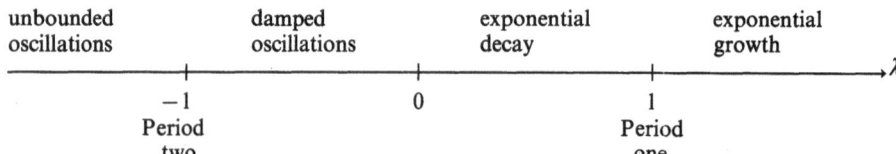

| unbounded oscillations | damped oscillations | exponential decay | exponential growth |

Fig. 2.2. Description of the asymptotic behavior of the first order difference equation $Y_{n+1} = \lambda Y_n$ as a function of λ

If λ satisfies either of the inequalities $-1 < \lambda < 0$ or $0 < \lambda < 1$ then E is locally *asymptotically stable*. If $|\lambda| > 1$ then the equilibrium is *unstable*. Translating this information to E_u (Fig. 2.1), since $r'(E_u) > 0$, this point is unstable. At E_s, some information about relative magnitude of $E_s r'(E_s)$ is needed to guarantee asymptotic stability. For example, if $E_s r'(E_s) > -2$ then linearization indicates that E_s is asymptotically stable.

When the actions of a previous generation determine the growth of a population, the resulting situation might be governed by the model

$$y_{n+1} = r(y_{n-1})y_n, \quad n = 1, 2, \ldots. \tag{2.4}$$

The Eq. (2.4) is a second order difference equation. As a first approximation to study the behavior of (2.4), suppose an equilibrium E exists; that is, $r(E) = 1$.

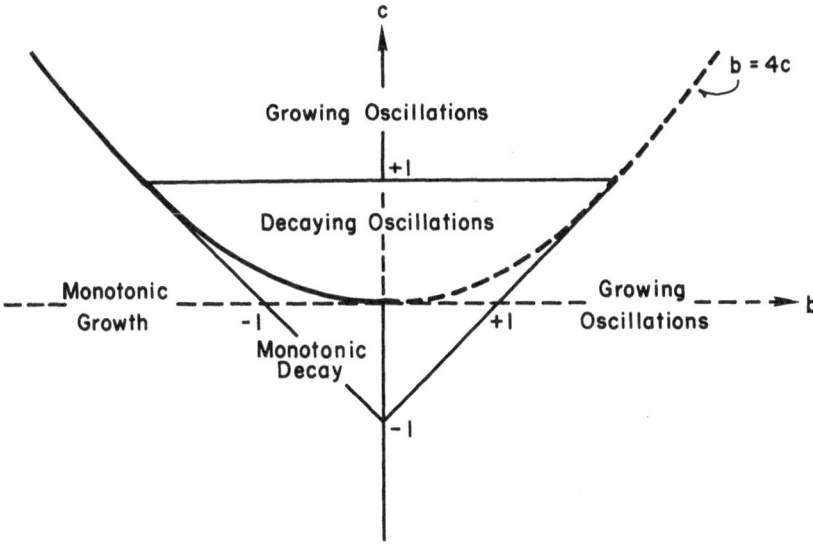

Fig. 2.3. The asymptotic behavior of the homogeneous second order linear difference equation $Y_{n+2}+bY_{n+1}+cY_n=0$ as a function of its coefficients

Linearization results in the equation

$$Y_{n+1}-Y_n-Er'(E)Y_{n-1}=0, \qquad n=1,2,\ldots . \tag{2.5}$$

The behavior of a general second order difference equation with constant coefficients is expressed in Fig. 2.3.

The numerical value of $Er'(E)$ again determines the growth or decay of solutions starting close to E. For example, if $r'(E)>0$, growth occurs; if $0<-Er'(E)\leq\frac{1}{4}$, decay results. Hence, stability can be determined in this manner.

Figure 2.4 compares the differences in behavior of solutions of difference equations due to delay effects in the growth rate. In general, this modelling exercise indicates that the population might be regulated more efficiently by using current data. It is also apparent that additional delays can influence the existence of oscillations.

One period delay in r	Monotone growth	Monotone decay		Oscillatory decay	Oscillatory growth	
Two period delay in r	Monotonic growth	Monotonic decay	Oscillatory decay	Oscillatory growth		$-Er'(E)$

Fig. 2.4. A comparison of the effects of delay in the difference equations $x_{n+1}=r(x_n)x_n$ and $x_{n+1}=r(x_{n-1})x_n$ in terms of the parameter $Er'(E)$

2.2 Discrete Logistic Equation

One of the simplest models that contains a formulation which represents effects of overcrowding is when r is a linear function of population size: $r(y_n) = 1 + r_0(E - y_n)/E$. Because of the name of the continuous analogue, this equation is often referred to as the discrete logistic equation. Scaling the variable y_n by the transformation $Z_n = y_n r_0/((1 + r_0)E)$ gives a canonical form

$$Z_{n+1} = \alpha Z_n(1 - Z_n), \quad \alpha = 1 + r_0. \tag{2.6}$$

The behavior of solutions of Eq. (2.6) is a function of the parameter α. This behavior, while it is complex and fascinating, also has several disturbing attributes. Only for $0 \leq \alpha \leq 4$ does (2.6) make sense with Z_n measuring a nonnegative quantity. If $\alpha < 0$, then small Z_0 results in $Z_1 < 0$; when $\alpha > 4$, then small Z_0 forces Z_2 to be negative. For $0 \leq \alpha \leq 4$ the solution sequence Z_n, $n = 0, 1, 2, \ldots$ does not change sign and remains less than or equal to one provided Z_0 satisfies $0 \leq Z_0 \leq 1$.

To study the sensitivity of the solutions to changes in the parameter α it is convenient to decompose the interval [0, 4] further. For $0 < \alpha < 1$, extinction of the population will occur since $Z_{n+1} \leq \alpha^n Z_0$; hence, $\lim_{n \to \infty} Z_n = 0$. When $\alpha = 1$, $Z_n = Z_0$, $n = 1, 2, \ldots$. For $1 < \alpha < 3$, as may be demonstrated by linearization, $Z_n = \dfrac{\alpha - 1}{\alpha}$ is an asymptotically stable equilibrium. The character of the approach is monotone decay if $1 < \alpha < 2$ and oscillatory decay if $2 < \alpha < 3$. The value $\alpha = 3$ results in a neutral oscillation about the equilibrium.

The solution behavior when $\alpha > 3$ becomes more interesting. There is no longer just a single stable equilibrium that governs the dynamics of the population; periodic solutions can emerge. The existence of a neutral oscillation of period two and its stability is now discussed. It is convenient to use (2.6) and write

$$Z_{n+2} = \alpha^2 Z_n[1 - (\alpha + 1)Z_n + 2\alpha Z_n^2 - \alpha Z_n^3]. \tag{2.7}$$

Suppose Z_n bounces between V_1 and V_2 then when $Z_n = V$ (either V_1 or V_2), $Z_{n+2} = V$. This leads to

$$\alpha V^3 - 2\alpha V^2 + (\alpha + 1)V + (\alpha^{-2} - 1) = 0.$$

One root must be $(\alpha - 1)/\alpha$, so the left side of this can be written as

$$\left(V - \frac{\alpha - 1}{\alpha}\right)(\alpha V^2 - (\alpha + 1)V + (1 + \alpha^{-1})) = 0.$$

The quadratic factors into roots

$$V = \frac{\alpha + 1 \pm [\alpha^2 - 2\alpha - 3]^{1/2}}{2\alpha}.$$

The assumption that $\alpha > 3$ implies both roots, V_1 and V_2, are real; in fact, it can be seen that they are positive. The oscillation exists but it is stable? For example, if Z_n is close to V, does this mean that Z_{n+2} is closer to V? For certain α, this is the case. From (2.7), the relationship

$$Z_{n+2}-Z_n = -\alpha^3 Z_n \left(Z_n - \frac{\alpha-1}{\alpha} \right)(Z_n - V_1)(Z_n - V_2)$$

results. Writing $Z_n = V_1 + P_n$, linearizing, and doing some algebra, we find that the equation

$$P_{n+2} = [1 - (\alpha+1)(\alpha-3)]P_n$$

determines the stability of V_1. Hence, α must satisfy $0 < (\alpha+1)(\alpha-3) < 2$ (or $3 < \alpha < 1 + \sqrt{6}$) for local asymptotic stability. A similar argument establishes the stability of V_2. The asymptotic behavior of Eq. (2.6) as discussed to the present is indicated in Fig. 2.5.

At $\alpha = 1 + \sqrt{6}$, the onset of double-double oscillations occurs; that is, a periodic solution of period 4 emerges. This is expressed in a classical bifurcation diagram in Fig. 2.6. Notice the traditional exchange of stability at the points of bifurcation. One can predict easily what happens next as α increases. A periodic solution of period 8 arises, then one of period 16, and, in general, one of period 2^n. This all transpires on a compact interval and there is a limit point on the α axis where this sequence of periodic solutions terminates. The value of $\alpha = 3.570$ seems prevalent for this behavior, although others have different numbers (e.g. Roughgarden,

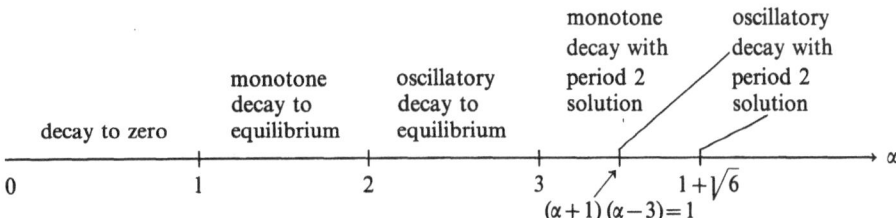

Fig. 2.5. Asymptotic behavior of the discrete logistic equation $Z_{n+1} = \alpha Z_n(1 - Z_n)$ as a function of the parameter α

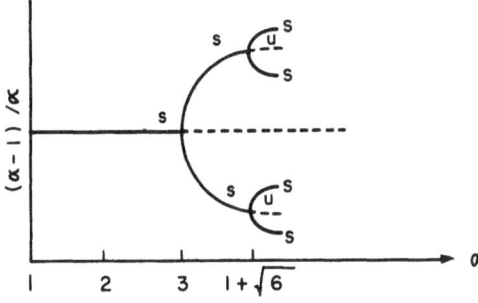

Fig. 2.6. A bifurcation diagram for the discrete logistic equation. s indicates an asymptotically stable solution while u indicates an unstable solution

1979). This value of α signals the onset of periodic solutions with odd period, beginning with very long periods, and as α increases, the period decreases to three. As α continues to increase, periodic solutions with arbitrary period and asymptotically aperiodic solutions occur. The behavior of this innocent looking equation has been termed "chaotic." Several authors have indicated the outcome of this deterministic population model should be regarded as stochastic (e.g. May, 1976).

There do seem to be some redeeming features of the model (2.6). Smale and Williams (1976) have demonstrated that for almost all $\alpha_1 \in (3, 4)$ there exists a single solution that is an attractor for almost all trajectories.

The remaining case $\alpha = 4$ is of interest also; not so much for its unusual behavior, but because it can be solved analytically. The function

$$Z_n = \frac{1}{2} - \frac{1}{2}\cos(2^n \arccos(1 - 2x_0))$$

is a solution of (2.6) with $\alpha = 4$. This solution oscillates, except for the initial conditions $x_0 = \frac{1}{2} - \frac{1}{2}\cos\frac{n\pi}{2^m}$, for n and m positive integers, in which case it tends to a limit.

2.3 Reflections on the Relevance of Discrete Models to Population Ecology

As May (1976) has pointed out, if models with chaotic behavior are relevant to populations, then wild oscillations of a population need not necessarily be the consequence of random environmental fluctuations but might be intrinsic to the population. An important question is "Do biological populations exhibit chaotic behavior?" Hassell et al. (1976) find little evidence to support an affirmative answer; certainly the question is unresolved at present.

The utility of chaotic type models has been questioned by several authors. Smith and Mead (1980) find that the spectral properties of the difference equation in the chaotic regime are, in practice, likely to be indistinguishable from those in the locally stable regime with a randomly fluctuating carrying capacity. O'Neill et al. (1982) use an error analysis approach to demonstrate that uncertainty in growth rate is as important in determining the regularity of the system as the mean value. In spite of the negative aspects of these works on models with a rich spectrum of behavior, the resulting discussions have been enriching to both population ecology and mathematics.

2.4 Summary

Discrete models, discussed here as difference equations, appear to be natural models to use to estimate the size of a population. Difference equations are innocent looking, but they can be dynamic monsters. While the behavior of some populations can be mimicked by these simple models, it is not clear if these models are extremely useful tools in population ecology. The behavioral repertoire of the simplest nonlinear model is extremely broad. Bifurcations can occur with extremely small perturbation of parameters. The phenomenon of chaotic behavior is not unique to the discrete logistic equation. There are, in the literature, models other than the logistic which exhibit chaotic motions and do not have some of the deficiencies, such as a restricted parameter range for biological significance, of the logistic. These include the widely applied Ricker model

$$X_{t+1} = X_t \exp\{r(1 - X_t/K)\},$$

among others (see May, 1976).

Chaotic behavior is not restricted to difference equations. It can occur in ordinary differential equations (Gilpin, 1978), but only for systems (i.e. not for scalar equations). However, this behavior also can be demonstrated for first order delay-differential equations and for partial differential equations.

3. Continuous Models of Populations

3.0 Introduction to Exponential Growth

Let $x(t)$ denote the population size (or biomass) at time t; let b and d denote the birth rate and depth rate, respectively (that is, b is the number of births per individual per unit time interval and analogously for d) on the time interval $[t, t + \Delta t]$, $\Delta t > 0$

$$x(t + \Delta x) - x(t) = bx(t)\Delta t + dx(t)\Delta t. \tag{3.1}$$

Dividing by Δt in (3.1) and letting Δt approach zero, we obtain

$$\frac{dx(t)}{dt} = rx(t), \tag{3.2}$$

where $r = b - d$ is the *intrinsic growth rate* of the population. The model (3.2) represents the traditional exponential growth $(r > 0)$ or decay $(r < 0)$ of a population. An important distinction between first order linear difference equations with constant coefficients and the analogous differential equation (3.2) is that (3.2) allows no oscillations. The deficiencies of (3.2) as a population model are as delineated in Sect. 2.0 where exponential growth is discussed.

3.1 A Density Dependent Growth Rate

In Sect. 2.1, for difference equations the assumption of density dependence was modelled by $r = r(x)$. The hypothesis that $r(x) \geq \delta > 0$ always leads to unbounded growth of the population, while $r(x) < -\delta < 0$ always leads to extinction. As a canonical form for $r(x)$, the function represented by the graph in Fig. 3.1 might serve as an initial approximation. It can be demonstrated by using linearization that the equilibrium E_u is unstable while the point E_s is locally asymptotically stable.

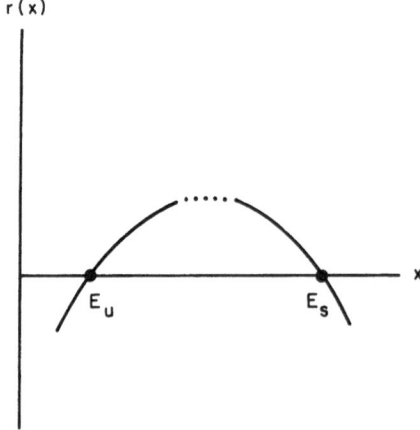

Fig. 3.1. The graph of a density dependent per capita growth rate

3.2 The Classical Logistic Equation

One of the first demographic studies that employed density dependent growth rates was that done by Pierre-Francois Verhulst (1838). In his book, Hutchinson (1978) eloquently presents the history of the logistic equation:

$$\frac{dx}{dt} = r(x)x, \tag{3.3}$$

where r is a linear function $r(x) = a - bx$. We proceed directly to the theory of the Eq. (3.3). The traditional way of writing the logistic equation is with $r(x) = r_0 x \left(1 - \frac{x}{K}\right)$, where r_0 (>0) is the *intrinsic growth* rate and K (>0) is called the *carrying capacity* of the population. When r_0 and K are constants, the equation may be solved by the variables separable method. This leads to

$$x(t) = \frac{Kx_0}{x_0 - (x_0 - K)e^{-r_0 t}}.$$

The geometrical structure of the positive quadrant of trajectories is given in Fig. 3.2. Each solution approaches the carrying capacity, K, as t approaches infinity, independent of initial population size, x_0. The sigmoid behavior represented by the bottom trajectory in Fig. 3.1 occurs for those solutions with x_0 satisfying $0 < x_0 < K/2$ ($K/2$ is the ordinate at which any inflection point of a solution occurs). When x is small, the solutions exhibit exponential growth for a period of time, then density dependent effects take over and the population saturates at carrying capacity. This saturation effect occurs monotonically for all trajectories.

Many of the populations whose graphs are shown in Sect. 2 exhibit the sigmoid characteristics of the logistic trajectories (e.g. Figs. 1.4–1.12). There are many hypotheses underlying the use of the logistic, and most of these are not valid for

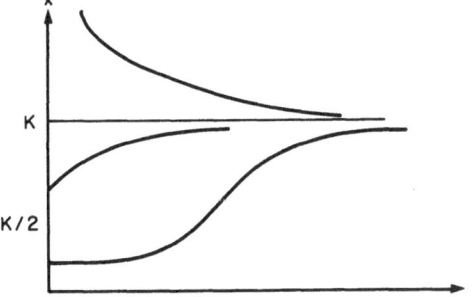

Fig. 3.2. Graphs of typical solutions of the logistic equation,

$$\frac{dx}{dt} = x\left(r_0 - \frac{r_0 x}{K}\right)$$

any population. Nevertheless, the logistic remains the most widely used population model. Some hypotheses that are employed in formulating the logistic equation are given below.

1. Biotic and abiotic parameters are constant for all time. In reality, birth and death rates are affected by many exogenous factors and will vary with these effects; a constant intrinsic growth rate and carrying capacity does not reflect this time variation.

2. Stochastic events are not considered in the model (of course not – it is a deterministic model).

3. All individuals of the population are treated equivalently. They are not differentiated by sex, age, social role, or physical location. This ecological homogeneity is probably not valid for any natural population.

4. The per capita growth rate responds instantaneously to changes in density. The assumption that no time delays occur in any process is probably invalid as well.

5. Resources are nonexpendable or are continuously renewed.

Each of these general objections to the logistic can be overcome through modification of the model. There are other changes in the traditional logistic equation that also need to be implemented. For example, if r_0 is not restricted to be positive, weird things can happen; when r_0 is negative, any solution with $x_0 > K$ is unbounded. How can a population, which cannot survive under the ideal conditions of no intraspecific competition, $(r_0 < 0)$ thrive and explode when the density dependence is strong? The logistic equation presents one way; but from a modelling perspective, it is not a feasible one.

The logistic equation has provided motivation for some of the early theoretical developments in evolutionary ecology. The idea of r- and K-selection are in direct reference to the parameters of the logistic (Roughgarden, 1979). Theoretical developments direct from this classical form of the logistic equation are fraught with difficulties. These difficulties are explored in Sect. 3.4 on the nonautonomous logistic equation.

3.3 The Logistic Equation with Harvesting

Suppose there is an exogeneous force which removes members of a population at a constant rate h. This physical process is called *harvesting* and can, for example, be

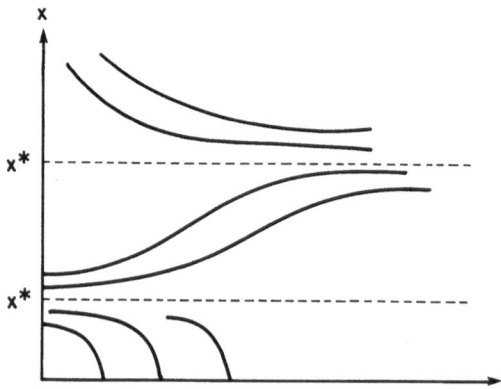

Fig. 3.3. The solution space of the logistic equation with harvesting

effected by hunters or fishermen. If the population is governed by the logistic equation, the model is (3.3) with harvesting included:

$$\frac{dx}{dt} = x(a-bx) - h.$$

(3.4)

Harvesting models have played an important role in the management of renewable resources (Clark, 1976); some interesting applications are developed later in this book by Dr. J. Conrad.

Equation (3.4) can be analyzed by employing a stability analysis. The equilibria are given as roots of $bx^2 - ax + h = 0$. The roots of this equation, x_*, x^* are real if and only if h satisfies the inequality $0 \le h \le a^2/4b$. The smaller root $x_* = [a - (a^2 - 4bh)]^{1/2}/2b$ is unstable and the root $x^* = [a + (a^2 - 4bh)^{1/2}]/2b$ is asymptotically stable. The solution space of (3.4) has a configuration given in Fig. 3.3.

The model has a threshold of extinction (x_*) below which extinction occurs in a finite time. If the initial population exceeds this threshold, the solution, $x(t)$, approaches the equilibrium population, x^*, as t tends to infinity.

Much will be said later about this situation but one concluding remark is given here. When the harvesting rate exceeds the critical value $h = a^2/4b$, extinction results independent of the initial population; hence, if (3.4) is biologically meaningful there is a critical harvesting rate. Certainly, for values of the harvest rate parameter near critical value, the model is interesting both from a mathematical and a bioeconomic viewpoint.

A discussion of this situation and an interesting application may be found in Brauer and Sanchez (1975).

3.4 The Nonautonomous Logistic Equation

Nearly Constant Coefficients

A hypothesis of the classical logistic model is that the parameters r and K are constant. The time-varying version of the logistic equation is

$$\frac{dx(t)}{dt} = r(t)x(t)\left[1 - \frac{x(t)}{K(t)}\right].$$

(3.5)

This equation is of Bernoulli-type and is solvable in closed form if r and K are piecewise continuous on $\mathbb{R}_+ = [0, \infty)$. The solution of (3.5) which passes through (t_0, x_0) is

$$x(t, t_0, x_0) = \frac{x_0 \exp\left[\int_{t_0}^{t} r(s)ds\right]}{1 + x_0 \int_{t_0}^{t} \exp\left[\int_{t_0}^{s} r(s_1)ds_1\right] \frac{r(s)}{K(s)} ds} . \tag{3.6}$$

When r and K satisfy the inequalities

$$0 < r_* \equiv \inf_{t \in \mathbb{R}_+} r(t) \leq r(t) \leq r^* \equiv \sup_{t \in \mathbb{R}_+} r(t) < \infty, \tag{3.7a}$$

$$0 < K_* \equiv \inf_{t \in \mathbb{R}_+} K(t) \leq K(t) \leq K^* \equiv \sup_{t \in \mathbb{R}_+} K(t) < \infty \tag{3.7b}$$

then the asymptotic behavior of (3.5) is much like the logistic with constant parameters (Coleman, 1979). There exists a solution of (3.5)

$$\tilde{x}(t) = \left[\int_{0}^{\infty} \exp\left[-\int_{0}^{s} r(t-\gamma)d\gamma\right] \frac{r(t-s)}{K(t-s)} ds\right]^{-1}$$

that is globally asymptotically stable. The solution $\tilde{x}$ depends upon the complete past history of r and K. If r and K are periodic then so is $\tilde{x}$.

A Deteriorating Environment

The behavior of the nonautonomous logistic equation is not always similar to the classical logistic model. This can be demonstrated by considering the deteriorating environment situation.

A *deteriorating environment* is modelled here by functions K with the properties that $K > 0$ on $\mathbb{R}_+$ and $\lim_{t \to \infty} K(t) = 0$. For convenience of illustration, assume for the present that $r > 0$ on $\mathbb{R}_+$.

Remark 1. If $\int_{t_0}^{\infty} r(s)ds = \infty$ then each solution of (3.5) satisfies $\lim_{t \to \infty} x(t, t_0, x_0) = 0$. Hence, a relatively large growth rate coupled with a deteriorating environment causes a population to track its environment to extinction. This plausible and expected result can be demonstrated directly from (3.6) by using L'Hospital's Rule.

Remark 2. If $\int_{0}^{\infty} r(t)dt < \infty$ then each solution of (3.5) converges; that is, $\lim_{t \to \infty} x(t, t_0, x_0) = x_\infty$. This result is valid even without the deteriorating environment hypothesis. In the case that K is constant, it can be shown that given any terminal value x_∞, with $x_\infty < K\left\{1 - \exp\left[-\int_{t_0}^{\infty} r(s)ds\right]\right\}^{-1} \equiv M$, there is a solution x with the behavior $\lim_{t \to \infty} x(t) = x_\infty$. Hence, if $\int_{t_0}^{\infty} r(s)ds$ is sufficiently small, M can be arbitrarily large and K is exceeded by the limit of any solution that is initially above K. For a deteriorating environment, there exist cases where the terminal value of every solution of (3.5) exceeds the terminal value of the carrying capacity.

If one attempts to interpret the above comments biologically, the following puzzle arises. A growth rate that is large can lead to extinction while a small growth rate can result in persistence, independent of initial population (and can lead to large terminal densities which exceed the terminal values of the carrying capacity). By interpreting r as the growth rate of the population in the absence of environmental stress, we reach a situation where a population, because of a small intrinsic growth rate, is barely able to persist under the best of conditions. However, it is able to survive and even fluorish in an intolerable environment. These conclusions again vividly indicate the inadequacy of (3.2) as a model of population growth. The difficulties here are easy to bypass in that a proper parametrization of the logistic equation does not lead to these dilemmas.

3.5 A Modified Logistic Equation

Modelling deficiencies of the classical logistic equation [(3.5), with constant coefficients], are promulgated by the role of both model parameters r and K. The model defects in the previous section are consequences of the domination of the intrinsic growth rate r in the equation. An equation of modified logistic type that does not allow solution behavior to be subjugated by r is

$$\frac{dx(t)}{dt} = x(t)\left(r - \frac{c}{K}x(t)\right). \tag{3.8}$$

In Eq. (3.8), the intrinsic growth rate, r, is expressed in units $(\text{time})^{-1}$ as is the positive parameter c. This new independent parameter c is a measure of the population response to environmental stress as represented by the ratio x/K.

A problem with the parameter K in the classical logistic equation is that it can be interpreted ambiguously as either a population carrying capacity or a steady state of the population. Since these interpretations need not be equivalent, it is convenient to reformulate the model as

$$\frac{dx(t)}{dt} = x(t)\left[r(t) - c\frac{x(t)}{B(t)}\right], \tag{3.9}$$

where B denotes the maximum population which the environment can support; that is, the environment can provide all necessary requirements for the maintenance of B individuals, but it will not support $B+1$ individuals. When r and B are constants, Eq. (3.9) has a stable equilibrium at $x = rB/c$. This motivates the definition of the (ultimate) population level parameter K as

$$K = \begin{cases} rB/c & \text{if } r > 0, \\ 0 & \text{if } r \leqq 0. \end{cases} \tag{3.10}$$

It must be the case that $r \leq c$ in Eq. (3.9). For species that have evolved in a manner which allows the population to exploit the full potential of the environment, one would expect to have $K = B$ and $r = c$; that is, the traditional logistic equation would be applicable.

Equation (3.9) can be shown to palliate some of the modelling difficulties associated with the classical logistic equation as it yields plausible results in many instances where the traditional logistic does not. For example, when $r < 0$, all solutions of the autonomous equation (3.9) approach zero as t approaches infinity while for the logistic some solutions blow up in a finite time.

In the nonautonomous case, it can be shown that the undesirable attributes exhibited in the deteriorating environment setting no longer hold (Hallam and Clark, 1981). For example, either a deteriorating growth rate or a deteriorating environment assures extinction of the population.

There have been many criticisms of the logistic equation (e.g. Gray, 1929; Kavanagh and Richards, 1934; Andrewartha and Birch, 1945; Pielou, 1977; Murray, 1979). However, despite the criticisms and obvious deficiencies, the equation continues to be the most frequently used continuous deterministic model of single species population growth in a limited environment. Its advantages are its analytical simplicity, the elementary interpretation of its biological parameters, and the fact that it often can be fitted to data.

The reparameterization suggested in (3.9) might prove beneficial for modelling populations with small growth rates or where the environment is changing in an unfavorable manner.

3.6 What Should a "Discrete" Logistic Model Look Like?

A comparison between the discrete logistic equation and the classical logistic equation indicates that the behaviors are considerably different. To find a difference equation that has the same behavior as the logistic, E. C. Pielou (1979) starts with the solution

$$x(t) = \frac{Kx_0}{x_0 - (x_0 - K)e^{-r_0 t}}$$

and considers this as the solution to the desired difference equation:

$$y_n = \frac{K}{1 + c\lambda^{-n}} \qquad (\lambda = e^{r_0}).$$

It can be shown that

$$y_{n+1} = \frac{\lambda y_n}{1 + \frac{\lambda - 1}{K} y_n}.$$

This difference equation has a behavior that is similar to the traditional logistic equation.

3.7 Density Dependent Representations

Smith's Modification of the Logistic Model

The per capita growth rate of the logistic equation is a linear function of the population density. Smith (1963), studying a population of *Daphnia magna* by an elegant set of experiments, found that his data did not support the linearity hypothesis (Fig. 3.4).

To find an adequate model, Smith argued as follows: The classical logistic equation contains the assumption that the growth rate is proportional to $(K-x)/K$, the proportion of maximal attainable population size still unrealized. This seems somewhat unrealistic and it might be more appropriate to have a growth rate that depends on the proportion of some limiting factor not yet consumed. A natural candidate for a limiting factor is the food supply not yet utilized. The hypothesis employed by Smith was that the per capita growth rate of a population is proportional to the rate of food supply not momentarily being used. This results in the model:

$$\frac{1}{x}\frac{dx}{dt} = r\left(1 - \frac{F}{T}\right),$$

where F is the rate at which a population of biomass x consumes resources, and T is the rate at which the population uses food when it is at the equilibrium K.

At this point, it is prudent to observe that this model shares some modelling difficulties with the logistic equation. The proportionality constant factor r can result in false feedback when r is negative, or in suspicious behavior if r is close to zero. Hence we assume that the per capita growth rate is a linear function of the ratio of the rates F/T; consequently,

$$\frac{1}{x}\frac{dx}{dt} = a - b\frac{F}{T}. \tag{3.11}$$

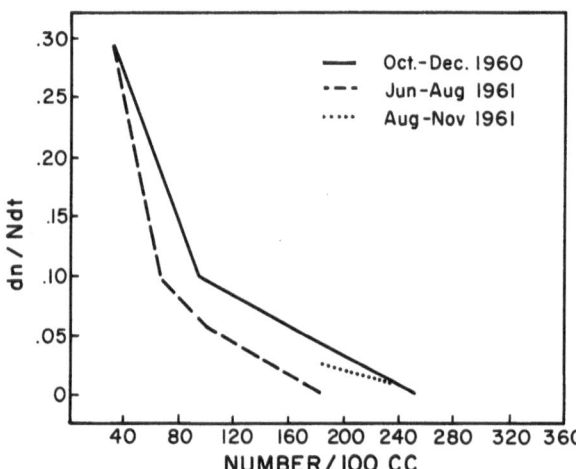

Fig. 3.4. Observed densities of *D. magna* at various specific rates of growth, using number of individuals as the measure of density. Data are combined from three sets of experiments (Smith, 1963)

The rates of consumption, F and T, depend, at least, upon the population biomass and upon the rate at which the population is growing. As a first approximation, let $F = c_1 x + c_2 dx/dt$.

As we remarked for the logistic equation earlier, the population carrying capacity, B, and the steady state of the population, K, need not be the same. The carrying capacity need not be an equilibrium so the role of saturation and equilibrium in Smith's development can be reinterpreted in the following way: When $x = B$, $F = T$ and consequently, $T = c_1 B + C_2 dx(B)/dt$. From (3.11), a computation shows that $T = c_1 B + c_2 B(a - b)$ and

$$\frac{1}{x}\frac{dx}{dt} = \frac{aB(c + a - b) - bcx}{B(c + a - b) + bx}.$$

This formulation, in an analogous manner to the logistic, yields a population carrying capacity K defined by

$$K = \begin{cases} \dfrac{aB(c + a - b)}{bc} & \text{if } b < c + a, \\ 0 & \text{if } b \geq c + a. \end{cases}$$

It can be demonstrated readily that each solution of this model with positive initial condition approaches K as t tends to infinity.

This model emphasizes a new saturation aspect to the role of density dependence in the theory of populations. There are in the literature many other representations of the per capita growth rate that reflect effects of density dependence. These include:

1. B. Gompertz (1825):

$$r(x) = \left(\ln \frac{B}{x} \right).$$

2. M. Rosenzweig (1971):

$$r(x) = \left(\left(\frac{B}{x} \right)^{-g} - 1 \right), \quad 0 < g \leq 1.$$

3. Goel, Maitra, and Montroll (1971) and Gilpin and Ayala (1973) use an opposite form of Rosenzweig's model:

$$r(x) = 1 - \left(\frac{x}{B} \right)^g, \quad 0 < g < 1.$$

4. T. Schoener (1973) takes $g = -1$ in the above formulation of Rosenzweig.

There are many other population models which have proved useful in application. Among these is the von Bertalanffy (1951) formulation that predicts the length and weight of an organism. Based upon a potential for growth much as the logistic and Smith's model, this model assumes that the rate of growth of the length is proportional to the length yet to be attained. The weight is then related to the length by an allometric law. The model of von Bertalanffy has been applied to fisheries by Beverton and Holt (1957).

4. Resource-Consumer Dynamics

4.0 Introduction

The quantity and quality of a population's resources determine many aspects of the growth of the population. Of the population models described previously, only the Smith model makes any pretense to consider explicitly the consumption rate as an important parameter. In this section, we will explore the dynamics of resource-consumer interactions from a modelling perspective. First, some general principles that govern most resource-consumer relationships are indicated:

1. When a resource is rare relative to the magnitude of the consumer population, the rate of consumption is determined solely by the amount of available resources.

2. When the resource is abundant, the rate of consumption is determined solely by the density of the consumer population.

3. When the resource is neither rare nor abundant, the rate of consumption is a function of both the amount of available resources and the density of the consumers.

4.1 A Model of a Resource-Consumer Interaction

Gallopin (1971) developed a rudimentary model that utilizes some of these principles to describe a single dynamic resource and a consumer population.

The Resource, A

The exogeneous rate of input of resource into the system is assumed to be $f(t)$; actually, in the analysis, f is taken to be constant. The amount of available resource A is, by conservative of mass, the difference between the resource that has entered the system and the resource that has been ingested by consumers. It is assumed that the excreted resource is not available for consumption.

The ingestion rate, c, should be a function of the amount of available food A; some measure of the population, either density or biomass, M; the rate of change of the population (growing organisms consume more resources per unit mass than nongrowing ones); and, perhaps, other factors. Thus,

$$c = c\left(A, M, \frac{dM}{dt}, \ldots\right).$$

In order to specify c further, it is convenient to introduce the parameter a, the ingestion rate per unit mass which corresponds to maintenance at complete satiation. If a population is satiated, an increase in food supply will not increase the ingestion rate. However, if the population is hungry, an increase in food will produce a significant increase in the ingestion rate. It is assumed that the change in the ingestion rate per unit change in available food is proportional to the hunger of the population:

$$\frac{\partial c}{\partial A} = h(M, A)(aM - c). \tag{4.1}$$

The proportionality parameter h depends only upon biomass M and available resources A. Note that aM is the total ingestion rate at satiation. A choice of h that results in many desirable properties of resource-consumer systems is that made by Gallopin: $h = \alpha/M$. For example, an increase in A has a greater effect on consumption when M is small than when M is large. Equation (4.1) with $h(A, M) = \alpha/M$ (in fact, for arbitrary h) is solvable as a first order differential equation in c; for $M > 0$, this yields

$$c(A, M) = aM + K(M)\exp(-\alpha A/M),$$

where $K(M)$ is a constant of integration that may depend upon M because the integration is with respect to A. Since $c(0, M) = 0$ for all $M > 0$, $K(M) = -aM$; hence

$$c(A, M) = aM(1 - \exp(-\alpha A/M)). \tag{4.2}$$

This formulation for rate of consumption has the following properties:
1. When food is abundant (A large), the ingestion rate is a function of M only:
$\lim\limits_{A \to \infty} c(A, M) = aM$.
2. When resources are scarce (A small), the ingestion rate is small:
$\lim\limits_{A \to \infty} c(A, M) = 0$.
3. When population biomass is small, the consumption rate is also small:
$\lim\limits_{\substack{M \to 0 \\ A \neq 0}} c(A, M) = 0$.
4. When the biomass is very large, the consumption rate will be proportional to available food: $\lim\limits_{M \to \infty} c(A, M) = a\infty A$.

Population Dynamics

Conservation of mass implies that

$$\frac{dM}{dt} = c(A, M) - \lambda(A, M), \tag{4.3}$$

where λ represents the losses of resource from the population. The losses include egestion, respiration, excretion, and mortality.

The following hypotheses are imposed: Egestion rate is proportional to consumption rate. The totality of the remaining losses are required to be proportional to the population biomass. The Eq. (4.3) with these hypotheses imposed becomes

$$\frac{dM}{dt} = (1-u)c - kM = \beta c - kM . \tag{4.4}$$

The conservation of mass law can be used to find a dynamic equation for A:

$$\frac{dA}{dt} = f - c .$$

When (4.2) is substituted in (4.4) the differential equations become a coupled system for A and M:

$$\frac{dM}{dt} = \beta a M \left(1 - \frac{k}{\beta a} - \exp(-\alpha A/M) \right),$$

$$\frac{dA}{dt} = f - aM(1 - \exp(-\alpha A/M)). \tag{4.5}$$

The system (4.5) can be analyzed by analytical methods if $f(t) \equiv f_0$ is a positive constant. In this case, there exists an equilibrium

$$M_0 = \frac{\beta f_0}{k},$$

$$A_0 = \frac{-\beta f_0}{k\alpha} \ln \left(1 - \frac{k}{a\beta} \right) \tag{4.6}$$

provided $k < a\beta$. If $k > a\beta$, then no (positive) equilibrium exists and extinction may be demonstrated by use of the comparison principle for differential inequalities (Lakshmikantham and Leela, 1969) applied to the M-equation in (4.5). The parameter k is a maintenance cost per unit mass of the population; extinction results if the maintenance cost exceeds the assimilation rate per unit mass ($a\beta$). (Most ecologists could have told us this before the mathematics started but it is nice to see that the mathematics confirms this belief.)

Persistence can occur for the population whose dynamics are modelled by (4.5). A linearization shows that the equilibrium (4.6) is locally asymptotically stable if and only if $k < a\beta$; that is, if it exists it is asymptotically stable.

The Dulac-Bendixson Nonexistence Criteria (with auxiliary function M^{-1}) can be used to demonstrate that no limit cycles exist for (4.5). Consequently, by the Poincaré-Bendixson Theorem, (A_0, M_0) is globally asymptotically stable.

There is a relationship between the hypotheses of the model of Smith (Sect. 3.7) and this one in that both formulations are concerned with consumption rates at a variable density and at equilibrium or saturation.

4.2 Other Model Mechanisms for Feeding Relationships

Representations Without Consumer Interference Mass Action

Among the earliest modelling efforts devoted to feeding mechanisms were important contributions of A. Lotka, an American biologist, and V. Volterra, an Italian mathematician. They assumed that the ingestion rate, F, is proportional to the product of the magnitudes of consumer density, x_c, and resource density, x_r:

$$F = f x_c x_r.$$

One criticism of this feeding representation is that when the resource is abundant, the rate of consumption does not depend solely upon the consumer population. Holling (1959) overcame this objection by maintaining the linear response for low density resource but imposed a saturation effect for large densities (Fig. 4.1).

Hyperbolic Responses

In a model of a fishery Ivlev (1961) used a formulation that allowed consumers to feed at a mass rate when food is abundant:

$$F = f x_c (1 - e^{-\gamma x_r}).$$

The models of Monod (1950), Watt (1959), and Holling (1959) all relate to the formulation

$$F = f \frac{x_c x_r}{c + x_r}.$$

This representation is referred to as a Michaelis-Menten-Monod formulation (M^3 for short; M^2 got into the act from enzyme kinetics) and is characteristic of hyperbolic type responses, Fig. 4.2.

Sigmoid Responses

Holling (1959) and Murdoch and Oaten (1975) described a sigmoid type functional response (Fig. 4.3).

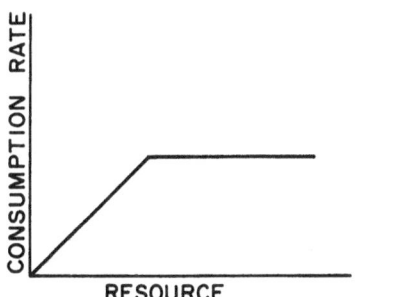

Fig. 4.1. A linear functional response (Holling)

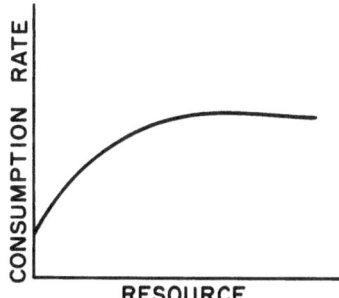

Fig. 4.2. A hyperbolic response for resource-consumer interactions

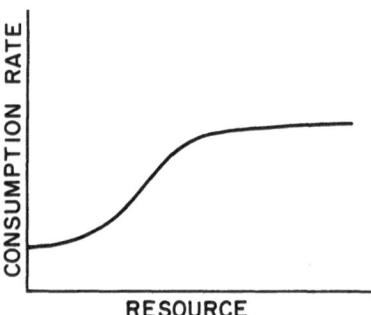

Fig. 4.3. A sigmoid response for resource-consumer interactions

The hyperbolic response seems to be most prevalent and occurs in some insects, some invertebrates (snails on mussels, starfish on snails) and some vertebrates (carp on bream roe); see Murdoch and Oaten (1975). Some phytoplankton respond in this way to various nutrients (Frost, 1974).

The sigmoid response has been observed in some parasitic insects and some vertebrates (deermouse on sawfly pupae). Hassell (1978) argued that this type of response is more common than originally believed.

The linear response type is claimed to occur for some invertebrates (crustacea on algae or yeast) although it is difficult to differentiate data fits to the linear response and the hyperbolic response.

4.3 Representations with Consumer Interference

Situations can occur in which consumer density increases but the feeding rate does not increase proportionately because of mutual interference between consumers. Consumption efficiency must decrease in these settings. Salt (1967), in studies of the effects of the ciliate *Woodruffia metabolica* on the species *Paramecium* and *Didinium nasutum* on *Paramecium aurelia*, found that feeding rates per unit consumer F/x_c react strongly to changes in density of consumers. Hassell (1971) has demonstrated that the searching efficiency of the insect parasite *Nemeritis canescens* decreases at high parasite densities.

With these examples in mind, DeAngelis, Goldstein, and O'Neill (1975) suggested that F/x_c should depend upon x_c and proposed a consumption formulation of

$$F = \frac{f x_r x_c}{b + x_c + d x_r}.$$

It should be recalled that Gallopin's formulation also has this consumer-interference feature with his formulation

$$F = a x_c (1 - \exp(-\alpha x_r)/x_c).$$

Resource-consumer mechanisms are formulations that attempt to emulate the consumption process. This biochemical process probably cannot be represented adequately without inclusion of some of the relevant biological and chemical interactions. Some progress on modelling the consumption process in this direction has been made by Lassiter (this volume).

References

Andrewartha, H.G., Birch, L.C. (1945). The Distribution and Abundance of Animals. University of Chicago Press, Chicago

Ayala, F.J. (1968). Genotype, environment, and population numbers. Science *162*: 1453–1459

Beverton, R.J.H., Holt, S.J. (1957). On the dynamics of exploited fish populations. Ministry of Agriculture, Fisheries, and Food (London). Fisheries Investigations Series 2 (19)

Bodenheimer, F.S. (1937a). Population problems of social insects. Biol. Rev. *12*: 393–340

Bodenheimer, F.S. (1937b). Studies in animal populations: II. Seasonal population trends of the honeybee. Quart. Rev. Biol. *12*: 406–425

Brauer, F., Sanchez, D.A. (1975). Constant rate population harvesting: equilibrium and stability. Theoret. Pop. Biol. *8*: 12–30

Braun, M. (1975). Differential Equations and Their Applications. Springer-Verlag, New York, 718 pp.

Carlson, T. (1913). Über Geschwindigkeit und Größe der Hefevermehrung in Würze. Biochem. Z. *57*: 313–334

Clark, C.W. (1976). Mathematical Bioeconomics: The Optimal Management of Renewable Resources. Wiley, New York, 352 pp.

Coddington, E.A., Levinson, N. (1955). Theory of Ordinary Differential Equations. McGraw-Hill, New York

Coleman, B.O. (1979). Nonautonomous logistic equations as models of the adjustment of populations to environmental change. Math. Biosci. *45*: 159–176

Connell, J.H. (1961). Effects of competition, predation by *Thais lapillus*, and other factors on natural populations of the barnacle *Balanus balanoides*. Ecol. Monogr. *31*: 61–104

Davidson, J. (1938). Trans. Roy. Soc. South Australia *62*: 342–346

DeAngelis, D.L., Goldstein, R.A., O'Neill, R.V. (1975). A model for trophic interaction. Ecology *56*: 881–982

Frost, B.W. (1974). Feeding processes at lower trophic levels in pelagic communities. In: The Biology of the Oceanic Pacific (C. B. Miller, ed.). Oregon State University Press, Corvallis, 59–77

Fujii, K. (1968). Studies on interspecies competition between the azuki bean weevil and the southern cowped weevil: III. Some characteristics of strains of two species. Researches on Population Ecology *10* (1): 87–98. Kyoto Univ., Kyoto, Japan

Gallopin, G.C. (1971). A generalized model of a resource-population system: I. General properties. II. Stability analysis. Oecologia *7*: 382–413; *7*: 414–432

Gause, G.F. (1934). The Struggle for Existence. Williams and Wilkins, Baltimore

Gilpin, M.E. (1979). Spiral chaos in a predator-prey model. Amer. Nat. *113*: 306–308

Gilpin, M.E., Ayala, F.J. (1973). Global models of growth and competition. Proc. Nat. Acad. Sci. USA *70*: 3590–3593

Goel, N.S., Maitra, S.C., Montroll, E.W. (1971). On the Volterra and other nonlinear models of interacting populations. Rev. Mod. Phys. *43*: 231–276

Gompertz, B. (1925). On the nature of the function expressive of the law of human mortality. Phil. Trans. *115*: 513–585

Gray, J. (1929). The kinetics of growth. Br. J. Exp. Biol. *6*: 248–274

Hallam, T.G., Clark, C.E. (1981). Nonautonomous logistic equations as models of population in a deteriorating environment. J. Theor. Biol. *93*: 303–311

Hassell, M.P. (1971). Mutual interference between searching insect parasites. J. Anim. Ecol. *40*: 473–486

Hassell, M.P. (1978). The Dynamics of Arthropod Predator-Prey Systems. Princeton University Press, Princeton

Hassell, M., Lawton, J., May, R. (1976). Pattern of dynamical behavior in single-species populations. J. Anim. Ecol. *45*: 471–486

Holling, C.S. (1959). The components of predation as revealed by a study of small-mammal predation of the European pine sawfly. Canad. Ent. *91*: 293–320

Hutchinson, G.E. (1978). An Introduction to Population Ecology. Yale University Press, New Haven, 260 pp.

Ivlev, U.S. (1961). Experimental ecology of the feeding of fishes. Yale University Press, New Haven

Kavanagh, A.J., Richards, W.O. (1934). The autocatalytic growth curve. Amer. Nat. *68*: 54–59

Lakshmikantham, V., Leela, S. (1969). Differential and Integral Inequalities. Academic Press, New York

Leunberger, D.G. (1979). Dynamic Systems. Wiley, New York, 446pp.

Levins, R. (1970). Extinction. Some Mathematical Questions in Biology, Vol. 2. Amer. Math. Soc. Providence R.I., 75–107

MacArthur, R.H., Connell, J.H. (1966). The biology of populations. Wiley, New York

MacLulich, D.A. (1937). Fluctuations in numbers of the varying Hare *(Lepus Americanus)*. University of Toronto Studies, Biol. Ser. *43*: 1–136

May, R.M. (ed.) (1976). Theoretical Ecology. Saunders. Philadelphia, 317pp.

M'Kendrick, A.G., Kesava Pai, M. (1911). The rate of multiplication of microorganisms: A mathematical study. Proc. Roy. Soc. Edinb. *31*: 649–655

Monod, J. (1950). La technique du culture continue: theorie et applications. Ann. Inst. Pasteur *79*: 390–410

Murdoch, W.W., Oaten, A. (1975). Predation and population stability. Adv. Ecol. Res. *9*: 1–131

Murray, B.G., Jr. (1979). Population Dynamics, Alternative Models. Academic Press, New York

Nicholson, A.J. (1954). An outline of the dynamics of animal populations. Australian J. Zool. *2*: 9–65

O'Neill, R.V., Gardner, R.H., Weller, D.E. (1982). Chaotic models as representations of ecological systems. Amer. Nat. *120*: 259–263

Pearl, R. (1982). The influence of density of population upon egg production in *Drosophila melanogaster*. J. Exp. Zool. *63*: 57–84

Pielou, E.C. (1977). Mathematical Ecology. Wiley-Interscience, New York

Riley, G.A. (1943). Physiological aspects of spring diatom flowerings. Bull. Bingham Oceanogr. Coll. *8*: 1–53

Rosenzweig, M. (1971). Paradox of enrichment: destabilization of exploitation ecosystems in ecological time. Science *171*: 385–387

Roughgarden, J. (1979). Theory of Population Genetics and Evolutionary Ecology: An Introduction. MacMillan, New York, 634pp.

Salt, G.W. (1967). Predation in experimental protozoan populations (Woodruffia-Paramecuim). Ecol. Monogr. *37*: 113–144

Schoener, T.W. (1973). Population growth regulated by intraspecific competition for energy or time. Theor. Pop. Biol. *4*: 56–84

Shelford, V.E. (1943). The relationship of snowy owl migration to the abundance of the collared lemming. Auk. *62*: 592–594

Smale, S., Williams, R.F. (1976). The qualitative analysis of a difference equation of population growth. J. Math. Biol. *3*: 1–4

Smith, F.E. (1963). Population dynamics in *Daphnia magna* and a new model for population growth. Ecology *44*: 651–663

Smith, R.H., Mead, R. (1980). The dynamics of discrete-time stochastic models of population growth. J. Theor. Biol. *86*: 607–627

Terao, A., Tanaka, T. (1928). Influence of temperature upon the rate of reproduction in the water-flea *Moina macrocopa*. Strauss. Proc. Imper. Acad. (Japan) *4*: 553–555

Utida, S. (1957). Cyclic fluctuations of population density intrinsic to the host-parasite system. Ecology *38*: 442–449

Utida, S. (1967). Damped oscillation of population density at equilibrium. Res. Pop. Ecol. *9*: 1–9

Verhulst, P.-F. (1938). Notice sur la loi que la population suit dans son accroissement. Correspondances Mathématiques et Physiques *10*: 113–121

von Bertalanffy, L. (1951). Theoretische Biologie, Vol. 2. Stoffwechsel, Wachstum, 2nd ed. Francke, Bern

Watt, K.E.F. (1959). A mathematical model for the effect of densities of attacked and attacking species on the numbers attacked. Canad. Ent. *91*: 129–144

The Formulation of Age-Structure Models

R. M. Nisbet and W. S. C. Gurney

1. The Mathematical Framework

1.1 Introduction – When Does Age-Structure Matter?

Not very long ago, two rash authors wrote that "although age-structure effects frequently influence the quantitative aspects of population dynamics, they are rather seldom responsible for qualitative changes in dynamic behaviour".* Although defensible in the context in which it was written, this assertion admits many exceptions – these are the subject of the present lectures.

Age-structure effects are likely to be ecologically important when we are studying a single population or a small number of populations over a time scale comparable with the lifetime of an individual member of the population(s). Thus, for example, they are vital to many of the epidemiological studies reported elsewhere in this volume. Their manifestation is frequently through population fluctuations whose structure reflects an element of synchronization of births and/or deaths in the population, as we show in Fig. 1 which illustrates examples of this phenomenon taken from studies of laboratory insect populations.

The dilemma facing the ecological modeller confronted with demographic changes associated with varying age structure in a population is how much detail to incorporate in his models. In this lecture, we shall outline the standard approaches to the mathematical descriptions of age (and size) structured populations and shall demonstrate the need for a compromise between the conflicting demands of mathematical tractability and biological realism. The two subsequent lectures explore a number of possible routes to such compromises.

1.2 Continuous Time Models

We assume we are modelling a closed population (no immigration or emigration) of a single species. We ignore all differences between individuals (even sex!) other than age and define an *age distribution* $f(a, t)$ by

$$f(a, t) = \lim_{da \to 0} \frac{\text{No. of individuals aged } a \to a + da \text{ at time } t}{da}. \tag{1.1}$$

---

* Nisbet and Gurney (1982)!

Biomathematics, Vol. 17, Mathematical Ecology
Edited by T. G. Hallam and S. A. Levin
© Springer-Verlag Berlin Heidelberg 1986

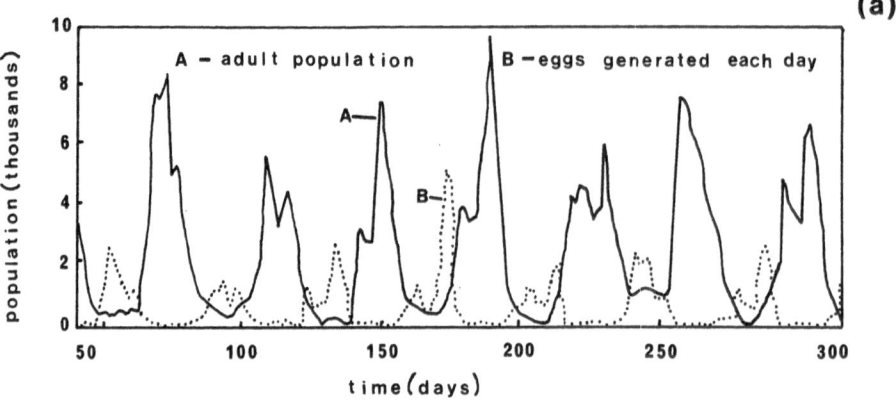

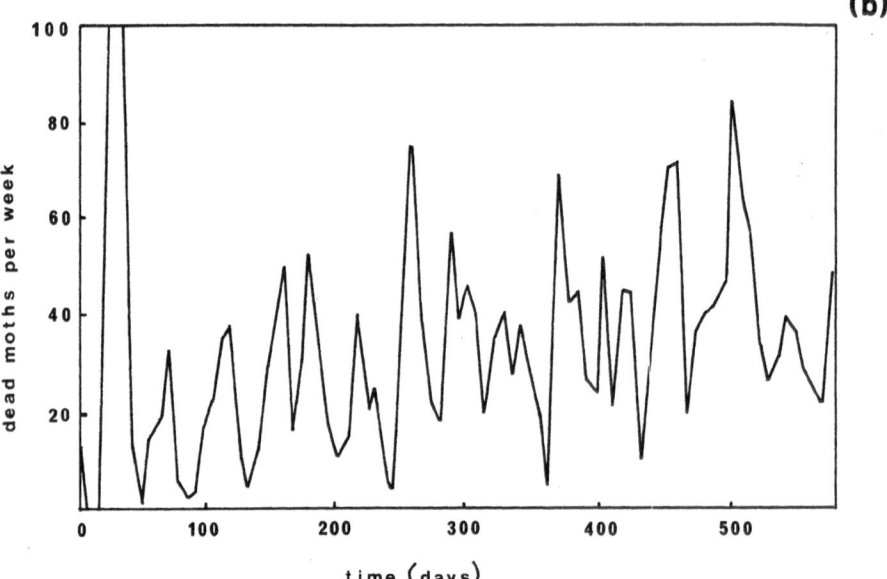

Fig. 1. Insect population fluctuations associated with age-structure effects. (**a**) Nicholson's Blowflies (Nicholson, 1954). The continuous curve is the total population, the broken curve the rate of egg production. The period of the cycles is 2–3 times the maturation time. (**b**) Lawton's *Plodia* (Gurney, Nisbet, and Lawton, 1983). The irregular fluctuations have a dominant period close to the generation time

We assume that the *per capita* death rate of individuals in the population depends only on age and time, and thus denote by $\delta(a, t)dt$ the proportion of individuals aged a who die during the infinitesimal time interval $t \rightarrow t + dt$. The probability that an individual, born at time t survives to age a is then

$$S(t, a) = \exp\left\{ - \int_{t}^{t+a} \delta(x - t, x)dx \right\}. \tag{1.2}$$

If $B(t)$ denotes the rate of addition of newborn individuals (aged zero) to the population at time t, then clearly

$$f(a,t)=B(t-a)S(t-a,a)=f(0,t-a)\exp\left\{-\int_{t-a}^{t}\delta(x-t+a,x)dx\right\},\qquad(1.3)$$

from which it is easy to demonstrate that the age distribution satisfies a partial differential equation (known as the McKendrick-Von Foerster equation)

$$\frac{\partial f(a,t)}{\partial t}+\frac{\partial f(a,t)}{\partial a}+\delta(a,t)f(a,t)=0.\qquad(1.4)$$

In this idealized population of androids, stamped with a date of manufacture but possessing no other distinguishing characteristics, we expect fecundity to depend only on age and time. We denote by $\beta(a,t)da$ the average rate of offspring production by an individual aged $a\to a+da$ at time t. Thus

$$B(t)=f(0,t)=\int_{0}^{\infty}\beta(a,t)f(a,t)da,\qquad(1.5)$$

an equation commonly referred to as a *renewal condition*, which can be used in two ways – as a boundary condition for the PDE (1.4), or [aided by Eq. (1.3)], as the basis for deriving an integral equation satisfied by $B(t)$, namely

$$B(t)=\int_{0}^{\infty}\beta(a,t)S(t-a,a)B(t-a)da,\qquad(1.6)$$

from the solution of which of course $f(a,t)$ can always be computed using (1.3).

The above formalism presupposes knowledge of the entire history (in principle back to $t=-\infty$!) of the recruitment rate $B(t)$ and the death rate $\delta(a,t)$, and although very valuable in "strategic" modelling, it is not directly applicable to models of specific populations. Either of two modifications enable us to circumvent this difficulty. The first possibility is to suppose that instead of knowing the entire history of $B(t)$ we know the initial age distribution, denoted by $\phi(a)$, at some time t (say $t=0$). Then Eq. (1.3) is replaced by

$$f(a,t)=\left[\begin{array}{ll}B(t-a)S(t-a,a)&\text{for}\quad t\geq a\\ \phi(a-t)\sigma(t,a)&\text{for}\quad t<a,\end{array}\right.\qquad(1.7)$$

where

$$\sigma(t,a)=\exp\left\{-\int_{0}^{t}\delta(a-t+x,x)dx\right\},\qquad(1.8)$$

the solution for $t<a$ representing "dying off" of individuals present at $t=0$. It is a routine exercise to verify that the Von Foerster-McKendrick equation (1.4) is still valid, though now only for $t>0$, and but that the integral equation (1.6) should be replaced by

$$B(t) = \int_0^t \beta(a, t)S(t-a, a)B(t-a)da + G(t) \qquad (1.9)$$

where

$$G(t) = \int_0^\infty \beta(t+x, t)\phi(x)\sigma(t, t+x)dx, \qquad (1.10)$$

and represents the production of offspring from the initial population. The integral equation (1.9) is often known as the Lotka renewal equation.

The second device for avoiding the need to know a long population history is particularly appropriate to laboratory populations which are "started" at some well defined time through intervention by the experimenter (e.g. by placing a few insect larvae in a bottle). We assume that prior to $t=0$ the system is *empty* [i.e. $f(a, t)=0$ for all a if $t \leq 0$], and that over some time interval after $t=0$ the experiment is started by the *innoculation* of newborn individuals at a rate $I(t)$. Equations (1.3) and (1.4) are still valid but the renewal condition (1.5) becomes

$$B(t) = \int_0^\infty \beta(a, t)f(a, t)da + I(t). \qquad (1.11)$$

The integral equation (1.9) for $B(t)$ retains its form, namely

$$B(t) = \int_0^t \beta(a, t)B(t-a)S(t-a, a)da + I(t). \qquad (1.12)$$

but the interpretation of the inhomogeneous term $(I(t))$ is different from that of $G(t)$ in (1.9).

1.3 Density-Independent Population Growth in an Unvarying Environment

We now specialize the discussion to the particular case where the vital rates depend only on age and not on time. The mathematical detail pertaining to this problem is presented in the notes by Frauenthal (this volume) so we shall omit it here. We do, however, note that if we denote the total population by $N(t)$, so that

$$N(t) = \int_0^\infty f(a, t)da, \qquad (1.13)$$

and define $F(a, t)da$ to be the *proportion* of the population with ages $a \rightarrow a+da$, then

$$F(a, t) = \frac{f(a, t)}{N(t)} \qquad (1.14)$$

We then expect (and Frauenthal proves) that after transients have died away, the normalized age distribution $F(a, t)$ approaches a stationary form $F^*(a)$, and the population grows or declines exponentially, i.e.

$$f(a, t) \propto F^*(a)e^{\lambda t}. \tag{1.15}$$

The growth (or decline) constant λ is the (only) real root of the equation

$$\int_0^\infty e^{-\lambda a} \beta(a) \exp\left\{-\int_0^a \delta(x)dx\right\} da = 1 \tag{1.16}$$

and it can be shown that this root is negative (implying ultimate decline of the population size towards zero) if

$$\int_0^\infty \beta(a) \exp\left\{-\int_0^a \delta(x)dx\right\} da < 1, \tag{1.17}$$

and positive (implying ultimate exponential growth) if this inequality is reversed.

1.4 Discrete-Time Models

A commonly used approach to modelling the dynamics of an age-structured population involves the use of *discrete age classes*, each of equal duration τ. This line of attack has proved particularly valuable in fisheries models purporting to describe changes in the stock of a species at a defined time each year.

Suppose there are $m+1$ age classes numbered $0, 1, 2, ..., m$, each covering an age interval τ. If $f_{i,t}$ denotes the number of individuals in age class i at time t and $P_{i,t}$ denotes the fraction of this population that survive to be in age class $i+1$ at time $t+\tau$, then

$$f_{i,t+\tau} = P_{i-1,t}f_{i-1,t}, \quad i = 1, 2, ..., m. \tag{1.18}$$

If $\beta_{i,t}$ is the average fecundity of a member of the ith age class at time t then the population of the first age class at time $t+\tau$ is clearly

$$f_{0,t+\tau} = \sum_{i=0}^m \beta_{i,t}f_{i,t}, \tag{1.19}$$

which is of course the discrete analog of the integral renewal condition (1.5).

The combined effects of Eqs. (1.18) and (1.19) can be recast in matrix notation if we define

$$\mathbf{f}_t \equiv \begin{bmatrix} f_{0,t} \\ f_{1,t} \\ \vdots \\ f_{m,t} \end{bmatrix}, \quad \mathbf{A}_t \equiv \begin{bmatrix} \beta_{0,t} & \beta_{1,t} & \beta_{2,t} & \cdots & \beta_{m,t} \\ P_{0,t} & 0 & 0 & \cdots & 0 \\ 0 & P_{1,t} & 0 & \cdots & 0 \\ \vdots & & & & \\ 0 & 0 & 0 & P_{m-1,t} & 0 \end{bmatrix} \tag{1.20}$$

implying

$$\mathbf{f}_{t+\tau} = \mathbf{A}_t \mathbf{f}_t. \tag{1.21}$$

This matrix formalism is particularly useful in the special case of density-independent growth in a constant environment, for which we can derive results analogous to those in Sect. 1.3. In this case the matrix $\mathbf{A}$ (normally called the *Leslie matrix*) does not vary with time and repeated application of (1.21) yields

$$\mathbf{f}_t = \mathbf{A}^k \mathbf{f}_0 \quad \text{where} \quad k = t/\tau. \tag{1.22}$$

The Perron-Frobenius theorem (see Frauenthal's notes) guarantees that the eigenvalue of maximum modulus of $\mathbf{A}$ is real and positive; if we denote this eigenvalue and the corresponding right eigenvector by y and $\mathbf{u}$ respectively, then (again appealing to Frauenthal for the supporting detail) we find that for large t

$$\mathbf{f}_t \propto \mathbf{u} e^{\lambda t} \quad \text{where} \quad \lambda = \tau^{-1} \ln y. \tag{1.23}$$

Thus the population ultimately grows or declines depending on whether $y \gtrless 1$, and it is found to decline if

$$\sum_{i=0}^{m} \beta_i \left\{ \sum_{j=0}^{i-1} P_j \right\} < 1, \tag{1.24}$$

a result which is the discrete analog of (1.17).

1.5 Elaboration of the Basic Models

The above models, in which vital rates depend only on age and time, are clearly of limited applicability. Suppose that per capita birth and death rates depend on some physiological factor like size or weight as well as age. If, for example, the factor is size (denoted by m), then clearly we require a distribution function $f(a, m, t)$ which describes the size distribution of population as well as that over age and time. We define a growth function $g(a, m, t)$ which represents the growth rate of an individual of mass m and age a at time t. There are then many published derivations (e.g. in our own book, Chap. 3, to take a not totally random example!) of the PDE obeyed by the distribution function, i.e.

$$\frac{\partial f}{\partial t} + \frac{\partial f}{\partial a} + \frac{\partial}{\partial m}(gf) + \delta f = 0, \tag{1.25}$$

which is coupled to a renewal condition

$$f(0, m, t) = \int_0^\infty \int_0^\infty \beta(a, m', m, t) f(a, m', t) dm' da. \tag{1.26}$$

The combination of (1.25) and (1.26) is normally analytically intractable (though a perfectly acceptable formalism for numerical work); there is however a special case which deserves comment. Suppose the functions β, δ, and g do not depend on age. Then m now represents a measure of the "physiological age" of an individual and we can describe the evolution of the population through time by a physiological age distribution $\varrho(m, t)$ defined by

$$\varrho(m, t) = \int_0^\infty f(a, m, t)da \tag{1.27}$$

which obeys the PDE

$$\frac{\partial \varrho}{\partial t} + \frac{\partial}{\partial m}(g\varrho) + \delta\varrho = \alpha(m, t) \tag{1.28}$$

with

$$\alpha(m, t) = \int_0^\infty \beta(m, m', t)\varrho(m', t)dm' . \tag{1.29}$$

Note that if we use m as a measure of physiological age there is no guarantee that all newborn individuals have the same physiological age. Thus the renewal condition (1.29) provides a driving term rather than a boundary condition for the PDE (1.28).

A variant on the above formalism [due we think to Sinko and Streifer (1971)] involves organisms which reproduce by binary fission. The PDE representing the balance of recruitment and loss of individuals of a particular size takes the distinctive form

$$\frac{\partial \varrho(x, t)}{\partial t} + \frac{\partial}{\partial m}(g(x, t)\varrho(x, t)) = -b(x)\varrho(x, t) + 4b(2x, t)\varrho(2x, t) \tag{1.30}$$

about whose formal mathematical properties very little is known except in a few special cases [the subject of the lecture by Diekmann (1983)]. Thus the description of age or size structure in even the simplest biological context (proliferation of microorganisms) is fraught with mathematical difficulty – a timely pointer to the difficulties to be expected in constructing tractable models of more complex ecological situations.

2. Models with Density Dependence

2.1 What does "Density-Dependent" Mean?

The size of a population can only be controlled if there is some mechanism that ensures that total births exceed total deaths when numbers are low and vice versa when they are high. The repertoire of known ways by which this happy state can be reached is virtually endless and the idealization of "density"- or "population"-dependent vital rates is hard to pin down in even the simplest population studies. For example in our paper to the symposium accompanying this course (Nisbet and Gurney, 1983c) we present a few models of laboratory moth populations regulated by the availability of larval food. We argue that shortage of food for the larvae could influence the overall population growth by

(a) increasing larval *mortality* (either uniformly or in "crowded" age cohorts).
(b) slowing larval *growth*, thereby delaying and reducing future adult recruitment.
(c) reducing the likelihood of successful *pupation*.
(d) reducing adult *fecundity* – in species where adults do not feed, eggs laid by adults are built from food consumed as larvae.

No doubt, the reader can add a few mechanisms of his own to the above list.

Given this virtually endless range of potential mechanisms resulting in "density-dependence" of vital rates, it is evident that there can be no general theory of population models with age and density dependent vital rates. In the remaining two sections, our aim is therefore to present adaptable techniques for *formulating* rigorous, but mathematically tractable models. The pointers to potential simplifications are found in the analysis of a few idealized cases which help elucidate the mathematical structures inherent in age structure problems.

2.2 An Extreme Idealization –
Vital Rates Controlled by Total Population

(A) *Discrete Time Formulation*

Just about the simplest defensible case (due to Beddington, 1974) to study is that in which the instantaneous vital rates are regulated by the *total population size*. In the discrete time formulation, this means that the elements of the Leslie matrix $\mathbf{A}$ become functions of the total population N_t where

$$N_t = \sum_{i=0}^{m} f_{i,t}, \tag{2.1}$$

i.e.

$$\beta_i = \beta_i(N_t); \quad P_i = P_i(N_t); \tag{2.2}$$

If this system has a steady state, then clearly this must be obtainable from

$$\sum_{i=0}^{m} \beta_i(N^*) \left\{ \prod_{j=0}^{i-1} P_j(N^*) \right\} = 1. \tag{2.3}$$

To investigate the local stability of this steady state, we set

$$\xi_t \equiv f_t - f^*; \quad n_t \equiv N_t - N^* \tag{2.4}$$

and find that to first order in ξ_t and n_t

$$
\begin{bmatrix} f_0^* + \xi_{0,t+\tau} \\ f_1^* + \xi_{1,t+\tau} \\ \vdots \\ f_m^* + \xi_{m,t+\tau} \end{bmatrix} = \begin{bmatrix} \beta_0^* + \left[\dfrac{d\beta_0}{dN}\right]^* n_t & \beta_1^* + \left[\dfrac{d\beta_1}{dN}\right]^* n_t \cdots\cdots & \\ P_0^* + \left[\dfrac{dP_0}{dN}\right]^* n_t & 0 & \cdots\cdots \\ 0 & P_1^* + \left[\dfrac{dP_1}{dN}\right]^* n_t \cdots\cdots \\ \vdots & \vdots \end{bmatrix} \begin{bmatrix} f_0^* + \xi_{0,t} \\ f_1^* + \xi_{1,t} \\ \vdots \\ f_m^* + \xi_{m,t} \end{bmatrix}
\tag{2.5}
$$

which, or multiplying up the R.H.S. and dropping second-order terms becomes

$$
\xi_{t+\tau} = A(N^*)\xi_t + n_t \left[\frac{dA}{dN}\right]^* f^* .
\tag{2.6}
$$

Thus

$$
\xi_{t+\tau} = A(N^*)\xi_t + \left[\sum_{j=0}^{m} \xi_{j,t}\right]\left[\frac{dA}{dN}\right]^* f^*
\tag{2.7}
$$

or in component form

$$
\xi_{i,t+\tau} = \sum_{j=0}^{m} A_{ij}(N^*)\xi_j + \sum_{j=0}^{m} \xi_{j,t} \sum_{k=0}^{m} \left[\frac{dA_{ik}}{dN}\right]^* f_k^*
\tag{2.8}
$$

i.e.

$$
\xi_{t+\tau} = B\xi_t \quad \text{where} \quad B_{ij} = A_{ij}(N^*) + \sum_{k=0}^{m} \left[\frac{dA_{ik}(N)}{dN}\right]^* f_k^* .
\tag{2.9}
$$

Thus local stability is controlled by the eigenvalues of this new matrix B with stability if $|\lambda_i \tau| < 1$ for all eigenvalues λ_i.

An interesting and useful variant of this formalism, intended for application to fish populations has been developed by Horwood and Shepherd (1981). It is argued that the population is regulated primarily by mortality during the (short) egg and larval stages and that all mortality thereafter can be regarded as density independent. It is assumed that there is a "juvenile" phase lasting n years after which adults can spawn and that spawning takes place at the start of the year (i.e. *before* significant mortality occurs). Finally, survival and fertility is assumed constant for all adults. The result of these assumptions is a model in which (using our notation)

$$
\begin{bmatrix} f_{1,t+1} \\ f_{2,t+1} \\ \vdots \\ \vdots \\ f_{n,t+1} \end{bmatrix} = \begin{bmatrix} 0 & 0 & 0 & \cdots\cdots & 0 & \beta P_0 \\ P_1 & 0 & 0 & \cdots\cdots & 0 & 0 \\ 0 & P_2 & 0 & \cdots\cdots & 0 & 0 \\ \vdots & & & & & \\ 0 & 0 & 0 & \cdots\cdots & P_{n-1} & P_n \end{bmatrix} \begin{bmatrix} f_{1,t} \\ f_{2,t} \\ f_{3,t} \\ \vdots \\ f_{n,t} \end{bmatrix}
\tag{2.10}
$$

in which f_1 now denotes the population of juveniles (excluding eggs and larvae) aged one, $f_2 \ldots f_{n-1}$ denotes the juvenile populations in the remaining year classes, and f_n represents the *total* adult population. The total egg production in one year is

$$E_t = \beta f_{n,t} \tag{2.11}$$

and a fraction P_0 of these are assured to survive to enter the first juvenile year class. The key assumption is now that P_0 is a function of E. Stability analysis very similar to that presented above can be performed and we refer the reader to the original paper for details. Here, we merely emphasize one special feature of the model, namely that the dynamics of the *adult* population obey a very simple *delay-difference equation* namely

$$f_{n,t} = \beta P_0(E_{t-n}) \left[\prod_{i=1}^{n-1} P_i \right] f_{n,t-n} + P_n f_{n,t-1} . \tag{2.12}$$

Since E_{t-n} is itself a function of $f_{n,t-n}$ [from (2.11)], Eq. (2.12) gives us the dynamics of the adult population explicitly without involving the other year classes.

(B) *Continuous Time Formulation*

We again postulate vital rates $\beta(a, N)$, $\delta(a, N)$ which depend on the total population $N(t)$, where

$$N(t) = \int_0^\infty f(a, t) da , \tag{2.13}$$

For notational convenience, we introduce a rather artificial survival function $\pi(a, N)$, defined by

$$\pi(a, N) = \exp \left\{ -\int_0^a \delta(x, N) dx \right\}, \tag{2.14}$$

representing the probability of surviving to age a in circumstances where the total population is held constant at N, throughout the aging process. The steady state population N^* (if there is one) is obtained from

$$\int_0^\infty \beta(a, N^*) \pi(a, N^*) da = 1 \tag{2.15}$$

and the stationary age distribution is

$$f^*(a) = \frac{N^* \pi(a, N^*)}{\int_0^\infty \pi(a, N^*) da} . \tag{2.16}$$

We linearize about the steady state by setting:

$$\xi(a, t) = f(a, t) - f^*(a); \quad n(t) = N(t) - N^* = \int_0^\infty \xi(a, t) da \tag{2.17}$$

and find that to first order

$$\frac{\partial \xi}{\partial t} + \frac{\partial \xi}{\partial a} + \delta(a, N^*)\xi + f^*(a)g(a)n(t) = 0, \tag{2.18a}$$

$$\xi(0, t) = \int_0^\infty \beta(a, N^*)\xi(a, t)da - Kn(t), \tag{2.18b}$$

where

$$\left[g(a) = \left[\frac{\partial \delta(a, N)}{\partial N} \right]^* \right.$$
$$\left. K = -\int_0^\infty f^*(a)\left[\frac{\partial \beta(a, N)}{\partial N} \right]^* da. \right. \tag{2.19}$$

It is possible – see Gurtin and MacCamy (1974) or Nisbet and Gurney (1982) to obtain formal conditions for the stability of the steady state by assuming

$$\xi(a, t) = L(a)e^{\lambda t}, \tag{2.20}$$

substituting in (2.18a) and (2.18b) and finding a gruesome equation, all of whose roots must have negative real parts. The interested reader is referred to the original paper cited for the details. We mention however two results of significance for strategic modelling

(a) With a *density-independent* death rate, a necessary and sufficient condition for neighborhood stability is $K > 0$, i.e. "on average", increasing the population must decrease the overall birth rate.
(b) With an *age-independent* death rate, sufficient conditions for local stability are $K \geq 0$ and $(d\delta(N)/dN)^* \geq 0$ (with equality permitted in at most one of these).

The proof of (a) is easy. The proof of (b) is messy and inelegant; details are in Nisbet and Gurney (1982).

As in the discrete time case, there is an important variant on the above formalism which produces a spectacular simplification. If instead of vital rates dependent on total population, we have vital rates which are regulated solely by the population of mature adults (aged τ or over), and if the adult fecundity and mortality are age-independent, then the changes in the adult population can be described by a delay-differential equation of the form

$$\frac{dN(t)}{dt} = R(N(t-\tau)) - D(N(t)) \tag{2.21}$$

in which $R(\cdot)$ and $D(\cdot)$ represent the rates of recruitment to and deaths from the adult population. This type of equation [which is manifestly the continuum analogue of (2.12)] has been extensively studied and has been applied to populations as different as blowflies (Gurney et al., 1980) and baleen whales (May, 1980).

To summarize this section, we have outlined the basic formalism for the unrealistic case of vital rates dependent on total population size, and illustrated one modification (involving "lumping" all the adults together) which produces a substantial simplification by yielding a single delay-differential or delay-difference equation. This is an example of the "stage structure" approach which is the subject of lecture 3, but more significantly for the present lecture it points to the possibility of obtaining compact reduced descriptions of non-linear age structure problems. This possibility is now explained in more detail.

2.3 A Route to Simplifying the Mathematics

(A) *The Gurtin-MacCamy Trick*

In the previous section we explored the consequences of a ludicrous idealization – vital rates controlled solely by total population – but saw that some of the resulting mathematics gave hints on useful simplifications when constructing more realistic models. In the same spirit we now stick with the idealisation and demonstrate a trick that, for certain fecundity functions, permits a full description of the population dynamics in terms of a few simple ordinary differential equations.

This key trick has an impressive pedigree including Gurtin and MacCamy (1974, 1979), Cushing (1977, 1980) MacDonald (1978) and many others. In its simplest form, in addition to assuming that vital rates are regulated by total population, we assume that

$$\delta(a, N) = \delta_0(N), \quad \beta(a, N) = \beta_0(N)G_s(a, c), \quad \text{(integer } s) \tag{2.22}$$

with

$$G_s(a, c) = \frac{c^{s+1}}{s!} a^s e^{-ca}, \quad c > 0 \tag{2.23}$$

i.e. that the per capita death rate is age independent and that the age dependence of the fecundity can be approximated by a gamma-distribution, the shape of which for a few values of s is illustrated in Fig. 2. We demonstrate the trick for the (silly but simple) case $s = 0$ and for the uncluttered case where the McKendrick-von Foerster equation (1.4) is valid for all t (before and after $t = 0$); the elaboration to different "start-up" regimes being straightforward.

We first derive an ODE for the total population by integrating equation (1.4) over all ages and obtaining

$$\frac{dN}{dt} = B(t) - \delta_0(N)N. \tag{2.24}$$

We then define an auxiliary variable $V(t)$ by

$$V(t) \equiv c \int_0^\infty f(a, t) \exp\{-ca\} da = B(t)/\beta_0(N) \tag{2.25}$$

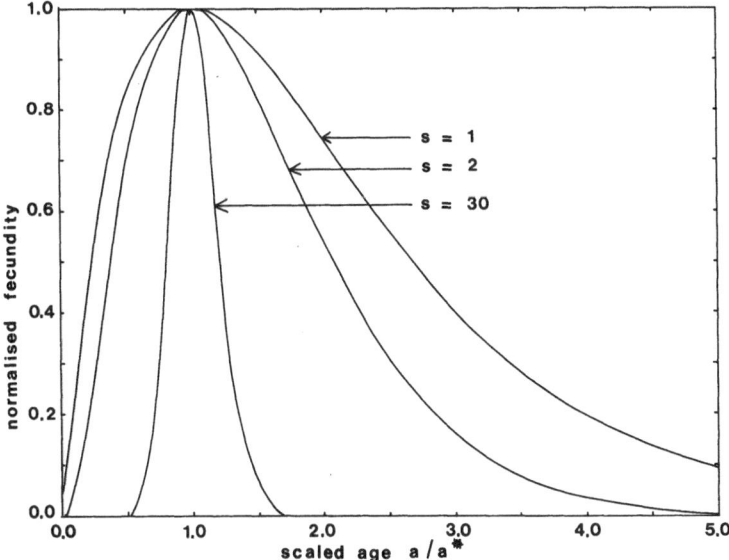

Fig. 2. The shape of the gamma distribution for a few values of s

and find after integrating by parts that

$$\frac{dV}{dt} = cB(t) - [\delta_0(N) + c]V(t).$$ (2.26)

Since the complete age distribution $f(a, t)$ is obtainable from $B(t)$ [or equivalently $V(t)$] and $N(t)$ with the aid of Eq. (1.3) it follows that the complete dynamics of the age-structured population are described by two ODE's – a spectacular simplification. With $s \geq 1$, we introduce further auxiliary variables and find that the reduction is still possible, the final description of the dynamics involving $s+1$ auxiliary equations.

(B) Using Realistic Fecundity Functions

The Gurtin-MacCamy trick has greatly facilitated a wide variety of "strategic" studies, but unfortunately the gamma distribution seldom provides a good fit to measured fecundities. The principal reason for this failure is that the life history of most species involves a distinct "juvenile" phase during which reproduction does not occur. We now outline a variant of the Gurtin-MacCamy trick which recognises the dynamic importance of this phase and yields a *small* set of delay-differential equations instead of the original age-structure equations.

We subdivide the population into two age classes:

$$0 \to a_1 \quad \text{AND} \quad a_1 \to \infty,$$
(immature) (matures)

and note that the *subpopulations* of immatures and matures at time t are given by

$$N_I(t) = \int_0^{a_1} f(a, t) da, \quad N_M(t) = \int_{a_1}^\infty f(a, t) da. \tag{2.27}$$

We permit density-dependence in the per capita death rates, but assume that all individuals within each age class experience the same mortality. Thus we assume that

$$\delta(a, t) = \begin{bmatrix} \delta_I(N_I(t), N_M(t)) & \text{for} & 0 < a < a_1 \\ \delta_M(N_I(t), N_M(t)) & \text{for} & a \geq a_1. \end{bmatrix} \tag{2.28}$$

We further assume that the fecundity has the form

$$\beta(a, t) = qg(N_I(t), N_M(t))h(a). \tag{2.29}$$

in which q represents the maximum possible fecundity, $h(a)$ is that fraction of q achieved by an adult of age a in the absence of density dependence, and $g(\cdot)$ represents the fractional reduction of the age specific fecundity due to density dependence. Essentially, we are assuming that the *shape* of the plot of fecundity against age is unaltered by density dependence, an assumption which of course restricts the applicability of the resulting models; in particular, for many anthropods the age of maturity is affected by density.

Nevertheless, it is often an *experimentally testable* assumption as we show in Fig. 3.

Our variant of the basic trick is produced by assuming that $h(a)$ is a *displaced* gamma distribution. We thus set

$$h(a) = \begin{bmatrix} 0 & \text{if} & a < a_1 \\ g_s(a - a_1, a_2) & \text{if} & a \geq a_1, \end{bmatrix} \tag{2.30}$$

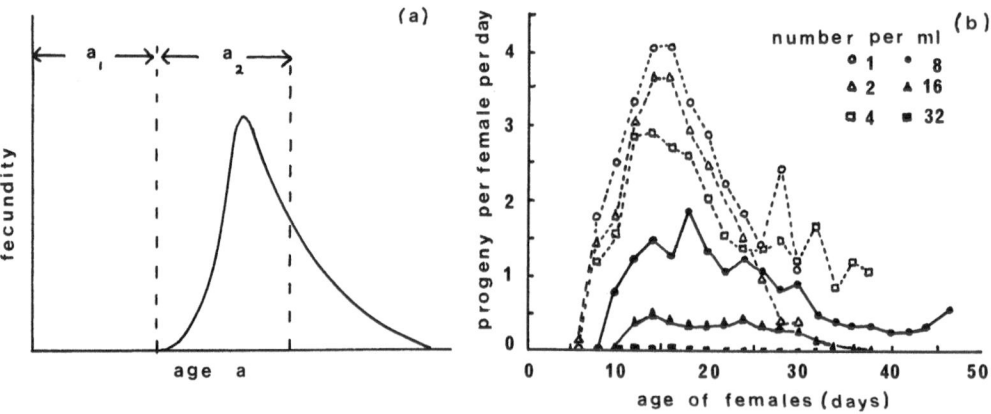

Fig. 3. (a) Parametrisation of the "displaced gamma" distribution used in Sect. 2.3. **(b)** Age and density dependence of fecundity for *Daphnia pulex* [from Ricklefs (1973) who quoted Warren (1971) who quoted Frank et al. (1957)!]

where

$$g_s(x,c) = A_s^{-1} x^s e^{-cx} \quad \text{and} \quad A_s = \max_{x \geq 0} \{x^s e^{-cx}\}. \tag{2.31}$$

To scale these functions we introduce a parameter $a_2 = s/c$: Fig. 3a then illustrates the roles played by the parameters a_1, a_2, and s in determining the shape of $h(a)$.

The reduced description is obtained by integrating the McKendrick equation over the age ranges $0 \rightarrow a_1$ and $a_1 \rightarrow \infty$, and using Eq. (1.3) to obtain

$$\frac{dN_I(t)}{dt} = B(t) - B(t-a_1)P(t) - \delta_I(N_I(t), N_M(t))N_I(t), \tag{2.32}$$

$$\frac{dN_M(t)}{dt} = B(t-a_1)P(t) - \delta_M(N_I(t), N_M(t))N_M(t), \tag{2.33}$$

with

$$P(t) \equiv \exp \left\{ - \int_{t-a_1}^{t} \delta_I(N_I(t'), N_M(t'))dt' \right\} \tag{2.34}$$

(These equations incidentially are a foretaste of the "stage structure" equations to be derived in the next lecture.) We see from (2.34) that

$$\frac{dP(t)}{dt} = P(t)[\delta_I(N_I(t-a_1), N_M(t-a_1)) - \delta_I(N_I(t), N_M(t))] \tag{2.35}$$

so we would have the dynamics expressed solely in terms of delay-differential equations but for the nasty integral equation defining $B(t)$. Fortunately, Blythe (1982) has shown that we can circumvent this problem by defining a new set of auxiliary variables

$$V_j(t) = \int_{a_1}^{\infty} g_{s+1-j}(a-a_1, c)f(a,t)da, \quad j = 1, 2, ..., s+1. \tag{2.36}$$

Then with a bit of slog he shows that

$$\frac{dV_j(t)}{dt} = \frac{sA_{s-1}}{A_s} V_{j+1}(t) - [c + \delta_M(N_I(t), N_M(t))]V_j(t)$$

$$\text{for} \quad j = 1, 2 ... s. \tag{2.37}$$

and

$$\frac{dV_{s+1}}{dt} = B(t-a_1)P(t) - [c + \delta_M(N_I(t), N_M(t))]V_{s+1}(t) \tag{2.38}$$

A full description of the dynamics is then embodied in equations (2.32), (2.33), (2.35), (2.37), and (2.38), which although intimidating analytically are readily integrated numerically.

3. Stage Structure Models

3.1 Introduction

Section 2 highlighted, on the one hand the lack of any general formalism for age- and density-dependent population dynamics, and on the other the possibilities for formulating realistic non-linear models representing particular regulatory mechanisms in specific populations. The key feature of these models was the definition of functionally distinct *stages* in the life of an individual (e.g. egg, larva, adult) and the recognition that the differences between individuals in the *same* stage could either be neglected [as in the models embodied in Eqs. (2.12) and (2.21)] or be treated in a rather simple minded way (as in Sect. 2.3 where we adopted a very simple representation of adult fecundity).

We have recently developed a general formalism to describe the changes of the "subpopulations" of distinct developmental stages in the extreme situation where all individuals within a stage are regarded as functionally identical, and where transitions between life history stages are triggered by a single factor such as age or size. The work has been reported in detail elsewhere (Gurney et al., 1983; Nisbet and Gurney, 1983a; Blythe et al., 1984) and presented in various conference papers so it would be inappropriate to repeat the detail here. Instead, we describe the main features (following very closely Nisbet and Gurney, 1983b) and refer the reader to the original papers for a fuller account.

3.2 Transitions Triggered by Age

Having assumed that we may approximate the life history of a species as a series of developmental stages within which all individuals are functionally identical, we now require to specify the "single factor" responsible for triggering transitions between instars. The natural first candidate is *age*, so we initially assume that transitions occur at fixed ages and derive expressions for the rates of recruitment, maturation and death in a particular development stage. The main results can be established intuitively using a "conveyer-belt" analogy, illustrated in Fig. 4, from which it is clear that the changes in the subpopulation $(N_i(t))$ of individuals in stage i are given by

$$\dot{N}_i(t) = R_i(t) - R_i(t-\tau_i)P_i(t) - \delta_i(t)N_i(t),\tag{3.1}$$

with the "through-stage-survival" $P_i(t)$ satisfying the differential equation

$$\dot{P}_i(t) = P_i(t)\left[\delta_i(t-\tau_i) - \delta_i(t)\right]\tag{3.2}$$

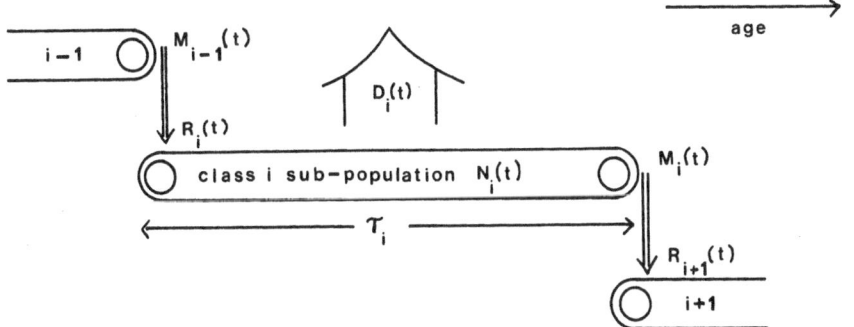

Fig. 4. A "conveyer-belt" analogy of the dynamics of the subpopulation of the ith development stage in a model where transitions between stages occur at fixed *ages*

and with

$$R_i(t) = R_{i-1}(t-\tau_i)P_i(t),$$
(3.3)

$$R_1(t) = \sum_i \beta_i(t)N_i(t).$$
(3.4)

As was emphasised in lecture 1, equations of this sort are formally valid if the population has run undisturbed (except via the t-dependence of the per capita vital rates) since $t = -\infty$! Of the two common situations emphasised in that lecture (a specified initial age distribution or innoculation of an empty system) the latter is more readily incorporated in this formalism.* Although we cannot retain the present "lumped" description *and* allow ourselves to add individuals of arbitrary age to the population at time t, if we assume that all individuals added to an age class are newly qualified for that class, then we can retain Eqs. (3.1) and (3.2) as the basic dynamic equations if we modify the definitions (3.3) and (3.4) to read

$$R_i(t) = R_{i-1}(t-\tau_{i-1})P_i(t) + I_i(t), \quad i > 1$$
(3.5)

$$R_1(t) = \sum_i \beta_i(t)N_i(t) + I_1(t).$$
(3.6)

Equations (3.1), (3.2), (3.5), (3.6) constitute the basic description for lumped age-class models with transitions triggered by age.

To illustrate the use of the recipe, we construct a very simple model of Lawton's experiments (Fig. 1b) on populations of the Indian meal-moth *Plodia interpunctella*. Adult *Plodia* do not feed, so the population was regulated by competition among larvae for food. We model this population by assuming that we need only consider two dynamically significant age classes:
(a) reproductively active adults of population $N_A(t)$
(b) a reproductively inactive larval stage of duration τ_L and population $N_L(t)$.
We assume a constant sex-ratio and that all females have a constant fecundity so that the net per capita adult fecundity is a constant (denoted by β_A), and we assume

---

* For details of how to deal with an initial age distribution, see Gurney, Blythe, and Nisbet (1985)

a constant, density-independent per capita adult death rate δ_A. We incorporate the resource limitation at the larval stage by assuming that a given larva competes equally with all other larvae (the "uniform competition" approximation) so that the per capita death rate of the larvae is a function of $N_L(t)$. For mathematical simplicity we choose

$$\delta_L(t) = \alpha N_L(t),\tag{3.7}$$

where α is a constant. We can now refer to the recipe and deduce that the dynamics of the system are described by three delay-differential equations

$$\dot{N}_L = R_L(t) - R_L(t - \tau_L)P_L(t) - \alpha N_L^2(t),\tag{3.8}$$

$$\dot{N}_A = R_A(t) - \delta_A N_A(t)\tag{3.9}$$

with

$$\dot{P}_L(t) = \alpha P_L(t)[N_L(t - \tau_L) - N_L(t)],\tag{3.10}$$

where, if the system is inoculated by adding a few eggs (new born "larvae") just after $t=0$ at a rate $I(t)$,

$$R_L(t) = \beta_A N_A(t) + I(t); \quad R_A(t) = R_L(t - \tau_L)P_L(t).\tag{3.11}$$

Figure 5 contains a "typical" solution of Eqs. (3.8)–(3.11) with parameters roughly appropriate to Lawton's *Plodia* – the main feature of the solution is a series of damped single-generation cycles with a period around 33 days. There is thus a possibility of explaining Lawton's observations as endogenous resonant quasicycles, the "driving" noise being demographic stochasticity.

More detailed discussion of this and a suite of related models is given in our paper to the research symposium (Nisbet and Gurney, 1983c).

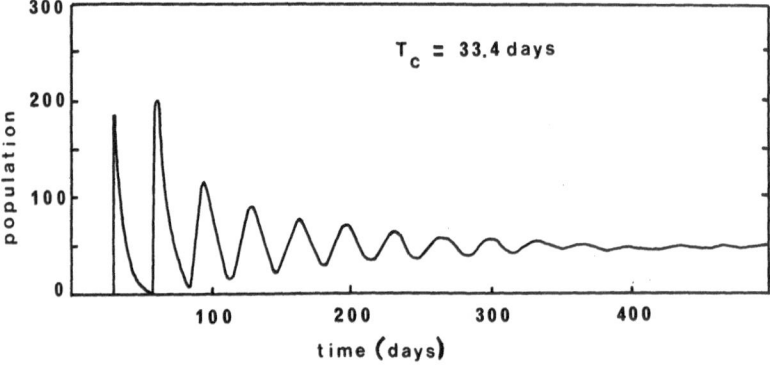

Fig. 5. A numerical solution of Eqs. (3.8)–(3.11) with $\beta_A = 9.4$ day^{-1}, $\delta_A = 0.2$ day^{-1}, $\tau_L = 28$ day. The constant α merely sets the scale of population – we set $\alpha = 5 \times 10^{-5}$ moth^{-1} day^{-1}

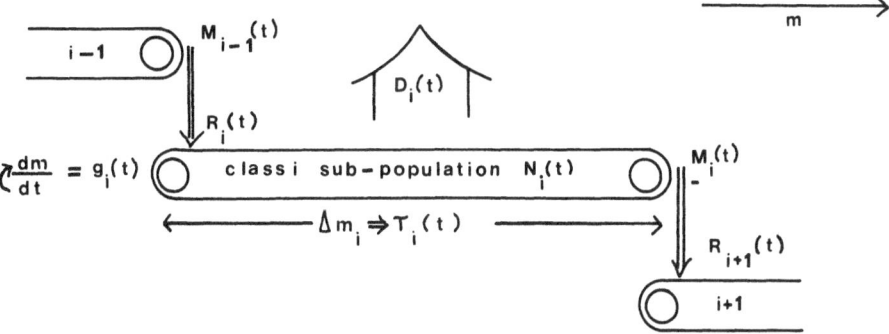

Fig. 6. A conveyer-belt analogy of the dynamics of the subpopulation of the ith developmental stage in a model where transitions between stages occur at fixed *sizes*

3.3 Growth Dependent Transitions

From physiological studies it is well known that for most insect species, it is not chronological age but weight gain that triggers the moult from one instar to its successor, a doubling of weight during an instar being typical (Dyar's "law"). The formalism outlined in the previous section now begins to earn its keep as it is readily extended to cover this situation. We stick with Oster's conveyor belt analogy but assume that the transition from one instar to its successor is triggered by achieving a weight gain Δm. The analogy is illustrated in Fig. 6. The dynamics of each of the various subpopulations are described by two delay differential equations

$$\dot{N}_i(t) = R_i(t) - R_i(t - \tau_i(t)) \frac{g_i(t)}{g_i(t - \tau_i(t))} P_i(t) - \delta_i(t) N_i(t), \tag{3.12}$$

$$\dot{P}_i(t) = P_i(t) \left[\frac{g_i(t) \delta_i(t - \tau_i(t))}{g_i(t - \tau_i(t))} - \delta_i(t) \right] \tag{3.13}$$

with the evaluation of the delay given by a third, namely,

$$\dot{\tau}_i(t) = 1 - g_i(t)/g_i(t - \tau_i(t)) \tag{3.14}$$

from which we can compute the lag at all times $t > 0$ if we know the value τ_{i0} of the lag at $t = 0$ and the "history" of the growth function at times $t \leq 0$. We can consistently interpret $g_i(t)$ at these times as the growth rate of a single individual introduced to the system at time $t(<0)$ and allowed to grow unimpeded by competition. This "unimpeded" growth rate will in many situations be constant (say g_{i0}); thus

$$\Delta m_i \equiv m_{i+1} - m_i = \int_{t - \tau_{i0}}^{t} g_i(t') dt' = g_{i0} \tau_{i0}. \tag{3.15}$$

We model the initial history in the same way as before, and assume an empty system to which a few individuals are added after $t=0$. Thus Eqs. (3.12)–(3.14) must be solved with

$$R_i(t) = R_{i-1}(t - \tau_{i-1}(t)) \frac{g_{i-1}(t)}{g_{i-1}(t - \tau(t))} P_{i-1}(t) + I_i(t), \tag{3.16}$$

$$R_1(t) = \sum_i \beta_i(t) N_i(t) + I_1(t). \tag{3.17}$$

With $N_i(t) = 0$ for all to ≤ 0, with the initial lag obtained from Eq. (3.15), and with

$$P_i(0) = \exp\left\{ -\int_{-\tau_{i0}}^{0} \delta_i(t') dt' \right\}. \tag{3.18}$$

Examples of the use of the formalism can be found in Nisbet and Gurney (1983a).

References

Beddington, J.R. (1974). Age distribution and the stability of simple discrete-time population models. Jour. Theor. Bio. *47*: 65–74

Blythe, S.P. (1982). Simple Age-Structure Models of Laboratory Insect Populations. Ph. D. Thesis, University of Strathclyde

Blythe, S.P., Nisbet, R.M., Gurney, W.S.C. (1984). The dynamics of population models with distributed maturation period. Theor. Pop. Biol. *25*: 289–311

Cushing, J.M. (1977). Integrodifferential Equations and Delay Models in Population Dynamics. Lect. Notes in Biomathematics, Vol. 20. Springer-Verlag, Germany

Cushing, J.M. (1980). Model stability and instability in age-structured populations. Jour. Theor. Biol. *86*: 709–730

Diekmann, O. (1983). The stable size distribution: an example in structured population dynamics. In: Mathematical Ecology, Lect. Notes in Biomathematics, Vol. 54, Springer-Verlag, Germany

Frank, P.W., Boll, C.D., Kelly, R.W. (1957). Vital statistics of laboratory cultures of *Daphnia pulex* Debeer as related to density. Physiol. Zool. *30*: 287–305

Gurney, W.S.C., Blythe, S.P., Nisbet, R.M. (1980). Nicholson's blowflies revisited. Nature, London *287*: 17–21

Gurney, W.S.C., Nisbet, R.M., Lawton, J.H. (1983). The systematic formulation of tractable single species models incorporating age-structure. Jour. of Animal Ecology *52*: 479–496

Gurney, W.S.C., Blythe, S.P., Nisbet, R.M. (1985). The systematic formulation of models of stage-structured populations. In: The Dynamics of Physiologically Structured Populations (O. Diekmann and J. A. J. Metz, Eds.). Springer-Verlag, Germany. In press

Gurtin, M.E., MacCamy, R.C. (1974). Non-linear age-dependent population dynamics. Archive for Rational Mechanics and Analysis *54*: 281–300

Gurtin, M.E., MacCamy, R.C. (1979): Some simple models for nonlinear age-dependent population dynamics. Math. Biosciences *43*: 199–211

Horwood, J.W., Shepherd, J.G. (1981). The sensitivity of age-structured populations to environmental variability. Math. Biosciences *57*: 59–82

MacDonald, N. (1978). Time Lags in Biological Models. Lect. Notes in Biomathematics, Vol. 27. Springer-Verlag, Germany

May, R.M. (1980). Mathematical models in whaling and fisheries management. In: Lectures on Mathematics in the Life Sciences, Vol. 13, pp. 1–63. American Mathematical Society

Nicholson, A.J. (1954). An outline of the dynamics of animal populations. Australian Jour. of Zoology 2: 9–65

Nisbet, R.M., Gurney, W.S.C. (1982). Modelling Fluctuating Populations. John Wiley & Sons Ltd., Chichester

Nisbet, R.M., Gurney, W.S.C. (1983a). The systematic formulation of population models for insects with dynamically varying instar duration. Theor. Pop. Bio. 23: 114–135

Nisbet, R.M., Gurney, W.S.C. (1983b). Tractable population models with age structure. In: Mathematics in Medicine and Biomechanics. (G. F. Roach, Ed.) Shiva Pub. Co.

Nisbet, R.M., Gurney, W.S.C. (1983c). Stage-structure models of uniform larval competition. In: Mathematical Ecology. Lect. Notes in Biomathematics, Vol. 54. Springer-Verlag, Germany

Ricklefs, R.C. (1973). Ecology. Nelson, London

Sinko, J.W., Streifer, W. (1971). A model for populations reproducing by fission. Ecology 52: 330–335

Warren, C.E. (1971). Biology and Water Pollution Control. Saunders, Philadelphia

Analysis of Age-Structure Models

James C. Frauenthal ⋆

Stable Population Theory – Discrete

Begin by developing a mathematical model for the growth of a population. Assume that the population is closed to migration, and that only the females are counted. Males are present for reproductive purposes, but are not specifically taken into consideration. In the case of human and other higher species, this makes sense for two reasons. Females know unequivocally who their offspring are; and (more importantly for the purposes of this model) females have a biologically well-defined beginning and end to their reproductive careers.

Although it will not be done here, the effects of migration could easily be added to the model. Further, its general structure makes the model an obvious candidate for realistic extrapolation computations (simulations). Here, however, the purpose is simply to investigate the implications of assuming that both the mortality and the fertility are functions of the age of individuals, but not functions of time. This first section will deal with discrete intervals of both age and time. The next section will repeat the development, but for a model which treats age and time as continuous variables.

For the sake of simplicity (and exactness), choose the intervals of age and of time to be identical in length. Typically, for models of the human population, these intervals are either one year or five. Without loss of generality, let t index time, with $t = 0, 1, 2, \ldots$ and i index age, with $i = 1, 2, 3, \ldots$. Also let

$k_{i,t}$ = number of females in the population in age group i at time t.

Conceptually, the model is composed of two relations, one of which results from the survival process and the other from the reproduction process. Consider first the survival of individuals from one interval to the next. Clearly, the females in age group i at time t will move into age group $i+1$ at time $t+1$, except for those lost due to death. Therefore, let

s_j = fraction of the females in age group j at time t who survive to be in age group $j+1$ at time $t+1$: $0 \leq s_j < 1$; $j = 1, 2, \ldots, n-1$ and $s_n = 0$.

It then follows that:

$$k_{j+1,t+1} = s_j k_{j,t}: \quad j = 1, 2, \ldots, n; \quad t = 1, 2, \ldots . \tag{1}$$

⋆ These notes were originally prepared with support from the Alfred P. Sloan Foundation while the author was at the State University of New York, Stony Brook, N.Y.

This set of equations describes the aging process for a particular group (cohort) of individuals as time passes. Notice the assumption that no individual lives for more than n time intervals.

The reproduction equation is a bit more complicated. Assume (since there is no migration) that all of the girls in the population who are in the youngest age group at time $t+1$ are the daughters of women in the population who were in the fertile age groups at time t. This simple idea is confounded slightly by the assumptions that the mothers survived until their daughters were born and that the daughters survived until time $t+1$. Therefore, let

> b_j = number of daughters per female in age group j that survive through the time interval in which they are born: $b_j \geq 0, j = 1, 2, ..., m-1, b_m \neq 0$, and $b_j = 0, j = m+1, m+2, ..., n$ $(n \geq m)$.

Thus,

$$k_{1,t+1} = b_1 k_{1,t} + b_2 k_{2,t} + \cdots + b_m k_{m,t}$$

$$= \sum_{j=1}^{m} b_j k_{j,t}. \tag{2}$$

These equations for survival and reproduction easily can be rewritten in matrix notation. To do so, let

$$\varrho_{ij} \equiv \begin{cases} b_j, & i=1, \quad j=1,2,...,m, \\ s_j, & i=j+1, \quad j=1,2,...,n-1, \\ 0, & \text{otherwise}, \end{cases}$$

$$\vec{K}_t \equiv \{k_{i,t}\} \quad \text{and} \quad \mathbf{L} \equiv \{\varrho_{ij}\},$$

where $\mathbf{L}$ is a Leslie Matrix, an $(n \times n)$ sparse matrix, all of the non-zero entries of which occur either in the first row (b's) or along the major sub-diagonal (s's). The matrix is thus positive semi-definite, or non-negative

$$\mathbf{L} = \begin{pmatrix} b_1 & b_2 & b_3 & \cdots & b_m & 0 & \cdots & 0 \\ s_1 & 0 & 0 & & 0 & & & \\ 0 & s_2 & 0 & & & & & \\ & 0 & & \ddots & & & & \vdots \\ \vdots & & & & s_m & & \ddots & \\ 0 & & \cdots & & 0 & 0 & s_{n-1} & 0 \end{pmatrix}.$$

Since there is a column of zeros, $\det \mathbf{L} = 0$. The basic equations (1) and (2) may now be written compactly:

$$\vec{K}_{t+1} = \mathbf{L}\vec{K}_t: \quad \vec{K}_0 \text{ given}, \quad t = 0, 1, 2, ... \tag{3}$$

The solution to (3) can be derived formally by induction; the result is:

$$\vec{K}_N = L^N \vec{K}_0, \quad \text{where} \quad \underbrace{L^N = L\,L\,L \cdots L.}_{N \text{ terms}} \tag{4}$$

It is convenient at this stage to observe that the matrices $\vec{K}_t$ and L can be partitioned.

Let

$$L \equiv \left(\frac{M \;\vline\; 0}{A \;\vline\; B}\right) \quad \text{and} \quad \vec{K}_t \equiv \left\{\begin{matrix} \vec{K}_t \\ \vec{D}_t \end{matrix}\right\},$$

where M is $(m \times m)$ with its only non-zero terms in the first row and along the major sub-diagonal;

$\quad$ 0 $\quad$ is $(m \times (n-m))$ and is all zeros;

$\quad$ A $\quad$ is $((n-m) \times m)$, all zeros except in the upper right corner;

$\quad$ B $\quad$ is $((n-m) \times (n-m))$ with its only non-zero terms along the major sub-diagonal;

$\quad$ $\vec{K}_t$ $\quad$ is $(1 \times m)$ with components $k_{i,t}$: $i = 1, 2, \ldots, m$;

$\quad$ $\vec{D}_t$ $\quad$ is $(1 \times (n-m))$ with components $k_{i,t}$: $i = m+1, m+2, \ldots, n$.

In addition, it is not hard to show that

$$L^N = \left(\frac{M^N \;\vline\; 0}{A_N \;\vline\; B^N}\right): \quad A_N = \sum_{i=0}^{n-1} B^i A M^{N-i-1}.$$

Substituting these relations into Eq. (4) leads to

$$\left\{\begin{matrix} \vec{K}_N \\ \vec{D}_N \end{matrix}\right\} = \begin{pmatrix} M & 0 \\ A & B \end{pmatrix}^N \left\{\begin{matrix} \vec{K}_0 \\ \vec{D}_0 \end{matrix}\right\} = \begin{pmatrix} M^N & 0 \\ A_N & B^N \end{pmatrix} \left\{\begin{matrix} \vec{K}_0 \\ \vec{D}_0 \end{matrix}\right\} \tag{5}$$

hence

$$\vec{K}_N = M^N \vec{K}_0 \quad \text{and} \quad \vec{D}_N = A_N \vec{K}_0 + B^N \vec{D}_0.$$

Notice that the components of $\vec{K}_N$ are independent of $\vec{D}_0$. This means that the population in the "reproductive interval" does not depend upon the population past the "reproductive interval." Clearly, the converse is not true. In fact, it is possible to show that

$$B^N = 0 \quad \text{for} \quad N > n - m.$$

This means that once the post-reproductive-age women initially present in the population have died (that is, after $n - m$ time intervals), their existence in the population is entirely forgotten.

The consequence is that women need only be counted up to the end of their reproductive years. It will therefore be adequate to consider as the governing equation:

$$\vec{K}_{t+1} = M\vec{K}_t. \tag{6}$$

We will refer to M as the "projection matrix" and L as the "complete projection matrix."

A Numerical Example

These ideas are commonly illustrated with a very simple example, often called the Fibonacci Rabbit Problem. The following facts are given: Rabbits survive exactly three time intervals, and each female produces one daughter in her first and one in her second interval of life. The problem is to trace the evolution of the rabbit population assuming that it is started by one female (and sufficient males) in the first interval of life. Based upon these facts, it follows that

$$\vec{K}_0 = \begin{Bmatrix} 1 \\ 0 \\ 0 \end{Bmatrix} \quad \text{and} \quad L = \begin{pmatrix} 1 & 1 & 0 \\ 1 & 0 & 0 \\ 0 & 1 & 0 \end{pmatrix}.$$

Next, project the population forward in time using matrix multiplication to get

Time, t		0	1	2	3	4	5	6	7	8	9	
Age	1	1	1	2	3	5	8	13	21	34	55	...
group,	2	0	1	1	2	3	5	8	13	21	34	...
i	3	0	0	1	1	2	3	5	8	13	21	...

Now look at the ratios $k_{i,t+1}/k_{i,t}$ of numbers of newborns at successive times, for $i=1$.

$t=$	0	1	2	3	4	5	...	$\to \infty$
$k_{1,t+1}/k_{1,t}$	1.0	2.0	1.5	1.66	1.60	1.62	...	$\to 1.618\ldots$

Note that the same limiting result is true for other age intervals i, and also for the ratio $k_{i,t}/k_{i+1,t}$ at $t=1, 2, \ldots$.

These observations can be summarized as follows:

$$\text{for all } i, \quad \lim_{t \to \infty} \frac{k_{i,t+1}}{k_{i,t}} = \lim_{t \to \infty} \frac{k_{i,t}}{k_{i+1,t}} = \lambda: \quad \text{a constant.}$$

The value of λ can be derived by observing that

$$k_{i,t}=k_{i,t-1}+k_{i,t-2}.$$

However, as $t\to\infty$, $k_{i,t-1}/k_{i,t-2}=\lambda$ and $k_{i,t}/k_{i,t-2}=\lambda^2$. Thus

$$\lambda^2-\lambda-1=0 \;\to\; \lambda=\frac{1\pm\sqrt{5}}{2}=1.618\ldots$$

(and the negative root is biologically irrelevant).

The Theorem of Perron and Frobenius

Returning to the theoretical development, look at the properties of the projection matrix, $\mathbf{M}$. Recall the theorem of Perron and Frobenius, which will be stated, but need not be proven in general here. Theorem (Perron-Frobenius):

> If $\mathbf{T}$ is a square matrix and $\mathbf{T}\geq 0$ with $\mathbf{T}^N>{}^0$ for some N, then $\mathbf{T}$ has a positive eigenvalue λ (multiplicity one) and corresponding column and row eigenvectors $\hat{u}>0$ and $\underset{\sim}{v}>0$. Further, λ is greater in absolute value than any other eigenvalue of $\mathbf{T}$.

Recall that both $\mathbf{L}$ and $\mathbf{M}$ are positive semi-definite square matrices. Since $\det\mathbf{L}=0$, $\mathbf{L}$ is reducible. It is not hard to show that $\det\mathbf{M}\neq 0$, therefore $\mathbf{M}$ is irreducible (that is, one can get from any vertex to any other vertex in the corresponding diagraph). Further, if any two successive age groups are fertile (as will be assumed henceforth), $\mathbf{M}$ is also primitive (that is, a sufficiently high power, N, of the matrix $\mathbf{M}$ will contain only positive elements).

Proceed now to look at some of the properties of $\mathbf{M}$, and to prove the associated special cases of the Perron-Frobenius theorem:

1. $\mathbf{M}$ has a positive eigenvalue λ_0 which is a simple root.

Proof. Recall that the eigenvalues of a matrix $\mathbf{M}$ satisfy the set of homogeneous equations

$$\mathbf{M}\hat{x}=\lambda\mathbf{I}\hat{x}: \quad \text{where} \quad \mathbf{I}\equiv \text{identity matrix}$$

thus

$$(\mathbf{M}-\lambda\mathbf{I})\hat{x}=0, \quad \text{so}, \quad \det(\mathbf{M}-\lambda\mathbf{I})=0.$$

As a consequence of the special form of $\mathbf{M}$, it follows that

$$\det(\mathbf{M}-\lambda\mathbf{I})=\lambda^m-b_1\lambda^{m-1}-s_1b_2\lambda^{m-2}-\cdots-s_1s_2\cdots s_{m-1}b_m=0.$$

Recall also that $s_i>0$: $i=1,2,\ldots,m$,

$$b_i\geq 0:\; i=1,2,\ldots,m-1 \quad \text{and} \quad b_m>0,$$

thus

$$s_1 s_2 \cdots s_{m-1} b_m > 0, \quad \text{so, } \mathbf{M} \text{ has no 0 eigenvalues } (\lambda \neq 0).$$

Dividing the characteristic equation through by λ^m and rearranging allows us to define the function $f(\lambda)$:

$$f(\lambda) = \frac{b_1}{\lambda} + \frac{s_1 b_2}{\lambda^2} + \frac{s_1 s_2 b_3}{\lambda^3} + \cdots + \frac{s_1 s_2 \cdots s_{m-1} b_m}{\lambda^m} = 1 .$$

But clearly, since all of the terms are positive and connected by plus signs

1) $\displaystyle \lim_{\lambda \to 0} f(\lambda) = \infty$,

2) $\displaystyle \lim_{\lambda \to \infty} f(\lambda) = 0$,

3) $f(\lambda)$ is a monotonic function .

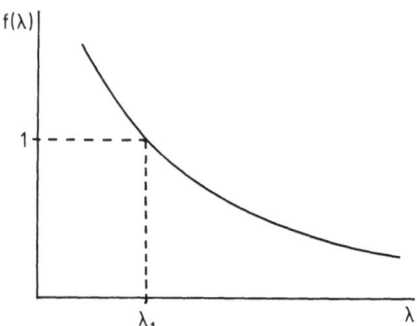

As illustrated in the figure to the right, these conditions together imply that there is only one real, positive eigenvalue λ_1 satisfying the equation $f(\lambda_1) = 1$. □

2. All the other eigenvalues of $\mathbf{M}$ are smaller in absolute value than the one real, positive eigenvalue.

Proof. Let λ_j represent any other eigenvalue: $j = 2, 3, \ldots, m$ and write

$$\lambda_j^{-1} = \exp\{\alpha + i\beta\} : \begin{cases} \alpha \text{ real}, \\ \beta \text{ real and positive}, \\ \beta \neq 2\pi, 4\pi, 6\pi, \ldots \end{cases}$$

thus

λ_j negative or complex .

Note that if $\beta = 2\pi, 4\pi, 6\pi, \ldots$, then λ_j is real and positive, which was just shown to be impossible.

Substitute the expression above into $f(\lambda_j) = 1$ and equate real parts to get

$$b_1 e^\alpha \cos\beta + s_1 b_2 e^{2\alpha} \cos 2\beta + \cdots + s_1 s_2 \cdots s_{m-1} b_m e^{m\alpha} \cos m\beta = 1 .$$

Now, for comparison, write $\lambda_1^{-1} = \exp\{\gamma\}$ and use this in $f(\lambda_1)$:

$$b_1 e^\gamma + s_1 b_2 e^{2\gamma} + \cdots + s_1 s_2 \cdots s_{m-1} b_m e^{m\gamma} = 1 .$$

Clearly, due to the restrictions on $\mathbf{M}$, at least 2 adjacent terms in the above two equations must be non-zero [say the nth and $(n+1)^{\text{st}}$]. But since $\beta \neq 2\pi, 4\pi, \ldots$ it follows that $\cos n\beta$ and $\cos(n+1)\beta$ cannot both equal unity. Thus some of the coefficients of the $\exp\{\cdot \alpha\}$ in the first equation must be smaller than the corresponding coefficients in the second equation. It therefore follows that

$$e^{\alpha} > e^{\gamma}, \quad \text{so} \quad |\lambda_j| < |\lambda_1| = \lambda_1 : \quad j = 2, 3, \ldots, m . \quad \Box$$

3. $\mathbf{M}$ has an eigenvector $\check{x}^{(1)}$ corresponding with λ_1 with all components positive.

   Proof. The eigenvector $\check{x}^{(1)} \equiv \{x_i^{(1)}\}$ satisfies the equations

$$\mathbf{M}\check{x}^{(1)} = \lambda_1 \check{x}^{(1)} : \quad \text{this is a set of } m \text{ simultaneous,} \\ \text{homogeneous linear equations} .$$

One of the components of $\check{x}^{(1)}$ may be selected freely. Without loss of generality, set

$$x_1^{(1)} = 1 .$$

But then from the second of the set of linear equations,

$$s_1 x_1^{(1)} = \lambda_1 x_2^{(1)} \quad \rightarrow \quad x_2^{(1)} = s_1/\lambda_1 .$$

Proceeding through the other equations in a similar manner, it follows that

$$x_j^{(1)} = \frac{s_1 s_2 \cdots s_{j-1}}{\lambda^{j-1}} : \quad j = 2, 3, \ldots, m .$$

But clearly, since $\lambda_1 > 0$ and $s_i > 0$: $i = 1, 2, \ldots, m$, then $\check{x}^{(1)} > 0$. These quantities are called the "components of the stable age distribution." $\quad \Box$

4. $\mathbf{M}$ has an eigenvector $y^{(1)}$ corresponding with λ_1 all components positive.

   Proof. Proceed exactly as above to define $y^{(1)} \equiv [y_i^{(1)}]$ and choose without loss of generality to set

$$y_1^{(1)} = 1 .$$

Then use the equations

$$y^{(1)}\mathbf{M} = \lambda_1 y^{(1)}$$

to work out the other components as done above. The result is

$$y_j^{(1)} = \sum_{n=j}^{m} (s_j s_{j+1} \cdots s_{n-1}) b_n \lambda_1^{j-n-1} .$$

Clearly, the $y_j^{(1)} > 0$ as above. These quantities are called the "reproductive values for women of age j." $\quad \Box$

5. For any not-identically-zero vector $\vec{K}_0$, there is a constant c which depends upon the vector, such that

$$\lim_{N \to \infty} \mathbf{M}^N \vec{K}_0 / \lambda_1^N = c \vec{x}^{(1)}.$$

Proof. The method employed to show that this is true is round-about, but introduces some useful ideas along the way. Realize that it is possible in theory to work out all of the eigenvalues and the associated row and column eigenvectors. These satisfy the equations

$$i = 1, 2, \ldots, m: \quad \mathbf{M}\vec{x}^{(i)} = \lambda_i \vec{x}^{(i)} \quad \text{and} \quad \underset{\sim}{y}^{(i)} \mathbf{M} = \lambda_i \underset{\sim}{y}^{(i)}.$$

Then define three matrices as follows:

$$\mathbf{X} \equiv [\vec{x}^{(1)} \cdots \vec{x}^{(m)}]: \quad \text{a row vector of column vectors,}$$

$$\mathbf{Y} \equiv \left\{ \begin{array}{c} \underset{\sim}{y}^{(1)} \\ \vdots \\ \underset{\sim}{y}^{(m)} \end{array} \right\}: \quad \text{a column vector of row vectors,}$$

$$\mathbf{\Lambda} \equiv \begin{pmatrix} \lambda_1 & & \\ & \ddots & \\ & & \lambda_m \end{pmatrix}: \quad \begin{array}{l} \text{the Jordan Canonical matrix} \\ \text{(all off-diagonal components are zero).} \end{array}$$

It has been assumed implicitly in the form of $\mathbf{\Lambda}$ that all of the eigenvalues are distinct. While this is very hard to show in any formal manner, it turns out always to be the case for human populations.

Without loss of generality, assume that the eigenvalues have been ordered such that

$$|\lambda_1| > |\lambda_2| \geq \cdots \geq |\lambda_m|.$$

In the new notation, the equations above may be rewritten

$$\mathbf{MX} = \mathbf{X\Lambda} \quad \text{and} \quad \mathbf{YM} = \mathbf{\Lambda Y}.$$

Thus

$$\mathbf{M} = \mathbf{X\Lambda X}^{-1} = \mathbf{Y}^{-1} \mathbf{\Lambda Y}$$

and as a consequence of uniqueness (and suitable normalization)

$$\mathbf{X} = \mathbf{Y}^{-1} \quad \text{and} \quad \mathbf{Y} = \mathbf{X}^{-1}, \quad \text{thus} \quad \mathbf{XY} = \mathbf{I}.$$

Consequently, observe that

$$\mathbf{M} = \mathbf{X}\mathbf{\Lambda}\mathbf{X}^{-1} = \mathbf{X}\mathbf{\Lambda}\mathbf{Y}$$
$$= \lambda_1 \vec{x}^{(1)} y^{(1)} + \lambda_2 \vec{x}^{(2)} y^{(2)} + \cdots + \lambda_m \vec{x}^{(m)} y^{(m)}$$
$$= \lambda_1 \mathbf{Z}^{(1)} + \lambda_2 \mathbf{Z}^{(2)} + \cdots + \lambda_m \mathbf{Z}^{(m)},$$

where

$$\mathbf{Z}^{(i)} \equiv \vec{x}^{(i)} y^{(i)} : \quad i = 1, 2, \dots, m.$$

The matrices $\mathbf{Z}^{(i)}$ are called the spectral components, and possess the special property that they are idempotent; that is, that they are powers of themselves.

$$[\mathbf{Z}^{(i)}]^N = \mathbf{Z}^{(i)} : \quad N = 1, 2, \dots, m.$$

To see that this is so, consider

$$\mathbf{M} = \mathbf{X}\mathbf{\Lambda}\mathbf{X}^{-1}, \quad \text{thus} \quad \mathbf{M}^2 = \mathbf{X}\mathbf{\Lambda}\mathbf{X}^{-1}\mathbf{X}\mathbf{\Lambda}\mathbf{X}^{-1} = \mathbf{X}\mathbf{\Lambda}^2\mathbf{X}^{-1}$$

and in general,

$$\mathbf{M}^N = \mathbf{X}\mathbf{\Lambda}^N\mathbf{X}^{-1}.$$

But $\mathbf{X}^{-1} = \mathbf{Y}$, so

$$\mathbf{M}^N = \mathbf{X}\mathbf{\Lambda}^N\mathbf{Y} = \lambda_1 \mathbf{Z}^{(1)} + \cdots + \lambda_m \mathbf{Z}^{(m)}.$$

By comparing this expression with

$$\mathbf{M}^N = [\lambda_1 \mathbf{Z}^{(1)} + \cdots + \lambda_m \mathbf{Z}^{(m)}]^N$$

it becomes apparent that for all i:

$$[\mathbf{Z}^{(i)}]^N = \mathbf{Z}^{(i)} \quad \text{and} \quad \mathbf{Z}^{(i)}\mathbf{Z}^{(j)} = 0 : \quad i \neq j.$$

Next, look at the expression

$$\mathbf{M}^N / \lambda_1^N = \mathbf{Z}^{(1)} + (\lambda_2/\lambda_1)^N \mathbf{Z}^{(2)} + \cdots + (\lambda_m/\lambda_1)^N \mathbf{Z}^{(m)}.$$

Clearly, since $|\lambda_1| > |\lambda_2| \geq \cdots \geq |\lambda_m|$, it follows that

$$\lim_{N \to \infty} (\lambda_i/\lambda_1)^N = 0 : \quad i = 2, 3, \dots, m.$$

Thus

$$\lim_{N \to \infty} \mathbf{M}^N / \lambda_1^N = \mathbf{Z}^{(1)} = \vec{x}^{(1)} y^{(1)}.$$

It therefore follows that since $y^{(1)} > 0$ and $\vec{K}_0 \not\equiv 0$ $(\vec{K}_0 \geq 0)$

$$\lim_{N \to \infty} \mathbf{M}^N \vec{K}_0 / \lambda_1^N = \vec{x}^{(1)} y^{(1)} \vec{K}_0 = c \vec{x}^{(1)}: \qquad y^{(1)} \vec{K}_0 = c > 0. \quad \square$$

This lengthy digression on the theorem of Perron and Frobenius allows for several useful conclusions to be drawn about the growth of an age-structured population whose vital rates are not functions of time. Recall that

$$\vec{K}_N = \mathbf{M}^N \vec{K}_0 \tag{7}$$

and thus from the asymptotic results found above, as $N \to \infty$

$$\vec{K}_N \to c \lambda_1^N \vec{x}^{(1)}. \tag{8}$$

Notice that the age structure and the temporal growth separate after the population has reached what is called the stable state (that is, when it is given by the asymptotic expression above).

1. The population eventually grows at a constant rate equal to λ_1, the dominant eigenvalue of the projection matrix $\mathbf{M}$. This rate is called the intrinsic rate of population growth. It is clear from the asymptotic expression that

$$\lim_{N \to \infty} \frac{k_{i,N+1}}{k_{i,N}} = \lambda_1: \qquad i = 1, 2, \ldots, m.$$

2. The population age distribution becomes proportional to $\vec{x}^{(1)}$, the eigenvector associated with the dominant root of the projection matrix $\mathbf{M}$. Notice that this is true regardless of the initial age distributed, $\vec{K}_0$. The age structure is referred to as the stable age distribution, and

$$\lim_{N \to \infty} \frac{k_{i,N}}{k_{i+1,N}} = \frac{x_i^{(1)}}{x_{i+1}^{(1)}} = \frac{\lambda_1}{s_i}: \qquad i = 1, 2, \ldots, m-1,$$

where the final equality follows from the definition of the components of $\vec{x}^{(1)}$ as found earlier.

It is interesting to reflect upon the consequences of the observation that the eventual age structure is independent of the initial age structure, so long as fertility and mortality are constant. What this means is that populations "forget" their past structure, a property called ergodicity. One consequence of ergodicity is that populations can be projected forward in time without trouble, but usually cannot be projected backwards.

Although it will not be demonstrated mathematically, it is possible to make a far more useful observation about ergodicity: It can be shown that if two populations which are initially different in their age structure are subjected to a sequence of identical mortality and fertility conditions which are changing with time, the age structures of the two populations eventually become the same. This usually is referred to as "weak ergodicity" and really extends beyond stable population theory.

Numerical Example – Continued

To conclude this section, I carry the Fibonacci Rabbit Problem a bit further, so as
to illustrate some of the results above. Recall that

$$
\vec{K}_0 = \begin{Bmatrix} 1 \\ 0 \\ 0 \end{Bmatrix} \quad \text{and} \quad \mathbf{L} = \begin{pmatrix} 1 & 1 & 0 \\ 1 & 0 & 0 \\ 0 & 1 & 0 \end{pmatrix}
$$

thus

$$
\vec{K}_0 = \begin{Bmatrix} 1 \\ 0 \end{Bmatrix} \quad \text{and} \quad \mathbf{M} = \begin{pmatrix} 1 & 1 \\ 1 & 0 \end{pmatrix}.
$$

1. Eigenvalues:

$$
\det(\mathbf{M} - \lambda \mathbf{I}) = \begin{vmatrix} 1-\lambda & 1 \\ 1 & -\lambda \end{vmatrix} = \lambda^2 - \lambda - 1 = 0
$$

thus

$$
\lambda_{1,2} = \frac{1 \pm \sqrt{5}}{2} = 1.618,\ -0.618.
$$

2. Eigenvectors:

$$
\mathbf{M}\vec{x}^{(i)} = \lambda_i \vec{x}^{(i)}, \quad \text{so,} \quad \begin{cases} x_1^{(i)} + x_2^{(i)} = \lambda_i x_1^{(i)}, \\ x_1^{(i)} = \lambda_i x_2^{(i)}. \end{cases}
$$

If one sets

$$
x_1^{(i)} = 1 \quad \rightarrow \quad x_2^{(i)} = 1/\lambda_i
$$

then

$$
\vec{x}^{(1)} = \begin{Bmatrix} 1 \\ 0.618 \end{Bmatrix} \quad \text{and} \quad \vec{x}^{(2)} = \begin{Bmatrix} 1 \\ -1.618 \end{Bmatrix}
$$

and

$$
\mathbf{X} = \begin{pmatrix} 1 & 1 \\ 0.618 & -1.618 \end{pmatrix}.
$$

The row eigenvectors could be found similarly; however, for a (2×2) matrix it is
easier to proceed as follows: Recall that

$$
\mathbf{Y} = \mathbf{X}^{-1} = \frac{1}{2.236} \begin{pmatrix} 1.618 & 1 \\ 0.618 & -1 \end{pmatrix} = \begin{pmatrix} 0.724 & 0.447 \\ 0.276 & -0.447 \end{pmatrix}.
$$

Thus

$$\underline{y}^{(1)} = [0.724 \quad 0.447].$$

3. Asymptotic Behavior:

$$\vec{K}_N = [\underline{y}^{(1)}\vec{K}_0]\lambda_1^N \vec{x}^{(1)} = 0.724\,(1.618)^N \begin{Bmatrix} 1 \\ 0.618 \end{Bmatrix}$$

so,

N	0	1	2	3	4	5	6	7
$k_{1,N}$	0.72	1.17	1.90	3.07	4.96	8.03	12.99	21.02

Stable Population Theory – Continuous

As in the previous section, consider a population which is closed to migration and in which only the females are counted. Again, ignore the males assumed to be present, as their reproductive behavior is difficult to quantify. Human females have a conveniently finite interval of roughly thirty years (from menarche to menopause) during which they are fertile. In addition, the females are always physically present at the birth of their children. In this section the objective is again to investigate the consequences of assuming that age-specific fertility and age-specific mortality are both time-independent, but now time t and age x will be treated as continuous variables.

Let

$$k(x,t)dx = \text{number of females in the population}$$
$$\text{who are between ages } x \text{ and } x+dx \text{ at time } t.$$

Since the mortality of a population is very important, we will derive the survivorship function (or life table function) rather carefully. In the process, it will be necessary to talk in probabilistic terms. Once these are understood, it will be sufficient to replace probabilities by fractions.

Let

$$f(x)dx = \text{Prob}\{\text{dying between age } x \text{ and } x+dx\}$$

and since everyone eventually dies, $f(x)$ is a proper density, so

$$\int_0^\omega f(x)dx = 1$$

and a distribution function for dying may be defined by

$$F(x) = \text{Prob}\{\text{dying prior to age } x\}: \quad \frac{dF(x)}{dx} = f(x).$$

It also is convenient to define a death rate, $\mu(x)$, called the force of mortality.

$$\mu(x) = \frac{f(x)}{1 - F(x)}$$

thus

$\mu(x)dx = \text{Prob}\{\text{dying between age } x \text{ and } x + dx | \text{survival to } x\}.$

Next, introduce a very important demographic variable, the survivorship, $\ell(x)$, which tells the number of individuals in a cohort of initial size $\ell(0)$ (usually taken to be 100,000) who survive to at least age x. In terms of the above definitions, and assuming the fractions and probabilities are interchangeable,

$$F(x) = 1 - \ell(x)/\ell(0).$$

Finally, define

$$p(x) = \frac{\ell(x)}{\ell(0)} = \text{fraction of females in the population}$$

who survive from birth to age x.

Properties of $p(x)$:
1. Continuous and differentiable;
2. Monotonically non-increasing;
3. $0 < p(x) \leq 1$ for $0 \leq x < \omega$
   and $p(\omega) = 0$: $\omega = $ terminal age.

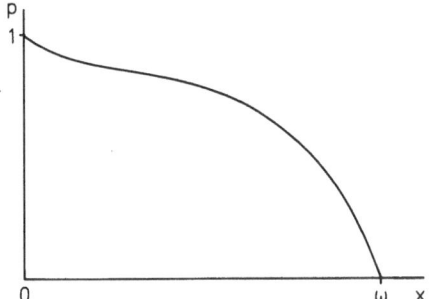

Specify a maternity function $m(x)$ such that

$m(x)dx = $ number of female babies born to a woman
when she is age x to $x + dx$.

While this definition makes no sense for any individual woman, it is perfectly reasonable as an averaged quantity when applied to a large population. In effect, it simply replaces probabilities for individuals with fractions for whole populations.

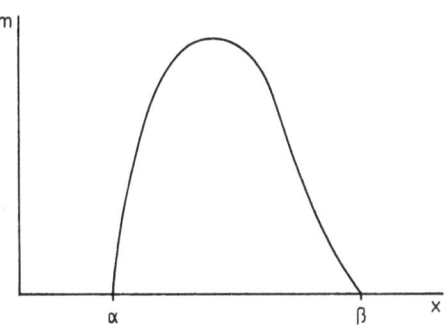

Properties of $m(x)$:
1. Continuous and differentiable;
2. $m(x) > 0$ for $\alpha < x < \beta$,
   $m(x) = 0$ for $\begin{cases} 0 \leq x \leq \alpha, \\ \beta \leq x \leq \omega, \end{cases}$
   where $0 \leq \alpha < \beta \leq \omega$;
3. $m(x)$ has a single maximum.

The second and third conditions actually follow from observations of human populations.

Notice that the average number of daughters born in the past to a woman who survives to age β is

$$\int_\alpha^\beta m(x)dx = \text{Gross Reproduction Rate (GRR)}.$$

Similarly, the average number of daughters to be born in the future to a woman herself just born is

$$\int_\alpha^\beta p(x)m(x)dx = \text{Net Reproduction Rate (NRR)} = R_0.$$

The product in the integrand of the net reproduction rate appears repeatedly and is called the net maternity, $\varphi(x) = p(x)m(x)$.

Let us calculate the total number of female births at time t, where:

$B(t)dt = $ number of female births
    during the time interval from t to $t+dt$.

Clearly, if we know the population age distribution at time t, $k(x, t)$, then

$$B(t) = \int_0^\omega k(x, t)m(x)dx. \tag{9}$$

Recognize that it is convenient to divide the population into two groups:
1. The females present at time $t=0$ (initial population).
2. The females born after time $t=0$ (daughters, grand-daughters, great-grand-daughters, ... of the initial population).
At any time t, all females older than age t are in the first group, while all females younger than age t are in the second. There is no overlap of ages between the two groups.

Next, imagine that a census is done at time $t=0$, so that we know the initial female population:

$$k_0(a) = k(a, 0): \quad 0 \leq a \leq \omega$$

then, at time t, the survivors of this group number

$$k(a+t, t) = k_0(a)\frac{p(a+t)}{p(a)}: \quad 0 \leq a \leq \omega, \quad t > 0$$

or in terms of a dummy variable $x = a+t$:

$$k(x, t) = k_0(x-t)\frac{p(x)}{p(x-t)}: \quad 0 \leq t \leq x \leq \omega.$$

This quantity counts the number of women in the first group noted above. The number of women in the second group at time t are the survivors of births into the population after $t=0$, thus

$$k(x, t) = B(t-x)p(x): \quad 0 \leq x \leq t.$$

Substitution into the integral expression (9) which defines $B(t)$ yields

$$B(t) = \int_0^t B(t-x)p(x)m(x)dx + \int_t^\omega k_0(x-t)\frac{p(x)}{p(x-t)}m(x)dx. \tag{10}$$

The first integral represents births to women born after $t=0$, while the second integral represents only the births to women in the initial population. It is customary to call the second integral $G(t)$, and to write it in the form

$$G(t) = \int_0^\omega k_0(a)\frac{p(a+t)}{p(a)}m(a+t)da. \tag{11}$$

Using this nomenclature, and recalling that $p(x)m(x) = \varphi(x)$, provides the form for the integral equation usually called the "renewal equation"

$$B(t) = G(t) + \int_0^t B(t-x)\varphi(x)dx: \quad t>0. \tag{12}$$

This is a non-homogeneous Volterra integral equation of the second kind with a band-limited kernel [meaning that $\varphi(x)$ is only non-zero for $\alpha < x < \beta$]. This equation may be solved in two ways.

I. Solution by Elementary Methods

Notice that since reproduction only takes place for women younger than age β, it follows that

$$G(t) \equiv 0 \quad \text{for} \quad t > \beta.$$

It therefore follows that for $t > \beta$, the renewal equation is homogeneous, and of the form

$$B(t) = \int_0^t B(t-x)\varphi(x)dx = \int_\alpha^\beta B(t-x)\varphi(x)dx: \quad t > \beta.$$

The limits on the integral have been changed because $t > \beta$ and

$$\varphi(x) \neq 0 \quad \text{only for} \quad \alpha < x < \beta.$$

It is now easy to confirm by direct substitution that the homogeneous form of the integral equation admits solutions of the form

$$B(t) = Q \exp\{rt\}.$$

Substituting the assumed form of the solution leads to

$$Q \exp\{rt\} = Q \exp\{rt\} \int_\alpha^\beta \exp\{-rx\}\varphi(x)dx.$$

For convenience of notation, define the integral to be:

$$\psi(r) \equiv \int_\alpha^\beta \exp\{-rx\}\varphi(x)dx$$

so the characteristic equation becomes:

$$\psi(r) = 1.$$

A number of observations may now be made about the nature of the solution to the integral equation.
1. The integral equation has exactly one real root $r=r_1$.

Proof. This follows simply from the observations
a. $\psi(r) \to 0$ as $r \to +\infty$;
b. $\psi(r) \to \infty$ as $r \to -\infty$;
c. $\psi(r)$ is monotonically decreasing as

$$\frac{d\psi(r)}{dr} = -\int_\alpha^\beta x \exp\{-rx\}\varphi(x)dx < 0.$$

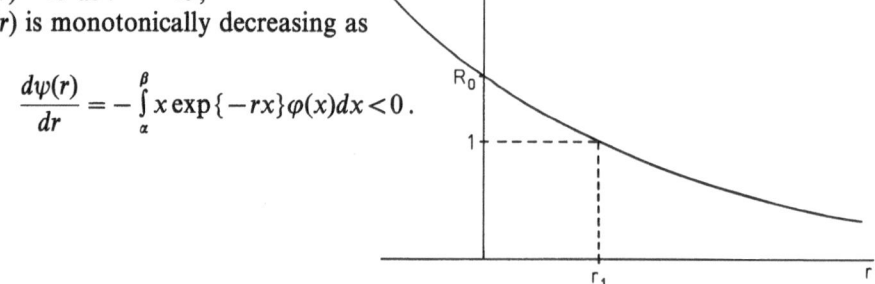

Thus for $r=r_1$, $\psi(r_1)=1$. This is the only real root. We may observe from the definitions that $\psi(0)=R_0$, the net reproduction rate for the population. So it follows that

$$r_1 \gtreqless 0 \quad \text{according to whether} \quad R_0 \gtreqless 1. \quad \square$$

2. All other roots $\{r_j\}: j=2,3,\ldots$ of the integral equation appear in complex conjugate pairs with $\mathrm{Re}\{r_j\} < r_1$.

Proof. For some $j \neq 1$, write $r_j = u + iv$: u, v real, $v > 0$. Substitute into the characteristic equation and equate real and imaginary parts

$$\int_\alpha^\beta \exp\{-ux\}\cos(vx)\varphi(x)dx = 1, \qquad \int_\alpha^\beta \exp\{-ux\}\sin(vx)\varphi(x)dx = 0.$$

Clearly, if the second equation holds for v, it also holds for $-v$ so $r_j^* = u - iv$ is also a root of the characteristic equation. Also, since for some arguments on the interval $\alpha \leq x \leq \beta$ it must be that $\cos(vx) < 1$, comparison of the first equation above with the equation $\psi(r_1) = 1$ provides $u < r_1$, thus $\mathrm{Re}\{r_j\} < r_1 : j = 2, 3, \ldots$.   □

Comments:
1. It is not hard to deduce that the real root r_1 is of multiplicity unity.
2. Cases in which there are conjugate pairs of complex roots with multiplicity greater than unity may certainly be constructed. However, multiple roots never seem to occur for real human populations. Therefore, assume that the roots are distinct, which simplifies the mathematics. It is possible to carry everything through with repeated roots present.
3. It can be shown that if $\alpha < \beta < \infty$ (that is, if the fertile interval is finite in width) then there are an infinite number of roots to the characteristic equation. Fortunately, only one is real, and all the rest appear in conjugate pairs of complex roots.
4. Since the renewal equation (12) is linear, it follows that solutions are linear combinations of the form

$$B(t) = Q_1 \exp\{r_1 t\} + Q_2 \exp\{r_2 t\} + \cdots$$
$$= \sum_j Q_j \exp\{r_j t\}: \quad t > \beta. \tag{13}$$

The next project is to evaluate the constants Q_j in order to analytically continue the solution so it is valid for $t > 0$. Evaluation of the constants, Q_j, proceeds as follows.

Define:

$$P_s = \int_0^\beta \exp\{-r_s t\} G(t) dt,$$

where $G(t)$ represents births to the initial population as before, and r_s is a particular root of the characteristic equation

$$\psi(r_s) = \int_\alpha^\beta \exp\{-r_s x\} \varphi(x) dx = 1. \tag{14}$$

Make use of the renewal equation to eliminate $G(t)$ in the definition of P_s

$$P_s = \int_0^\beta \exp\{-r_s t\} \left\{ B(t) - \int_0^t B(t-x)\varphi(x) dx \right\} dt.$$

Next, write the formal solution in the form

$$B(t) = Q_s \exp\{r_s t\} + \sum_{j \neq s} Q_j \exp\{r_j t\} \tag{15}$$

and substitute in the definition of P_s

$$P_s = \int_0^\beta \exp\{-r_s t\}\left\{Q_s \exp\{r_s t\} - Q_s \int_0^t \exp\{r_s(t-x)\}\varphi(x)dx\right\}dt + R_s$$

$$= Q_s \int_0^\beta \left\{1 - \int_0^t \exp\{-r_s x\}\varphi(x)dx\right\}dt + R_s,$$

where

$$R_s = \sum_{j \neq s} R_{s,j}$$

with

$$R_{s,j} = \int_0^\beta \exp\{-r_s t\}\left\{Q_j \exp\{r_j t\} - Q_j \int_0^t \exp\{r_j(t-x)\}\varphi(x)dx\right\}dt.$$

Next, use the characteristic equation (14)

$$\int_0^\beta \exp\{-r_s x\}\varphi(x)dx = 1$$

to replace the "1" in the curly brackets in the expression for P_s, so that

$$P_s = Q_s \int_0^\beta \int_t^\beta \exp\{-r_s x\}\varphi(x)dxdt + R_s$$

$$= Q_s \int_0^\beta \int_0^x \exp\{-r_s x\}\varphi(x)dtdx + R_s$$

$$= Q_s \int_0^\beta x \exp\{-r_s x\}\varphi(x)dx + R_s.$$

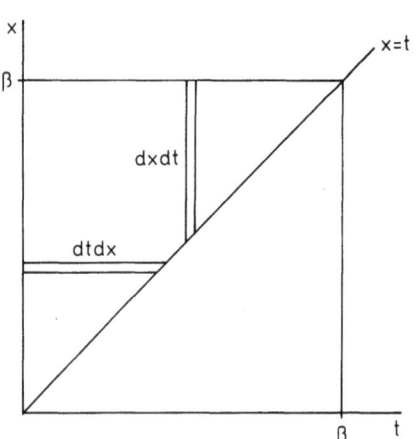

We could proceed in exactly the same way to show that $R_{s,j} \equiv 0$, thus $R_s \equiv 0$ (so long as roots are all distinct). Thus

$$Q_s = \frac{P_s}{\int_0^\beta x \exp\{-r_s x\}\varphi(x)dx} = \frac{\int_0^\beta \exp\{-r_s t\}G(t)dt}{\int_\alpha^\beta x \exp\{-r_s x\}\varphi(x)dx}. \tag{16}$$

II. Solution by Laplace Transforms

A more direct method of solution is available using Laplace transforms. Begin by defining the Laplace transform of a function $h(t)$:

$$\mathcal{L}\{h(t)\} = h^*(r) = \int_0^\infty \exp\{-rt\}h(t)dt.$$

It will be necessary to make use of one of the elementary properties of Laplace transforms, the transform of the convolution of two functions

$$\mathcal{L}\{h(t)*k(t)\} = \mathcal{L}\left\{\int_0^t h(t-x)k(x)dx\right\} = h^*(r)k^*(r).$$

Start with the renewal equation (12)

$$B(t) = G(t) + \int_0^t B(t-x)\varphi(x)dx: \quad t>0.$$

Multiply through by $\exp\{-rt\}$ and integrate with respect to t from zero to infinity. The result is

$$B^*(r) = G^*(r) + B^*(r)\varphi^*(r), \quad \text{where} \quad \mathcal{L}\{B(t)\} = B^*(r),$$
$$\mathcal{L}\{G(t)\} = G^*(r),$$

Thus:
$$\mathcal{L}\{\varphi(t)\} = \varphi^*(r).$$

$$B^*(r) = \frac{G^*(r)}{1-\varphi^*(r)}.$$

Next, invert the transform to get back $B(t)$ from $B^*(r)$. This is done with the aid of some observations:

Look at the denominator in the definition of $B^*(r)$:

$$\varphi^*(r) = \int_0^\infty \exp\{-rx\}\varphi(x)dx = \psi(r) = 1.$$

This is precisely the characteristic equation (14) which was studied with some care earlier. Therefore, the nature of the root $r = r_1, r_2, \ldots$ is known.

It is possible to write

$$1 - \varphi^*(r) = (r-r_1)(r-r_2)(r-r_3)\cdots$$

and if all of the roots are distinct (as assumed earlier), a partial fraction expansion may be used to write

$$B^*(r) = \frac{G^*(r)}{1-\varphi^*(r)} = \frac{Q_1}{r-r_1} + \frac{Q_2}{r-r_2} + \frac{Q_3}{r-r_3} + \cdots.$$

Evaluation of the Q_j:

$$Q_j = \lim_{r \to r_j} \left\{ \frac{(r-r_j)G^*(r)}{1-\varphi^*(r)} - (r-r_j) \sum_{i \neq j} \frac{Q_i}{r-r_i} \right\}.$$

The second term goes to zero; the first must be evaluated by l'Hopital's Rule:

$$Q_j = \frac{G^*(r_j)}{-\dfrac{d\varphi^*}{dr}\bigg|_{r=r_j}} = \frac{\displaystyle\int_0^\infty \exp\{-rt\}G(t)dt}{\displaystyle\int_0^\infty x\exp\{-rx\}\varphi(x)dx} = \frac{\displaystyle\int_0^\beta \exp\{-rt\}G(t)dt}{\displaystyle\int_\alpha^\beta x\exp\{-rx\}\varphi(x)dx}.$$

Notice that this result is the same as the one found earlier in (16). The final observation needed before inverting $B^*(r)$ to find $B(t)$ is that

$$\mathscr{L}\{Q_j \exp\{r_j t\}\} = Q_j \int_0^\infty \exp\{-rt\} \exp\{r_j t\}dt = \frac{Q_j}{r-r_j}$$

hence

$$\mathscr{L}^{-1}\left\{ \frac{Q_j}{r-r_j} \right\} = Q_j \exp\{r_j t\}.$$

It therefore follows as in (13) that

$$B(t) = \mathscr{L}^{-1}\{B^*(r)\} = \mathscr{L}^{-1}\left\{ \sum_j \frac{Q_j}{r-r_j} \right\} = \sum_j Q_j \exp\{r_j t\}$$
$$= Q_1 \exp\{r_1 t\} + Q_2 \exp\{r_2 t\} + \cdots .$$

Successive Generations Model

It is instructive to think of a self-renewing population in a slightly different way. As before, at $t=0$ there is a group of females present who constitute the initial population. As time passes, these women give birth to a group of daughters who are accounted for by the function $G(t)$, defined exactly as in (11). For notational convenience, choose to set $G(t) = B_0(t)$.

The daughters of the initial population grow up, experiencing mortality $p(x)$, and bear their daughters, the grand-daughters of the initial population. It follows that this last group of women are determined by the expression

$$B_1(t) = \int_0^t B_0(t-x)\varphi(x)dx.$$

Following this line of reasoning inductively leads to the general expression for the nth generation

$$B_n(t) = \int_0^t B_{n-1}(t-x)\varphi(x)dx: \quad n=1,2,3,\dots . \tag{17}$$

Notice that each successive generation is just the convolution of the previous generation with the net maternity function. We therefore can formally Laplace transform the successive generations:

$$\mathcal{L}\{B_0(t)\} = B_0^*(r) = G^*(r),$$

$$\mathcal{L}\{B_1(t)\} = \mathcal{L}\{B_0(t) * \varphi(t)\} = B_0^*(r)\varphi^*(r) = G^*(r)\varphi^*(r),$$

$$\mathcal{L}\{B_2(t)\} = \mathcal{L}\{B_1(t) * \varphi(t)\} = B_1^*(r)\varphi^*(r) = G^*(r)[\varphi^*(r)]^2,$$

$$\vdots$$

$$\mathcal{L}\{B_n(t)\} = \mathcal{L}\{B_{n-1}(t) * \varphi(t)\} = B_{n-1}^*(r)\varphi^*(r) = G^*(r)[\varphi^*(r)]^n.$$

But the birth trajectory $B(t)$ at any time t is just the sum of the births in all generations at time t

$$B(t) = B_0(t) + B_1(t) + \cdots = \sum_{j=0}^{\infty} B_j(t).$$

Since the Laplace transform of a sum is the sum of the Laplace transforms,

$$B^*(r) = \sum_{j=0}^{\infty} B_j^*(r) = G^*(r) \sum_{j=0}^{\infty} [\varphi^*(r)]^j.$$

Assuming there is some region of the complex plane where $|\varphi^*(r)| < 1$ we may sum the geometric series to get

$$B^*(r) = \frac{G^*(r)}{1 - \varphi^*(r)}.$$

From here on, the solution proceeds as before.

Asymptotic Behavior of the Birth Trajectory

Proceed now to look at the long-term behavior of the solution (13)

$$B(t) = \sum_j Q_j \exp\{r_j t\}.$$

First, rewrite the solution in the form

$$B(t) = Q_1 \exp\{r_1 t\} \left\{ 1 + \sum_{j \neq 1} \frac{Q_j}{Q_1} \exp\{(r_j - r_1)t\} \right\}.$$

Recall that $\mathrm{Re}\{r_j\} < r_1 : j = 2, 3, \ldots$, thus

$$\lim_{t \to \infty} \exp\{(r_j - r_1)t\} = 0: \quad j = 2, 3, \ldots .$$

It therefore follows that

$$B(t) \rightarrow Q_1 \exp\{r_1 t\} \quad \text{as} \quad t \rightarrow \infty, \tag{18}$$

where

$$Q_1 = \frac{\int_0^\beta \exp\{-r_1 t\} G(t) dt}{\int_\alpha^\beta x \exp\{-r_1 x\} \varphi(x) dx}.$$

The asymptotic behavior of the population follows directly from the definition

$$k(x, t) = B(t - x) p(x)$$

thus

$$k(x, t) \rightarrow Q_1 \exp\{r_1 t\} [\exp\{-r_1 x\} p(x)] \quad \text{as} \quad t \rightarrow \infty. \tag{19}$$

Notice that the time behavior and the age behavior have separated. It is apparent that the age-structure assumes a fixed shape which is the product of an exponential and the (scaled) Life Table survivorship function. As time passes, the age-structure simply grows exponentially at all ages.

If the net reproduction rate $R_0 = 1$ then the dominant root $r_1 = 0$. This case is referred to as a "stationary" population as $t \rightarrow \infty$. Note that the age-structure is proportional to the Life Table survivorship and the total population size becomes constant.

If the net reproduction rate $R_0 > 1$ (< 1) then the dominant root $r_1 > 0$ (< 0). This situation is referred to as a "stable" population in the limit as $t \rightarrow \infty$. It is seen that relative to the stationary case, the population age-structure has an excess of young (old) people, and the population as a whole grows (shrinks) as time passes.

Momentum of Population Growth

To conclude this section, consider a sample computation which makes use of stable population theory. The purpose is to discover what happens to a stably growing population when there is a shift in the maternity behavior.

Imagine that a population has been growing in a stable fashion for a long time so that

$$k(x, t) = B(t - x) p(x) = B_0 \exp\{rt\} [\exp\{-rx\} p(x)].$$

Next, assume that at $t = 0$ there is an abrupt shift in the maternity behavior so that $\bar{R}_0 = 1$. Assume further that this maternity shift occurs by a proportional scaling at all ages, so that

$$\bar{m}(x) = m(x)/R_0: \quad R_0 = \int_\alpha^\beta m(x) p(x) dx, \quad \text{so}, \quad \bar{R}_0 = 1.$$

After a while, the population will reestablish stability, with the new stable form given by

$$\bar{B}(t) = \bar{Q}\exp\{\bar{r}t\} = B_\infty : \quad \text{since} \quad \bar{R}_0 = 1, \quad \bar{r} = 0.$$

As previously discovered,

$$\bar{Q} = B_\infty = \frac{\int_0^\infty G(t)\exp\{-\bar{r}t\}dt}{\int_\alpha^\beta x\exp\{-\bar{r}t\}p(x)\bar{m}(x)dx} = \frac{1}{\mu}\int_0^\infty G(t)dt,$$

where use has been made of the definition of the mean age of childbearing in a stationary population:

$$\mu \equiv \frac{\int_\alpha^\beta xp(x)m(x)dx}{\int_\alpha^\beta p(x)m(x)dx} = \int_\alpha^\beta xp(x)\bar{m}(x)dx \tag{20}$$

and $G(t)$ is the part of the birth trajectory which is due to the females in the population at time $t=0$ when the fertility shift occurs.

Recall from (11) that

$$G(t) = \int_0^\infty k_0(a)\frac{p(a+t)}{p(a)}\bar{m}(a+t)da,$$

where

$$k_0(a) = B_0\exp\{-ra\}p(a).$$

Thus

$$B_\infty = \frac{B_0}{\mu}\int_0^\infty\int_0^\infty\exp\{-ra\}p(a+t)\bar{m}(a+t)dadt.$$

Make the substitution $x = a+t$ in the inner integral to get

$$\frac{B_\infty}{B_0} = \frac{1}{\mu R_0}\int_0^\infty\int_t^\infty\exp\{-r(x-t)\}p(x)m(x)dxdt$$

$$= \frac{1}{\mu R_0}\int_0^\infty\exp\{-rx\}p(x)m(x)\int_x^\infty\exp\{rt\}dtdx$$

$$= \frac{1}{\mu r R_0}\int_0^\infty[1-\exp\{-rx\}]p(x)m(x)dx$$

$$= \frac{R_0-1}{\mu r R_0}.$$

It is also not hard to find the ratio of the population size at time $t=0$ to the eventual population size as $t\to\infty$ from the relations

$$k_0(x) = B_0 \exp\{-rx\}p(x),$$

$$k_\infty(x) = B_\infty p(x).$$

Thus

$$\frac{K_\infty}{K_0} = \frac{B_\infty \int_0^\infty p(x)dx}{B_0 \int_0^\infty \exp\{-rx\}p(x)dx} = \frac{R_0-1}{\mu r R_0}b\mathring{e}_0,$$

where the integral in the numerator is the expectation of life at birth, $\mathring{e}_0$, and the integral in the denominator is the reciprocal of the intrinsic birth rate, b. All of the quantities in this expression refer to the population before the shift in fertility.

To interpret the result, notice that one part of it comes from the shift in the birth trajectory and the other part from the redistribution of the age structure.

While the above result is exact, it is not very easily visualized; therefore, look for an approximation in terms of R_0. Recall that to a rather good approximation $R_0 \simeq \exp\{r\mu\}$, hence

$$\frac{B_\infty}{B_0} = \frac{R_0-1}{\mu r R_0} = R_0^{-1/2}\frac{R_0^{1/2} - R_0^{-1/2}}{\mu r}$$

$$\simeq R_0^{-1/2}[\exp\{r\mu/2\} - \exp\{-r\mu/2\}]/\mu r$$

$$\simeq R_0^{-1/2}[1 + (r\mu)^2/24 + \cdots].$$

Typically, $r\simeq 0.01$ and $\mu \simeq 30$, so that $(r\mu)^2/24 \simeq 0.004$; hence

$$\frac{B_\infty}{B_0} \simeq R_0^{-1/2}.$$

This means that if $R_0 > 1$, then $B_\infty < B_0$, as shown below:

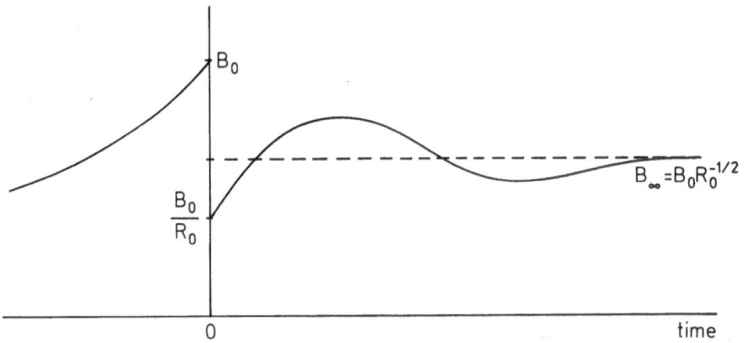

In order to determine the asymptotic behavior of the total population size, it is also necessary to look at the product $b\mathring{e}_0$. To do so, define

$$S(\varrho) = \frac{\int\limits_0^\infty p(x)dx}{\int\limits_0^\infty \exp\{-\varrho x\}p(x)dx} : \quad \begin{cases} S(0) = 1, \\ S(r) = b\mathring{e}_0. \end{cases}$$

Differentiate with respect to ϱ to get

$$\frac{dS}{d\varrho} = S\frac{\int\limits_0^\infty x\exp\{-\varrho x\}p(x)dx}{\int\limits_0^\infty \exp\{-\varrho x\}p(x)dx} \simeq S(A_0 - \varrho\sigma^2),$$

where the approximation results from expanding the exponentials within the integrals, and using the definitions of the mean age A_0 and variance σ^2 of the life table (stationary) population $p(x)$. Separating the resulting differential equation and integrating

$$\int\limits_1^{b\mathring{e}_0} \frac{dS}{S} = \int\limits_0^r (A_0 - \varrho\sigma^2)d\varrho$$

leads to

$$b\mathring{e}_0 \simeq \exp\left\{A_0 r - \frac{\sigma^2}{2}r^2\right\} \simeq \exp\left\{\left[\left(A_0 - \frac{\sigma^2}{2}r\right)\Big/\mu\right]\ln R_0\right\}$$

$$= R_0^{[(A_0 - \sigma^2 r/2)/\mu]}.$$

Although it cannot be demonstrated in any satisfactory mathematical manner, it turns out that the quantity in square brackets is about unity for all populations; hence

$$b\mathring{e}_0 \simeq R_0.$$

Thus,

$$\frac{K_\infty}{K_0} = \frac{R_0 - 1}{\mu r R_0} b\mathring{e}_0 \simeq R_0^{-1/2} \cdot R_0 = R_0^{1/2}.$$

In words, this approximation says that if a population is growing in a stable manner with a net reproduction rate of R_0 and then abruptly scales its maternity to the bare replacement level, the eventual population will number about $\sqrt{R_0}$ times the population size at the time of the maternity adjustment. Notice further that, although the birth rate declines after the fertility adjustment, the ultimate population growth is brought about by readjustment of the age structure, as shown below.

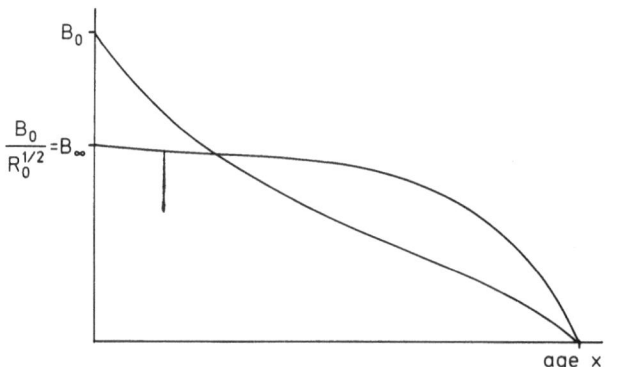

Ratio of areas:

$$\frac{K_\infty}{K_0} \simeq \sqrt{R_0}\,.$$

The abrupt shift in maternity discussed above provides a limiting case to how quickly maternity can be adjusted, but it is not very realistic. A more acceptable model which admits a simple solution is the case where maternity is adjusted as follows. All females alive at $t=0$ continue to reproduce with maternity unaffected, while all females born after $t=0$ shift their maternity to the bare replacement level by a proportional scaling at all ages. Thus if the original net reproduction rate is R_0 and the maternity $m(x)$, then at time t:

$$\bar{m}(x) = \begin{cases} m(x): & x > t, \\ m(x)/R_0: & x \leq t. \end{cases}$$

The effect of this modification is to scale the results for B_∞/B_0 by a factor R_0 and to leave everything else unchanged.

Continuous Age-Structured Models

The models considered thus far have been restricted to those in which mortality and fertility are taken as functions of the age of individuals but not of time. In this section, those models will be generalized to consider mortality as $\mu(x, t)$ and fertility as $\beta(x, t)$.

Obviously, it is hoped that these new models will reduce to those considered earlier if $\mu(x, t) = \mu(x)$ and $\beta(x, t) = m(x)$.

Choose as the dependent variable the population age density at time t, $k(x, t)$, defined so that $k(x, t)dx$ counts the number of individuals between the ages of x and $x + dx$ at time t. Clearly, the total population, $P(t)$, is given by the integral

$$P(t) = \int_0^\omega k(x, t)dx, \tag{21}$$

where ω is the oldest age that anyone attains.

The assumption that there is a finite maximum age is physically realistic, and simplifies certain mathematical details.

Next, derive the crucial equation: a conservation equation for individuals in the population as time passes. It depends on the fact that in one unit (say, year) of time, everyone in the population ages by one unit (year) of age, or else they drop out of the group under study.

Define the "cohort derivative," D, of the population age density, $k(x, t)$, by the formal expression

$$Dk(x, t) = \lim_{h \to 0} \frac{k(x+h, t+h) - k(x, t)}{h}.$$

To rewrite this expression in more familiar notation, add and subtract $k(x+h, t)$ on the right-hand side, so

$$Dk(x, t) = \lim_{h \to 0} \frac{k(x+h, t+h) - k(x+h, t) + k(x+h, t) - k(x, t)}{h}$$

$$= \lim_{h \to 0} \frac{k(x+h, t+h) - k(x+h, t)}{h} + \lim_{h \to 0} \frac{k(x+h, t) - k(x, t)}{h}$$

$$= \frac{\partial k(x, t)}{\partial t} + \frac{\partial k(x, t)}{\partial x}.$$

The conditional probability of dying between ages $x+r$ and $x+r+dr$, given that one is alive at age $x+r$ at time $t+r$, is $\mu(x+r, t+r)dr$, and the cohort population at risk of dying at age $x+r$ at time $t+r$ is $k(x+r, t+r)$; thus the number of individuals who die between age x at time t and age $x+h$ at time $t+h$ is given by

$$\int_0^h k(x+r, t+r)\mu(x+r, t+r)dr.$$

Simple accounting thus provides the balance between the population age density at two nearby instants of time an amount h apart:

$$k(x+h, t+h) = k(x, t) - \int_0^h k(x+r, t+r)\mu(x+r, t+r)dr.$$

Rearranging, dividing through by h and taking the limit as $h \to 0$ provides the relation

$$Dk(x, t) + \mu(x, t)k(x, t) = 0$$

or, in its more usual form,

$$\frac{\partial k(x, t)}{\partial x} + \frac{\partial k(x, t)}{\partial t} + \mu(x, t)k(x, t) = 0. \tag{22}$$

This equation is called either the McKendrick equation or the von Foerster equation. It is a partial differential equation of the hyperbolic type, with characteristics given by $t = x + (\text{constant})$. This means that if the point (x, t) is on a characteristic, then so is the point $(x + h, t + h)$, or in other words, that cohorts age along characteristics. In the demographic literature, an age/time plot of the characteristics is called a Lexis diagram.

Notice that the McKendrick equation (as it should properly be called, since McKendrick's work is dated more than thirty years before von Foerster's) describes the aging process, but not the birth process. To complete the formulation, two side conditions are needed. The first is the usual initial condition on a partial differential equation of the form:

$$k(x, 0) = k_0(x) \tag{23}$$

and the second is a boundary condition

$$k(0, t) = \int_0^\infty \beta(x, t) k(x, t) dx \equiv B(t) \tag{24}$$

which can be seen to be just a generalization of the integral equation (9) discussed at length in the previous section.

The set of equations is in its most general form, but not its most common one. Typically, it is assumed that the time dependence of both the mortality and the fertility functions is related to the total population size at time t and not to time itself. This has the effect of making the equations non-linear. For the remainder of this section, we consider mortality and fertility functions of the form $\mu(x, P)$ and $\beta(x, P)$.

The partial differential equation can now be integrated along characteristics to give the formal solution:

$$k(x, t) = k_0(x - t)\pi(x - t, x, 0), \quad x \geq t,$$
$$k(x, t) = B(t - x)\pi(0, x, t - x), \quad t > x,$$

where the function π is a generalization of the survivorship discussed earlier, and is defined by the expression

$$\pi(x_0, x, t) = \exp\left\{ -\int_{x_0}^x \mu(y, P(y - x_0 + t)) \right\} dy.$$

The important thing to notice about this result is that a knowledge of P and B (along with the initial age distribution and the mortality and fertility functions) completely determines the age distribution, $k(x, t)$, for all future time.

Transformation to a Set of Ordinary Differential Equations

The procedure to be discussed is due to Gurtin and MacCamy, though the general idea is often called the method of moments. In order to make the transformation from a partial differential equation to a set of ordinary differential equations, it will be necessary to assume that the mortality is age independent, and that the fertility function is separable. In particular, assume

$$\mu(x, P) = \mu(P) \quad \text{only}$$

$$\beta(x, P) = \beta(P)x^n \exp\{-\gamma x\}: \quad n = 0, 1, 2, \ldots, \quad \gamma > 0.$$

The number of simultaneous equations which result depend upon n; there will be $n+2$ equations. Notice that the major change occurs as n increases from zero. When $n=0$, the age dependent portion of the fertility function achieves its maximum at age zero, while for $n>0$, the fertility is zero at age zero, rises to a single maximum at age n/γ, and then dies away with increasing age.

The Transformation Procedure

Conceptually, what will be done is to multiply the partial differential equation by a weighting function, $g(x)$, and integrate from $x=0$ to $x=\infty$. The weighting function successively assumes the forms

$$g(x) = x^k \exp\{-\gamma x\}: \quad k = 0, 1, \ldots, n.$$

The resulting set of equations will depend upon $B(t)$ and $P(t)$, as well as n auxiliary variables which represent moments of the age density, $k(x, t)$ with respect to x, though their precise physical meaning is obscure.

The Two Equation Model ($n=0$)

Begin by multiplying the partial differential equation (22) by the weighting factor $g(x)=1$ and integrating from $x=0$ to $x=\infty$:

$$\int_0^\infty \frac{\partial k(x, t)}{\partial t} dx + \int_0^\infty \frac{\partial k(x, t)}{\partial x} dx + \mu(P) \int_0^\infty k(x, t) dx = 0$$

but this can be rewritten

$$\frac{d}{dt} \int_0^\infty k(x, t) dx + k(\infty, t) - k(0, t) + \mu(P) \int_0^\infty k(x, t) dx = 0$$

and since $k(\infty, t) = 0$ and $k(0, t) = B(t)$, this may be written

$$\dot{P} + \mu(P)P - B = 0. \tag{25a}$$

This is, of course, one equation in two unknowns. Next, go on to multiply the partial differential equation by $g(x) = \exp\{-\gamma x\}$ and proceed exactly as before. The result is:

$$\dot{B} + [\beta(P) - \mu(P) - \gamma]B = 0. \tag{25b}$$

This is the second equation in the same two unknowns. The initial conditions follow from

$$P(0) = \int_0^\infty k_0(x)dx,$$

$$B(0) = \beta(P(0)) \int_0^\infty \exp\{-\gamma x\} k_0(x)dx.$$

Illustrative Example

It is often the case that mortality is more sensitive to population size variations than fertility. Consider the situation where $\beta(P) = \beta_0$, a constant independent of population size. The O.D.E.'s (25) then become

$$\dot{P} = -\mu(P)P + B,$$

$$\dot{B} = [\kappa - \mu(P)]B: \quad \kappa = \beta_0 - \gamma.$$

It will now be demonstrated that this system of equations admits no periodic solutions. First, observe that all equilibrium points lie on the line (invariant set) $B = \kappa P$. Any closed orbit surrounds at least one equilibrium point, and thus must cut the invariant set. But this is a contradiction since, once on the invariant set, it can never be left.

The Three Equation Model ($n = 1$)

Proceed exactly as before, assuming now that $\mu(x, P) = \mu(P)$ only, and $\beta(x, P) = \beta(P)x \exp\{-\gamma x\}$. Successively weight the P.D.E. by $g(x) = 1$, $g(x) = \exp\{-\gamma x\}$ and $g(x) = x \exp\{-\gamma x\}$; the resulting equations following integration from $x = 0$ to $x = \infty$ are:

$$\dot{P} + \mu(P)P - k(0, t) = 0,$$

$$\dot{G} + [\mu(P) + \gamma]G - k(0, t) = 0, \tag{26}$$

$$\dot{B} + [\mu(P) + \gamma]B - \beta(P)G = 0,$$

where $G(t)$ is an auxiliary function defined by

$$G(t) = \int_0^\infty \exp\{-\gamma x\} k(x, t)dx.$$

Notice that $k(0, t)$ is written instead of $B(t)$ in the first two equations. Normally it would be written $B(t)$; however, the $k(0, t)$ was left because it allows for an interesting application to be considered.

Cannibalism

Consider an age-structured model in which the adults consume their own egges or newborns. Consider particularly the case in which mortality of the species is age-independent, and fertility as a function of age starts at zero, rises to a single maximum, and then dies away. The three-equation O.D.E. model (26) satisfies these restrictions.

If $B(t)$ is thought of as the rate of production of eggs (or newborns), and $k(0, t)$ as the number of newborns who survive cannibalism, then the difference is the number of eggs cannibalized.

Observe that it is no longer the case that $B(t) = k(0, t)$. One possible model assumes that egg consumption depends upon the product of the number of eggs and the number of adults; hence

$$B(t) - k(0, t) = B(t)P(t)$$

so one may write $k(0, t) = B(t)[1 - P(t)]$, or, since this allows $k(0, t) < 0$,

$$k(0, t) = \max\{B(t)[1 - P(t)], 0\}.$$

This idea opens up a great many possible models, but this section will go no further. It is sufficient to note that a substantial literature has developed in this area.

References

Coale, A.J. (1972). The Growth and Structure of Human Populations: A Mathematical Investigation. Princeton University Press

Frauenthal, J.C. (1975). Birth trajectory under changing fertility conditions. Demography *12*: 447–454

Frauenthal, J.C. (1983). Some simple models of cannibalism. Math. Biosci. *63*: 87–98

Gurtin, M.E., MacCamy, R.C. (1979). Some simple models of nonlinear age dependent population dynamics. Math. Biosci. *43*: 199–211

Keyfitz, N. (1977). Introduction to the Mathematics of Population with Revisions. Addison-Wesley, Reading, Mass.

Keyfitz, N. (1977). Applied Mathematical Demography. Wiley-Interscience, New York

Leslie, P.H. (1945). On the use of matrices in certain population mathematics. Biometrica *33*: 183–212

Lopez, A. (1961). Weak ergodicity. *In:* Problems in Stable Population Theory. Office of Population Research, Princeton

McKendrick, A.C. (1926). Applications of mathematics to medical problems. Proc. Edinburgh Math. Soc. *44*: 98–130

Parlett, B. (1970). Ergodic properties of populations I: The one sex model. Theor. Pop. Biol. *1*: 191–207

Pollard, J.H. (1973). Mathematical Models for the Growth of Human Populations. Cambridge University Press

Sharpe, F.R., Lotka, A.J. (1911). A problem in age distribution. Phil. Mag. *21*: 435–438

von Foerster, H. (1959). Some remarks on changing populations. *In:* The Kinetics of Cellular Proliferation. Grune and Stratton, New York

Random Walk Models of Movement and Their Implications

Simon A. Levin

I. Introduction

Biologists long have sought quantitative models to describe the process of dispersal: to aid understanding, to guide experimentation, and to facilitate prediction. The most common such models are of the random walk type, deriving from the assumption that individuals move in a series of discrete steps with probabilities totally determined by positional information. Learning is ignored.

The simplest random walk model may be motivated by the following illustration. Let an organism be located at position 0 at time 0, as in Fig. 1. Assume further that at discrete times kt, the organism jumps either forward or backward λ units,

and that either event has probability 1/2. All of these assumptions may be relaxed, but will be observed here to make the presentation clearer.

If m and n both are even integers, what is the probability that at time nt the organism is at position $m\lambda$ (after its latest jump)? This is just the probability that $(n+m)/2$ forward steps have been taken, and $(n-m)/2$ backward steps, and thus is given by the appropriate term of the Bernoulli (binomial) distribution:

$$\text{Prob} = (\tfrac{1}{2})^n \frac{n!}{\left(\dfrac{n+m}{2}\right)! \left(\dfrac{n-m}{2}\right)!}. \tag{1}$$

For n large, this converges to the Gaussian (normal) distribution

$$Ce^{-m^2/2n}, \tag{2}$$

where

$$C = \sqrt{\frac{2}{\Pi n}}.$$

Biomathematics, Vol. 17, Mathematical Ecology
Edited by T. G. Hallam and S. A. Levin
© Springer-Verlag Berlin Heidelberg 1986

In terms of $x=\lambda m$ and $t=\tau n$, (2) may be written:

$$C \exp\left(-\frac{x^2}{4t} \cdot \frac{2\tau}{\lambda^2}\right).$$

(3)

This tends to the limiting form

$$P(x,t)=C \exp\left(-\frac{x^2}{4Dt}\right), \quad \text{where} \quad C=\frac{1}{2\sqrt{\Pi Dt}},$$

(4)

provided λ and τ both are allowed to shrink to zero in such a way that the limit

$$\lim_{\lambda,\tau \to 0} \frac{\lambda^2}{\tau}=2D$$

(5)

exists. Similar forms also can be obtained in higher dimensions (see Okubo, 1980; Lin and Segel, 1974). D is termed the *diffusion* coefficient.

(4) represents the approximate limiting distribution to be expected when individuals spread according to the random walk model proposed above. As Okubo (1980) observes, the limiting approximation is valid only when "the time of observation t is much greater than the duration time τ of each random step, and when the scale of observation x is much greater than the length of each random step." In other words, the use of the diffusion approximation is justified only on scales that involve a great many individual steps. The limiting form (4) can be derived under more general conditions: it does not depend, for example, on the assumption that each individual moves at each time step.

Note that the population described by (4) always is normally distributed (for $t>0$), and has a variance $(2Dt)$ which increases linearly with time. Note further (by direct differentiation) that at any point x, the rate of change of population density is proportional to the second derivative of population density with respect to x; that is,

$$\frac{\partial P}{\partial t}=D\frac{\partial^2 P}{\partial x^2}.$$

(6)

This equation is known as the diffusion equation (or the heat equation), and also could be derived from first principles. It describes the spread of a diffusing population with any distribution (not just the normal). Moreover, the property that the variance increases linearly with time at the rate $2D$ also is true in general.

Let N be the total population size; that is,

$$N= \int_{-\infty}^{\infty} P(x,t)dx.$$

(7)

Then

$$\frac{\partial N}{\partial t} = \int_{-\infty}^{\infty} \frac{\partial P}{\partial t}dx= \int_{-\infty}^{\infty} D\frac{\partial^2 P}{\partial x^2}dx = D\frac{\partial P}{\partial x}\bigg|_{-\infty}^{\infty}.$$

(8)

Thus, provided we assume that $\partial P/\partial x$ tends to zero at $\pm\infty$, N doesn't change with time.

The mean of the distribution is given by

$$m = \int_{-\infty}^{\infty} xP(x,t)dx/N . \tag{9}$$

By application of the chain rule, we see that m also is a constant, provided P and $\partial P/\partial x$ vanish sufficiently rapidly at $\pm\infty$. Similarly, if the variance V is defined by

$$V = \int_{-\infty}^{\infty} (x-m)^2 P(x,t)dx/N , \tag{10}$$

we see (again using integration by parts) that

$$\frac{dV}{dt} = 2D . \tag{11}$$

Thus, in general,

$$V = V_0 + 2Dt . \tag{12}$$

Property (12), that the variance increases linearly with time, is a characteristic of the simplest (constant coefficient) model of dispersal. Several researchers (e.g. Kareiva, 1983) have used this property to test the adequacy of the simplest model. For data on the foraging movements of phytophagous insects, Kareiva estimated D from the slope of the regression of V on t. He then used this estimate of D to generate a series of probability distributions for the spread of the insects, and compared it with what was actually observed. He found that the agreement was excellent in most cases, but that in some cases a habitat-dependent diffusion model,

$$\frac{\partial P}{\partial t} = \frac{\partial^2 (D(x)P)}{\partial x^2} , \tag{13}$$

provided better agreement. His conclusion: the basic diffusion model is an excellent starting point, but must be modified when there is substantial habitat variability.

Another major modification is necessary when growth and spread occur simultaneously. Here, Eq. (6) becomes

$$\frac{\partial P}{\partial t} = D\frac{\partial^2 P}{\partial x^2} + F(P,x,t) , \tag{14}$$

where $F(P,x,t)$ is the local population growth rate. In the next section, I shall discuss some uses of this equation and its higher dimensional versions in studying the rates of spread of species entering new habitats. In a later chapter, the extension of (14) to interacting populations is discussed, as are the associated problems of the generation and maintenance of spatial pattern for both the scalar and vector versions.

II. Waves of Advance of Species and Genes

One of the earliest applications of diffusion models of movement was to the rate of advance of advantageous alleles entering novel habitats. Fisher (1937), considering the case of selection operating on two alleles at a single autosomal locus, proposed the model

$$\frac{\partial P}{\partial t} = D\frac{\partial^2 P}{\partial x^2} + rP(1-P),\tag{15}$$

where P is the frequency of the advantageous allele. Later authors (Aronson and Weinberger, 1975; Hoppensteadt, 1975; Hadeler and Rothe, 1975) have more correctly considered the equations for genotype frequencies; but the basic insights available in Fisher's original model remain essentially unchanged (see review by Hadeler, 1976).

Fisher conjectured that an advancing wave would relax asymptotically to a front, with a characteristic speed of advance, and gave a formula for that speed. Kolmogorov et al. (1937), in a classic paper in mathematical biology, formalized these results for the more general equation,

$$\frac{\partial P}{\partial t} = D\frac{\partial^2 P}{\partial x^2} + f(P),\tag{16}$$

where

$$f(0) = f(1) = 0, \quad f > 0 \text{ on } (0,1)\tag{17}$$

and

$$f'(0) > f'(P) \text{ on } [0,1].\tag{18}$$

By looking for non-negative wave solutions of the form

$$P = H(x - ct), \quad c > 0.\tag{19}$$

Kolmogorov et al. showed that there exist monotone wave solutions for all wave speeds c greater than or equal to the critical value

$$c^* = 2\sqrt{D \cdot f'(0)},\tag{20}$$

and none for $c < c^*$. They showed further that for an initial Heaviside distribution,

$$P_0(x) = \begin{cases} 1 & \text{for} \quad x < 0 \\ 0 & \text{for} \quad x > 0, \end{cases} \tag{21}$$

the wave corresponding to $c = c^*$ is attracting (see Hadeler, 1976). A recent and thorough monograph on the subject by Bramson (1983) provides an excellent treatment. Bramson discussed more general initial distributions (not all of which converge to the c^*-wave), and shows that the rate of relaxation is logarithmic. Such problems remain objects of intense mathematical interest. Recent work (Fife, 1984) summarizes the current state of knowledge, considers more general functions $f(P)$, and treats the problem of spatially slowly varying waves.

Among biologists, such results have been of considerable interest, perhaps more for the ecological analogues than for the population genetic situations which motivated Fisher's model. Skellam (1951) applied models of the form (16) to the study of species invading new habitats. Indeed, even for the linear growth model $f(P) = rP$, interesting results obtain. In this case, there is no true front, because the solutions are spreading normal distributions which also increase exponentially in amplitude. Thus, the speed of propagation is infinite. However, if one superimposes the requirement of a detection threshold – a level of density below which the species is not observed – and considers the observed spread of the species, one finds again that the "front" propagates at the speed given by (20).

Skellam applied the model to examine the rates of spread of a variety of species, from oaks to muskrats, and to consider the adequacy of simple diffusion in explaining rates of spread. The technique remains a potentially powerful tool to this day in the examination of the rates of spread of species entering novel habitats. One can examine observed patterns of spread to see how well they fit the predictions of a diffusion model, and try to relate the observed rates of spread to independently measured estimates of diffusion (D) and growth at low densities ($f'(0)$). Each of the latter, in theory, may be estimated independently of data on the spread of fronts, and several current studies seek to identify the appropriate parameters and make the critical comparisons.

III. Further Considerations

The examination of fronts is not restricted to the scalar case. Similar investigations have been carried out for the three equations describing genotype frequencies (Aronson and Weinberger, 1975), and for spreading epidemics (see, for example, Hadeler, 1984). Kendall (1965), and numerous investigators since, have studied the coupled diffusion equations describing the spread and change in the susceptible and infective classes within a population, and determined the asymptotic speed of advance.

Considerable attention also has been showered upon the study of stable spatial patterns, in genetics as well as in ecology. These are treated in a later chapter in this volume.

References

Aronson, D.G., Weinberger, H.F. (1975). Nonlinear diffusion in population genetics, combustion, and nerve propagation. pp. 5–49, *In*: J. Goldstein (ed.), Partial Differential Equations and Related Topics. Lecture Notes in Mathematics 445. Springer-Verlag, Heidelberg

Bramson, M. (1983). Convergence of solutions of the Kolmogorov equation to travelling waves. Mem. Amer. Math. Soc. #285

Fife, P.C. (1984). Current topics in reaction-diffusion systems. *In*: M. G. Velarde (ed.), Proceedings of NATO Conference on Nonequilibrium Phenomena in Physics and Related Fields. Plenum

Fisher, R.A. (1937). The wave of advance of advantageous genes. Ann. Eugen. London 7:355–369

Hadeler, K.P. (1976). Nonlinear diffusion equations in biology. *In*: W. N. Everett and B. D. Sleeman (eds.), Ordinary and Partial Differential Equations. Lect. Notes in Mathematics 564, Springer-Verlag, Heidelberg

Hadeler, K.P. (1984). Spread and age structure in epidemic models in Perspectives in Mathematics. Anniversary of Oberwolfach, 1984. pp. 295–320. Birkhäuser-Verlag, Basel

Hadeler, K.P., Rothe, E. (1975). Travelling fronts in nonlinear diffusion equations. J. Math. Biol. 2:251–263

Hoppensteadt, F. (1975). Mathematical Theories of Populations: Demographics, Genetics and Epidemics. SIAM Reg. Conf. Series 20, Philadelphia

Kareiva, P. (1983). Local movement in herbivorous insects: applying a passive diffusion model to mark-recapture field experiments. Oecologia 57:322–327

Kendall, D.G. (1965). Mathematical models of the spread of infection. Mathematics and Computer Science in Biology and Medicine, London H.M.S.O. pp. 213–225

Kolmogorov, A., Petrovskij, I., Piskunov, N. (1937). Étude de l'équation de la diffusion avec croissance de la quantité de la matière et son application à un problème biologique. Bull. Univ. Moscou Ser. Internation., Sec. A, 1 (6) 1–25

Lin, C.C., Segel, L.A. (1974). Mathematics Applied to Deterministic Problems in The Natural Sciences. MacMillan, New York. 604 + xix pages

Okubo, A. (1980). Diffusion and Ecological Problems: Mathematical Models. Biomathematics 10, Springer-Verlag, New York

Skellam, J.G. (1951). Random dispersal in theoretical populations. Biometrika 38: 196–218

Stochastic Population Theory: Birth and Death Processes

*Luigi M. Ricciardi**

1. Introduction

Birth and death processes were introduced by Feller (1939) and have since been used as models for population growth, queue formation, in epidemiology and in many other areas of both theoretical and applied interest. From the standpoint of the theory of stochastic processes they represent an important special case of Markov processes with countable state spaces and continuous parameters.

In the sequel we shall provide in a rather straightforward fashion a description of birth and death processes, without referring to the general theory of stochastic processes. Furthermore, we shall start with a discussion of the simple birth process and the simple death process to gain some insight into the mathematical techniques to be used successively in more complicated situations.

The simple birth process was first studied by Yule (1924) in connection with the mathematical theory of evolution; successively, and independently, it was proposed by Furry (1937) in order to describe the multiplication of particles in cosmic-ray showers. For this reason, the simple birth process is often referred to as the Yule-Furry process.

Yule's starting point was to consider the number of species within a genus and to conceive the creation of a new species by mutation as a random event whose probability of occurrence is proportional to the number of existing species. The creation of a new species at a rate proportional to the number of existing species and independent of their ages and sizes may be taken as somewhat plausible if one assumes that each species rapidly reaches its largest size (determined by the environmental carrying capacity) and thereafter does not change size in an appreciable manner; moreover, the assumption that different species do not interact strongly with one another has to be added. All this is suggested by Darwin's view that the process of species modification will generally affect only a few species at the same time, the variability of each species being quite independent of that of all others.

The main flaw of Yule's approach clearly is that it neglects the differences in species sizes and completely ignores the possibility of a species becoming extinct so that its conclusions appear to be rather inexact. Nevertheless, the Yule-Furry process allows one to gain some insight into the stochastic behavior of populations and can be taken as a basis for developing more realistic extensions.

* Work performed under CNR-JSPS Cooperation Programme, Contracts No. 83.00032.01 and No. 84.00227.01 and under MPI financial support

2. The Simple Birth Process

Let us denote by $X(t)$ the total number of individuals of a population at time t and by

$$p_n(t)=P\{X(t)=n\}, \quad (n=0,1,2,...) \tag{2.1}$$

the probability that the random variable $X(t)$ takes the value n. We shall assume that all individuals are able to give birth to new individuals. If we are modelling the growth of a population of unicellular organism, such as bacteria, we must take into account the fact that whenever two new daughter organisms come into being the reproducing individual ceases to exist as such. Instead, in the case of the higher animals we might suppose that the parent organism coexists with the newly generated individual. In either case the total population size increases by exactly one unit as a result of a single reproduction. We now make the assumption that the probability for *a given* individual to give origin to a new one in the time interval $(t, t+\Delta t]$ is $\lambda\Delta t$, where λ denotes a positive constant. For single cell organisms this assumption is equivalent to assuming that the time τ elapsing between birth and the division into the two daughter cells is exponentially distributed:

$$P\{\tau\leq T\}=1-e^{-\lambda T} \quad (0\leq T<\infty).$$

The probability that the whole population consisting of $X(t)$ individuals at time t gives rise to a new individual in $(t, t+\Delta t]$ is then $\lambda X(t)\Delta t+o(\Delta t)$, where $o(\Delta t)$ denotes quantities such that

$$\lim_{\Delta t\uparrow 0}\frac{o(\Delta t)}{\Delta t}=0.$$

Therefore, assuming that Δt is very small, we can henceforth disregard the term $o(\Delta t)$ and write the above probability directly as $\lambda X(t)\Delta t$.

In order to determine the probabilities (2.1), which characterize the growth process, we proceed as follows. We consider the probability $p_n(t+\Delta t)$ that $X(t+\Delta t)=n$ and write it down as the sum of the probabilities of two mutually exclusive events: i) the probability that at time t n individuals are already present in the population and that no birth takes place in $(t, t+\Delta t]$; ii) the probability that $n-1$ individuals are present at time t and that one birth takes place in $(t, t+\Delta t]$. Disregarding the probabilities that multiple births take place in $(t, t+\Delta t]$, which are quantities $o(\Delta t)$, we thus have:

$$p_n(t+\Delta t)=p_n(t)(1-\lambda\Delta t)^n$$
$$+p_{n-1}(t)\lambda(n-1)\Delta t+o(\Delta t). \tag{2.2}$$

Disregarding the quantities $o(\Delta t)$ in the first term of the right hand side of (2.2) after a simple algebra we obtain:

$$\frac{p_n(t+\Delta t)-p_n(t)}{\Delta t}=p_{n-1}(t)\lambda(n-1)-p_n(t)\lambda n+\frac{o(\Delta t)}{\Delta t}.$$

Passing to the limit as $\Delta t \to 0$ we are finally led to the following system of differential equations in the unknowns $p_n(t)$:

$$\frac{dp_n(t)}{dt} = \lambda(n-1)p_{n-1}(t) - \lambda n p_n(t). \tag{2.3}$$

Since new individuals can only be generated by previously existing individuals, we must assume that the initial population size consists of a certain nonzero number j of individuals, i.e., $X(0) = j$. Therefore, the initial conditions

$$p_m(0) \equiv P\{X(0) = m\} = \begin{cases} 1, & m = j \\ 0, & m \neq j \end{cases} \tag{2.4}$$

must be associated with Eq. (2.3) and these must be considered only for $n \geq j$, with $p_{j-1}(t) \equiv 0$ in the first equation. For $n = j$, from (2.3) we have:

$$\frac{dp_j(t)}{dt} = -\lambda j p_j(t). \tag{2.5}$$

Therefore, it follows

$$p_j(t) = e^{-\lambda j t} \tag{2.6}$$

where use of (2.4) has been made to determine the arbitrary constant arising from integration of (2.5). We can then set $n = j+1$ in (2.3) and substitute in it the expression (2.6):

$$\frac{dp_{j+1}(t)}{dt} = \lambda j e^{-\lambda j t} - \lambda(j+1)p_{j+1}(t). \tag{2.7}$$

From (2.7) it easily follows:

$$p_{j+1}(t) = j e^{-\lambda j t}(1 - e^{-\lambda t}),$$

where use of initial condition (2.4) has been made. Equation (2.3) can thus be solved in succession and the general form for $p_{j+k}(t)$ ($k = 0, 1, 2, \ldots$) can be inferred after a few iterations. The result is:

$$p_{j+k}(t) = \binom{j+k-1}{j-1} e^{-\lambda j t}(1 - e^{-\lambda t})^k \quad (k = 0, 1, \ldots) \tag{2.8}$$

or, equivalently,

$$p_n(t) = \binom{n-1}{j-1} e^{-\lambda j t}(1 - e^{-\lambda t})^{n-j} \quad (n = j, j+1, \ldots), \tag{2.9}$$

showing that the population size at time t has a negative binomial distribution in which the probability of success in a single trial is $e^{-\lambda t}$. Therefore, the mean population size, $\langle X(t)|X(0)=j\rangle$, and the variance, $\text{Var}\{X(t)|X(0)=j\}$, of the population size are given by

$$\langle X(t)|X(0)=j\rangle=je^{\lambda t}, \tag{2.10a}$$

$$\text{Var}\{X(t)|X(0)=j\}=je^{\lambda t}(e^{\lambda t}-1), \tag{2.10b}$$

respectively. It is interesting to remark that under the simple birth process model the mean population size coincides with the size of a deterministically growing population in a Malthusian scheme. The variance of the population size also increases steadily and, for large t, is approximately given by $je^{2\lambda t}$. From (2.10a) and (2.10b) we see that the coefficient of variation V is given by:

$$V\equiv\frac{[\text{Var}\{X(t)|X(0)=j\}]^{1/2}}{\langle X(t)|X(0)=j\rangle}=\frac{[je^{\lambda t}(e^{\lambda t}-1)]^{1/2}}{je^{\lambda t}}.$$

This approaches the value $1/\sqrt{j}$ as $t\to\infty$. Hence, we see that this variation coefficient is asymptotically very small if the initial number of individuals is very large.

3. The Simple Death Process

The simple birth process considered in Sect. 2 entails a monotonically increasing population size. We shall now define and study the converse process in which the population size $X(t)$ is monotonically decreasing. To this purpose, let us refer to a population of individuals which is subject to random death in the sense that the probability for a given individual to die in the time interval $(t, t+\Delta t]$ is $\mu\Delta t$, with μ a positive constant. Furthermore, we make the assumption that at time $t=0$ the population consists of $X(0)=j$ individuals. From these assumptions, it follows that the probability of one death occurring in $(t, t+\Delta t]$ in the whole population is $\mu X(t)\Delta t+o(\Delta t)$. Proceeding as in Sect. 2, we are easily led to the following system of differential equations for the probability $p_n(t)$ of having n individuals at time t:

$$\frac{dp_n(t)}{dt}=\mu(n+1)p_{n+1}(t)-\mu np_n(t) \tag{3.1}$$

which must be solved with the initial conditions

$$p_m(0)=\begin{cases}1, & m=j\\0, & m\neq j.\end{cases} \tag{3.2}$$

It is a simple matter to solve equations (3.1) recursively by making use of conditions (3.2). The result is:

$$p_n(t)=\binom{j}{n}e^{-n\mu t}(1-e^{-\mu t})^{j-n} \quad (n=0,1,...,j) \tag{3.3}$$

which shows that the population size $X(t)$ is binomially distributed with mean and variance given by

$$\langle X(t)\,|\,X(0)=j\rangle = je^{-\mu t}$$

and

$$\mathrm{Var}\{X(t)\,|\,X(0)=j\} = je^{-\mu t}(1-e^{-\mu t}),$$

respectively. We thus see that the mean population exponentially decreases with time in a fashion identical to the population size of a deterministically declining population such that the relative number of deaths per unit time remains a constant μ.

In conclusion, it should be pointed out that in the simple death process individual life time is again exponentially distributed and that asymptotic extinction is now a sure event since $p_0(t)\to 1$ as $t\to\infty$.

4. The Simple Birth and Death Process

We shall now combine the assumptions underlying the two models discussed in the foregoing – one involving birth only, the other death only – and assume that two types of transitions can take place in construction of the simple birth and death process. Namely, we shall assume that the probability for any given individual to give birth in the time interval $(t, t+\Delta t]$ is $\lambda\Delta t$ while the probability of dying in that time interval is $\mu\Delta t$. Equivalently, denoting by $\Delta X(t, t+\Delta t)$ the increment $X(t+\Delta t)-X(t)$ of the population size in $(t, t+\Delta t]$ we can make the following assumptions:

$$P\{\Delta X(t, t+\Delta t)=1\,|\,X(t)=n\} = \lambda n\Delta t + o(\Delta t)$$

$$P\{\Delta X(t, t+\Delta t)=-1\,|\,X(t)=n\}\mu n\Delta t + o(\Delta t) \qquad (4.1)$$

$$P\{|\Delta X(t, t+\Delta t)|>1\,|\,X(t)=n\} = o(\Delta t).$$

From (4.1) it follows

$$P\{\Delta X(t, t+\Delta t)=0\,|\,X(t)=n\} = 1-(\lambda+\mu)n\Delta t + o(\Delta t). \qquad (4.2)$$

From (4.1) and (4.2) by a procedure similar to that employed in Sects. 2 and 3 we easily obtain

$$p_n(t+\Delta t)=p_n(t)\left[1-(\lambda+\mu)n\Delta t\right]$$
$$+ p_{n-1}(t)(n-1)\lambda\Delta t + p_{n+1}(t)\mu(n+1)\Delta t + o(\Delta t)$$

or, in the limit as $\Delta t\to 0$:

$$\frac{dp_n(t)}{dt} = -n(\mu+\lambda)p_n(t) + \lambda(n-1)p_{n-1}(t)$$
$$+ \mu(n+1)p_{n+1}(t) \qquad (n=0, 1, 2, \dots) \qquad (4.3)$$

with the initial condition

$$p_n(0) = \begin{cases} 1, & n=j \\ 0, & n \neq j \end{cases} \tag{4.4}$$

and $p_{-1}(t) \equiv 0$ in the first equation. Before showing how Eq. (4.3) can be solved to determine the functions $p_n(t)$, let us calculate the mean population size $\langle X(t) \rangle$ and its variance $\mathrm{Var}\{X(t)\}$ by a straightforward procedure. To this purpose we make use of the definition

$$\langle X(t) \rangle = \sum_{n=0}^{\infty} n p_n(t) \tag{4.5}$$

and differentiate both sides with respect to t:

$$\frac{d}{dt} \langle X(t) \rangle = \sum_{n=0}^{\infty} n \frac{dp_n(t)}{dt}$$

$$= - \sum_{n=0}^{\infty} n^2 (\mu + \lambda) p_n(t) + \lambda \sum_{n=1}^{\infty} n(n-1) p_{n-1}(t)$$

$$+ \mu \sum_{n=0}^{\infty} n(n+1) p_{n+1}(t), \tag{4.6}$$

where the last equality follows from (4.3). Setting $n-1=n$ and $n+1=n$ in the second and third sum on the right hand side of (4.6), respectively, after some straightforward algebra we are led to the following differential equation:

$$\frac{d}{dt} \langle X(t) \rangle = (\lambda - \mu) \langle X(t) \rangle. \tag{4.7}$$

On the other hand from (4.4) and (4.5) we obtain the initial condition

$$\langle X(0) \rangle = j. \tag{4.8}$$

From (4.7) and (4.8) we immediately get

$$\langle X(t) | X(0) = j \rangle = j e^{(\lambda - \mu)t}. \tag{4.9}$$

By a similar procedure an equation for

$$\mathrm{Var}\{X(t)\} = \langle X^2(t) \rangle - [\langle X(t) \rangle]^2$$

can be derived and solved with the initial condition

$$\mathrm{Var}\{X(0)\} = 0.$$

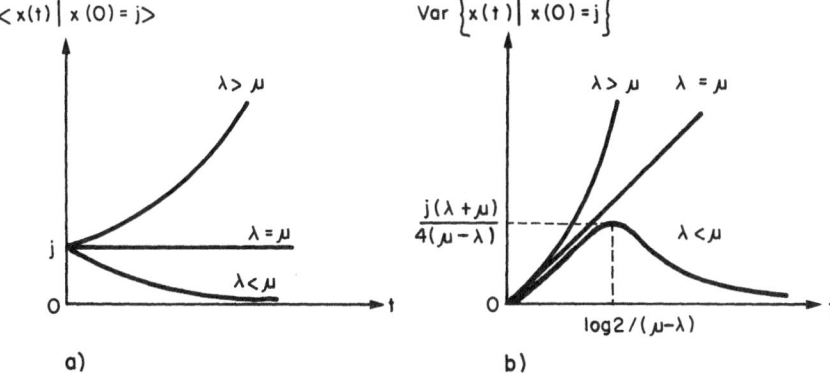

Fig. 1. a) The mean population size, conditional upon $X(0)=j$, exponentially grows for $\lambda > \mu$, exponentially decreases if $\lambda < \mu$ and remains constant if $\lambda = \mu$. b) The variance of the population size is sketched for the case $X(0) = j$. It grows exponentially if $\lambda > \mu$ and linearly if $\lambda = \mu$. When $\lambda < \mu$ it first increases to reach its maximum value at $t = \log 2/(\mu - \lambda)$, then decreases exponentially after exhibiting an inflection point at $t = 2\log 2/(\mu - \lambda)$

The result can be seen to be the following:

$$\text{Var}\{X(t)\,|\,X(0)=j\} = \begin{cases} 2j\mu t, & \lambda = \mu \\[2mm] \dfrac{j(\lambda+\mu)}{\lambda-\mu}\,e^{(\lambda-\mu)t}[e^{(\lambda-\mu)t}-1], & \lambda \neq \mu. \end{cases} \tag{4.10}$$

Note that in the case $\lambda \neq \mu$ this variance not only depends on the intrinsic "growth rate" $\lambda - \mu$ but also on the sum $\lambda + \mu$. Figure 1 is a sketch of mean and variance for the simple birth and death process.

We now turn to the solution of Eq. (4.3) with the initial conditions (4.4). To this purpose it is convenient to change Eq. (4.3) into a unique partial differential equation of the first order whose general solution can be obtained by standard techniques. The starting point consists of introducing the so called "moment generating function" $M(\theta, t)$ defined as follows in terms of the p_n's:

$$M(\theta, t) = \sum_{n=0}^{\infty} e^{\theta n} p_n(t), \tag{4.11}$$

where θ is a "dummy" variable. Multiplying both sides of (4.3) by $e^{\theta n}$ and summing over n from 0 to ∞, one then easily obtains:

$$\frac{\partial M}{\partial t} = [\lambda(e^\theta - 1) + \mu(e^{-\theta} - 1)]\frac{\partial M}{\partial \theta}. \tag{4.12}$$

The initial condition to be associated with (4.12) is readily obtained from (4.4) and (4.11):

$$M(\theta, 0) \equiv \sum_{n=0}^{\infty} e^{\theta n} p_n(0) = e^{\theta j}. \tag{4.13}$$

The general solution of Eq. (4.12) can be found by the so-called "method of characteristics". To see how this can be used, let us refer to a general semi-linear partial differential equation of the first order, that we write in the form:

$$\sum_{n=1}^{n} P_k(x_1, x_2, ..., x_n, f) \frac{\partial f}{\partial x_k} = R(x_1, x_2, ..., x_n, f) \tag{4.14}$$

with the coefficients P_k's and R functions of the n variable x_i as well as of the unknown solution f. From (4.14) one first forms the subsidiary equations

$$\frac{dx_1}{P_1} = \frac{dx_2}{P_2} = ... = \frac{dx_n}{P_n} = \frac{df}{R} \tag{4.15}$$

and then finds n independent integrals of those equations that can be written in the form

$$u_1(x_1, x_2, ..., x_n, f) = c_1$$
$$u_2(x_1, x_2, ..., x_n, f) = c_2$$
$$................$$
$$u_n(x_1, x_2, ..., x_n, f) = c_n \tag{4.16}$$

with $c_1, c_2, ..., c_n$ arbitrary constants. The most general solution of (4.14) can then be expressed as:

$$\Phi(c_1, c_2, ..., c_n) = 0 \tag{4.17}$$

or, equivalently, as

$$c_i = \Psi(c_1, c_2, ..., \hat{c}_i, ..., c_n), \tag{4.18}$$

where the notation $\hat{c}_i$ means that on the right hand side of (4.18) the constant c_i is missing and where Ψ denotes an arbitrary function. For the case of Eq. (4.12) the subsidiary equations are

$$\frac{dt}{1} = \frac{d\theta}{-[\lambda(e^{\theta} - 1) + \mu(e^{-\theta} - 1)]} = \frac{dM}{0}. \tag{4.19}$$

The first and third expressions yield the integral

$$M(\theta, t) = c_1 \tag{4.20}$$

while the first and second give:

$$dt = \frac{-e^{\theta} d\theta}{(e^{\theta} - 1)(\lambda e^{\theta} - \mu)}. \tag{4.21}$$

By integration, from (4.21) we obtain:

$$
c_2 = \begin{cases} t + \dfrac{1}{\lambda - \mu} \log \left| \dfrac{e^\theta - 1}{\lambda e^\theta - \mu} \right|, & \text{for } \lambda \neq \mu \\[4mm] t - \dfrac{1}{\lambda(e^\theta - 1)}, & \text{for } \lambda = \mu \end{cases}
$$

(4.22)

or:

$$
c_2' = \begin{cases} \dfrac{(e^\theta - 1)e^{(\lambda - \mu)t}}{\lambda e^\theta - \mu}, & \text{for } \lambda \neq \mu \end{cases}
$$

(4.23a)

$$
\qquad\quad \lambda t - \dfrac{1}{e^\theta - 1}, \qquad \text{for } \lambda = \mu
$$

(4.23b)

where $c_2' = e^{c_2(\lambda - \mu)}$ when $\lambda \neq \mu$ and $c_2' = \lambda c_2$ when $\lambda = \mu$.

Let us now study separately the cases $\lambda \neq \mu$ and $\lambda = \mu$.

i) *Assume* $\lambda \neq \mu$. From (4.18), (4.20), and (4.23a) one obtains:

$$
M(\theta, t) = \Psi \left[\frac{(e^\theta - 1)e^{(\lambda - \mu)t}}{\lambda e^\theta - \mu} \right].
$$

(4.24)

The arbitrary function Ψ can be specified by making use of (4.13):

$$
\Psi \left(\frac{e^\theta - 1}{\lambda e^\theta - \mu} \right) = e^{\theta j}.
$$

(4.25)

Hence:

$$
\Psi(u) = \left(\frac{\mu u - 1}{\lambda u - 1} \right)^j
$$

(4.26)

having set $u = (e^\theta - 1)/(\lambda e^\theta - \mu)$. From (4.24) and (4.26) one can find:

$$
M(\theta, t) = \left[\frac{\mu v(\theta, t) - 1}{\lambda v(\theta, t) - 1} \right]^j
$$

(4.27)

where we have set

$$
v(\theta, t) = \frac{(e^\theta - 1)e^{(\lambda - \mu)t}}{\lambda e^\theta - \mu}.
$$

(4.28)

ii) *Assume now* $\lambda = \mu$. From (4.18), (4.20), and (4.23b) we get:

$$
M(\theta, t) = \Psi \left(\lambda t - \frac{1}{e^\theta - 1} \right).
$$

(4.29)

Hence:

$$M(\theta, t) = \left[\frac{1 - (\lambda t - 1)(e^\theta - 1)}{1 - \lambda t(e^\theta - 1)} \right]^j. \tag{4.30}$$

We are now in the position to obtain the probabilities $p_n(t)$ $(n = 0, 1, 2, ...)$, solutions of (4.3) with conditions (4.4). Indeed, from (4.11) it follows

$$F(s, t) \equiv M(\log s, t) = \sum_{n=0}^{\infty} p_n(t) s^n \tag{4.31}$$

so that these probabilities identify with the coefficients of the Taylor series expansion of $F(s, t) = M(\log s, t)$. For this reason, the function $F(s, t)$ is also named the "probability generating function". Let us again treat two cases separately.

i) $\lambda \neq \mu$. From (4.27) and (4.31) we obtain

$$F(s, t) = \left[\frac{\mu(1 - \alpha) - (\lambda - \mu\alpha)s}{\mu - \lambda\alpha - \lambda(1 - \alpha)s} \right]^j \tag{4.32}$$

having set

$$\alpha = e^{(\lambda - \mu)t}. \tag{4.33}$$

If the population initially consists of one individual (i.e. $j = 1$), from (4.32) one readily finds:

$$F(s, t) \equiv \frac{\mu(1 - \alpha) - (\lambda - \mu\alpha)s}{\mu - \lambda\alpha} \frac{1}{1 - \lambda\varrho s} = \frac{\mu(1 - \alpha) - (\lambda - \mu\alpha)s}{\mu - \lambda\alpha} \sum_{n=0}^{\infty} (\lambda\varrho)^n s^n$$

$$= \mu\varrho s^0 + \sum_{n=1}^{\infty} \left[\mu\varrho(\lambda\varrho)^n - \frac{\lambda - \mu\alpha}{\lambda\varrho(\mu - \lambda\alpha)} (\lambda\varrho)^n \right] s^n \tag{4.34}$$

having set

$$\varrho = \frac{1 - \alpha}{\mu - \lambda\alpha} \equiv \frac{1 - e^{(\lambda - \mu)t}}{\mu - \lambda e^{(\lambda - \mu)t}}. \tag{4.35}$$

Combining the expressions (4.32) and (4.34) we thus obtain:

$$p_0^{(1)}(t) = \mu\varrho$$

$$p_n^{(1)}(t) = (\lambda\varrho)^n \left[\mu\varrho - \frac{\lambda - \mu\alpha}{(\mu - \lambda\alpha)\lambda\varrho} \right] \quad (n = 1, 2, ...) \tag{4.36}$$

with the upper index reminding us that the population initially consisted of one individual. The case $j > 1$ leads to somewhat more complicated expressions for the probabilities $p_n(t)$. By straightforward calculations one is led to the following result:

$$p_0(t) = (\mu\varrho)^j$$

$$p_n(t) = \sum_{r=0}^{\min(j,n)} \binom{j}{r} \binom{j+n-r-1}{j-1} (\mu\varrho)^{j-r}(\lambda\varrho)^{n-r}[1-(\mu+\lambda)\varrho]^r \qquad (4.37)$$

$$(n = 1, 2, \ldots).$$

with ϱ given by (4.35).

ii) $\lambda = \mu$. From (4.30) and (4.31) we now obtain:

$$F(s,t) = \left[\frac{1-(\lambda t - 1)(s-1)}{1 - \lambda t(s-1)}\right]^j. \qquad (4.38)$$

Assuming again that initially only one individual exists, setting $j=1$ in (4.38) we easily obtain:

$$F(s,t) \equiv \frac{\lambda t + (1-\lambda t)s}{1+\lambda t} \; \frac{1}{1 - \dfrac{\lambda t}{1+\lambda t}\, s}$$

$$= \frac{\lambda t}{1+\lambda t}\, s^0 + \sum_{n=1}^{\infty} \frac{(\lambda t)^{n-1}}{(1+\lambda t)^{n+1}}\, s^n. \qquad (4.39)$$

Hence:

$$p_0^{(1)}(t) = \frac{\lambda t}{1+\lambda t},$$

$$p_n^{(1)}(t) = \frac{(\lambda t)^{n-1}}{(1+\lambda t)^{n+1}}, \qquad (n = 1, 2, \ldots). \qquad (4.40)$$

More complicated expressions hold if $j > 1$.

We conclude this section by stressing that the simple birth and death process studied in the foregoing not only provides an interesting model for simple growth processes but also possesses the remarkable feature of being solvable. Indeed, we have established simple closed form expressions for the probabilities $p_n(t)$ $(n = 1, 2, \ldots)$ in all cases (i.e., $\lambda \gtrless \mu$, $j \geq 1$); furthermore, we also have been able to calculate explicitly mean and variance of the population's size. All this allows one to draw an accurate picture of the population's time evolution, which is often impossible for other models. In particular, we can easily get information about the extinction probability of the population. Indeed, from (4.37) [or directly taking the limit of (4.32) as $s \to 0$] one obtains:

$$p_0(t) = \left\{\frac{\mu[e^{(\lambda-\mu)t}-1]}{\lambda e^{(\lambda-\mu)t}-\mu}\right\}^j \qquad (\lambda \neq \mu). \qquad (4.41)$$

Furthermore, the case $\lambda = \mu$ can also be treated either by taking the limit of (4.38) as $s \to 0$ or by applying l'Hospital rule to (4.41). The result is:

$$p_0(t) = \left(\frac{\lambda t}{1 + \lambda t} \right)^j \quad (\mu = \lambda).$$

(4.42)

Therefore, by taking the limits of (4.41) and (4.42) for $t \to \infty$, we obtain for the probability that asymptotically the population goes to extinction:

$$\lim_{t \to \infty} p_0(t) = \begin{cases} 1, & \text{for } \lambda \le \mu \\ \left(\dfrac{\mu}{\lambda} \right)^j, & \text{for } \lambda > \mu. \end{cases}$$

(4.43)

Hence, the population is doomed to extinction unless the birth rate exceeds the death rate. In the latter case the probability for the population to be still existing after infinitely long times is given by $1 - \left(\dfrac{\mu}{\lambda} \right)^j$, a quantity that becomes unity only for infinitely large birth rates or vanishing death rates. Note that asymptotic extinction is certain also in the case $\lambda = \mu$ in which, as shown by (4.9), the mean population size stays constantly equal to the initial population size $j > 0$. This should be of no surprise; it merely indicates the inadequacy of the mean value above to represent the features of the population growth process. As we shall see, a similar situation is also encountered in other population growth models.

As a final comment we mention that simple birth and death processes are sometimes referred to as the Feller-Arley process since their study has been essentially performed by these two authors (see Feller, 1940; Arley, 1943).

5. Simple Birth and Death Processes with Immigration

In order to improve the birth and death population model so far considered we must take into account the realistic circumstance that many populations are not isolated so that immigration effects may take place. In the following, along the lines of Kendall (1948), we shall assume that immigration can be accounted for by a suitable adjustment of the population's death rate; indeed, it is reasonable to postulate that the probability of a single loss in the time interval $(t, t + \Delta t]$ due to both emigrations and deaths is proportional to the population size $X(t)$ and to Δt. Immigration effects will instead be accounted for by assuming that they occur according to a Poisson process independent of the population size. In mathematical terms, we assume that

$$P\{\Delta X(t, t + \Delta t] = 1 \mid X(t) = n\} = \lambda n \Delta t + v \Delta t + o(\Delta t)$$

$$P\{\Delta X(t, t + \Delta t] = -1 \mid X(t) = n\} = \mu n \Delta t + o(\Delta t)$$

(5.1)

$$P\{|\Delta X(t, t + \Delta t]| > 1 \mid X(t) = n\} = o(\Delta t).$$

By a procedure similar to that of Sect. 4 it is found that the moment generating function $M(\theta, t)$ now satisfies the following equation:

$$\frac{\partial M}{\partial t} = [\lambda(e^\theta - 1) + \mu(e^{-\theta} - 1)]\frac{\partial M}{\partial \theta} + \nu(e^\theta - 1)M \tag{5.2}$$

with the usual initial condition

$$M(\theta, 0) = e^{\theta j}. \tag{5.3}$$

Note that in the limit as $\nu \to 0$ (i.e. if the immigration rate vanishes) Eq. (5.2) identifies with (4.12). To solve (5.2), we consider the subsidiary equations

$$\frac{dt}{1} = \frac{d\theta}{-[\lambda(e^\theta - 1) + \mu(e^{-\theta} - 1)]} = \frac{dM}{\nu(e^\theta - 1)M}. \tag{5.4}$$

For convenience, we shall assume that $\lambda \neq \mu$ since the case $\lambda = \mu$ can then be obtained from this by a limit procedure. Since first and second expressions of (5.4) are identical to those occurring in the absence of immigration, from (4.23a) we obtain:

$$\frac{(e^\theta - 1)e^{(\lambda - \mu)t}}{\lambda e^\theta - \mu} \equiv \text{const}. \tag{5.5}$$

From the second and third expressions in (5.4) it follows:

$$\frac{dM}{M} = -\frac{\nu e^\theta}{\lambda e^\theta - \mu}\, d\theta. \tag{5.6}$$

Hence:

$$(\lambda e^\theta - \mu)^{\nu/\lambda}M = \text{const}. \tag{5.7}$$

Making use of (4.18), (5.5), and (5.7) we can write the usual general solution of (5.2) as:

$$M(\theta, t) = (\lambda e^\theta - \mu)^{-\nu/\lambda}\Psi\left[\frac{(e^\theta - 1)e^{(\lambda - \mu)t}}{\lambda e^\theta - \mu}\right]. \tag{5.8}$$

As before, the arbitrary function Ψ is specified by the initial condition (5.3). Indeed, this yields:

$$e^{\theta j}(\lambda e^\theta - \mu)^{\nu/\lambda} = \Psi\left(\frac{e^\theta - 1}{\lambda e^\theta - \mu}\right) \tag{5.9}$$

or:

$$\Psi(u) = \left(\frac{\mu u - j}{\lambda u - 1}\right)^j \left(\frac{\mu - \lambda}{\lambda u - 1}\right)^{v/\lambda} \tag{5.10}$$

having set again

$$u = \frac{e^\theta - 1}{\lambda e^\theta - \mu} \tag{5.11}$$

so that:

$$e^\theta = \frac{\mu u - 1}{\lambda u - 1}. \tag{5.12}$$

Making use of (5.10), (5.11), and (5.12), from (5.8) we finally obtain:

$$M(\theta, t) = \frac{(\lambda - \mu)^{v/\lambda}\{\mu[e^{(\lambda - \mu)t} - 1] - [\mu e^{(\lambda - \mu)t} - \lambda]e^\theta\}^j}{\{[\lambda e^{(\lambda - \mu)t} - \mu] - \lambda[e^{(\lambda - \mu)t} - 1]e^\theta\}^{j + v/\lambda}}. \tag{5.13}$$

As a special case, let us suppose that initially no individuals are present so that the first contribution to the population size comes from the immigration phenomenon. Setting $j = 0$ and $\theta = \log s$ in (5.13) we obtain the probability generating function [see Eq. (4.31)]:

$$F(s, t) = \left[\frac{\lambda - \mu}{\lambda e^{(\lambda - \mu)t} - \mu}\right]^{v/\lambda} \left\{1 - \frac{\lambda[e^{(\lambda - \mu)t} - 1]}{\lambda e^{(\lambda - \mu)t} - \mu} s\right\}^{-v/\lambda} \tag{5.14}$$

which (see, for instance, Parzen 1960) is the probability generating function of a negative binomial distribution with mean

$$\langle X(t) | X(0) = 0 \rangle = \begin{cases} \dfrac{v}{\lambda - \mu} [e^{(\lambda - \mu)t} - 1] & \text{for} \quad \lambda \neq \mu \\ vt, & \text{for} \quad \lambda = \mu. \end{cases} \tag{5.15}$$

Hence, having started with zero individuals at time zero, the mean population size grows exponentially for large t with rate $\lambda - \mu$ if $\lambda > \mu$; it grows linearly with rate equal to the immigration rate if birth and death rates are equal; it asymptotically tends to the constant value $v/(\mu - \lambda)$ if the death rate is larger than the birth rate (see Fig. 2). It is interesting to point out that in the latter case a limiting stable distribution exists. Indeed, from (5.14) it follows:

$$F(s, \infty) \equiv \lim_{t \to \infty} F(s, t) = \left(\frac{\mu - \lambda}{\mu - \lambda s}\right)^{v/\lambda}, \qquad \mu > \lambda. \tag{5.16}$$

$<x(t)\,|\,x(0)=0>$

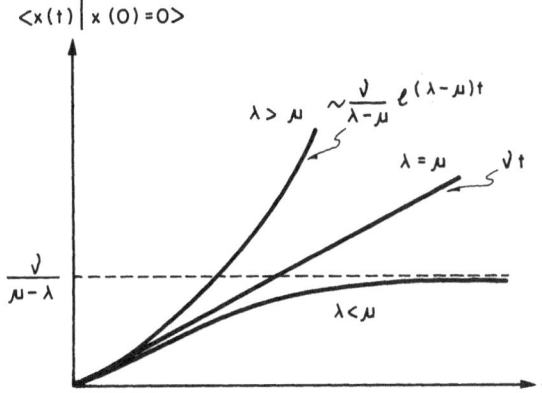

Fig. 2. Mean value, conditional on $X(0)=0$, for a population modeled as a simple birth and death process with immigration rate v

When the birth rate is zero, from (5.16) one has:

$$\lim_{\lambda\to 0} F(s,\infty) = \lim_{\lambda\to 0} \exp\left\{-\frac{v}{\lambda}\log\left[1+\frac{\lambda(1-s)}{\mu-\lambda}\right]\right\}$$

$$= \lim_{\lambda\to 0} \exp\left\{-\frac{v}{\lambda}\left[\frac{\lambda(1-s)}{\mu-\lambda}-\frac{1}{2}\frac{\lambda^2(1-s)^2}{(\mu-\lambda)^2}+\cdots\right]\right\} = e^{v/\mu(s-1)}$$

which is the probability generating function of a Poisson process with parameter v/μ. In other words, if the only contribution to population growth is due to random immigration, asymptotically the population size is a random variable Poisson distributed with mean value given by the ratio of immigration and death rates.

6. The General Birth and Death Process

The simple birth and death process discussed in Sect. 4 may be viewed as a stochastic generalization of the simple Malthusian growth law which is expressed by the differential equation

$$\frac{dx}{dt}=(\lambda-\mu)x. \tag{6.1}$$

If $x(0)=j$, that is j individuals are present at time t, from (6.1) it follows

$$x(t)=je^{(\lambda-\mu)t}. \tag{6.2}$$

Therefore, the mean population size (4.9) coincides with the deterministic population size in a Malthusian growth model in which the fertility is given by the difference $\lambda-\mu$ between birth and death rates. However, such an equality between the stochastic mean and the solution of the analogous deterministic model is by no means general. As we shall see in future examples, a discrepancy between stochastic means and deterministic solutions is usually to be expected, even though the functional form of those two quantities may at times be the same (say, both exponential, both logistic, etc.).

Let us now consider a deterministic growth model for which the fertility (i.e. the relative change of population size per unit time) is not constant as in the Malthusian law. For instance, we may suppose that the fertility is proportional to the population size $x(t)$ and write:

$$\begin{cases} \dfrac{dx}{dt} = \alpha x^2 \\ x(0) = j. \end{cases} \tag{6.3}$$

Hence, we obtain

$$x(t) = \frac{j}{1 - \alpha jt} \tag{6.4}$$

which implies

$$\lim_{t \to 1/(\alpha j)} x(t) = +\infty. \tag{6.5}$$

Due to the property (6.5) the deterministic process expressed by growth Eq. (6.3) is said to be *divergent* or *explosive* as it entails infinite growth in finite time. Is there a stochastic analogue of a divergent growth process? To answer this question, we shall now define a generalization of the simple birth process of Sect. 2. To this purpose, we use the following assumptions:

$$P\{\Delta X(t, t + \Delta t] = 1 \mid X(t) = n\} = \lambda_n \Delta t + o(\Delta t)$$

$$P\{|\Delta X(t, t + \Delta t]| > 1 \mid X(t) = n\} = o(\Delta t), \tag{6.6}$$

where $\lambda_0, \lambda_1, \lambda_2, \ldots$ are arbitrary (positive) real numbers. Proceeding as before, from (6.6) we obtain the differential equations

$$\left. \begin{array}{l} \dfrac{dp_n(t)}{dt} = \lambda_{n-1} p_{n-1}(t) - \lambda_n p_n(t) \quad (n = 1, 2, \ldots) \\[3mm] \dfrac{dp_0(t)}{dt} = -\lambda_0 p_0, \end{array} \right\} \tag{6.7}$$

where $p_n(t) = P\{X(t) = n\}$ $(n = 0, 1, 2, \ldots)$. Note that if we take $\lambda_n = \lambda n$ we recover the simple birth process of Sect. 2. In general the process described by Eq. (6.7) may turn out to be "explosive" in the sense that

$$\sum_k p_{hk}(s, t) < 1 \quad (t > s), \tag{6.8}$$

where

$$p_{hk}(s, t) = P\{X(t) = k \mid X(s) = h\} \quad (t > s). \tag{6.9}$$

Inequality (6.8) means that there is a non-zero probability, equal to $1 - \sum_k p_{hk}(s, t)$, that the process undergoes an infinite number of transitions from state h in the time interval $(t - s)$.

In the sequel we shall study under what conditions the process is not explosive. To this purpose, let us assume that in the time interval $(t, t + \Delta t]$ the population size changes from h to $h + 1$ with probability $\lambda_h \Delta t + o(\Delta t)$ and from h to $h - 1$ with probability $\mu_h \Delta t + o(\Delta t)$, the probability of any other changes being $o(\Delta t)$. Since the process is time homogeneous we have:

$$p_{hh}(\Delta t) = 1 - (\lambda_h + \mu_h)\Delta t + o(\Delta t) \tag{6.10}$$

and

$$p_{hk}(\Delta t) = \begin{cases} \lambda_h \Delta t + o(\Delta t), & \text{if } k = h+1 \\ \mu_h \Delta t + o(\Delta t), & \text{if } k = h-1 \\ o(\Delta t), & \text{if } k \neq h, k \neq h-1, k \neq h+1, \end{cases} \tag{6.11}$$

where we have set:

$$p_{hk}(\Delta t) = P\{X(t + \Delta t) = k \mid X(t) = h\}, \quad \forall h, k \in N^+. \tag{6.12}$$

Making use of first principles, the following two equations can be immediately seen to hold:

$$p_{hk}(t + \Delta t) = p_{hk}(t)[1 - (\lambda_k + \mu_k)\Delta t] + p_{h,k-1}(t)\lambda_{k-1}\Delta t$$
$$+ p_{h,k+1}(t)\mu_{k+1}\Delta t + o(\Delta t) \tag{6.13}$$

and

$$p_{hk}(t + \Delta t) = [1 - (\lambda_h + \mu_h)\Delta t]p_{hk}(t) + \lambda_h \Delta t p_{h+1,k}(t)$$
$$+ \mu_h \Delta t p_{h-1,k}(t) + o(\Delta t) \tag{6.14}$$

in which we take $p_{h,-1}(t) = p_{-1,k}(t) = 0$.

Equations (6.13) are said to be of the "forward" type since the initial state h is considered as a parameter while the next state k is variable. In Eq. (6.14), instead, the arrival state k is fixed while the starting state h is viewed as variable. For this reason (6.14) are "backward" equations.

By simple manipulations equations (6.13) and (6.14) can be re-written in the following differential form:

$$\frac{d}{dt} p_{hk}(t) = -(\lambda_k + \mu_k)p_{hk}(t) + \lambda_{k-1}p_{h,k-1}(t) + \mu_{k+1}p_{h,k+1}(t) \quad (k \geq 0)$$

$$\tag{6.15}$$

$$\frac{d}{dt} p_{hk}(t) = -(\lambda_h + \mu_h)p_{hk}(t) + \mu_h p_{h-1,k}(t) + \lambda_h p_{h+1,k}(t) \quad (h \geq 0).$$

Equations (6.15) are the basic equations for a general birth and death process as they describe the time evolution of the transition probabilities $p_{hk}(t)$. In the sequel we shall concentrate on the study of the general birth process. To this purpose we set $\mu_r = 0$ $(r \geq 0)$ in equations (6.15) while assuming $\lambda_r \geq 0$ $(r \geq 0)$. We then obtain:

$$\frac{dp_{hk}(t)}{dt} = -\lambda_k p_{hk}(t) + \lambda_{k-1} p_{h,k-1}(t) \qquad (k \geq 0), \tag{6.16}$$

and

$$\frac{dp_{hk}(t)}{dt} = -\lambda_h p_{hk}(t) + \lambda_h p_{h+1,k}(t) \qquad (h \geq 0), \tag{6.17}$$

where (6.16) are the *forward equations* and (6.17) the *backward equations*. Note that if $\lambda_0 = 0$, then if for some t it is $X(t) = 0$ the process terminates because no births occur in $(t, t + \Delta t]$. In this sense the state 0 is an "absorbing" state. Along the lines of Prabhu (1965) we shall now state and prove an important theorem which also gives a necessary and sufficient condition for the non-explosiveness of the general birth process.

Theorem 1. *The forward equations (6.16) with the λ_r all distinct have the unique solution given by*

$$p_{hk}(t) = \begin{cases} 0, & \text{for } k < h \\ \sum_{n=h}^{k} A_{hk}^{(n)} e^{-\lambda_n t}, & \text{for } k \geq h, \end{cases} \tag{6.18}$$

where the quantities $A_{hk}^{(n)}$ are specified as follows:

$$A_{hk}^{(n)} = \frac{\lambda_h \lambda_{h+1} \ldots \lambda_{k-1}}{(\lambda_h - \lambda_n)(\lambda_{h+1} - \lambda_n) \ldots (\lambda_{n-1} - \lambda_n)(\lambda_{n+1} - \lambda_n) \ldots (\lambda_k - \lambda_n)}. \tag{6.19}$$

Furthermore $\sum_k p_{hk}(t) = 1$ if and only if

$$\sum_{r=h}^{\infty} \frac{1}{\lambda_r} = +\infty. \tag{6.20}$$

Proof. We denote by $\Pi_{hk}(\tau)$ the (always existing) Laplace transform of $p_{hk}(t)$:

$$\Pi_{hk}(\tau) = \int_0^{\infty} dt\, e^{-\tau t} p_{hk}(t) \qquad (\tau > 0). \tag{6.21}$$

Since

$$\int_0^{\infty} dt\, e^{-\tau t} \frac{dp_{hk}(t)}{dt} = e^{-\tau t} p_{hk}(t) \big|_0^{\infty} + \tau \Pi_{hk}(\tau) = -\delta_{hk} + \tau \Pi_{hk}(\tau) \tag{6.22}$$

with $\delta_{hk}=0$ for $h\neq k$ and $\delta_{hk}=1$ for $h=k$, by taking the Laplace transform of the forward equation (6.16) we obtain:

$$(\tau+\lambda_k)\Pi_{hk}(\tau)=\delta_{hk}+\lambda_{k-1}\Pi_{h,k-1}(\tau) \qquad (k\geqq 0).\tag{6.23}$$

Solving these equations by iteration for $k=0,1,2,\dots$ one easily finds

$$\Pi_{hk}(\tau)=0, \quad \text{for} \quad k<h \tag{6.24}$$

and

$$\Pi_{hk}(\tau)=\frac{\lambda_h\lambda_{h+1}\dots\lambda_{k-1}}{(\tau+\lambda_h)(\tau+\lambda_{h+1})\dots(\tau+\lambda_k)}, \quad \text{for} \quad k\geqq h.\tag{6.25}$$

A partial fraction expansion of the right hand side of (6.25) then yields:

$$\Pi_{hk}(\tau)=\sum_{n=h}^{k}\frac{A_{hk}^{(n)}}{\tau+\lambda_n} \qquad (k\geqq h),\tag{6.26}$$

where the coefficients $A_{hk}^{(n)}$ are given by

$$A_{hk}^{(n)}=\lim_{\tau\to-\lambda_n}(\tau+\lambda_n)\Pi_{hk}(\tau).\tag{6.27}$$

Making use of (6.25), from (6.27) we obtain

$$\Pi_{hk}(\tau)=\sum_{n=h}^{k}\frac{\lambda_h\lambda_{h+1}\dots\lambda_{k-1}}{(\lambda_h-\lambda_n)(\lambda_{h+1}-\lambda_n)\dots(\lambda_{n-1}-\lambda_n)(\lambda_{n+1}-\lambda_n)\dots(\lambda_k-\lambda_n)}$$
$$\cdot\left(\frac{1}{\tau+\lambda_n}\right).\tag{6.28}$$

From (6.24) and (6.28) the result (6.18) then immediately follows by recalling that $(\tau+\lambda_n)^{-1}$ is the inverse Laplace transform of $e^{-\lambda_n t}$. This concludes the proof of the first part of Theorem 1. To prove the second part we sum both sides of Eq. (6.16) over k from h to an integer, that we re-name k, to obtain:

$$\frac{dQ_{hk}(t)}{dt}=-\lambda_k P_{hk}(t)\tag{6.29}$$

where we have set:

$$Q_{hk}(t)=\sum_{r=h}^{k}P_{hr}(t).\tag{6.30}$$

Now we define the Laplace transform of $Q_{hk}(t)$:

$$\tilde{Q}_{hk}(\tau)\equiv\int_0^\infty dt\,e^{-\tau t}Q_{hk}(t) \qquad (\tau>0).\tag{6.31}$$

Taking the Laplace transform of both sides of (6.29) and using the fact that $Q_{hk}(0) = 1$ we then obtain:

$$\tau \tilde{Q}_{hk}(\tau) = 1 - \lambda_k \tilde{\Pi}_{hk}(\tau). \tag{6.32}$$

On the other hand, from (6.25), we have

$$\lambda_k \tilde{\Pi}_{hk}(\tau) = \frac{1}{\left(1 + \dfrac{\tau}{\lambda_h}\right)\left(1 + \dfrac{\tau}{\lambda_{h+1}}\right) \cdots \left(1 + \dfrac{\tau}{\lambda_k}\right)}. \tag{6.33}$$

Therefore,

$$\lim_{k \to \infty} \lambda_k \tilde{\Pi}_{hk}(\tau) = \begin{cases} 0, & \text{if } \sum_{r=h}^{\infty} \dfrac{1}{\lambda_r} = \infty \\[2mm] \sum_{r=h}^{\infty} \left(1 + \dfrac{\tau}{\lambda_r}\right)^{-1}, & \text{if } \sum_{r=h}^{\infty} \dfrac{1}{\lambda_r} < \infty. \end{cases} \tag{6.34}$$

From (6.34) and (6.32) it follows that $\lim_{k \to \infty} \tau \tilde{Q}_{hk}(\tau)$ equals 1 or is less than 1 according as the series $\sum_{r=h}^{\infty} \dfrac{1}{\lambda_r}$ is divergent or convergent. Since τ^{-1} is the Laplace transform of the unit function, it follows that for all finite t

$$\lim_{k \to \infty} Q_{hk}(t) \equiv \sum_{k=h}^{\infty} P_{hk}(t) \tag{6.35}$$

equals 1 in the first case whereas it is less than 1 in the second case. This completes the proof of Theorem 1. It should be mentioned that the result (6.18) is due to Feller (1940) while criterion (6.20) is due both to Feller (1940) and Lundberg (1940). Another important theorem also due to Feller (1957) is the following.

Theorem 2. *The solution (6.18) of the forward equations of the general birth process also satisfies the backward equations. Let us denote this solution as $F_{hk}(t)$ and let $B_{hk}(t)$ be an arbitrary nonnegative solution of the backward equations. Then*

$$B_{hk}(t) \geq F_{hk}(t). \tag{6.36}$$

Proof. Proceeding as in the proof of Theorem 1 from (6.17) we obtain:

$$(\tau + \lambda_h)\tilde{B}_{hk}(\tau) = \delta_{hk} + \lambda_h \tilde{B}_{h+1,k}(\tau) \quad (h \geq 0), \tag{6.37}$$

where $\tilde{B}_{hk}(\tau)$ is the Laplace transform of $B_{hk}(t)$. Let us now impose that $\tilde{B}_{hk}(\tau) = 0$ $= \tilde{F}_{hk}(\tau)$ for $h > k$, where $\tilde{F}_{hk}(\tau)$ is the Laplace transform of $F_{hk}(t)$. Solving successively (6.37) for $h = k, k-1, \ldots$, we obtain for $h \leq k$:

$$\tilde{B}_{hk}(\tau) = \frac{\lambda_{k-1}\lambda_{k-2}\cdots\lambda_h}{(\tau + \lambda_k)(\tau + \lambda_{k-1})\cdots(\tau + \lambda_h)} = \tilde{F}_{hk}(\tau), \tag{6.38}$$

where the last equality follows from (6.25) after a straightforward interpretation of the used notation. Therefore, $\tilde{F}_{hk}(\tau)$ is also a solution of the backward equations. Now, by integration of equations (6.17) written for $p_{hk}(t) \equiv B_{hk}(t)$ we obtain:

$$B_{hk}(t) = \delta_{hk} e^{-\lambda_h t} + \lambda_h \int_0^t d\theta B_{h+1,k}(t-\theta) e^{-\lambda_h \theta}. \tag{6.39}$$

For $h > k$ we have $B_{hk}(t) \geq 0$ whereas we know that $F_{hk}(t) = 0$. Hence (6.36) is satisfied for $h > k$. On the other hand, setting $h = k$ in (6.39) we obtain

$$B_{kk}(t) \geq e^{-\lambda_k t} = F_{kk}(t) \tag{6.40}$$

with the last equality following from (6.18). Using (6.40) in (6.39) with $h = k-1$ we then obtain:

$$B_{k-1,k}(t) \geq \lambda_{k-1} \int_0^t d\theta F_{kk}(t-\theta) e^{-\lambda_{k-1}\theta} = F_{k-1,k}(t), \tag{6.41}$$

where the last equality follows from the fact that $F_{hk}(t)$ satisfies (6.39). By iteration of this procedure (6.36) follows for all h, which concludes the proof of Theorem 2.

Finally, we have the following corollary.

Corollary. *If the series (6.20) diverges, the forward and the backward equations have the unique solution given by (6.18).*

Proof. From (6.36) it follows

$$\sum_{k=0}^{\infty} B_{hk}(t) \geq \sum_{k=0}^{\infty} F_{hk}(t). \tag{6.42}$$

By Theorem 1 the divergence of the series (6.20) implies $\sum_{k=0}^{\infty} F_{hk}(t) = 1$ so that from (6.42) it follows $\sum_{k=0}^{\infty} B_{hk}(t) \geq 1$. Therefore, it must be $\sum_{k=0}^{\infty} B_{hk}(t) = 1$ which is possible if and only if the equal sign holds in (6.36) for all k and finite t.

The above considerations are limited to the general birth process. In the case when deaths also occur the forward and backward equations are given by (6.15). In this case the analysis of the uniqueness of their solutions is more cumbersome. Here we limit ourselves to mentioning the following conditions that are sufficient for the uniqueness of the solution of forward and backward equations *and* for $\sum_{k=h}^{\infty} F_{hk}(t) = 1$:

   i) $\lambda_n = 0$ for some $n \geq 1$.
This means that an upper limit to the population size is automatically placed.
   ii) $\lambda_n > 0$ for $n \geq j$ (where j is the initial population size) and

$$\sum_{n=j}^{\infty} \frac{1}{\lambda_n} = \infty. \tag{6.44}$$

This condition essentially insures that the number of births taking place in any future time interval cannot be infinitely large.

iii) $\lambda_n > 0$ for $n \geq j$ and

$$\sum_{n=j}^{\infty} \frac{\mu_n \mu_{n-1} \cdots \mu_j}{\lambda_n \lambda_{n-1} \cdots \lambda_j} = \infty . \tag{6.45}$$

Here the finiteness of the population is achieved by a suitable balance of birth and death rates.

iv) $\lambda_n > 0$ for $n \geq j$ and

$$\sum_{n=j}^{\infty} \left[\frac{1}{\lambda_n} + \frac{\mu_n}{\lambda_n \lambda_{n-1}} + \cdots + \frac{\mu_n \mu_{n-1} \cdots \mu_{j+1}}{\lambda_n \lambda_{n-1} \cdots \lambda_j} + \frac{\mu_n \mu_{n-1} \cdots \mu_j}{\lambda_n \lambda_{n-1} \cdots \lambda_j} \right] = \infty \tag{6.46}$$

which involves a combination of the factors in ii) and iii).

In conclusion, we point out explicitly that in the case of the simple birth and death process of Sect. 4 one has for all $n \geq 1$

$$\lambda_n = \lambda n$$

$$\mu_n = \mu n$$

so that $\sum \lambda_n^{-1} = \infty$ and therefore the uniqueness of the solution as well as its non-explosiveness is secured by ii).

A detailed analysis of the properties of birth and death process is due to Ledermann and Reuter (1954) and to Karlin and McGregor (1957a, b). A lucid presentation of the main results can also be found in Bailey (1964), Prabhu (1965), and Bharucha-Reid (1960).

7. Non Homogeneous Processes

In the foregoing we have consistently assumed that birth and death rates are time independent, their only possible form of dependence being on the population size. In this section we shall allow birth and death rates to be explicitly time dependent. The ensuing stochastic process will therefore be non homogeneous. First, however, we shall study into some details the so called Polya process. This is a non homogeneous birth process in which the probability of having a new individual born in time $(t, t + \Delta t]$ in a population consisting of $X(t)$ individuals is given by

$$P\{\Delta X(t) = 1 \mid X(t) = n\} = \frac{\lambda(1 + \mu n)}{1 + \lambda \mu t} \Delta t + o(\Delta t) \qquad (n = 0, 1, 2, \ldots)$$

$$P\{\Delta X(t) > 1 \mid X(t) = n\} = o(\Delta t) \tag{7.1}$$

with λ and μ (the latter not to be taken as a death rate) positive parameters. While this model has been proved useful in the treatment of electron-photon cascade problems, it is somewhat questionable whether it is of great interest for the study of the dynamics of populations of biological organisms. However, it undoubtedly provides a good example of a non homogeneous stochastic process of the birth type, so that we will study it in some detail.

Similar to what we did in Sect. 4, from (7.1) a first order partial differential equation can be derived for the moment generating function (4.11). We now have:

$$\frac{\partial M}{\partial t} = \frac{\lambda(e^{\theta}-1)}{1+\lambda\mu t}\left(M+\mu\frac{\partial M}{\partial \theta}\right). \tag{7.2}$$

Equation (7.2) can be thrown into a more convenient form by setting

$$\lambda\mu t = \tau, \quad e^{\theta}-1 = \varrho, \quad \mu\log M = S. \tag{7.3}$$

Indeed, we obtain:

$$(1+\tau)\frac{\partial S}{\partial \tau} - \varrho(1+\varrho)\frac{\partial S}{\partial \varrho} = \varrho \tag{7.4}$$

which is an equation of the type (4.14). The subsidiary equations are

$$\frac{d\tau}{1+\tau} = -\frac{d\varrho}{\varrho(1+\varrho)} = \frac{dS}{\varrho}. \tag{7.5}$$

From the first two expressions in (7.5) we obtain:

$$\frac{(1+\tau)\varrho}{1+\varrho} = c_1 \tag{7.6}$$

while from the second and third we get

$$(1+\varrho)e^{S} = c_2, \tag{7.7}$$

with c_1 and c_2 arbitrary constants. Recalling (4.18) we can write down the general solution of (7.4) in the form:

$$(1+\varrho)e^{S} = \Psi\left[\frac{(1+\tau)\varrho}{1+\varrho}\right], \tag{7.8}$$

where the arbitrary function Ψ can be determined by the initial condition. Let us assume, for simplicity, $X(0) = 0$. Then:

$$M(\theta, 0) \equiv \sum_{n=0}^{\infty} e^{\theta n}p_n(0) = 1 \tag{7.9}$$

so that for $t=0$ from (7.3) and (7.9) we obtain $\tau=S=0$. From (7.8) it then follows:

$$\Psi\left(\frac{\varrho}{1+\varrho}\right)=1+\varrho \qquad (7.10)$$

i.e.

$$\Psi(u)=\frac{1}{1-u}, \qquad (7.11)$$

where we have set

$$\frac{\varrho}{1+\varrho}=u, \qquad \varrho=\frac{u}{1-u}. \qquad (7.12)$$

From (7.8), (7.11), and (7.12) we thus obtain

$$(1+\varrho)e^S=\frac{1+\varrho}{1-\varrho\tau}, \qquad (7.13)$$

or

$$S=-\log(1-\varrho\tau). \qquad (7.14)$$

Passing to the old variables via (7.3) we get the moment generating function

$$M(\theta,t)=[1-\lambda\mu t(e^\theta-1)]^{-1/\mu} \qquad (7.15)$$

from which the probability generating function $F(s,t)$ can also be written down:

$$F(s,t)\equiv M(\log s,t)=[(1+\lambda\mu t)-\lambda\mu t s]^{-1/\mu} \qquad (7.16)$$

showing that $X(t)$ has again a negative binomial distribution with individual probabilities

$$p_n(t)\equiv P\{X(t)=n\,|\,X(0)=0\}$$
$$=\frac{(\lambda t)^n}{n!}(1+\lambda\mu t)^{-(n+1/\mu)}\prod_{r=1}^{n-1}(1+r\mu) \qquad (n=1,2,\ldots). \qquad (7.17)$$

Therefore,

$$\langle X(t)\,|\,X(0)=0\rangle=\lambda t$$
$$\mathrm{Var}\{X(t)\,|\,X(0)=0\}=\lambda t(1+\lambda\mu t) \qquad (7.18)$$

showing that the mean population size linearly increases with time. Instead the variance only for small times grows roughly in a linear fashion being otherwise steadily above such levels. Hence, for small times this growth process behaves as a Poisson process. In the more interesting case $X(0)=j>0$ one can also describe the growth process, the only difference now being that (7.11) should read:

$$\Psi(u) = \left(\frac{1}{1-u}\right)^{1+j\mu}.$$

(7.19)

Therefore:

$$S = \log \frac{(1+\varrho)^{j\mu}}{(1-\varrho\tau)^{1+j\mu}}$$

(7.20)

and hence:

$$M(\theta, t) = e^{\theta j}[1 - \lambda\mu t(e^\theta - 1)]^{-\frac{1+j\mu}{\mu}}$$

(7.21)

and:

$$F(s, t) \equiv M(\log s, t) = s^j[(1 + \lambda\mu t) - \lambda\mu t s]^{-\frac{1+j\mu}{\mu}}.$$

(7.22)

By direct differentiation of (7.21) it is easy to show that mean and variance of the population size are now given by

$$\langle X(t) \mid X(0)=j\rangle = j + (1+j\mu)\lambda t,$$

(7.23)

and

$$\text{Var}\{X(t) \mid X(0)=j\} = (1+j\mu)(1+\lambda\mu t)\lambda t,$$

(7.24)

respectively. Hence, the difference between population size and its mean is approximately Poisson distributed when t is small.

We now turn to a generalization of the simple birth and death process considered in Sect. 4 and assume that birth and death rates are both time dependent:

$$\lambda_n = \lambda(t)n,$$
$$\mu_n = \mu(t)n.$$

(7.25)

It is then straightforward to see that in place of Eqs. (4.3) we now have:

$$\frac{dp_n(t)}{dt} = -n[\mu(t)+\lambda(t)]p_n(t) + (n-1)\lambda(t)p_{n-1}(t)$$
$$+ (n+1)\mu(t)p_{n+1}(t) \quad (n=1, 2, \ldots),$$
$$\frac{dp_0(t)}{dt} = \mu(t)p_1(t)$$

(7.26)

with the initial condition

$$p_n(0) = \begin{cases} 1, & n=j \\ 0, & n \neq j. \end{cases} \tag{7.27}$$

Hence, the equation for the moment generating function reads:

$$\frac{\partial M}{\partial t} = [\lambda(t)(e^\theta - 1) + \mu(t)(e^{-\theta} - 1)] \frac{\partial M}{\partial \theta} \tag{7.28}$$

which is an obvious generalization of Eq. (4.12). From (7.28) we can immediately write the equation for the probability generating function $F(s, t)$ defined in (4.31):

$$\frac{\partial F}{\partial t} = (s-1)[\lambda(t)s - \mu(t)] \frac{\partial F}{\partial s} \tag{7.29}$$

with initial condition

$$F(s, 0) \equiv \sum_{n=0}^{\infty} p_n(0)s^n = s^j. \tag{7.30}$$

To solve (7.29) we consider the subsidiary equations

$$\frac{dt}{1} = -\frac{ds}{(s-1)[\lambda(t)s - \mu(t)]} = \frac{dF}{0}. \tag{7.31}$$

The first and third quantities yield

$$F = c_1 \qquad (c_1 \text{ a constant}) \tag{7.32}$$

while from the first and the second we obtain the differential equation

$$\frac{ds}{dt} = (s-1)[\mu(t) - \lambda(t)s], \tag{7.33}$$

or:

$$\frac{d\sigma}{dt} = \lambda(t) - [\mu(t) - \lambda(t)]\sigma \tag{7.34}$$

having performed the transformation

$$s - 1 = \frac{1}{\sigma}. \tag{7.35}$$

Multiplying both sides of (7.34) by the integrating factor $e^{\phi(t)}$, where

$$\phi(t) = \int_0^t d\tau[\mu(\tau) - \lambda(\tau)], \tag{7.36}$$

and integrating with respect to t we easily obtain:

$$\sigma e^{\phi(t)} - \int_0^t d\tau \lambda(\tau) e^{\phi(\tau)} = c_2 \quad (c_2 \text{ a constant}).$$ (7.37)

Switching back to s via (7.35), from (7.37) we obtain the second integral

$$\frac{e^{\phi(t)}}{s-1} - \int_0^t d\tau \lambda(\tau) e^{\phi(\tau)} = c_2 .$$ (7.38)

From (7.32) and (7.38) the general solution of (7.29) follows:

$$F(s,t) = \Psi \left\{ \frac{e^{\phi(t)}}{s-1} - \int_0^t d\tau \lambda(\tau) e^{\phi(\tau)} \right\},$$ (7.39)

where Ψ is the arbitrary function to be determined by the initial condition (7.30). It is a simple matter to show that the functional form of Ψ is the following:

$$\Psi(u) = \left(1 + \frac{1}{u} \right)^j .$$ (7.40)

Hence, from (7.39) we obtain the probability generating function:

$$F(s,t) = \left\{ 1 + \left[\frac{e^{\phi(t)}}{s-1} - \int_0^t d\tau \lambda(\tau) e^{\phi(\tau)} \right]^{-1} \right\}^j$$ (7.41)

with $\phi(t)$ given by (7.36). It is now easy to see that the right hand side of (7.41) can be re-written as the right hand side of (4.32). It follows that an expansion of $F(s,t)$ in powers of s is again possible so that the individual probabilities $p_n(t)$ can be explicitly written down. One thus obtains:

$$p_0(t) = \left[1 - \frac{1}{e^{\phi(t)} + H(t)} \right]^j ,$$ (7.42a)

$$p_n(t) = \sum_{r=0}^{\min(j,n)} \binom{j}{r} \binom{j+n-r-1}{j-1} \left[1 - \frac{1}{e^{\phi(t)} + H(t)} \right]^{j-r}$$

$$\cdot \left[\frac{H(t)}{e^{\phi(t)} + H(t)} \right]^{n-r} \left[\frac{1 - H(t)}{e^{\phi(t)} + H(t)} \right]^r \quad (n = 1, 2, \ldots),$$ (7.42b)

where we have set

$$H(t) = \int_0^t d\tau \lambda(\tau) e^{\phi(\tau)}$$ (7.43)

and where $\phi(t)$ is given by (7.36).

Equation (7.42a) can be used to investigate the chance of extinction of the population. To this purpose we note that

$$e^{\phi(t)} + H(t) \equiv e^{\phi(t)} + \int_0^t d\tau \lambda(\tau) e^{\phi(\tau)}$$

$$= e^{\phi(t)} + \int_0^t d\tau \mu(\tau) e^{\phi(\tau)} - \int_0^t d\tau [\mu(\tau) - \lambda(\tau)] e^{\phi(\tau)}$$

$$= e^{\phi(t)} + \int_0^t d\tau \mu(\tau) e^{\phi(\tau)} - e^{\phi(\tau)} \big|_0^t$$

$$= 1 + \int_0^t d\tau \mu(\tau) e^{\phi(\tau)} \tag{7.44}$$

so that (7.42a) can be re-written as:

$$p_0(t) = \left[\frac{\int_0^t d\tau \mu(\tau) e^{\phi(\tau)}}{1 + \int_0^t d\tau \mu(\tau) e^{\phi(\tau)}} \right]^j . \tag{7.45}$$

Hence

$$\lim_{t \to \infty} p_0(t) = 1 \tag{7.46}$$

if and only if

$$\lim_{t \to \infty} \int_0^t d\tau \mu(\tau) e^{\phi(\tau)} = \infty . \tag{7.47}$$

We conclude by mentioning that the moments of the population size can be obtained directly by differentiation of the moment generating function $M(\theta, t) \equiv F(e^\theta, t)$ given by (7.41). A straightforward calculation yields:

$$\langle X(t) | X(0) = j \rangle = j \exp \left\{ \int_0^t d\tau [\lambda(\tau) - \mu(\tau)] \right\} \tag{7.48}$$

and

$$\text{Var}\{X(t) | X(0) = j\} = j \exp \left\{ 2 \int_0^t d\tau [\lambda(\tau) - \mu(\tau)] \right\}$$

$$\cdot \int_0^t d\tau [\lambda(\tau) + \mu(\tau)] \exp \left\{ \int_0^\tau d\sigma [\mu(\sigma) - \lambda(\sigma)] \right\} . \tag{7.49}$$

8. The Logistic Process

We shall conclude our discussion of birth and death processes with some remarks on the so called "logistic" process. This was introduced by Prendville (1948) as an example of a regulated birth and death process in which the state space is confined to a preassigned strip $0 \leq n_1 \leq n \leq n_2$ by a suitable dependence of birth and death rates on the population size n. This process was thoroughly analyzed by Takashima (1957); a brief survey of it is also present in Iosifescu and Tautu (1973) but unfortunately their text is marred by some misleading misprints. Hence, we shall derive anew the quantities of interest by making use of the procedure already implemented in the previous sections.

The logistic process can be defined by assuming that birth rate b_n per *individual* and death rate d_n per *individual* in a population consisting of n organisms are given by

$$b_n = \begin{cases} \alpha \left(\dfrac{n_2}{n} - 1 \right), & \text{for} \quad 0 < n_1 \leq n \leq n_2 \\ \\ 0, & \text{otherwise} \end{cases} \tag{8.1}$$

and

$$d_n = \begin{cases} \beta \left(1 - \dfrac{n_1}{n} \right), & \text{for} \quad 0 < n_1 \leq n \leq n_2 \\ \\ 0, & \text{otherwise} \end{cases} \tag{8.2}$$

respectively, where α and β are positive constants and $n_1, n_2 \in N^+$ and $n_1 < n_2$. Assumptions (8.1) and (8.2) express the statement that as the population size n approaches the upper limit n_2 the birth rate per individual decreases to zero while the death rate per individual approaches its maximum value given by $\beta \left(1 - \dfrac{n_1}{n_2} \right)$. Vice versa, as n decreases to the lower limit n_1 birth and death rates per individual tend to $\alpha \left(\dfrac{n_2}{n_1} - 1 \right)$ and 0, respectively.

Let us now denote by $p_{jn}(t)$ the transition probability of the process $X(t)$ representing the population size, with $X(0) = j$ being the initial number of individuals in the population:

$$p_{jn}(t) = P\{X(t) = n \mid X(0) = j\} . \tag{8.3}$$

Making use of the first of Eqs. (6.15), setting in it $h = j$, $k = n$, $\lambda_n = nb_n$, and $\mu_n = nd_n$, and recalling that it must be $p_{jn}(t) = 0$ for $n < n_1$ and $n > n_2$ we obtain the following evolution equations for the transition probabilities of the logistic process:

$$\frac{dp_{j,n_1}(t)}{dt} = -\alpha(n_2 - n_1)p_{j,n_1}(t) + \beta p_{j,n_1+1}(t), \tag{8.4a}$$

$$\frac{dp_{jn}(t)}{dt} = -[\alpha n_2 - \beta n_1 + (\beta - \alpha)n]p_{jn}(t)$$
$$+ \beta(n+1-n_1)p_{j,n+1}(t) + \alpha(n_2-n+1)p_{j,n-1}(t)$$
$$(n = n_1+1, n_1+2, ..., n_2-1), \tag{8.4b}$$

$$\frac{dp_{j,n_2}(t)}{dt} = -\beta(n_2 - n_1)p_{j,n_2}(t) + \alpha p_{j,n_2-1}(t). \tag{8.4c}$$

Let us denote by $F(s, t)$ the probability generating function:

$$F(s, t) = \sum_{n=n_1}^{n_2} P_{jn}(t) s^n \tag{8.5}$$

and write a partial differential equation for it. To this purpose, we multiply both sides of (8.4b) by s^n and sum from $n_1 + 1$ to $n_2 - 1$. Using the identity

$$\sum_{n=n_1+1}^{n_2-1} \frac{dp_{jn}(t)}{dt} s^n = \sum_{n=n_1}^{n_2} \frac{dp_{j,n}(t)}{dt} s^n - s^{n_1} \frac{dp_{j,n_1}(t)}{dt}$$

$$- s^{n_2} \frac{dp_{j,n_2}(t)}{dt} \tag{8.6}$$

and Eqs. (8.4a) and (8.4c), a rather cumbersome procedure finally yields

$$\frac{\partial F(s, t)}{\partial t} + (s-1)(\alpha s + \beta) \frac{\partial F(s, t)}{\partial s} = (s-1)\left(\alpha n_2 + \frac{\beta n_1}{s}\right) F(s, t). \tag{8.7}$$

To determine $F(s, t)$ we consider the subsidiary equations:

$$\frac{dt}{1} = \frac{ds}{(s-1)(\alpha s + \beta)} = \frac{dF}{(s-1)\left(\alpha n_2 + \dfrac{\beta n_1}{s}\right) F}. \tag{8.8}$$

From the first and second quantities we easily obtain:

$$dt = \frac{1}{\alpha + \beta} \left[\frac{1}{s-1} - \frac{\alpha}{\alpha s + \beta} \right] ds. \tag{8.9}$$

Hence:

$$\frac{s-1}{\alpha s + \beta} e^{-(\alpha+\beta)t} = c_1 \quad (c_1 \text{ a constant}). \tag{8.10}$$

From the second and third quantities in (8.8) we then get

$$\frac{dF}{F} = \frac{1}{\alpha s + \beta} \left[\alpha n_2 + \frac{\beta n_1}{s} \right] ds \tag{8.11}$$

or, more conveniently,

$$\frac{dF}{F} = \left[n_2 \frac{\alpha}{\alpha s + \beta} + \frac{n_1}{s} - n_1 \frac{\alpha}{\alpha s + \beta} \right] ds \tag{8.12}$$

whose general solution is:

$$s^{-n_1} (\alpha s + \beta)^{n_1 - n_2} F(s, t) = c_2 \quad (c_2 \text{ a constant}). \tag{8.13}$$

Making use of (8.10) and (8.13) we can now write down the general solution of Eq. (8.7):

$$F(s,t) = s^{n_1}(\alpha s + \beta)^{n_2 - n_1} \Psi \left[\frac{s-1}{\alpha s + \beta} e^{-(\alpha+\beta)t} \right].$$ (8.14)

To determine the arbitrary function Ψ we make use of the initial condition

$$F(s,0) = \sum_{n=n_1}^{n_2} p_{jn}(0) s^n = s^j$$ (8.15)

to obtain:

$$\Psi(u) = (1 + \beta u)^{j-n_1}(1 - \alpha u)^{n_2 - j}(\alpha + \beta)^{n_1 - n_2},$$ (8.16)

where we have set

$$u = \frac{s-1}{\alpha s + \beta}, \qquad s = \frac{1 + \beta u}{1 - \alpha u}.$$ (8.17)

From (8.14) and (8.16) it finally follows:

$$\begin{aligned} F(s,t) = \;& \frac{s^{n_1}}{(\alpha + \beta)^{n_2 - n_1}} \\ & \cdot \{[\alpha + \beta e^{-(\alpha+\beta)t}]s + \beta[1 - e^{-(\alpha+\beta)t}]\}^{j-n_1} \\ & \cdot \{\alpha[1 - e^{-(\alpha+\beta)t}]s + [\beta + \alpha e^{-(\alpha+\beta)t}]\}^{n_2 - j}. \end{aligned}$$ (8.18)

We shall now make use of expression (8.18) to determine mean and variance of the population size. To this purpose, we remark that from (8.5) it follows:

$$\langle X(t) | X(0) = j \rangle \equiv \sum_{n=n_1}^{n_2} n p_{jn}(t) = \left[\frac{\partial F(s,t)}{\partial s} \right]_{s=1}$$ (8.19)

and

$$\begin{aligned} \mathrm{Var}\{X(t) | X(0) = j\} &\equiv \sum_{n=n_1}^{n_2} n^2 p_{jn}(t) - \langle X(t) | X(0) = j \rangle^2 \\ &= \left\{ \frac{\partial^2 F(s,t)}{\partial s^2} + \frac{\partial F(s,t)}{\partial s} - \left[\frac{\partial F(s,t)}{\partial s} \right]^2 \right\}_{s=1}. \end{aligned}$$ (8.20)

From (8.18), (8.19), and (8.20) after a fairly large amount of calculations the following expressions for the mean and the variance are found:

$$\begin{aligned} \langle X(t) | X(0) = j \rangle = \;& \frac{1}{\alpha + \beta} \{(\alpha n_2 + \beta n_1) \\ & - [(\alpha n_2 + \beta n_1) - j(\alpha + \beta)] e^{-(\alpha+\beta)t}\} \end{aligned}$$ (8.21)

and

$$\text{Var}\{X(t)\,|\,X(0)=j\} = \frac{\alpha\beta(n_2-n_1)}{(\alpha+\beta)^2}$$
$$+ \frac{\beta-\alpha}{\alpha+\beta}\left(j - \frac{\alpha n_2 + \beta n_1}{\alpha+\beta}\right)e^{-(\alpha+\beta)t}$$
$$+ \left[\frac{\beta^2 n_1 - \alpha^2 n_2}{(\alpha+\beta)^2} - \frac{j(\beta-\alpha)}{\alpha+\beta}\right]e^{-2(\alpha+\beta)t}. \tag{8.22}$$

We can now take the limit of (8.18) as $t\to\infty$ to investigate whether an equilibrium regime is asymptotically reached by the population. The answer is yes since

$$\lim_{t\to\infty} F(s,t) = \frac{s^{n_1}}{(\alpha+\beta)^{n_2-n_1}}(\alpha s + \beta)^{n_2-n_1}$$
$$\equiv \frac{s^{n_1}}{(\alpha+\beta)^{n_2-n_1}}\sum_{k=0}^{n_2-n_1}\binom{n_2-n_1}{k}\alpha^k\beta^{n_2-n_1-k}s^k. \tag{8.23}$$

Re-writing the last expression on the right hand side of (8.23) as

$$F(s,\infty) = \sum_{n=n_1}^{n_2}\left\{\frac{1}{(\alpha+\beta)^{n_2-n_1}}\binom{n_2-n_1}{n-n_1}\alpha^{n-n_1}\beta^{n_2-n}\right\}s^n \tag{8.24}$$

and comparing it with (8.5) written for $t=\infty$ it follows:

$$P_{jn}(\infty)\equiv P_n = \frac{1}{(\alpha+\beta)^{n_2-n_1}}\binom{n_2-n_1}{n-n_1}\alpha^{n-n_1}\alpha^{n_2-n} \qquad (n_1\leq n\leq n_2), \tag{8.25}$$

whereas it is $p_n=0$ for $n<n_1$ and $n>n_2$. Note that the probabilities (8.25) are independent of the initial population size j. From (8.25), or more directly from (8.21) and (8.22), the asymptotic mean and variance of the population can be obtained:

$$\langle X(\infty)\rangle \equiv \lim_{t\to\infty}\langle X(t)\,|\,X(0)=j\rangle = \frac{\alpha n_2 + \beta n_1}{\alpha+\beta} \tag{8.26}$$

and

$$\text{Var}\{X(\infty)\} = \lim_{t\to\infty}\text{Var}\{X(t)\,|\,X(0)=j\} = \frac{\alpha\beta(n_2-n_1)}{(\alpha+\beta)^2}. \tag{8.27}$$

9. Concluding Remarks

In the previous sections we have considered birth and death processes and we have shown by simple straightforward arguments that equations can be written down for the transition probabilities $p_{hk}(t)$. Actually all this is part of a more general theory about which we want to say a little here (see also Fisz, 1963).

Let $\{X(t); 0 \le t < \infty\}$ be a Markov process with at most a countable number of states, say $0, 1, 2, \ldots$ and let

$$p_{hk}(\tau, t) = P\{X(t) = k \mid X(\tau) = h\} \quad (h, k = 0, 1, 2, \ldots) \tag{9.1}$$

be the transition probabilities from state h at time τ to state k at time t $(t > \tau)$. We then have:

i) $\quad p_{hk}(\tau, t) \ge 0$, \hfill (9.2a)

ii) $\quad p_{hk}(\tau, t) = \sum_r p_{hr}(\tau, s) p_{rk}(s, t)$, \hfill (9.2b)

where s is any arbitrary instant such that $\tau < s < t$. Equation (9.2b) is known as the Chapman-Kolmogorov equation expressing the statement that the probability of the transition from state h at time τ to state k at time t can be written, due to the Markov property, as the sum of the probabilities of transitions from h at time τ to any other state r at an intermediate time s, followed by the transition from state r at time s to state k at time t. Let us now assume that

$$\lim_{\Delta t \to 0} \frac{1}{\Delta t} [1 - p_{hh}(t, t + \Delta t)] = q_h(t) \quad (h = 0, 1, 2, \ldots) \tag{9.3}$$

and

$$\lim_{\Delta t \to 0} \frac{1}{\Delta t} p_{hk}(t, t + \Delta t) = q_{hk}(t) \quad (h, k = 0, 1, 2 \ldots) \tag{9.4}$$

the convergence being uniform in t. For small Δt we then have the following asymptotic equalities:

$$p_{hh}(t, t + \Delta t) \sim 1 - q_h(t) \Delta t + o(\Delta t) \tag{9.5}$$

and

$$p_{hk}(t, t + \Delta t) \sim q_{hk}(t) \Delta t + o(\Delta t). \tag{9.6}$$

The functions $q_h(t)$ and $q_{hk}(t)$ are often called "intensity functions". The following theorem then holds.

Theorem. *Let $p_{hk}(\tau, t)$ be the probability transition function of a Markov process with at most a countable number of states $h = 0, 1, 2, \ldots$. Suppose that the intensity functions $q_h(t)$ and $q_{hk}(t)$ exist and are continuous. Then:*

i) *The functions* $p_{hk}(\tau, t)$ *satisfy the system of differential equations*

$$\frac{\partial p_{hk}(\tau, t)}{\partial \tau} = q_h(\tau) p_{hk}(\tau, t) - \sum_{r \neq h} q_{hr}(\tau) p_{rk}(\tau, t) \tag{9.7}$$

with

$$p_{hk}(t, t) = \begin{cases} 1, & h = k \\ 0, & h \neq k. \end{cases} \tag{9.8}$$

ii) *If, moreover, the convergence in* (9.4) *is uniform in h for fixed k, then the functions* $p_{hk}(\tau, t)$ *satisfy the system of differential equations*

$$\frac{\partial p_{hk}(\tau, t)}{\partial t} = -p_{hk}(\tau, t) q_k(t) + \sum_{r \neq k} p_{hr}(\tau, t) q_{rk}(t) \tag{9.9}$$

with

$$p_{hk}(\tau, \tau) = \begin{cases} 1, & h = k \\ 0, & h \neq k. \end{cases} \tag{9.10}$$

Here we shall limit ourselves to proving ii). To this purpose, we re-write (9.2b) setting in it $t = t + \Delta t$ and $s = t$:

$$p_{hk}(\tau, t + \Delta t) = \sum_r p_{hr}(\tau, t) p_{rk}(t, t + \Delta t). \tag{9.11}$$

Hence:

$$\frac{p_{hk}(\tau, t + \Delta t) - p_{hk}(\tau, t)}{\Delta t} = \frac{1}{\Delta t} \left[\sum_r p_{hr}(\tau, t) p_{rk}(t, t + \Delta t) - p_{hk}(\tau, t) \right]$$

$$= \sum_{r \neq k} p_{hr}(\tau, t) \frac{p_{rk}(t, t + \Delta t)}{\Delta t} - p_{hk}(\tau, t) \frac{1 - p_{kk}(t, t + \Delta t)}{\Delta t}. \tag{9.12}$$

Passing to the limit as $\Delta t \to 0$ and making use of (9.3) and (9.4) from (9.12) Eq. (9.9) follows. Note that use of the absolute convergence of $\sum_r p_{hr}$ has been made. Similar considerations lead to the proof of (9.7). For obvious reasons equation (9.7) is called "backward" while Eq. (9.9) is said to be "forward" [see also Eq. (6.15)]. Now, it can be shown that backward and forward equations have identical solutions $p_{hk}(\tau, t)$ satisfying (9.2a), (9.2b), (9.8), and (9.10) and that the corresponding intensity functions exist and are continuous. The proof is in Kolmogorov (1931) for the homogeneous case, i.e., when $p_{hk}(\tau, t) = p_{hk}(t - \tau)$. The nonhomogeneous case is instead considered in Feller (1936). Note that if the process is homogeneous (such as the birth and death processes introduced in Sect. 1 through 7) one has

$$q_h(t) = q_h$$
$$q_{hk}(t) = q_{hk} .$$

(9.13)

Equations (9.7) and (9.9) then read:

$$-\frac{dp_{hk}(t)}{dt} = q_h p_{hk}(t) - \sum_{r \neq h} q_{hr} p_{rk}(t)$$

(9.14)

and

$$\frac{dp_{hk}(t)}{dt} = -p_{hk}(t)q_k + \sum_{r \neq k} p_{hr}(t)q_{rk} ,$$

(9.15)

respectively, while initial conditions (9.8) and (9.10) become:

$$p_{hk}(0) = \begin{cases} 1, & h = k \\ 0, & h \neq k . \end{cases}$$

(9.16)

Moreover, for homogeneous processes existence and finiteness of the intensity functions q_h and q_{hk} does not need to be assumed in all cases. For instance, if

$$\lim_{t \to 0} p_{hh}(t) = 1 \quad (h = 0, 1, ...)$$

(9.17)

(continuity at zero) then the existence of q_h and q_{hk} is insured. Moreover, if convergence is uniform in h, then q_h and q_{hk} are finite. In the particular case when the number of states of $X(t)$ is finite, from (9.17) the finiteness of q_h and q_{hk} follows. Further considerations and rigorous proofs of the statements can be found, for instance, in Doob (1942).

Acknowledgements

I wish to thank Dr. Amelia G. Nobile for several interesting discussions.

References

Arley, N. (1940). On the Theory of Stochastic Processes and their Application to the Theory of Cosmic Ray Radiation. New York: Wiley

Bailey, N. (1964). The Elements of Stochastic Processes with Applications to the Natural Sciences. New York: Wiley

Bharucha-Reid, A.T. (1960). Elements of the Theory of Markov Processes and Their Applications. New York: McGraw

Doob, J.L. (1942). Topics on the theory of Markov chains, Trans. Amer. Math. Soc. *52*, 37–64

Feller, W. (1936). Zur Theorie des stochastischen Prozesses, Math. Ann. *113*, 113–160

Feller, W. (1938). Die Grundlagen der Volterraschen Theorie des Kampfes ums Dasein in wahrscheinlichkeitstheoretischer Behandlung, Acta Biotheoretica *5*, 11–40

Feller, W. (1945). On the integro-differential equations of purely discontinuous Markoff processes, Trans. Amer. Math. Soc. *48*, 488–515. Errata ibid. *58*, 474

Feller, W. (1957). An Introduction to Probability Theory and its Applications, Vol. 1 (2nd Ed.). New York: Wiley

Fisz, M. (1963). Probability Theory and Mathematical Statistics. New York: Wiley

Furry, W.H. (1937). On fluctuation phenomena in the passage of high energy electrons through lead, Phys. Rev. *52*, 569

Iosifescu, M., Tautu, P. (1973). Stochastic Processes and Applications in Biology and Medicine, Vol. II. Berlin: Springer-Verlag

Karlin, S., McGregor, J.L. (1957a). The differential equations of birth and death processes and the Stiltjes moment problem, Trans. Amer. Math. Soc. *85*, 489–546

Karlin, S., McGregor, J.L. (1957b). The classification of birth and death processes, Trans. Amer. Math. Soc. *86*, 366–400

Kendall, D.G. (1948). On the generalized birth and death process, Ann. Math. Statist. *19*, 1–15

Kolmogorov, A. (1931). Über die analytischen Methoden in der Wahrscheinlichkeitsrechnung, Math. Ann. *104*, 415–458

Ledermann, W., Reuter, G.E.H. (1954). Spectral theory for the differential equations of simple birth and death processes, Phil. Trans. Roy. Soc. London A*246*, 321–369

Lundberg, O. (1940). On Random Processes and their Application to Stickness and Accident Statistics. Stockholm: Uppsula

Parzen, E. (1965). Modern Probability Theory and its Applications. New York: The Macmillan Company

Prabhu, N.U. (1965). Stochastic Processes. Basic Theory and its Applications. New York: The Macmillan Company

Prendville, B.J. (1949). Discussion (Symposium on Stochastic Processes), J. Roy. Statist. Soc. *11*, 273

Takashima, M. (1957). Note on evolutionary processes, Bull. Math. Statist. *7*, 18–24

Yule, U. (1924). A mathematical theory of evolution based on the conclusions of Dr. J.C.Willis, F.R.S., Phil. Trans. B*213*

Stochastic Population Theory: Diffusion Processes

*Luigi M. Ricciardi**

1. Introduction

In the previous contribution we discussed birth and death processes as models of populations subject to random growth. There, the population size at each instant was represented as a discrete random variable labeled by the considered instant. Since the probabilistic description was characterized by a straightforward integration, we purposely avoided spending time on definitions and mathematical preliminaries. However, it is sometimes convenient to model population growth by continuous differential equations and by their stochastic counterparts. This implies that the population size at each instant can be any nonnegative real number or else a continuous space-continuous time stochastic process. It is to be stressed that this is evidently an approximation which, however, may prove useful in making inferences about global properties of the population dynamics such as stability, extinction, etc.

In the case of continuous space-continuous state approximations the stochastic framework is no longer self-explanatory. Hence, some mathematical preliminaries are necessary. In the sequel we shall briefly review some definitions and properties of stochastic processes with a special view of the description of Markov and diffusion processes. This is motivated by the particularly important role played by these processes in population dynamics modeling. The literature on this subject is too vast to be recalled here. We limit ourselves to mentioning the books by Karlin and Taylor (1975, 1981) where a comprehensive treatment of stochastic processes can be found as well as numerous applications to biologically interesting instances.

2. Stochastic Processes: Preliminaries

Consider an experiment specified by:
a) the experiment's outcomes, ξ, forming the space S;
b) certain subsets of S (called *events*) and by the probabilities of these events.
Let us now associate with every outcome ξ, according to a certain rule, a real function, $X(t, \xi)$, of the time variable. Thus doing we have constructed a family of

---

* Work performed under CNR-JSPS Cooperation Programme, Contracts No. 83.00032.01, 85.00002.01, and No. 84.00227.01 and under MPI financial suffort

functions, one for each ξ. This family of functions is called a *stochastic process*, or a *random function*.

A stochastic process can thus be viewed as a function of two variables, t and ξ. The domain of ξ is the set S of all possible outcomes or states of the experiment, often called the state space of the process. This space may consist of a discrete set of points, in which case our process is denoted as a discrete state process; or of a continuum of states, in which case we have a continuous state process. The domain of t is a set of real numbers, which likewise may be discrete or continuous. Unless specified otherwise, we shall assume that the domain of t is the entire time axis, i.e., the real line. ·

In summary, the stochastic process $X(t, \xi)$ has the following properties:
1. For a specific outcome ξ_i, $X(t, \xi_i)$ is a single deterministic function of time. This time function is called a *realization* or *sample path* of the stochastic process.
2. For a specific time t_i, $X(t_i, \xi)$ is a quantity depending on ξ, i.e., it is a random variable.
3. $X(t_i, \xi_j)$ is a number.

In the following, a stochastic process will simply be denoted by $X(t)$, i.e., we shall omit the explicit specification of the dependence on ξ. However, this dependence is to be always implicitly assumed.

From the above considerations we see (cf. Fig. 1) that $X(t)$ may denote four different things:
1. A family of functions of time (t and ξ variables)
2. A single function of time (t variable and ξ fixed)
3. A random variable (t fixed and ξ variable)
4. A single number (t fixed and ξ fixed).

In the foregoing we have considered *one-dimensional* stochastic processes. However, the above definition can be easily extended to the general case of an

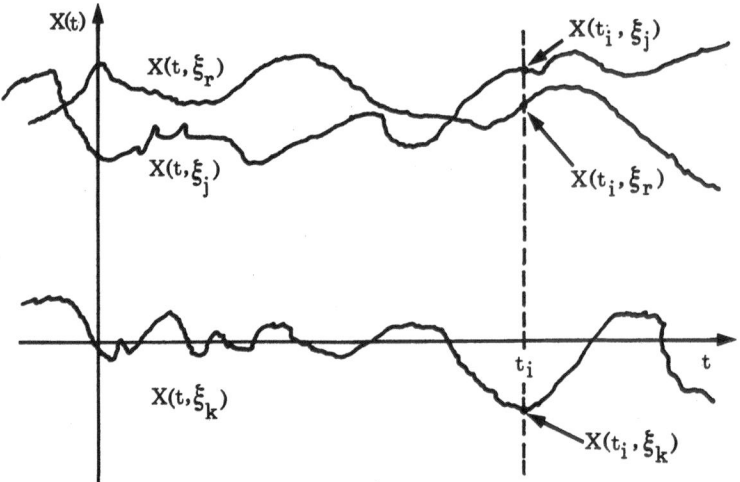

Fig. 1. Illustrating the definition of stochastic process. The time functions $X(t, \xi_r)$, $X(t, \xi_j)$, and $X(t, \xi_k)$ are sample paths. $X(t_i, \xi)$, where ξ takes the values ξ_r, ξ_j, and ξ_k, is a random variable labelled by the chosen instant t_i

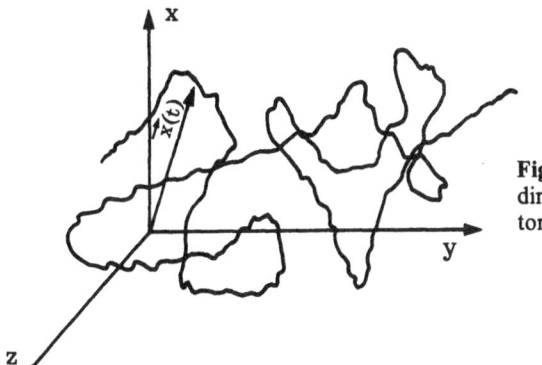

Fig. 2. One sample path of the three-dimensional process depicting the trajectory of a colloidal particle

n-dimensional stochastic process. Instead of $X(t, \xi)$, we would consider the *vector function* $\mathbf{X}(t, \xi)$, or simply $\mathbf{X}(t)$, where each component of this vector is a one-dimensional stochastic process.

Example 1. Let $\mathbf{X}(t)$ be the displacement (from an arbitrarily fixed origin) of a colloidal particle subject to the impacts of molecules in the surrounding medium. A specific outcome of the underlying experiment is the selection of a colloidal particle and $\mathbf{X}(t)$ determines the trajectory of the particle. This trajectory looks very irregular and cannot be described by a formula. Moreover, if one knows $X(t)$ up to a certain time t_1, one cannot determine the process at times $t > t_1$ (cf. Fig. 2). We shall return to this particular process later on. For now, it suffices to remark that this is a three-dimensional stochastic process in continuous time with a continuum of states.

It is easy to come up with other examples of stochastic processes. Probably the most commonly encountered ones deal with fluctuations due to *noise* in physical devices and biological structures such as thermal noise, shot noise, membrane noise, etc. Depending upon the physical experiment and our model of it, we may have one of four types of one-dimensional stochastic processes: discrete or continuous time variable, and discrete or continuous state variable.

3. Markov Processes

How much information about a stochastic process do we need before we can, at least in principle, answer all questions concerning it? We now address this problem, known as "process specification" or "probability law determination."

Consider a stochastic process $X(t)$ and, for sake of argument, let $X(t)$ be continuous in both state and time (the other cases follow similarly). For n fixed instants $t_1, t_2, ..., t_n$, the values of $X(t_i)$, $(i = 1, 2, ..., n)$ determine a family of random variables whose joint, or n^{th} order, p.d.f. $f_n[X(t_1); X(t_2); ...; X(t_n)]$ is assumed to be known if $X(t)$ is completely specified. Conversely, if we know the nth order p.d.f., f_n, for *all* n and *all* t_i, then $X(t)$ is specified since, for instance, any statistical averages of $X(t)$ can be calculated. By choosing the instants t_i sufficiently close to one another, we can replace $X(t)$ by a sequence of random variables, $X(t_i)$, with an accuracy satisfactory for most practical purposes. Thus $X(t)$ can be characterized by an infinite sequence of p.d.f.'s of increasing order, namely

$$f_1[X(t_1)], \quad f_2[X(t_1); X(t_2)], \dots, f_n[X(t_1); X(t_2); \dots; X(t_n)], \dots .$$

Of course, these densities are not all independent since the joint p.d.f.'s corresponding to different values of n are connected via the *compatibility relations* (marginal densities):

$$\int dX(t_j) f_n[X(t_1); X(t_2); \dots; X(t_j); \dots; X(t_n)] = f_{n-1}[X(t_1); \dots; X(t_{j-1}),$$
$$X(t_{j+1}); \dots; X(t_n)], \quad (1 \leq j \leq n), \tag{3.1}$$

where the integral on the l.h.s. is over the entire domain of the random variable $X(t_j)$. Thus, in general, we are confronted with the impossible task of determining the functions $f_n[X(t_1); \dots; X(t_n)]$ for *all* n and t_i. However, the situation improves a great deal when dealing with the class of processes known as *Markov processes*, also called *processes without aftereffect*. A process $X(t)$ is Markov if for all n and $t_1 < t_2 < \dots < t_n$ we have:

$$f[X(t_n)|X(t_{n-1}); X(t_{n-2}); \dots; X(t_1)] = f[X(t_n)|X(t_{n-1})], \tag{3.2}$$

where the conditional p.d.f. $f[\dots|\dots]$ is defined (cf. Bayes theorem) as

$$f[X(t_n)|X(t_{n-1}); \dots; X(t_1)] = \frac{f_n[X(t_n); X(t_{n-1}); \dots; X(t_1)]}{f_{n-1}[X(t_{n-1}); X(t_{n-2}); \dots; X(t_1)]}. \tag{3.3}$$

The function on the r.h.s. of (3.2) is called, for evident reasons, the *transition* p.d.f. of the process $X(t)$. In general, this will be a function of four variables: the pair $x_n = X(t_n)$ and $x_{n-1} = X(t_{n-1})$, as well as t_n and t_{n-1}. Note that (3.2) will certainly be satisfied by a "zero-memory" process such as the Poisson process. In fact, the Markov property may be viewed as a restriction on the process' memory: Given the "present," $X(t_{n-1})$, the "future," $X(t_n)$, is independent of the "past," $X(t_i)$ ($i < n-1$).

A consequence of property (3.2) is that a Markov process is completely specified (in the sense discussed earlier) if one knows the first order or *univariate* p.d.f., $f_1[X(t)]$, *and the transition p.d.f.*, $f[X(t)|X(\tau)]$. Indeed,

$$f_n[X(t_1); X(t_2); \dots; X(t_n)] = f_{n-1}[X(t_{n-1}); X(t_{n-2}); \dots; X(t_1)]$$
$$\times f[X(t_n)|X(t_{n-1}); \dots; X(t_1)]$$
$$= f[X(t_n)|X(t_{n-1})] f[X(t_{n-1})|X(t_{n-2})] \dots f[X(t_2)|X(t_1)] f_1[X(t_1)],$$

where the last equality follows from the assumed Markov character of the process. It is worth remarking explicitly that the knowledge of $f_1[X(t)]$ and $f[X(t)|X(\tau)]$ is equivalent to the knowledge of the second order p.d.f. or *bivariate* function $f_2[X(t); X(\tau)]$.

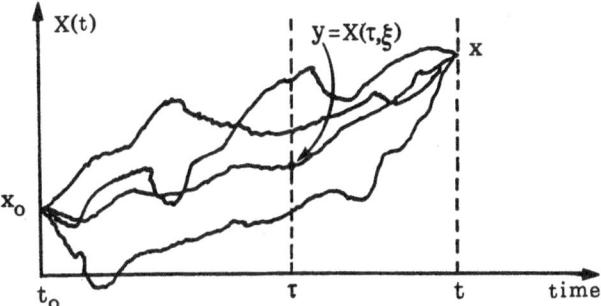

Fig. 3. Illustrating the unfolding of the transition p.d.f. $f(x, t|x_0, t_0)$ as expressed by Eq. (3.4)

Another very important class of processes, also completely specified by the bivariate function $f_2[X(t); Y(\tau)]$ are *Gaussian* or *normal* processes. A process is called Gaussian if the functions $f_n[X(t_1); X(t_2); \dots; X(t_n)]$ are *normal* for all n and $t_1, t_2, \dots, t_n$. As we shall see later on, such a multidimensional Gaussian density function is completely specified by the mean values $E[X(t_i)]$ $(i = 1, 2, \dots, n)$ and by the covariance function,

$$K(t, \tau) = E\{[X(t) - m(t)] \cdot [X(\tau) - m(\tau)]\} = R(t, \tau) - m(t)m(\tau),$$

where $R(t, \tau) = E[X(t)X(\tau)]$ is the so-called correlation function. On the other hand, to determine $E[X(t_i)]$ and $K(t, \tau)$ all one needs to know is the process' bivariate p.d.f. It is worth recalling that this definition does not imply that the higher order moments of a Gaussian process are all zero. We shall have more to say about Gaussian processes in the future.

While both Markov and Gaussian processes are specified by their bivariate density function, the two properties, Markovity and normality, are quite distinct: a process may be both, neither, or either. Unless stated otherwise, from now on we shall deal exclusively with Markov processes.

The above considerations indicate that a problem of primary interest in describing a stochastic process is the determination of its transition p.d.f. A first step in this direction consists of writing an integral equation for it, known as the Smolukowski or Chapman-Kolmogorov equation. To derive it (see Fig. 3) we let $t_0 < \tau < t$ be three otherwise arbitrary instants and set $x_0 = X(t_0)$, $y = X(\tau)$, and $x = X(t)$. Then we have, via the compatibility relations (3.1):

$$f_2(x, t; x_0, t_0) = \int dy f_3(x, t; y, \tau; x_0, t_0),$$

where the customary more explicit notations $f_2[X(t); X(t_0)] = f_2(x, t; x_0, t_0)$ and $f_3[X(t); X(\tau); X(t_0)] = f_3(x, t; y, \tau; x_0, t_0)$ have been adopted. Using Bayes' theorem (3.3), we thus have with similar notations:

$$f(x, t|x_0, t_0) f_1(x_0, t_0) = \int dy f(x, t|y, \tau; x_0, t_0) f_2(y, \tau; x_0, t_0).$$

Using finally the Markov property (3.2), and simplifying, we obtain the Smolukowski equation

$$f(x, t|x_0, t_0) = \int dy f(x, t|y, \tau) f(y, \tau|x_0, t_0).$$ (3.4)

As Eq. (3.4) shows, we can view the transition from state x_0 at time t_0 to state x at time t as a two stage transition passing through any intermediate state y at an arbitrarily selected time τ. The probability for the process to pass from $X(t_0)$ to $X(t)$ is thus written as the sum of the probabilities pertaining to the transitions $X(t_0) \rightarrow X(\tau)$ and, subsequently, $X(\tau) \rightarrow X(t)$.

Equation (3.4) has been derived for a process continuous in time and state. However, it is not difficult to prove, by quite similar arguments, that analogous equations hold for the other three types of processes. In all cases, such equations are called Smolukowski or Chapman-Kolmogorov equations. For instance, for a process discrete both in state and time, for all $n > m > 0$ we have:

$$P(X_n = k|X_0 = j) = \sum_i P(X_m = i|X_0 = j)P(X_n = k|X_m = i),$$ (3.5)

where, for example, $P\{X_n = k|X_m = i\}$ denotes the probability for the process to attain state k at time n, conditional upon being in state i at time m. For a process discrete in time and continuous in state starting in state x_0 at time 0, the Smolukowski equation instead reads:

$$f(x, n|x_0, 0) = \int dy f(x, n|y, m) f(y, m|x_0, 0),$$ (3.6)

where the notation used is self-explanatory and $n > m > 0$. Finally, for a process discrete in state and continuous in time starting in state j at time u, we have:

$$p_{jk}(u, t) = \sum_i p_{ji}(u, v)p_{ik}(v, t) \quad (t > v > u),$$ (3.7)

where $p_{jk}(u, t)$ is the conditional probability of going from state j at time u to state k at time t. The latter case contains, for instance, the birth and death processes considered earlier.

It should be noted that these Smolukowski equations have been written by assuming that the underlying processes originate at some "initial" state with certainty. Consequently, the corresponding transition functions are expected to satisfy suitable initial conditions reflecting the specification of the process' initial value. For the discrete-state cases, described by Eqs. (3.5) and (3.7), the initial conditions are:

$$P(X_0 = k|X_0 = j) = p_{jk}(u, u) = \delta_{jk},$$

δ_{jk} is the Kronecker symbol defined as:

$$\delta_{jk} = \begin{cases} 1, & \text{for } j = k \\ 0, & \text{for } j \neq k. \end{cases}$$

In the continuous state case described by Eq. (3.4) the appropriate initial condition is instead:

$$\lim_{t \to t_0} f(x, t|x_0, t_0) = \delta(x - x_0),$$
(3.8)

where $\delta(x)$ is the Dirac delta-function:

$$\delta(x) = \begin{cases} 0, & x \neq 0 \\ \infty, & x = 0 \end{cases}, \quad \text{and} \quad \int_a^b \delta(x)dx = 1, \quad \forall a < 0 < b.$$

Note that for $t_0 = 0$, relation (3.8) yields the appropriate initial condition for the continuous state-discrete time process described by Eq. (3.6).

While the interpretation of the initial condition for the discrete state cases is straightforward, initial condition (3.8) is probably best understood by thinking of $\delta(x)$ as the limit case of an infinitely peaked Gaussian p.d.f. so that at the initial time the probability for the process to take any other value other than x_0 is zero.

4. Kinetic Equations

Consider a Markov process with a continuum of state values in continuous time. Its transition p.d.f., $f(x, t|x_0, t_0)$, was shown in Sect. 3 to satisfy the Smolukowski equation:

$$f(x, t|x_0, t_0) = \int dy f(x, t|y, \tau) f(y, \tau|x_0, t_0)$$
(4.1)

with $t > \tau > t_0$ arbitrary instants and $X(t) = x$, $X(\tau) = y$, $X(t_0) = x_0$. Equation (4.1) is to be looked at as a compatibility relation holding for any Markov process, but it is not sufficient to determine the process' transition p.d.f. To accomplish this task, further assumptions besides the Markov assumption are necessary, as we shall soon see.

However, first let us re-write Eq. (4.1) in a differential form. With this purpose in mind, let us change in Eq. (4.1) t into $t + \Delta t$ and τ into t, where Δt is a small, but finite, temporal increment. Subtracting $f(x, t|x_0, t_0)$ from both sides, we then obtain:

$$f(x, t + \Delta t|x_0, t_0) - f(x, t|x_0, t_0) = \int dy f(x, t + \Delta t|y, t) f(y, t|x_0, t_0)$$
$$- f(x, t|x_0, t_0).$$
(4.2)

Let us now consider an arbitrary function $R(x)$ vanishing at the end points of the state space sufficiently rapidly together with its derivatives of all orders. Multiplying both sides of Eq. (4.2) by $R(x)/\Delta t$ and integrating over the state space, we obtain:

$$\int dx R(x) \frac{f(x, t + \Delta t|x_0, t_0) - f(x, t|x_0, t_0)}{\Delta t}$$

$$= \frac{1}{\Delta t} \int dx\, R(x) \int dy f(x, t + \Delta t|y, t) f(y, t|x_0, t_0)$$

$$- \frac{1}{\Delta t} \int dx\, R(x) f(x, t|x_0, t_0).$$
(4.3)

Substituting the Taylor expansion about the point y for $R(x)$ in the first integral on the r.h.s. of Eq. (4.3):

$$R(x) = R(y) + \sum_{n=1}^{\infty} \frac{d^n R(y)}{dy^n} \frac{(x-y)^n}{n!} \tag{4.4}$$

and taking the limit as $\Delta t \to 0$, we obtain:

$$\int dx\, R(x) \frac{\partial f}{\partial t} = \lim_{\Delta t \to 0} \frac{1}{\Delta t} \int dy\, R(y) f(y, t|x_0, t_0) \int dx f(x, t + \Delta t|y, t)$$

$$+ \sum_{n=1}^{\infty} \frac{1}{n!} \int dy \left\{ \frac{d^n R(y)}{dy^n} f(y, t|x_0, t_0) \right.$$

$$\times \lim_{\Delta t \to 0} \frac{1}{\Delta t} \int dx (x-y)^n f(x, t + \Delta t|y, t) \Big\}$$

$$- \lim_{\Delta t \to 0} \frac{1}{\Delta t} \int dx\, R(x) f(x, t|x_0, t_0) \tag{4.5}$$

or:

$$\int dx\, R(x) \frac{\partial f}{\partial t} = \sum_{n=1}^{\infty} \frac{1}{n!} \int dx \frac{d^n R(x)}{dx^n} f(x, t|x_0, t_0) A_n(x, t) \tag{4.6}$$

having used the "normalization condition":

$$\int dx f(x, t + \Delta t|y, t) = 1 \tag{4.7}$$

and having set:

$$A_n(x, t) = \lim_{\Delta t \to 0} \frac{1}{\Delta t} \int dy (y-x)^n f(y, t + \Delta t|x, t), \quad (n = 1, 2, \ldots). \tag{4.8}$$

An integration by parts of the r.h.s. of Eq. (4.6), in which the vanishing of $R(x)$ and its derivatives at the ends of integration interval is used, shows that:

$$\int dy \frac{d^n R(y)}{dy^n} f(y, t|x_0, t_0) A_n(y, t)$$

$$= (-1)^n \int dx\, R(x) \frac{\partial^n}{\partial x^n} [A_n(x, t) f(x, t|x_0, t_0)]. \tag{4.9}$$

Equation (4.6) thus yields:

$$\int dx\, R(x) \left\{ \frac{\partial f}{\partial t} - \sum_{n=1}^{\infty} \frac{(-1)^n}{n!} \frac{\partial^n}{\partial x^n} [A_n(x, t) f(x, t|x_0, t_0)] \right\} = 0. \tag{4.10}$$

Now due to the arbitrariness of the function $R(x)$ the bracketed terms must be identically zero, and we have our desired result:

$$\frac{\partial f(x, t | x_0, t_0)}{\partial t} = \sum_{n=1}^{\infty} \frac{(-1)^n}{n!} \frac{\partial^n}{\partial x^n} [A_n(x, t) f(x, t | x_0, t_0)]. \tag{4.11}$$

This is the so-called *kinetic equation* or differential form of the Smolukowski equation (4.1), holding under the sole assumption that the process under consideration is Markov.

The functions $A_n(x, t)$ defined by (4.8) are called *infinitesimal moments* of the process. The process is temporally homogeneous if the transition p.d.f. only depends on the difference between the present time and the initial time:

$$f(y, t + \Delta t | x, t) = f(y, \Delta t | x, 0). \tag{4.12}$$

We thus see from (4.8) that in this case the infinitesimal moments do not depend on the time variable.

Upon setting:

$$y - x \equiv X(t + \Delta t) - X(t) = \Delta x \tag{4.13}$$

we see from (4.8) that for Δt sufficiently small

$$E\{(\Delta x)^n | X(t) = x\} \simeq A_n(x, t) \Delta t \quad (n = 1, 2, \ldots). \tag{4.14}$$

This relation tells that the moments of the process' increment over a small time interval Δt, conditional upon $X(t) = x$, are proportional to the time interval itself. In particular,

$$\begin{aligned} E\{\Delta x | X(t) = x\} &= A_1(x, t) \Delta t, \\ E\{(\Delta x)^2 | X(t) = x\} &= A_2(x, t) \Delta t \end{aligned} \tag{4.15}$$

so that the conditional variance of the process' increment is:

$$\sigma^2\{\Delta x | X(t) = x\} \simeq A_2(x, t) \Delta t - [A_1(x, t)]^2 (\Delta t)^2. \tag{4.16}$$

Therefore:

$$\lim_{\Delta t \to 0} \frac{\sigma^2\{\Delta x | X(t) = x\}}{\Delta t} = A_2(x, t). \tag{4.17}$$

This is the reason why the second order infinitesimal moment $A_2(x, t)$ is customarily called *infinitesimal variance*. The first order moment

$$A_1(x, t) = \lim_{\Delta t \to 0} \frac{E\{\Delta x | X(t) = x\}}{\Delta t} \tag{4.18}$$

is the so-called *drift* of the process.

The differential form (4.11) of the Smolukowski equation involves the derivatives of the transition p.d.f. with respect to the present time t and the present state x, whereas the initial time t_0 and initial state x_0 appear as parameters. However, as we shall now see, the role of the present and initial variables can be interchanged in the sense that one can derive another differential form of the Smolukowski equation in which the present variables appear as parameters. With this in mind, after setting in (4.1) $t_0 = t_0 - \Delta t$ and $\tau = t_0$, we obtain:

$$f(x, t|x_0, t_0 - \Delta t) = \int dy f(y, t_0|x_0, t_0 - \Delta t) f(x, t|y, t_0),\tag{4.19}$$

where now $x = X(t)$, $x_0 = X(t_0 - \Delta t)$, $y = X(t_0)$ and $\Delta t > 0$ is a small increment. Making use of the identity

$$f(x, t|x_0, t_0) = \int dy f(x, t|x_0, t_0) f(y, t_0|x_0, t_0 - \Delta t)\tag{4.20}$$

Equation (4.19) can be re-written as:

$$
\begin{aligned}
&f(x, t|x_0, t_0 - \Delta t) - f(x, t|x_0, t_0)\\
&\quad = \int dy f(y, t_0|x_0, t_0 - \Delta t)\left[f(x, t|y, t_0) - f(x, t|x_0, t_0)\right],
\end{aligned}\tag{4.21}
$$

or:

$$
\begin{aligned}
&f(x, t|x_0, t_0 - \Delta t) - f(x, t|x_0, t_0)\\
&\quad = \frac{\partial^n f(x, t|x_0, t_0)}{\partial x_0^n} \sum_{n=1}^{\infty} \frac{1}{n!} \int dy f(y, t_0|x_0, t_0 - \Delta t)(y - x_0)^n
\end{aligned}\tag{4.22}
$$

having expanded $f(x, t|y, t_0)$ in the r.h.s. of Eq. (4.21) as a Taylor series about x_0. Dividing both sides of (4.22) by $(-\Delta t)$ and taking the limit as $\Delta t \to 0$, we finally obtain:

$$\frac{\partial f(x, t|x_0, t_0)}{\partial t_0} + \sum_{n=1}^{\infty} \frac{A_n(x_0, t_0)}{n!} \frac{\partial^n f(x, t|x_0, t_0)}{\partial x_0^n} = 0,\tag{4.23}$$

where $A_n(x_0, t_0)$, $(n = 1, 2, ...)$ are the infinitesimal moments earlier defined. Equation (4.23) is the preannounced alternative differential expansion of the Smolukowski equation.

In conclusion, the transition p.d.f. of a Markov process satisfies the kinetic equations (4.11) and (4.23). The former is a *forward* equation as the initial variables are fixed; the latter is a *backward* equation in the sense that it describes the unfolding of the process that leads to a preassigned state at the present time.

5. Diffusion Equations

We now raise the question: What do we do with the above kinetic equations derived for a Markov process? The answer is simple enough: Not much! Because of the presence of arbitrarily high order derivatives with respect to the state variable, Eqs. (4.11) and (4.23) are not very helpful in determining the transition p.d.f. of the process. This situation is greatly improved if from a certain n all the infinitesimal moments vanish. Indeed, when this is the case we are confronted with partial differential equations whose solutions can be searched by standard analytical or numerical techniques. An important, and a priori unsuspected, result due to Pawula (1967) provides a specification of the order of the equation when only a finite number of infinitesimal moments are non vanishing. Indeed, Pawula proved that if the infinitesimal moments $A_n(x, t)$ exist for all n, the vanishing of any even order infinitesimal moments implies $A_n(x, t) = 0$ for $n \geq 3$.

This implies that whenever the kinetic equation contains a finite number of derivatives, it is an equation of order 2 at most. Let us first examine the case in which $A_2 = A_3 = \ldots = 0$, i.e., when only the first order derivative with respect to x survives. In this case the transition p.d.f. of the process satisfies

$$\frac{\partial f}{\partial t} + \frac{\partial}{\partial x}[A_1(x, t)f] = 0 \tag{5.1}$$

as well as

$$\frac{\partial f}{\partial t_0} + A_1(x_0, t_0)\frac{\partial f}{\partial x_0} = 0 \tag{5.2}$$

with the initial condition (3.8). One thus finds:

$$f(x, t|x_0, t_0) = \delta\left[x - \int_{t_0}^{t} A_1(x_0, \tau)d\tau - x_0\right], \tag{5.3}$$

telling us that the process is deterministic as all sample paths coincide with the function

$$x(t) = \int_{t_0}^{t} A_1(x_0, \tau)d\tau + x_0. \tag{5.4}$$

We now come to the case of when $A_2(x, t) > 0$ and $A_r(x, t) = 0$ for $r > 2$. In this case (4.11) becomes the so-called *Fokker-Planck*, or *forward*, equation:

$$\frac{\partial f}{\partial t} = -\frac{\partial}{\partial x}[A_1(x, t)f] + \frac{1}{2}\frac{\partial^2}{\partial x^2}[A_2(x, t)f] \tag{5.5}$$

whereas Eq. (4.23) simplifies into the so-called *Kolmogorov* or *backward* equation:

$$\frac{\partial f}{\partial t_0} + A_1(x_0, t_0)\frac{\partial f}{\partial x_0} + \frac{1}{2}A_2(x_0, t_0)\frac{\partial^2 f}{\partial x_0^2} = 0. \tag{5.6}$$

Equations (5.5) and (5.6) are also called *diffusion equations*. In numerous instances they can be used to determine the transition p.d.f. of the Markov process under study, and hence (cf. Sect. 3) to achieve its description. However, at a first glance one may feel that we are trapped in a tautological argument. Indeed, we have derived equations (5.5) and (5.6) with a view to solve them and thus obtain the unknown transition p.d.f. $f(x, t|x_0, t_0)$. On the other hand, these equations are useful only if the drift and infinitesimal variance of the process are known which, due to definitions (4.8), seems to require that we already possess the transition p.d.f. However, this is not the case. All we really need to know in order to determine the drift and infinitesimal variance is the probability density function of the increment of the process over a small time interval Δt, which can often be easily accomplished as we shall see later in some specific instances. Furthermore, in many cases Eqs. (5.5) and (5.6) directly arise as the limit of difference equations and in others the drift and infinitesimal variance can be directly evaluated when the process is due to random fluctuations of a parameter appearing in an ordinary differential equation. We can therefore safely regard Eqs. (5.5) and (5.6) as completely specified. The problem is then how to solve them in order to determine the unknown function $f(x, t|x_0, t_0)$. Clearly this cannot be identified with any solutions of the diffusion equations. Being a transition p.d.f. it must be non negative for all choices of initial and present variables; as $t \to t_0$ it must become a delta-function [recall that at the initial time the whole probability mass is concentrated at $x_0 = X(t_0)$] and, unless there is a "leak" somewhere, the probability mass must be conserved over the state space, i.e.:

$$\int f(x, t|x_0, t_0)dx = 1 , \tag{5.7}$$

where the integral is over the entire state space interval.

Let us now re-write the Fokker-Planck equation (5.5) in the form

$$\frac{\partial f}{\partial t} + \frac{\partial j}{\partial x} = 0 , \tag{5.8}$$

having set:

$$j(x, t|x_0, t_0) = A_1(x, t)f(x, t|x_0, t_0)$$

$$-\tfrac{1}{2}\frac{\partial}{\partial x}[A_2(x, t)f(x, t|x_0, t_0)] . \tag{5.9}$$

Equation (5.8) can be viewed as the equation of "conservation of probability" because it has the form of the equations of mass transport or diffusion involving a flow of mass. Indeed, interpreting $j(x, t)$ as the "probability current" density, i.e., as the amount of probability crossing the abscissa x in the positive direction per unit time, we have for any interval (y, z):

$j(y, t|x_0, t_0)\Delta t \simeq$ amount of probability *entering* (y, z) across y in Δt,

$j(z, t|x_0, t_0)\Delta t \simeq$ amount of probability *leaving* (y, z) across z in Δt.

Therefore, $[j(y, t|x_0, t_0) - j(z, t|x_0, t_0)]\Delta t$ is roughly the increment of the total probability in the interval (y, z):

$$[j(y, t|x_0, t_0) - j(z, t|x_0, t_0)]\Delta t$$
$$\simeq \int_y^z f(x, t + \Delta t|x_0, t_0)dx - \int_y^z f(x, t|x_0, t_0)dx . \tag{5.10}$$

Hence:

$$\frac{j(y, t|x_0, t_0) - j(z, t|x_0, t_0)}{z - y}$$
$$\simeq \frac{1}{\Delta t} \int_y^z \frac{f(x, t + \Delta t|x_0, t_0) - f(x, t|x_0, t_0)}{z - y} dx$$
$$\simeq \frac{f(v, t + \Delta t|x_0, t_0) - f(w, t|x_0, t_0)}{\Delta t}$$

with v and w in (y, z). In the limit as $z \to y$ and $\Delta t \to 0$, we thus find

$$-\frac{\partial j(y, t|x_0, t_0)}{\partial y} = \frac{\partial f(y, t|x_0, t_0)}{\partial t}$$

which coincides with Eq. (5.8) after setting $y = x$. Therefore, just as the equations of heat conduction or transport of mass involve a flow of heat or mass, Eq. (5.8), and therefore the Fokker-Planck equation, involves a flow of "probability mass." We can thus picture a diffusion process $X(t)$ as the ensemble of trajectories (of a very irregular form) described by representative points undergoing diffusion. At time $t = t_0$ all points are at x_0; then they start spreading about in such a way that the relative number of points in any interval (y, z) at any time t corresponds to the quantity

$$\int_y^z f(x, t|x_0, t_0)dx .$$

Let us now assume that f is non zero over an interval (the "diffusion interval") (a, b), finite or infinite, and zero outside it. Then, the whole probability mass is confined to this interval:

$$\int_a^b f(x, t|x_0, t_0)dx = 1, \quad \forall t . \tag{5.11}$$

Equation (5.8) then yields:

$$\int_a^b \frac{\partial j(x, t|x_0, t_0)}{\partial x} dx = 0 \tag{5.12}$$

or:

$$j(a, t|x_0, t_0) = j(b, t|x_0, t_0) . \tag{5.13}$$

If, however, there is no flow of representative points across the boundaries, the stronger condition

$$j(a, t|x_0, t_0) = j(b, t|x_0, t_0) = 0 \tag{5.14}$$

holds. This means that whenever a representative point reaches an end of the diffusion interval its trajectory terminates and no representative points can enter the diffusion interval. This situation quite naturally arises if $(a, b) \equiv (-\infty, +\infty)$ in which case one usually also expects that

$$\lim_{x \to \pm \infty} f(x, t|x_0, t_0) = 0. \tag{5.15}$$

6. The "Stationary" Case

Let us consider a homogeneous diffusion process, i.e. one with a stationary p.d.f.:

$$f(x, t|x_0, t_0) = f(x, t - t_0|x_0, 0) \equiv f(x, t - t_0|x_0) \tag{6.1}$$

and ask whether this transition p.d.f. has limit as $t \to \infty$. Usually this limit is a probability density function, $W(x)$, which does not depend on the initial state x_0. Therefore, the process attains a condition of statistical equilibrium whereby time does not play any role and the dependence on the initial state is forgotten. $W(x)$ is commonly called *steady state* distribution, even though it is a p.d.f. Eq. (5.8) now becomes:

$$\frac{\partial W}{\partial t} \equiv 0 = \frac{\partial j_0}{\partial x}. \tag{6.2}$$

This implies that the stationary probability density current j_0 is a constant. Therefore, (5.9) becomes an ordinary differential equation for $W(x)$:

$$\frac{d}{dx}[A_2(x)W(x)] - 2A_1(x)W(x) + 2j_0 = 0, \tag{6.3}$$

whose general solution can be written as:

$$W(x) = \frac{c}{A_2(x)} \exp\left\{2\int^x \frac{A_1(y)}{A_2(y)} dy\right\}$$
$$- \frac{2j_0}{A_2(x)} \int_z^x \exp\left\{-2\int^w \frac{A_1(y)}{A_2(y)} dy\right\} dw \cdot \exp\left[2\int_z^x \frac{A_1(y)}{A_2(y)} dy\right], \tag{6.4}$$

where c and z are arbitrary. Note, however, that changing z is equivalent to changing c so that only c and j_0 can be regarded as arbitrary constants. If there is no flow of probability at the boundaries, then $j_0 = 0$ [cf. (5.14)] and the constant c is determined by imposing

$$\int\limits_a^b W(x)dx = 1,$$ (6.5)

where (a, b) is the diffusion interval. The result is:

$$W(x) = \frac{c}{A_2(x)} \exp\left\{ 2\int\limits_z^x \frac{A_1(y)}{A_2(y)} dy \right\}$$ (6.6)

with

$$c = \left\{ \int\limits_a^b dx [A_2(x)]^{-1} \exp\left\{ 2\int\limits_z^x \frac{A_1(y)}{A_2(y)} dy \right] \right\}^{-1}.$$ (6.7)

If $c \neq 0$, (6.6) gives the steady state distribution. Note that this is expressed in terms of integrals involving the drift and infinitesimal variance which are known if the Fokker-Planck or the Kolmogorov equations are given. Without having to solve these equations, we can thus determine the stationary properties of the process. To determine its "transient" behaviour one instead needs to solve the diffusion equations in order to find the transition p.d.f., which is, in general, a rather complicated task.

We now notice that due to Eq. (6.1) we can write

$$\frac{\partial f}{\partial t_0} = -\frac{\partial f}{\partial t}.$$ (6.8)

Then setting

$$a(x) = \tfrac{1}{2} A_2(x),$$
$$b(x) = A_1(x),$$ (6.9)

the diffusion equations (5.6) and (5.7) read:

$$\frac{\partial f}{\partial t} = \frac{\partial^2}{\partial x^2} [a(x)f] - \frac{\partial}{\partial x} [b(x)f],$$ (6.10a)

$$\frac{\partial f}{\partial t} = a(x_0) \frac{\partial^2 f}{\partial x_0^2} + b(x_0) \frac{\partial f}{\partial x_0}.$$ (6.10b)

The problem of integrating these equations is complicated by the fact that $a(x)$ may vanish or $b(x)$ may become singular as x approaches some value x^*. This value is said to be a singular point, in either case, for the underlying diffusion process. If no such point exists (which occurs, for instance, if the infinitesimal variance and drift are constants), then Eqs. (6.10) are defined for x ranging over the entire interval $(-\infty, \infty)$. The transition p.d.f. can then be determined by solving either equation with the initial condition

$$\lim_{t \downarrow 0} f(x, t|x_0) = \delta(x - x_0). \tag{6.11}$$

This *uniquely* determines the transition p.d.f. However, when the process is singular, condition (6.11) does not suffice and, in fact, one is not even free to choose which of the two diffusion equations to solve in order to find f. The integration problem for singular diffusion equations has been solved by Feller (1952; 1954), but with the aid of too high a level of mathematics to be reproduced here. Nevertheless, the conclusions of Feller's study can be simply enough reported. With this objective in mind, let us consider an interval $I \equiv (r_1, r_2)$ with $-\infty \leqq r_1 < r_2 \leqq \infty$ and let us assume (which is safe for all the cases of practical interest) that the functions $a(x)$, $a'(x)$, and $b(x)$ are continuous for $x \in I$, with the infinitesimal variance $a(x) > 0$ inside (r_1, r_2). This means that the only possible singular points are the end points r_1 and r_2 of I, the "diffusion interval". Denoting by r_i the generical end point, it may occur that $X(t)$ never attains, for finite t, the value r_i. In this case r_i is said to be an *inaccessible boundary*. Otherwise, r_i is said to be *accessible*. A finer classification partitions accessible boundaries into *exit* boudaries and *regular* boundaries, whereas inaccessible ones can be either *entrance* boundaries or *natural* boundaries. Also this finer classification reflects specific features of $X(t)$. When r_i is regular, nothing particular happens to $X(t)$ in the neighborhood of r_i. Thus if we wish to confine the whole probability mass to (r_1, r_2), the initial condition (6.11) must be complemented with boundary conditions, one for each regular boundary. An exit boundary is, instead, one such that a probability flow exists from the inside of the diffusion interval onto the point r_i whereas no probability flow takes place from r_i into the diffusion interval. In other words, r_i acts as an "absorbing boundary" in the sense that the process' sample paths starting at x_0 ($r_1 < x_0 < r_2$) terminate as soon as they attain the value r_i. As for inaccessible boundaries, r_i is entrance if r_i is not attainable from the inside of (r_1, r_2) whereas starting from r_i the process can take values in (r_1, r_2). Finally, natural boundaries are the ones characterized by zero probability flow. Therefore, they are inaccessible from the inside of the diffusion intervals and if some probability mass is initially assigned to them, it stays there forever without being able to spread into the diffusion interval. The following diagram summarizes the boundary classification.

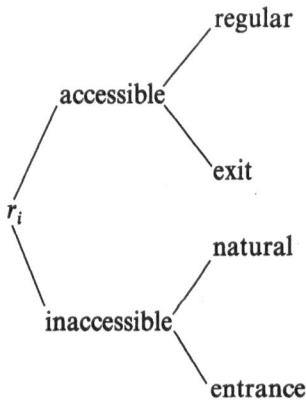

The interest in this classification rests upon the fact, proved by Feller, that if one knows the nature of the boundaries of the diffusion interval one can decide what kind of boundary conditions, if any, have to be associated with the diffusion equations in order to determine the transition p.d.f. of the process. Of course, if such a procedure is to be of practical usefulness, one should be able to establish the nature of the boundaries without having to know preliminarily the process' transition p.d.f. This is actually the case. Indeed, as proved by Feller, the classification of the boundaries depends solely upon certain integrability properties of the coefficients $a(x)$ and $b(x)$ of the diffusion equations. More specifically, let x' be a point contained inside the diffusion interval $I=(r_1, r_2)$ and let us introduce the following four functions:

$$f(x) = \exp\left[-\int_{x'}^{x} \frac{b(z)}{a(z)} dz \right],$$

$$g(x) = [a(x)f(x)]^{-1},$$

$$h(x) = f(x) \int_{x'}^{x} g(z) dz,$$

$$k(x) = g(x) \int_{x'}^{x} f(z) dz.$$

(6.12)

Let us further denote by $\psi(x) \in L(I_i)$ the (Lebesgue) integrability of a non negative function $\psi(x)$ in the interval (x', r_i), i.e.

$$\psi(x) \in L(I_i) \leftrightarrow \int_{I_i} dx \psi(x) < \infty$$

(6.13)

and by $\psi(x) \notin L(I_i)$ the opposite case, i.e., the case when the integral in (6.13) is divergent. Feller's criterion for the classification of the boundary r_i is then the following:

$$r_i \text{ is} \begin{cases} regular, & \text{if } f(x) \in L(I_i) \text{ and } g(x) \in L(I_i) \\ exit, & \text{if } g(x) \notin L(I_i) \text{ and } h(x) \in L(I_i) \\ entrance, & \text{if } f(x) \notin L(I_i) \text{ and } k(x) \in L(I_i) \\ natural, & \text{otherwise.} \end{cases}$$

Some interesting and often encountered situations are the following:
a. *Both boundaries are natural.* Then the initial condition (6.11) alone uniquely determines the transition p.d.f. as a solution of the diffusion equations (6.10). Either equation may thus be used to find f. No boundary conditions, other than $f(r_i, t|x_0) = 0$, can be imposed.
b. *One-boundary is natural and the other is exit.* The initial condition (6.11) uniquely specifies the transition p.d.f. as the solution of the Fokker-Planck equation (6.10a). The Kolmogorov equation (6.10b) has, instead, infinitely many solutions all satisfying condition (6.11). In this case use of the Fokker-Planck equation is advisable.

c. *Both boundaries are exit.* As in b. Again, we can determine $f(x, t|x_0)$ as the solution of Fokker-Planck equation (6.10a) satisfying initial condition (6.11).
d. *One or both boundaries are regular.* In this case neither Eqs. (6.10) can be solved with the initial condition (6.11) alone. Boundary conditions also have to be imposed.

Of the various boundary conditions one can associate with the diffusion equations at a regular boundary r_i, the following two are particularly interesting.

1. *Total reflection at r_i.* By this one means that there is no flow of probability at the boundaries. Therefore the probability density current, j_0, must be zero at the boundary r_i (and, therefore, everywhere due to the assumed stationarity of the p.d.f.). From (5.9), using the stationarity assumption, we thus obtain the following boundary condition for the Fokker-Planck equation:

$$\lim_{x \to r_i} \left\{ \frac{\partial}{\partial x} [a(x) f(x, t|x_0)] - b(x) f(x, t|x_0) \right\} = 0. \tag{6.14}$$

2. *Total absorption at r_i.* In this case the boundary condition imposed for the solution of the Fokker-Planck equation is:

$$\lim_{x \to r_i} [a(x) f(x) f(x, t|x_0)] = 0, \tag{6.15}$$

where $f(x)$ is the function defined in (6.12). In the particular case of non singular diffusion equations, $a(x)$ and $b(x)$ being bounded and $a(x) > 0$ over the whole interval (r_1, r_2) imply that $f(x)$ is also bounded. Condition (6.15) then becomes

$$\lim_{x \to r_i} f(x, t|x_0) = 0. \tag{6.16}$$

Conditions (6.15), and (6.16) for the non-singular case, insure that r_i acts as an absorbing boundary in the sense that all sample paths terminate as they reach r_i. Condition (6.14), instead, makes r_i act as reflecting: upon reaching r_i, either sample paths are immediately reflected backward or they stay there for a while but then have to turn back. The boundary plays much the same role as the rigid walls of a container in which the molecules of a gas undergo random motion.

We conclude this section with a few considerations on some particular diffusion processes that will be encountered in the sequel.

A. The Wiener Process

Let us consider the diffusion interval obtained by setting $b(x) = 0$ and $a(x) = \sigma^2/2$ in (6.10), with σ^2 an arbitrary constant. The forward equation (6.10a) then reads

$$\frac{\partial f}{\partial t} = \frac{\sigma^2}{2} \frac{\partial^2 f}{\partial x^2}. \tag{6.17}$$

Since this is nonsingular, the diffusion interval is the entire real line with the points at infinity natural boundaries. Hence, to determine the transition p.d.f. we have to look for a solution of (6.17) satisfying the delta-condition. If we assume that the process originates at $x=0$, this transition p.d.f. is easily recognized to be

$$f(x, t|0) = \frac{1}{\sqrt{2\pi t}\,\sigma} \exp\left\{-\frac{x^2}{2\sigma^2 t}\right\}. \tag{6.18}$$

If instead we solve equation (6.17) under the initial condition

$$\lim_{t\downarrow t_0} f(x, t|x_0, t_0) = \delta(x-x_0) \tag{6.19}$$

in place of (6.18) we obtain:

$$f(x, t|x_0, t_0) = \frac{1}{\sqrt{2\pi(t-t_0)}\,\sigma} \exp\left\{-\frac{(x-x_0)^2}{2\sigma^2(t-t_0)}\right\} \tag{6.20}$$

which is still a normal density. However, now its mean value is x_0 and its variance is $\sigma^2(t-t_0)$.

An important feature of the process $X(t)$ whose transition p.d.f. is given by (6.20) is that, besides being Markov, it has independent increments. Indeed, let $t_0 < t_1 < t_2 < \ldots < t_n$ be arbitrary instants and let $Y_1 = X(t_1) - x_0$, $Y_2 = X(t_2) - X(t_1)$, ..., $Y_n = X(t_n) - X(t_{n-1})$ be the n random varaibles representing the increments of the process over the time intervals (t_0, t_1), (t_1, t_2), ..., (t_{n-1}, t_n). Then

$$\begin{aligned}
\Phi_n(Y_1, Y_2, \ldots, Y_n) \\
= f(x_0 + Y_1, t_1|x_0, t_0) f(x_1 + Y_2, t_2|x_1, t_1) \ldots f(x_{n-1} + Y_n|x_{n-1}, t_{n-1}) \\
= f(Y_1, t_1 - t_0|0) f(Y_2, t_2 - t_1|0) \ldots f(Y_n, t_n - t_{n-1}|0)
\end{aligned} \tag{6.21}$$

where the left hand side is the joint p.d.f. of the n increments and where for all y and $\tau < t$ the transition p.d.f.

$$f(x, t|y, \tau) = \frac{1}{\sqrt{2\pi(t-\tau)}\,\sigma} \exp\left\{-\frac{(x-y)^2}{2\sigma^2(t-\tau)}\right\} \equiv f(x-y, t-\tau|0) \tag{6.22}$$

is obtained from (6.20) by identifying x_0 with y and t_0 with τ. Since the right hand side of (6.21) is the product $\Phi_1(Y_1)$, $\Phi_2(Y_2) \ldots \Phi_n(Y_n)$ of the univariate p.d.f.'s of the n considered increments, the independence of these increments follows. Moreover, since the increment $X(t_{k+1}) - X(t_k)$ has the same p.d.f. as $X(t_{k+1} + \alpha) - X(t_k + \alpha)$, whatever α and t_k, it also follows that these increments are stationary. Note that the stationary-independent-increment property of the process with transition p.d.f. (6.22) is a consequence of the invariance of this function under both space and time shifts.

The diffusion process whose transition p.d.f. is given by (6.20) is said to be a *Wiener process* conditioned upon $X(0) = x_0$ while the process $W(t)$ having transition p.d.f. (6.18) is often referred to as the *standard Wiener process*. This can also be directly defined as the process $W(t)$ such that

1. $W(0)=0$
2. $W(t)$ has stationary independent increments.
3. For every $t>0$, $W(t)$ is normally distributed
4. For all $t>0$, $E[W(t)]=0$.

It can be seen that these four properties uniquely specify the probability law of $W(t)$ up to the parameter σ^2 appearing in (6.18).

An important feature of the Wiener process $W(t)$ is its being Gaussian. A stochastic process is said to be nonsingular *Gaussian* or *normal* if for any finite number of instants $t_1, t_2, ..., t_n$ the joint n-dimensional density of the random variables $X(t_1), X(t_2), ..., X(t_n)$ is given by

$$f_n(x_1, t_1; x_2, t_2; ...; x_n, t_n)$$
$$=(2\pi)^{-n/2}|\Delta|^{-1/2}\exp\{-1/2(X-m)^T K^{-1}(X-m)\},\tag{6.23}$$

where T denotes the transpose of a vector,

$$X^T=[x_1 x_2 ... x_n],$$
$$m^T=[E\{X(t_1)\}E\{X(t_2)\} ... E\{X(t_n)\}],$$
$$K=\|E\{X(t_i)-E[X(t_i)]\}\{X(t_j)-E\{X(t_j)\}\}\|,$$

and $\Delta=\det(K)\neq0$. From (6.23) we see that to specify the probability law of a Gaussian process it is sufficient to know the vector m and the covariance matrix K, which can be calculated if one knows the transition and the univariate p.d.f.'s of the process, or equivalently, the second order joint p.d.f. of the process. Now, for the Wiener process one has:

$$f_2(x_1, t_1; x_2, t_2)=f(x_2, t_2|x_1, t_1)f(x_1, t_1|0),\tag{6.24}$$

where the functions on the right hand side are immediately obtained from (6.18) and (6.22). One can then prove that

$$K(t_1, t_2)=\sigma^2\min(t_1, t_2).\tag{6.25}$$

Since for the Wiener process m is the null vector, from (6.23) and (6.25) the n-dimensional joint p.d.f. can immediately be written down.

Another important diffusion process often met in the applications is the so-called *Wiener process with drift*. Its transition p.d.f. is

$$f(x, t|x_0, t_0)=\frac{1}{\sqrt{2\pi(t-t_0)}\sigma}\exp\left\{-\frac{[x-x_0-\mu(t-t_0)]^2}{2\sigma^2(t-t_0)}\right\},\tag{6.26}$$

where the constant μ denotes the drift. The function (6.26) is the solution of the diffusion equation

$$\frac{\partial f}{\partial t}=\frac{\sigma^2}{2}\frac{\partial^2 f}{\partial x^2}-\mu\frac{\partial f}{\partial x}\tag{6.27}$$

with the initial condition (6.19). The function (6.26) is again a normal density. However, now its mean value $x_0+\mu(t-t_0)$ linearly shifts in time, the direction being determined by the sign of μ. As one can directly see from (6.26), the Wiener process with drift is also defined over the whole real line, with the points at infinity natural boundaries. However, one can restrict it to other intervals by making use of suitable boundary conditions. For instance, one can restrict the Wiener process with drift to the interval $(0, \infty)$ by imposing the reflection condition (6.16) at the zero (regular) boundary, i.e., by requiring that:

$$\left[\frac{\sigma^2}{2} \frac{\partial f(x, t|x_0)}{\partial x} - \mu f(x, t|x_0) \right]_{x=0} = 0 \tag{6.28}$$

which expresses the conservation of the probability mass in the interval $(0, \infty)$. Thus doing, one can prove that the transition p.d.f. $f_r(x, t|x_0)$ is given by:

$$f_r(x, t|x_0) = \frac{1}{\sqrt{2\pi t}\sigma} \left[\exp\left\{ -\frac{(x - x_0 - \mu t)^2}{2\sigma^2 t} \right\} \right.$$

$$+ \exp\left\{ -\frac{4\mu x_0 t + (x + x_0 - \mu t)^2}{2\sigma^2 t} \right\} \right]$$

$$\left. - \frac{\mu}{\sigma^2} \exp\left[\frac{2\mu x}{\sigma^2} \right] \left\{ 1 - \mathrm{Erf}\left(\frac{x + x_0 + \mu t}{\sqrt{2t}\sigma} \right) \right\} \right\}, \tag{6.29}$$

where $\mathrm{Erf}(z)$ is the error function defined as:

$$\mathrm{Erf}(z) = \frac{2}{\sqrt{\pi}} \int_0^z dx\, e^{-x^2}. \tag{6.30}$$

B. The Ornstein-Uhlenbeck Process

We now consider a diffusion equation with constant infinitesimal variance and linear drift with negative slope, say:

$$\frac{\partial f}{\partial t} = \frac{\sigma^2}{2} \frac{\partial^2 f}{\partial x^2} - \frac{\partial}{\partial x}[(\mu - \delta x)f] \tag{6.31}$$

with $\delta > 0$, μ, and σ^2 arbitrary constants. A straightforward calculation again shows that the points at infinity are natural boundaries so that the transition p.d.f. can be obtained as the solution of (6.31) that satisfies the initial delta-condition. The diffusion process whose transition p.d.f. satisfies equation (6.31) is called the *Ornstein-Uhlenbeck* process. Clearly, for $\delta = \mu = 0$ one recovers the Wiener process. Solving equation (6.31) with condition (6.11) one obtains (cf. Ricciardi, 1977):

$$f(x, t|x_0, t_0) = \frac{1}{\left[\pi \frac{\sigma^2}{\delta} (1 - e^{-2\delta t}) \right]^{1/2}}$$

$$\times \exp\left\{ -\frac{\left[x - \frac{\mu}{\delta} + \left(\frac{\mu}{\delta} - x_0 \right) e^{-\delta t} \right]^2}{\frac{\sigma^2}{\delta}(1 - e^{-2\delta t})} \right\}. \tag{6.32}$$

Therefore, f is a normal density with conditional mean $E[(X(t)]$ and variance $\text{Var}[X(t)]$ given by

$$E[X(t)] = \frac{\mu}{\delta} - \left(\frac{\mu}{\delta} - x_0\right)e^{-\delta t}$$

$$\text{Var}[X(t)] = \frac{\sigma^2}{2\delta}(1-e^{-2\delta t}).$$

(6.33)

Note that, distinct from the case of the Wiener process, the Ornstein-Uhlenbeck process admits of a steady state distribution. As one can see directly from (6.32), this is a normal density with mean μ/δ and variance $\sigma^2/(2\delta)$. As is well known, also the Ornstein-Uhlenbeck process is Gaussian with correlation function $R(t,\tau)$ $\equiv E[X(t)X(\tau)]$ given by

$$R(t,\tau) = \frac{\mu^2}{\delta^2} + \frac{\mu}{\delta}\left(x_0 - \frac{\mu}{\delta}\right)(e^{-\delta t} + e^{-\delta t'})$$

$$+ \left[\left(x_0 - \frac{\mu}{\delta}\right)^2 - \frac{\sigma^2}{2\delta}\right]e^{-\delta(t+t')} + \frac{\sigma^2}{2\delta}e^{-\delta|t-t'|}.$$

(6.34)

The expression of the correlation function does not change if we assume that $\delta < 0$. Of course, in this case $X(t)$ is no longer an Ornstein-Uhlenbeck process (see also Sect. 7). Use of (6.34) will be made in the next section to obtain the covariance of $X(t)$, i.e. the function

$$\Phi(t,t') = R(t,t') - E[X(t)]E[X(t')],$$

(6.35)

where $E[X(t)]$ and $E[X(t')]$ can be obtained via the first of (6.33).

7. Exponential Growth with Random Harvesting and Immigration

As an example of construction of a diffusion process let us consider a population initially subject to a Malthusian growth. The population size $x(t)$ is then given by

$$x(t) = x_0 e^{\alpha t},$$

(7.1)

where $\alpha > 0$ is the fertility of the population and x_0 is the initial population size. As we already know (see Sect. 6 of previous contribution) a stochastic analogue of (7.1) is provided by a simple birth and death process with parameters λ and μ such that $\lambda - \mu = \alpha$. We shall now consider such a birth and death process and assume that random immigration and random harvesting also take place according to Poisson schemes. Denoting by ν and ε the rates of immigration and harvesting, respectively, and assuming that changes of size Δx can take place in time Δt, from Eq. (6.1) we easily obtain:

$$
\begin{aligned}
f(x, t+\Delta t|x_0) = & \{1-[(\lambda+\mu)x+\nu+\varepsilon]\Delta t\} f(x, t|x_0) \\
& +\lambda(x-\Delta x)\Delta t f(x-\Delta x, t|x_0) \\
& +\mu(x+\Delta x)f(x+\Delta x, t|x_0)+\nu\Delta t f(x-\Delta x, t|x_0) \\
& +\varepsilon\Delta t f(x+\Delta x, t|x_0)+o(\Delta t).
\end{aligned} \tag{7.2}
$$

Expanding the functions $f(x+\Delta x, t|x_0)$ and $f(x-\Delta x, t|x_0)$ as Taylor series about the point x, from (7.2), it follows:

$$
\begin{aligned}
& \frac{f(x, t+\Delta t|x_0)-f(x, t|x_0)}{\Delta t} \\
& = -\Delta x\left\{[(\lambda-\mu)x+\nu-\varepsilon]\frac{\partial f(x, t|x_0)}{\partial x}+(\lambda-\mu)f(x, t|x_0)\right\} \\
& \quad +\tfrac{1}{2}(\Delta x)^2\left\{[\lambda+\mu)x+\nu+\varepsilon]\frac{\partial^2 f(x, t|x_0)}{\partial x^2}+(\lambda+\mu)\frac{\partial f(x, t|x_0)}{\partial x}\right\}+o(\Delta t).
\end{aligned} \tag{7.3}
$$

In order to obtain a diffusion equation, we must take the limit of (7.3) as $\Delta t \to 0$ with a suitable choice of Δx. At the same time, we must take λ, ν and ε infinitely large to avoid the trivial case $\partial f/\partial t = 0$ and hence $f(x, t|x_0)=\delta(x-x_0)$. In conclusion, we make in (7.3) the following substitutions:

$$
\lambda \to \frac{\lambda}{\Delta x}+\frac{\varrho}{2(\Delta x)^2}, \qquad \mu \to \frac{\mu}{\Delta x}+\frac{\varrho}{2(\Delta x)^2},
$$

$$
\nu \to \frac{\nu}{\Delta x}+\frac{\eta}{2(\Delta x)^2}, \qquad \varepsilon \to \frac{\varepsilon}{\Delta x}+\frac{\eta}{2(\Delta x)^2}, \tag{7.4}
$$

$$
(\Delta x)^2 = \sigma^2 \Delta t,
$$

where ϱ, η, and σ^2 are arbitrary constants. Passing to the limit as $\Delta t \to 0$, we then obtain:

$$
\begin{aligned}
\frac{\partial f(x, t|x_0)}{\partial t} = & -[(\lambda-\mu)x+\nu-\varepsilon]\frac{\partial f(x, t|x_0)}{\partial x}+(\lambda-\mu)f(x, t|x_0) \\
& +\tfrac{1}{2}(\varrho x+\eta)\frac{\partial^2 f(x, t|x_0)}{\partial x^2}+\varrho\frac{\partial f(x, t|x_0)}{\partial x}
\end{aligned} \tag{7.5}
$$

or

$$
\frac{\partial f(x, t|x_0)}{\partial t} = -\frac{\partial}{\partial x}[A_1(x)f(x, t|x_0)]+\tfrac{1}{2}\frac{\partial^2}{\partial x^2}[A_2(x)f(x, t|x_0)], \tag{7.6}
$$

where we have set:

$$A_1(x) = (\lambda - \mu)x + v - \varepsilon$$
$$A_2(x) = \varrho x + \eta .$$
(7.7)

Note that due to the arbitrariness of σ^2 in (7.4) the parameters ϱ and η appearing in (7.7) can be taken as arbitrary with the only constraint $\varrho \geq 0$, $\eta \geq 0$. The diffusion process obtained by the above sketched limit procedure will be discussed in detail elsewhere together with some more general birth and death schemes (Giorno, Nobile, and Ricciardi, 1985). Here we limit ourselves to the following remarks.

i) Assume that $\eta = 0$. We are then led to the well known Feller process (Feller, 1951). The diffusion interval is $(0, \infty)$ with 0 an exit boundary if $v \leq \varepsilon$, a regular boundary if $\varepsilon < v < \varepsilon + \varrho/2$ and an entrance boundary if $v \geq \varepsilon + \varrho/2$. The point at infinity is always a natural boundary. This is "attracting" if $\lambda > \mu$. For this diffusion population model the transition p.d.f. $f(x, t|x_0)$ is known (Feller, loc. cit.) when 0 is an exit boundary or a regular boundary with an absorption condition superimposed. This is given by (cf. Ricciardi, 1977):

$$f(x, t|x_0) = \frac{2(\lambda - \mu)}{\varrho[e^{(\lambda - \mu)t} - 1]} \exp\left\{ -\frac{2(\lambda - \mu)[x + x_0 e^{(\lambda - \mu)t}]}{\varrho[e^{(\lambda - \mu)t} - 1]} \right\}$$

$$\times [e^{-(\lambda - \mu)t} x/x_0]^{\left(v - \varepsilon - \frac{\varrho}{2}\right)/\varrho}$$

$$\times I_{1 - \frac{2(v - \varepsilon)}{\varrho}} \left\{ \frac{4(\lambda - \mu)}{\varrho[1 - e^{(\lambda - \mu)t}]} [e^{-(\lambda - \mu)t} xx_0]^{1/2} \right\},$$
(7.8)

where $I_k(x)$ is the Bessel function defined as:

$$I_k(x) = \sum_{n=0}^{\infty} \frac{(x/2)^{2n+k}}{n! \Gamma(n+k+1)}.$$
(7.9)

For completeness, we mention that the transition p.d.f. can also be obtained in the special case $\varrho = 4$, $\lambda - \mu = -2$, $v - \varepsilon = 1, 2, \ldots$ in which the origin is a regular boundary if $v - \varepsilon = 1$ and an entrance boundary if $v - \varepsilon = 2, 3, \ldots$. Indeed, such a diffusion process can be transformed into the so-called Rayleigh process for which the transition p.d.f. is known (Giorno et al., 1985).

ii) Assume that $\varrho = 0$. Then, the diffusion interval is the whole real line. Hence, if $\lambda < \mu$ the limit diffusion process is an Ornstein-Uhlenbeck process, i.e., a normal Markov process. The transition p.d.f. is then a Gaussian density whose (conditional) mean and variance can be immediately written down by recalling (6.33):

$$E[X(t)|X(0) = x_0] = \frac{v - \varepsilon}{\mu - \lambda} - \left(\frac{v - \varepsilon}{\mu - \lambda} - x_0 \right) e^{-(\mu - \lambda)t}$$
(7.10)

and

$$\text{Var}\{X(t)|X(0) = x_0\} = \frac{\eta}{2(\mu - \lambda)} [1 - e^{-2(\mu - \lambda)t}].$$
(7.11)

Since the point $x=0$ is a regular boundary, to make this process suitable to describe the dynamics of a population a reflecting condition must be imposed at zero. However, the transition p.d.f. is not known if $v \neq \varepsilon$. Let us now suppose $\lambda > \mu$. Then we obtain a diffusion process again defined over the entire real line. The points at infinity are still natural boundaries (although of an "attracting" type). Now the process is again normal and Markov. The difference with respect to the case $\lambda < \mu$ is disclosed by the calculation of the covariance $\Phi(t, t')$. Indeed, for $\lambda \neq \mu$ from (6.34) and (6.35) one obtains:

$$\Phi(t, t') = \frac{\eta}{2(\lambda - \mu)} \left[e^{(\lambda - \mu)(t + t')} - e^{(\lambda - \mu)|t - t'|} \right]. \tag{7.12}$$

Hence, we have for all h:

$$\lim_{t \to \infty} \Phi(t, t+h) = \begin{cases} \dfrac{\eta}{2(\mu - \lambda)} e^{-(\mu - \lambda)|h|}, & \lambda < \mu \\ +\infty, & \lambda > \mu \end{cases} \tag{7.13}$$

showing that as t approaches infinity no steady state process is obtained if $\lambda > \mu$. The transition p.d.f. in the presence of a reflecting condition at $x=0$ should be calculated to model a population growth process. However, this seems to be unknown in the general case $v \neq \varepsilon$. If $v = \varepsilon$ (i.e. when harvesting and immigration occur with equal rates) the transition p.d.f. of the population size with a reflecting condition at $x=0$ can be seen to have the following form:

$$f(x, t | x_0) = \left\{ \frac{\mu - \lambda}{\pi \eta [1 - e^{-2(\mu - \lambda)t}]} \right\}^{1/2} \left[\exp \left\{ -\frac{[x - x_0 e^{-(\mu - \lambda)t}]^2}{\frac{\eta}{\mu - \lambda} [1 - e^{-2(\mu - \lambda)t}]} \right\} \right.$$

$$\left. + \exp \left\{ -\frac{[x + x_0 e^{-(\mu - \lambda)t}]^2}{\frac{\eta}{\mu - \lambda} [1 - e^{-2(\mu - \lambda)t}]} \right\} \right], \quad \lambda \neq \mu. \tag{7.14}$$

iii) Let us finally assume $\lambda = \mu$. Then from (7.7) we see that the limit diffusion is a Wiener process with drift $v - \varepsilon$ and infinitesimal variance η. Hence, the population transition p.d.f. in the presence of a reflecting boundary at $x=0$ is immediately obtained from (6.29):

$$f_r(x, t | x_0) = \frac{1}{\sqrt{2\pi\eta t}} \left[\exp \left\{ -\frac{[x - x_0 - (v - \varepsilon)t]^2}{2\eta t} \right\} \right.$$

$$+ \exp \left\{ -\frac{4(v - \varepsilon)x_0 t + [x + x_0 - (v - \varepsilon)t]^2}{2\eta t} \right\} \right]$$

$$\left. - \frac{v - \varepsilon}{\eta} \exp \left[\frac{2(v - \varepsilon)x}{\eta} \right] \left\{ 1 - \mathrm{Erf} \left[\frac{x + x_0 + (v - \varepsilon)t}{\sqrt{2\eta t}} \right] \right\} \right].$$

8. Stochastic Differential Equations

In Sect. 7 we have considered an example of a diffusion process arising as the limit of a birth and death process with random immigration and random harvesting. Now we shall sketch heuristically a different procedure, of high conceptual and practical interest, to obtain diffusion processes. Successively, use of this procedure will be made to discuss some diffusion models of population growth.

Let us consider the (one dimensional) motion of a colloidal particle in a viscous medium (Brownian motion). The equation describing such motion is the *deterministic* Newton's law:

$$m\frac{dv}{dt} = -\beta v + F(t), \tag{8.1}$$

where $-\beta v$ is the elastic restoring force due to the friction experienced by the colloidal particle and $F(t)$ is the force acting on the particle due to the impacts from the molecules in the medium. In this equation everything is deterministic, and the velocity $v(t)$ is uniquely determined if we know the velocity at some arbitrarily chosen initial time and if we know the force $F(t)$. Alternatively, if we could observe the movement of the particle and measure the velocity $v(t)$ at all times, we could determine the force $F(t)$ acting on the particle:

$$F(t) = m\frac{dv}{dt} + \beta v. \tag{8.2}$$

Similarly, if we could perform the same observation on a (countable) ensemble of colloidal particles, we would determine uniquely an *ensemble* of functions $F_i(t)$:

$$F_i(t) = m\frac{dv_i}{dt} + \beta v_i \quad (i = 1, 2, \ldots), \tag{8.3}$$

where i refers to the ith measurement.

Although in principle possible, such a procedure is not useful. Indeed, what one is interested in is not the precise description of the movement of a single particle but rather some global information of the type: how many particles at a given time have an energy above a preassigned value? How many particles are likely to be found in an assigned region at a given time? Etc. Equivalently formulated, these questions become: What is the probability that at a given time a particle has an energy above a preassigned value? What is the probability that a particle will be found in an assigned region at a given time? Etc. In other words, we are interested in the global properties of an ensemble of macroscopically identical particles. This means that instead of considering the equations:

$$m\frac{dv_i}{dt} + \beta v_i = F_i(t) \quad (i = 1, 2, \ldots), \tag{8.4}$$

where everything is deterministic, we prefer to consider a unique *stochastic equation*:

$$m\frac{dv}{dt} + \beta v = F(t) \tag{8.5}$$

with the following interpretation[*]: $F(t)$ is a random function, i.e., a stochastic process; each and every one of its (continuum of) sample paths, say $F(\xi, t)$, is a deterministic function representing the actual force acting on a particle of the ensemble, thus generating the function $v(\xi, t)$. This, in turn, can be looked upon as the sample path of a stochastic process $v(t)$. Thus Eq. (8.5) can be viewed as a transformation mapping a stochastic process $F(t)$ into another stochastic process $v(t)$. The latter is specified if one knows $F(t)$ and $v(0) = v_0$. In general v_0 is not unique but is described by an initial p.d.f. $\psi(v_0)$.

Extending the situation above, we can consider an equation of the type

$$\frac{dx}{dt} = G[x, t, \Lambda(t)], \tag{8.6}$$

where $\Lambda(t)$ is a stochastic process and G is an assigned function. Then $x(t)$, the "solution" of Eq. (8.6), is itself a random function. Clearly, its determination cannot be accomplished unless $\Lambda(t)$ is specified.

In the following we shall confine ourselves to considering the more modest, although in practice very interesting, case when Eq. (8.6) has the form:

$$\frac{dx}{dt} = h(x) + k(x)\Lambda(t), \tag{8.7}$$

where h and k are deterministic functions, sufficiently smooth to make permissible the mathematical steps we are going to need, and where $k(x)$ is a positive function. If $\Lambda(t)$ is a "well behaved" stochastic process, in the sense that its sample paths are smooth functions, Eq. (8.7) can be handled with the classical methods of the theory of ordinary differential equations and one can look for a solution $x(t)$ such that $x(0) = x_0$. Such a solution will be a stochastic process with sample paths all originating at x_0. However, with the exception of the case when $\Lambda(t)$ is a Gaussian process, a simple description of the solution $x(t)$ is not possible because $x(t)$ is expected to be non-Markov due to the assumed smoothness of the sample paths of $\Lambda(t)$. For $x(t)$ to be a Markov process we should instead deprive the sample paths of $\Lambda(t)$ of memory in the sense, that, for instance, the random variables $\Lambda(t_1)$ and $\Lambda(t_2)$ should be uncorrelated for all $t_2 \neq t_1$, a situation clearly in contrast with the smoothness property of the sample paths. On the other hand, assuming such a lack of correlation creates serious mathematical difficulties

---

[*] To simplify the notation, we shall denote by the same symbol the stochastic process and any of its realizations. For instance, $F(t)$ is either the force acting on a specific particle or the stochastic process $F(\xi, t)$ with ξ varying over the ensemble. The interpretation will be clear from the context.

concerning the interpretation of Eq. (8.7) because the sample paths of $x(t)$ may not admit of derivatives in the conventional sense. Actually, one suspects that the whole machinery of traditional calculus may be inadequate to handle equations such as (8.7). This is, indeed, the case. To overcome such difficulties new calculi have been devised, the most familiar ones being that of Ito and that of Stratonovich. While referring to any of the several specialized texts on this subject (for instance, those of Arnold, Wong, Jazwinski, or Soong) for a precise formulation of the integration problem of stochastic differential equations and for an exposition of the necessary mathematical tools, we shall proceed heuristically to illustrate some instances frequently mentioned in the biological literature. But first we need to introduce the notion of "white noise". To this purpose, let us return to the stochastic differential equation

$$\frac{dx}{dt} = h(x) + k(x)\Lambda(t) \tag{8.8}$$

earlier mentioned, but let us now assume that $\Lambda(t)$ is a stationary process with zero mean and with a rather narrow and peaked correlation function:

$$E[\Lambda(t)] = g_1 = 0, \tag{8.9a}$$

$$E[\Lambda(t_1)\Lambda(t_2)] = g_2(t_1, t_2) = g_2(t_2 - t_1), \tag{8.9b}$$

where $g_2(\tau)$ is appreciably non zero only in the neighborhood of $\tau = 0$ with a very sharp maximum at $\tau = 0$. More generally, for any group of instants $t_1, t_2, ..., t_n$ all lying close to each other we set:

$$E[\Lambda(t_1)\Lambda(t_2) ... \Lambda(t_n)] = g_n(t_1, t_2, ..., t_n), \tag{8.10}$$

and, again, assume that the nth order correlation function g_n has a sharp maximum at $t_1 = t_2 = ... = t_n$, being otherwise effectively zero. Finally, we assume that when $t_1, t_2, ..., t_r$ are proximal to each other, and also when $t_{r+1}, t_{r+2}, ..., t_s$ are proximal but far from the group $t_1, t_2, ..., t_r$ and so on, then

$$E[\Lambda(t_1) ... \Lambda(t_r)\Lambda(t_{r+1}) ... \Lambda(t_s)\Lambda(t_{s+1}) ... \Lambda(t_p) ...]$$

$$= E[\Lambda(t_1) ... \Lambda(t_r)]E[\Lambda(t_{r+1}) ... \Lambda(t_s)]E[\Lambda(t_{s+1}) ... \Lambda(t_p)] ...$$

$$= g_r(t_1, ..., t_r)g_s(t_{r+1}, ..., t_s)g_p(t_{s+1}, ..., t_p) ..., \tag{8.11}$$

where the functions g_n have already been qualitatively specified.

All these assumptions about the stochastic process $\Lambda(t)$ appearing in Eq. (8.8) may look rather artificial at this stage, but the motivation for them will soon emerge. With this in mind, let us perform a change of variable in (8.8) by setting:

$$y = \Phi(x), \quad x = \Phi^{-1}(y) \tag{8.12}$$

with

$$\Phi(x) = \int^x \frac{dz}{k(z)} \ . \tag{8.13}$$

Then, Eq. (8.8) is changed into:

$$\frac{dy}{dt} = \frac{H(y)}{K(y)} + \Lambda(t) \tag{8.14}$$

upon setting:

$$H(y) = h[\Phi^{-1}(y)] \tag{8.15}$$
$$K(y) = k[\Phi^{-1}(y)] \ .$$

The advantage of this procedure is that we have constructed a stochastic process $y(t)$ defined by the simpler equation (8.14) in which $\Lambda(t)$ appears in a purely *additive way*. Note that Eq. (8.14) is of the same type as Eq. (8.5) for the velocity of a Brownian particle. Due to the above assumptions on $\Lambda(t)$, we now expect $y(t)$, and hence $x(t)$, to be Markov. Its transition p.d.f. $f_y(y, t|y_0)$ thus satisfies the differential form (4.11) of the Smolukowski equation that we now write as:

$$\frac{\partial f_y}{\partial t} = \sum_{n=1}^{\infty} \frac{(-1)^n}{n!} \frac{\partial^n}{\partial y^n} [B_n f_y] \ . \tag{8.16}$$

Let us evaluate the infinitesimal moments B_n by using Eq. (8.14). First, we express the increment of y over a small time interval Δt by means of the approximation:

$$\Delta y \equiv y(t + \Delta t) - y(t) \simeq \frac{H(y)}{K(y)} \Delta t + \int_t^{t+\Delta t} \Lambda(\tau) d\tau , \tag{8.17}$$

where, here and in the following, the value of the process at time t is considered as fixed. Note that Eq. (8.17) requires that $H(y)$ and $K(y)$ are smooth; however, the smoothness of the sample paths of $\Lambda(t)$ is not implied. Taking the expectation of both sides of (8.17), due to (8.9a), in the limit as $\Delta t \to 0$ we obtain

$$B_1(y) = \lim_{\Delta t \to 0} \frac{1}{\Delta t} E[\Delta y] = \frac{H(y)}{K(y)} \ . \tag{8.18}$$

To calculate B_2 we now square both sides of (8.17) and obtain:

$$(\Delta y)^2 \simeq O[(\Delta t)^2] + 2\Delta t \frac{H(y)}{K(y)} \int_t^{t+\Delta t} \Lambda(\tau) d\tau$$
$$+ \int_t^{t+\Delta t} \Lambda(\tau) d\tau \int_t^{t+\Delta t} \Lambda(\theta) d\theta . \tag{8.19}$$

Upon taking the expectations and after dividing by Δt, for small Δt we are left with:

$$\Delta t B_2 \simeq E[(\Delta y)^2] = \int_t^{t+\Delta t} d\tau \int_t^{t+\Delta t} d\eta\, g_2(\tau - \eta), \tag{8.20}$$

after making use of Eq. (8.9). Using the earlier specified qualitative behavior of g_2, it then follows

$$B_2 \simeq \frac{E[(\Delta y)^2]}{\Delta t} = \sigma^2, \tag{8.21}$$

with

$$\sigma^2 = \int_{-\infty}^{\infty} g_2(s)\,ds \tag{8.22}$$

and with the result becoming exact in the limit as $\Delta t \to 0$.

Proceeding along similar lines, it is not difficult to become convinced that due to the assumed properties (8.9) and (8.11) of $A(t)$ the following relationship holds:

$$\Delta t B_n(y) \simeq E[(\Delta y)^n] = o(\Delta t) \quad (n=3,4,\ldots). \tag{8.23}$$

Equation (8.16) thus becomes the Fokker-Planck equation

$$\frac{\partial f_y}{\partial t} = -\frac{\partial}{\partial y}[B_1(y)f_y] + \frac{\sigma^2}{2}\frac{\partial^2 f_y}{\partial y^2}, \tag{8.24}$$

with $B_1(y)$ and σ^2 given by (8.18) and (8.21), respectively. The conclusion is that Eq. (8.14) can be thought of as defining a diffusion process $y(t)$ whose drift equals the deterministic part of the r.h.s. of the equation while its infinitesimal variance depends exclusively on the characteristics $(g_2(\tau))$ of the random part of the equation. Furthermore, if we impose that $P\{y(0)=y_0\}=1$, with y_0 non random, then the specification of such a diffusion process is unique.

Let us now examine the infinitesimal moments $A_n(x)$ of the Markov process $x(t)$ defined by Eq. (8.8). Denoting its transition p.d.f. by $f_x(x,t|x_0)$, we know that for small Δt we have:

$$\begin{aligned}
A_n(x)\Delta t &\simeq \int (x'-x)^n f_x(x',\Delta t|x)\,dx' \\
&= \int [\Phi^{-1}(y')-x]^n f_x[\Phi^{-1}(y'),\Delta t|x]\,d[\Phi^{-1}(y')] \\
&= \int [\Phi^{-1}(y')-x]^n f_y[y',\Delta t|\Phi(x)]\,dy' \quad (n=1,2,\ldots),
\end{aligned} \tag{8.25}$$

having made use of the one-to-one transformation (8.12) and of the relation

$$f_x(x',t|x) = \left[\frac{f_y(y',t|y)}{|dx'/dy'|}\right]_{y'=\Phi(x')} \tag{8.26}$$

between the transition p.d.f.'s of the processes $x(t)$ and $y(t)$. Let us now expand $\Phi^{-1}(y)$ as a Taylor series about the point $y = \Phi(x)$:

$$\Phi^{-1}(y) = \Phi^{-1}[\Phi(x)] + \alpha_1(x)[y' - \Phi(x)] + \tfrac{1}{2}\alpha_2(x)[y' - \Phi(x)]^2$$
$$+ \sum_{n=3}^{\infty} \frac{\alpha_n(x)}{n!}[y' - \Phi(x)]^n. \tag{8.27}$$

with

$$\alpha_n(x) = \frac{d^n \Phi^{-1}(y)}{dy^n}\bigg|_{y=\Phi(x)}. \tag{8.28}$$

Since $\Phi^{-1}[\Phi(x)] = x$ and (8.15) holds a straightforward application of the theorem on the derivative of the inverse function yields:

$$\alpha_1 = k(x)$$
$$\alpha_2 = k'(x)k(x), \tag{8.29}$$

where $k'(x)$ denotes $dk(x)/dx$. Making use of (8.29) and of (8.27), from (8.25) we obtain:

$$\Delta t A_1(x) \simeq k(x) \int [y' - \Phi(x)] f_y[y', \Delta t | \Phi(x)] dy'$$
$$+ \tfrac{1}{2}k'(x)k(x) \int [y' - \Phi(x)]^2 f_y[y', \Delta t | \Phi(x)] dy'$$
$$+ \sum_{j=3}^{\infty} \frac{\alpha_j(x)}{j!} \int [y' - \Phi(x)]^j f_y[y', \Delta t | \Phi(x)] dy'. \tag{8.30}$$

Therefore, in the limit as $\Delta t \to 0$, we find:

$$A_1(x) = k(x)B_1[\Phi(x)] + \tfrac{1}{2}k'(x)k(x)B_2[\Phi(x)]$$
$$+ \sum_{j=3}^{\infty} \frac{\alpha_j(x)}{j!} B_j[\Phi(x)], \tag{8.31}$$

where the $B_n(y)$'s are given by (8.18), (8.21), and (8.23). Recalling (8.15) we thus find:

$$A_1(x) = h(x) + \frac{\sigma^2}{4} \frac{dk^2(x)}{dx}. \tag{8.32}$$

Using the same procedure one can prove that

$$A_2(x) = \sigma^2 k^2(x),$$
$$A_n(x) = 0 \quad (n = 3, 4, \ldots). \tag{8.33}$$

By virtue of (8.32) this shows that the Markov process $x(t)$ defined by Eq. (8.8) is a diffusion process with drift and infinitesimal variance given by

$$A_1(x) = h(x) + \frac{\sigma^2}{4} \frac{dk^2(x)}{dx} \equiv h(x) + \tfrac{1}{4}\frac{dA_2}{dx} \tag{8.34a}$$

$$A_2(x) = \sigma^2 k^2(x). \tag{8.34b}$$

Clearly, if k is a constant, $A_1(x)$ and $A_2(x)$ coincide with the corresponding functions determined earlier for the purely additive random noise case after rescaling the correlation function. We should, however, emphasize the determinant role played above by our assumptions on $\Lambda(t)$ required for (8.34) to hold. The main trouble with such assumptions is that they do not uniquely characterize $\Lambda(t)$. Any other stationary random function $\Lambda'(t)$ on the r.h.s. of Eq. (8.8), with zero mean and with correlation functions $g'_n(t_1, t_2, ..., t_n) \neq g_n(t_1, t_2, ..., t_n)$, but with the same qualitative behavior, would describe the same diffusion process $x(t)$ provided only that

$$\int_{-\infty}^{\infty} g'_2(\tau)d\tau = \int_{-\infty}^{\infty} g_2(\tau)d\tau. \tag{8.35}$$

To obtain a one-to-one correspondence between $x(t)$ and $\Lambda(t)$, the latter should be better specified. This is usually done by considering the limiting case of $\Lambda(t)$, when

$$g_2(t_1, t_2) = g_2(t_2 - t_1) = \sigma^2 \delta(t_2 - t_1), \tag{8.36}$$

where $\delta(t)$ is the Dirac delta function and σ^2 is a constant, the only (harmless) element of indeterminacy. Note that with the choice (8.36), relation (8.22) is satisfied. Let us further assume that the functions g_n appearing in (8.10) are such that for all integers $n \geq 0$ and for all $t_1, t_2, ...$

$$g_{2n+1}(t_1, t_2, ..., t_{2n+1}) = 0$$
$$g_{2n}(t_1, t_2, ..., t_{2n}) = \sum g_2(t_i, t_j)g_2(t_k, t_h) ..., \tag{8.37}$$

where the sum is taken over all the different ways in which one can divide the $2n$ instants $t_1, t_2, ..., t_{2n}$ into n pairs.

As Wang and Uhlenbeck (1945) proved, assuming that $\Lambda(t)$ is a stationary process with correlation functions g_n satisfying (8.36) and (8.37) implies that $\Lambda(t)$ is a *normal* process. However, this is a typical example in which the covariance matrix is singular due to (8.36) and (8.37). In particular, we have

$$\text{Var}[\Lambda(t)] = g_2(0) = \infty,$$

which indicates that the sample paths of $\Lambda(t)$ are totally memoryless (and thus fluctuate infinitely rapidly) due to the assumed total lack of correlation between the r.v.'s $\Lambda(t)$ and $\Lambda(t + \Delta t)$, no matter how small Δt is or what t is. This peculiar stochastic process is called *white noise*.

Quite evidently, there is interest in stochastic differential equations such as (8.8) for the purpose of modelling biological or physical phenomena. The deterministic part of the equation

$$\frac{dx}{dt} = h(x) \tag{8.38}$$

can be taken as describing the time course of the state of the system under study when all the relevant environmental and internal parameters are fixed. Adding the term $k(x)\Lambda(x)$ to the r.h.s. of (8.38) can then be viewed as equivalent to switching from the description of one single system to that of an ensemble of macroscopically identical systems, all starting at $x(0)=x_0$, the state of each of which varies in time as a sample path of the diffusion process specified by (8.34). The random term on the r.h.s. of Eq. (8.8) is usually interpreted as a perturbation due to the overall effect of the numerous microscopic, unknown or only partially known, environmental or internal fluctuations of some of the parameters appearing in Eq. (8.38), for which a complete deterministic description is impossible or useless.

Before discussing a few examples it should be emphasized that in any actual biological or physical situation the noise affecting the system's evolution would not be exactly "white", whereas it is often reasonable to assume that it is stationary and that its correlation functions possess the qualitative features mentioned in the foregoing. It thus appears reasonable to conclude (Stratonovich, 1963, 1968) that for such realistic noise, denoted by $\Lambda_R(t)$, the differential equation

$$\frac{dx_R}{dt}=h(x_R)+k(x_R)\Lambda_R(t) \tag{8.39}$$

leads us to a stochastic process $x_R(t)$ which is not exactly a diffusion process (because the infinitesimal moments $A_3, A_4, \ldots$ are not exactly zero). However, while Eq. (8.8) is meaningful [because for such types of noise, we may safely assume the differentiability of the sample paths of $x_R(t)$] the diffusion process $x(t)$ specified by (8.34) can be looked upon as the limiting case of a sequence of processes $x_R(t)$ in which the actual "colored" noise becomes whiter and whiter. Therefore, writing Eq. (8.8) and interpreting $\Lambda(t)$ as white noise is a useful expedient for determining the limiting process $x(t)$. The goodness of the representation of the state of the system under study by $x(t)$, instead of $x_R(t)$, depends on how close the actual noise is to the white noise. Much more on this subject can be found in the quoted texts by Stratonovich, and in a short, but important, paper by Wong and Zakai (1965).

According to Eq. (8.34) and to the above remarks, the Wiener process $W(t)$ can be thought of as the limit of the process $z(t)$ defined by the stochastic differential equation

$$\frac{dz}{dt}=\Lambda_R(t) \tag{8.40}$$

when $\Lambda_R(t)\to\Lambda(t)$. In this sense (and only in this sense) can $\Lambda(t)$ be formally interpreted as the derivative of the Wiener process. Therefore, we can write:

$$\frac{dx}{dt}=h(x)+k(x)\Lambda_R(t)$$
$$=h(x)+k(x)\frac{dz}{dt} \tag{8.41}$$

where $z(t)$ is the solution of (8.40). From (8.41) we also find

$$dx=h(x)dt+k(x)dz\neq h(y)dt+k(y)dW=dy, \tag{8.42}$$

where $y(t)$ *is not* equal to $x(t)$. The arbitrary identification of dy with dx in Eq. (8.42) and of $x(t)$ with the diffusion process with infinitesimal moments (8.34) is responsible for what is often incorrectly referred to as the Ito versus Stratonovich controversy (cf., for instance, Mortensen, 1969; Gray and Caughey, 1965; Arnold, 1974, Chap. 10).

It should be mentioned that an equation such as

$$dx = h(x)dt + k(x)dW, \tag{8.43}$$

which we have just derived heuristically, is called an Ito equation; this can be handled by means of a somewhat ad hoc calculus (Ito' Calculus) which differs in several fundamental ways from the one to which we are accustomed. We will not elaborate on this here; we only mention that the solution, $x(t)$, of (8.43) is also a diffusion process. Its drift and infinitesimal variance are, however, given by

$$\begin{aligned} B_1(x) &= h(x) \\ B_2(x) &= \sigma^2 k^2(x). \end{aligned} \tag{8.44}$$

Comparing (8.44) with (8.34) we thus see that $B_2(x) = A_2(x)$, while $A_1(x)$ differs from $B_1(x)$ by the term

$$\frac{\sigma^2}{4} \frac{dk^2(x)}{dx} \equiv \tfrac{1}{4} \frac{dB_2}{dx} = \tfrac{1}{4} \frac{dA_2}{dx}.$$

The only case where this difference vanishes is clearly when $k(x) = \text{const}$, i.e., when we are dealing with a purely additive noise.

9. Diffusion Models in Population Growth

In Sect. 7 an example has been provided to show how a diffusion process can be constructed from a birth and death process with random immigration and harvesting to model population growth. In the sequel we shall make use of the considerations of Sect. 8 to provide various diffusion models to population growth by suitably parameterizing first order differential equations expressing the population's growth law in the absence of random effects. We shall start with the simplest growth law, i.e. the Malthusian one, and thus consider a population, initially consisting of $x(0) = x_0$ individuals, whose fertility αx is proportional to the population size $x(t)$. Hence, the growth law is expressed by the differential equation

$$\begin{aligned} \frac{dx}{dt} &= \alpha x \\ x(0) &= x_0. \end{aligned} \tag{9.1}$$

In order to include random effects such as environmental variability, we now set in Eq. (9.1) $\alpha = a + \Lambda(t)$, where $\Lambda(t)$ is the white noise defined in Sect. 8. Hence, we obtain the stochastic differential equation

$$\frac{dx}{dt} = ax + x\Lambda(t)$$ (9.2)

with the initial condition

$$P\{x(0) = x_0\} = 1.$$ (9.3)

Equation (9.2) has been considered repeatedly in the literature. Here we limit ourselves to mentioning the works by Lewontin and Cohen (1969), Levins (1969), May (1972, 1973), Capocelli and Ricciardi (1974). According to the discussion of Sect. 8, Eq. (8.2) defines a diffusion process $X(t)$ having drift $A_1(x)$ and infinitesimal variance $A_2(x)$ given by

$$A_1(x) = \left(\alpha + \frac{\sigma^2}{2}\right)x$$ (9.4)

and

$$A_2(x) = \sigma^2 x^2,$$ (9.5)

respectively. Hence the diffusion interval of interest is $(0, \infty)$, with the end points natural boundaries. As shown in Capocelli and Ricciardi (1976), the transition p.d.f. $f(x, t|x_0)$ is given by:

$$f(x, t|x_0) = \frac{1}{\sigma x \sqrt{2\pi t}} \exp\left\{-\frac{\left[\ln\frac{x}{x_0} - \alpha t\right]^2}{2\sigma^2 t}\right\}, \qquad 0 < x < \infty$$ (9.6)

showing that $X(t)$ is a lognormal process whose mean $M(t|x_0)$ and variance $V(t|x_0)$ are

$$M(t|x_0) = x_0 \exp\left[\left(\alpha + \frac{\sigma^2}{2}\right)t\right]$$

$$V(t|x_0) = x_0^2 \exp\left[2\left(\alpha + \frac{\sigma^2}{2}\right)t\right][e^{\sigma^2 t} - 1].$$ (9.7)

From (9.6) one can also easily obtain the mode $\mu(t|x_0)$ of the population size. This is given by

$$\mu(t|x_0) = x_0 \exp[(\alpha - \sigma^2)t].$$ (9.8)

From (9.8) and the first of (9.7) we thus draw the following conclusions:

i) if $\alpha > \sigma^2$ the most likely population size increases without bounds as $t \to \infty$; if $\alpha < \sigma^2$ the most likely population size tends to zero as $t \to \infty$; it finally stays constantly equal to the initial population size if $\alpha = \sigma^2$.

ii) The mean population size grows to infinity with time if $\alpha > -\sigma^2/2$ whereas it goes to zero or stays constantly equal to x_0 if $\alpha \leqq -\sigma^2/2$.

In particular, from i) and ii) it follows that if $-\sigma^2/2 < \alpha < \sigma^2$ the mean population size grows without bounds in time while the most likely population size shrinks to zero. Hence, for $-\sigma^2/2 < \alpha < 0$ contrary to the deterministic behavior (which entails an exponentially decreasing population size) the mean population size grows to infinity. However, the most likely population size tends to zero.

Let us now substitute Eq. (9.2) with the stochastic equation involving the differential of the Wiener process [cf. Eq. (8.43)]. The resulting diffusion process has the drift $B_1(x)$ and infinitesimal variance $B_2(x)$ given by

$$B_1(x) = \alpha x$$
$$B_2(x) = \sigma^2 x^2 .$$

(9.9)

This implies that we still have a lognormal diffusion process. However, its mean and mode are obtained from (9.7) and (9.8) by the substitution $\alpha \to \alpha - \sigma^2/2$. Hence, now negative values of the fertility never imply a growing mean. In this sense, the process (9.9) has been viewed by some authors as more suitable to be taken as the stochastic counterpart of the deterministic Malthusian growth law (see, for instance, May, 1972, 1973). Whether this is really the case will be discussed in some detail later on.

Equation (9.1) can be viewed as an approximation to modelling populations that grow in an environment characterized by infinite resources. However, due to the finiteness of the latter, to the accumulation of toxic products, to competition, etc. some regulation effects have to be taken into account in order to impose an upper bound to the number of individuals that can coexist in a preassigned environment. Such regulation can be expressed by changing Eq. (9.1) into an equation of the form

$$\frac{dx}{dt} = \alpha x [1 - \Phi(x)],$$

(9.10)

where $\Phi(x)$ is a suitable function of the population size. The Malthusian case is then contained in (9.10) as it corresponds to the choice $\Phi \equiv 0$.

Two cases well known in the population biology literature arise for $\Phi(x) = \beta x/\alpha$ and $\Phi(x) = \beta \log x/\alpha$, where $\beta > 0$ and α are arbitrary parameters. The former is denoted as the *logistic* growth process and the latter as the *Gompertz* growth process. A third case that we shall consider is when $\Phi(x) = 1 + x(x - \gamma)$, with $\gamma > 0$ an arbitrary parameter. This case will be referred to as "logistic-like". From (9.10), with the proper choice of $\Phi(x)$ we obtain the logistic equation

$$\frac{dx}{dt} = \alpha x - \beta x^2 ,$$

(9.11)

the Gompertz equation

$$\frac{dx}{dt} = \alpha x - \beta x \log x \tag{9.12}$$

and the logistic-like equation

$$\frac{dx}{dt} = -\alpha x(x-\gamma). \tag{9.13}$$

Denoting by x_0 the initial population size, the solutions of (9.11) and (9.12) can be easily seen to be:

$$x(t) = \alpha x_0 [\alpha e^{-\alpha t} + \beta x_0 (1-e^{-\alpha t})]^{-1} \tag{9.14}$$

and

$$x(t) = \exp\left[\frac{\alpha}{\beta} - \left(\frac{\alpha}{\beta} - \log x_0\right)e^{-\beta t}\right], \tag{9.15}$$

respectively. As for the solution of (9.13), it can be expressed in the following implicit form:

$$|x(t) - \gamma| = \frac{|x_0 - \gamma|}{x_0} x(t) \exp\left\{\frac{\gamma}{x_0} - \frac{\gamma}{x(t)} - \alpha \gamma^2 t\right\}. \tag{9.16}$$

It is not difficult to prove that functions (9.14)–(9.16) are qualitatively alike (Nobile and Ricciardi, 1984) in the sense that they all exhibit an S-shaped behavior as t increases; in all cases the number of individuals tend to a finite saturation value given by α/β for the logistic growth, by $e^{\alpha/\beta}$ for the Gompertz case and by γ for the logistic-like growth law. From Eqs. (9.11) to (9.13) one sees that the growth rate is symmetric for $x_0 \in (0, \alpha/\beta)$ in the logistic case while it is skewed in the other two cases with a positive skewness for the Gompertz law and a negative skewness for the logistic-like growth law.

By using the machinery of Sect. 8 we shall now construct the diffusion counterparts of Eqs. (9.11)–(9.13). To this purpose, we set in (9.11) $\alpha = \alpha + \Lambda(t)$ and thus obtain the stochastic differential equation

$$\frac{dX}{dt} = \alpha X - \beta X^2 + X\Lambda(t). \tag{9.17}$$

This defines a diffusion process $X(t)$ whose drift $A_1(x)$ and infinitesimal variance $A_2(x)$ are given by

$$A_1(x) = \left(\alpha + \frac{\sigma^2}{2}\right)x - \beta x^2$$

$$A_2(x) = \sigma^2 x^2. \tag{9.18}$$

The diffusion interval to be considered is $(0, \infty)$ with the end points natural boundaries. A feature of this diffusion process is that it admits of a steady distribution whatever the parameters α, β, and σ^2 are. This can be seen easily by making use of (6.6) since the constant given by (6.7) turns out to be nonzero. Alternatively, we can model the logistic growth process in random environment by means of the equation

$$dX = (\alpha X - \beta X^2)dt + XdW(t).$$ (9.19)

Drift $B_1(x)$ and infinitesimal variance $B_2(x)$ of the resulting diffusion processes are then given by

$$B_1(x) = \alpha x - \beta x^2$$
$$B_2(x) = \sigma^2 x^2.$$ (9.20)

In this case, by making use again of (6.6) and (6.7) we are led to the conclusion that the steady state distribution exists only if $\alpha > \sigma^2/2$, i.e. only for sufficiently large intrinsic growth rates, or equivalently, only if the intensity of the noise is small enough. Furthermore, one can see that if $\alpha < \sigma^2/2$, the point $x = 0$ is a natural boundary of an attracting type. By this we mean that with probability one asymptotically the population size attains arbitrarily small values. Such diversity of behavior of the models defined by (9.18) and (9.20) more pressingly than in the Malthusian case raises the question of what stochastic equation should be taken as the counterpart of Eq. (9.11). While postponing the answer to this question, we here limit ourselves to mentioning that such diversity of behavior is discussed in detail by Feldman and Roughgarden (1975).

Let us now turn to Gompertz equation (9.12) and proceed as was done for the logistic case (see also Capocelli and Ricciardi, 1974). By the same procedure we are led to the stochastic equations

$$\frac{dX}{dt} = \alpha X - \beta X \log X + X \Lambda(t)$$ (9.21)

and

$$dX = (\alpha X - \beta X \log X)dt + XdW(t).$$ (9.22)

Equations (9.21) and (9.22) define diffusion process on $(0, \infty)$, with the endpoints natural boundaries. The associated drifts $C_1(x)$ and infinitesimal variances $C_2(x)$ can be written as

$$C_1(x) = (m - \beta \log x)x,$$
$$C_2(x) = \sigma^2 x^2,$$ (9.23)

where $m = \alpha + \sigma^2/2$ in the case of (9.21) and $m = \alpha$ in the case of (9.22). In either case, the steady state distribution always exists irrespective of the magnitude of α and σ^2, and the asymptotic behavior of the population sizes in the two models are identical. In this sense, we can conclude that Eqs. (9.21) and (9.22) are both equally suited to represent the stochastic analogue of the Gompertz model (see also Nobile and Ricciardi, 1980).

Stochastic diffusion processes can also be constructed by suitably parameterizing the logistic-like Eq. (9.13). As shown in Nobile and Ricciardi (1984a) use of equations involving white noise leads to a diffusion process that behaves differently from that which is obtained by writing the equation involving the differential of the Wiener process, so that again a question of modelling naturally arises. Such a question is the object of Sect. 10. In conclusion, we would like to mention that alternative diffusion models of population growth in a random environment have been discussed recently elsewhere (Nobile and Ricciardi, 1984b).

10. Discrete Approximations

In order to overcome the question of what stochastic equation should be associated with an assigned deterministic growth model we shall now outline a straightforward procedure to construct diffusion models starting from difference equations. For the sake of simplicity, we shall refer to the logistic equation (9.11); however, our considerations can be extended to other models.

Let us start by remarking that two simple discrete approximations can be constructed for a logistic population growth model. First of all, we can approximate equation (9.11) by the difference equation

$$y_{(n+1)\tau} - y_{n\tau} = \alpha \tau y_{n\tau} - \beta \tau y_{n\tau}^2 \quad (n=0, 1, ...)$$

$$y_0 = x_0,$$

(10.1)

where $\tau > 0$ is an arbitrarily fixed quantity. In order to construct from (10.1) a stochastic model, we consider a simple scheme by which the quantity $\alpha\tau$ in (10.1) is taken as a sequence $Z_{n\tau}$ of random variables mutually independent with probabilities:

$$P\{Z_{n\tau} = \sigma\sqrt{\tau}\} = \tfrac{1}{2} + \frac{\alpha\sqrt{\tau}}{2\sigma}$$

$$P\{Z_{n\tau} = -\sigma\sqrt{\tau}\} = \tfrac{1}{2} - \frac{\alpha\sqrt{\tau}}{2\sigma} \quad (n=0, 1, ...).$$

(10.2)

In other terms, we substitute (10.1) with the stochastic difference equations

$$Y_{(n+1)\tau} - Y_{n\tau} = Z_{n\tau} Y_{n\tau} - \beta\tau Y_{n\tau}^2 \quad (n=0, 1, ...),$$

$$P\{Y_0 = x_0\} = 1.$$

(10.3)

Of course, more general schemes could be considered, but this would only add unnecessary complexity to the forthcoming calculations without giving any better insight into the procedure. From (10.3) one easily obtains:

$$\tau^{-1} E\{Y_{(n+1)\tau} - Y_{n\tau} | Y_{n\tau} = x\} = \alpha x - \beta x^2,$$

$$\tau^{-1} E\{[Y_{(n+1)\tau} - Y_{n\tau}]^2 | Y_{n\tau} = x\} = \sigma^2 x^2 + \beta^2 x^4 \tau^2 - 2\alpha\beta x^3 \tau, \tag{10.4}$$

$$\tau^{-1} E\{[Y_{(n+1)\tau} - Y_{n\tau}]^{2+p} | Y_{n\tau} = x\} = o(\tau) \quad (p=1, 2, \ldots).$$

Hence, as $\tau \to 0$, $Y_{n\tau}$ converges to the diffusion process $Y(t)$ whose infinitesimal moments are given by (9.20).

We now construct a second discrete approximation to the process described by (9.11) in the following way. We start from the solution (9.14) of (9.11) and calculate it at the instants $t = n\tau$ ($n = 0, 1, \ldots$). By a simple procedure we are then led to the following system of difference equations:

$$\frac{x_{(n+1)\tau} - x_{n\tau}}{x_{n\tau}} + 1 = \left[\frac{1 - e^{-\alpha\tau}}{\alpha\tau} \beta\tau x_{n\tau} + e^{-\alpha\tau} \right]^{-1} \quad (n=0, 1, \ldots), \tag{10.5}$$

where we have set $x_{n\tau} = x(n\tau)$ ($n = 0, 1, \ldots$). As before, we now substitute $\alpha\tau$ with the random variables $Z_{n\tau}$ defined by (10.2) so that (10.5) is changed into the system of stochastic difference equations

$$\frac{X_{(n+1)\tau} - X_{n\tau}}{X_{n\tau}} + 1 = \left[\beta\tau X_{n\tau} \frac{1 - e^{-Z_{n\tau}}}{Z_{n\tau}} + e^{-Z_{n\tau}} \right]^{-1} \quad (n=0, 1, \ldots). \tag{10.6}$$

Proceeding as before, we calculate the conditional moments of the increment of $X_{n\tau}$ over the time interval τ. However, the calculations are now more complex. We start introducing the random variables

$$\zeta_{n\tau} = \left\{ \left[\beta\tau X_{n\tau} \frac{1 - e^{-Z_{n\tau}}}{Z_{n\tau}} + e^{-Z_{n\tau}} \right]^{-1} | X_{n\tau} = x \right\} \quad (n=0, 1, \ldots). \tag{10.7}$$

Then upon setting $\varepsilon = \sigma\sqrt{\tau}$ we have:

$$P\left\{ \zeta_{n\tau} = \varepsilon \left[\frac{\beta x \varepsilon^2}{\sigma^2} (1 - e^{-\varepsilon}) + \varepsilon e^{-\varepsilon} \right]^{-1} \right\} = \frac{1}{2} + \frac{\alpha\varepsilon}{2\sigma^2}$$

$$\hspace{8cm} (n=0, 1, \ldots). \tag{10.8}$$

$$P\left\{ \zeta_{n\tau} = \varepsilon \left[\frac{\beta x \varepsilon^2}{\sigma^2} (e^{\varepsilon} - 1) + \varepsilon e^{\varepsilon} \right]^{-1} \right\} = \frac{1}{2} - \frac{\alpha\varepsilon}{2\sigma^2}.$$

Hence, we have:

$$E\{[X_{(n+1)\tau} - X_{n\tau}]^r | X_{n\tau} = x\} = x^r E\{[\zeta_{n\tau} - 1]^r\} \quad (r=1, 2, \ldots; n=0, 1 \ldots) \tag{10.9}$$

so that the moments of $\zeta_{n\tau} - 1$ have to be determined in order to obtain the conditional moments of the change in population size over the time interval τ.

Making use of (10.8) one obtains:

$$E[\zeta_{n\tau}-1]=\varepsilon^2\sigma^{-2}[\varDelta(\varepsilon)]^{-1}[\varepsilon(\alpha+\beta x)\sinh\varepsilon+\varepsilon^2\sigma^{-2}\alpha\beta x(\cosh\varepsilon-1)$$
$$+\sigma^2\cosh\varepsilon-\sigma^2\varepsilon^{-2}\varDelta(\varepsilon)]\quad(n=0,1,...),\tag{10.10}$$

where

$$\varDelta(\varepsilon)=[\sigma^{-2}\beta x\varepsilon^2(1-e^{-\varepsilon})+\varepsilon e^{-\varepsilon}][\sigma^{-2}\beta x\varepsilon^2(e^\varepsilon-1)+\varepsilon e^\varepsilon].\tag{10.11}$$

Let us now write $\varDelta(\varepsilon)$ as:

$$\varDelta(\varepsilon)=\varepsilon^2+2\sigma^{-2}\beta x\varepsilon^4+o(\varepsilon^4)\tag{10.12}$$

and expand the exponentials appearing in (10.10) to the second order in ε. After some algebra we then obtain:

$$E[\zeta_{n\tau}-1]=\frac{-\beta\alpha\varepsilon^2+\alpha\varepsilon^2+\sigma^2\varepsilon^2/2+o(\varepsilon^2)}{\sigma^2+2\beta x\varepsilon^2+o(\varepsilon^2)}.\tag{10.13}$$

From (10.13) and (10.9), recalling that $\varepsilon=\sigma\sqrt{\tau}$, one gets:

$$\lim_{\tau\to0}\tau^{-1}E[X_{(n+1)\tau}-X_{n\tau}|X_{n\tau}=x]=\left(\alpha+\frac{\sigma^2}{2}\right)x-\beta x^2.\tag{10.14}$$

We now come to the calculation of the infinitesimal variance. To this purpose it is advantageous to make use of the identity

$$E[(\zeta_{n\tau}-1)^2]=\{E[\zeta_{n\tau}^2]-1\}-2E[\zeta_{n\tau}-1]\quad(n=0,1,...).\tag{10.15}$$

The second term on the right hand side of (10.15) is given by (10.13). On the other hand, by a procedure similar to that used before, one can prove that (Ricciardi, 1979):

$$E[\zeta_{n\tau}^2]-1=\frac{2\sigma^2\varepsilon^2+2\alpha\varepsilon^2-2\beta x\varepsilon^2+o(\varepsilon^2)}{\sigma^2+4\beta\varepsilon^2x+o(\varepsilon^2)}.\tag{10.16}$$

From (10.9), (10.13), (10.15) and (10.16) we finally obtain:

$$\lim_{\tau\to0}\tau^{-1}E\{[X_{(n+1)\tau}-X_{n\tau}]^2|X_{n\tau}=x\}=\sigma^2x^2.\tag{10.17}$$

Following similar lines one can also prove that all higher order conditional moments of the change in population size over the time interval τ are quantities $o(\tau)$ so that the higher order infinitesimal moments of the limit of $X_{n\tau}$, when $\tau\to0$, vanish. This proves via approximation (10.6) in the limit when $\tau\to0$, that $X_{n\tau}$ converges to the diffusion process having drift and infinitesimal variance given by (9.18).

The above considerations have led us to a direct interpretation of the diffusion models of logistic growth generated by equations (9.17) and (9.19). The former is suitable to describe the diffusion process that arises as the limit when $\tau \to 0$ of the difference equations (10.6) constructed by discretizing and then randomizing the solution of (9.11); the latter is instead appropriate to depict the diffusion model obtained by discretizing the Eq. (9.11). Therefore, it appears that Eq. (9.17) is the stochastic analogue of an intrinsically continuous population growth model whereas Eq. (9.19) is appropriate to model an intrinsically discontinuous growth process by means of a stochastic continuous approximation.

To shed some more light on the question of what diffusion process should be taken as the stochastic analogue of a deterministic population model, let us reason as follows. Suppose we start with the Stratonovich equation (8.8):

$$\frac{dx}{dt} = h(x) + k(x)\Lambda(t). \tag{10.18}$$

Then, we obtain the diffusion process characterized by the infinitesimal moments

$$A_1(x) = h(x) + \tfrac{1}{4} \frac{dA_2(x)}{dx}$$
$$A_2(x) = \sigma^2 k^2(x). \tag{10.19}$$

Therefore, in order to be led to the same process the following Ito equation must be considered:

$$dx = \left[h(x) + \frac{\sigma^2}{2} k(x) \frac{dk(x)}{dx} \right] dt + k(x)dW(t). \tag{10.20}$$

Vice versa, if the random growth of the population is modeled by the Ito equation

$$dy = h(y)dt + k(y)dW(t) \tag{10.21}$$

due to (8.44) the equivalent Stratonovich equation must be written in the form:

$$\frac{dy}{dt} = \left[h(y) - \frac{\sigma^2}{2} k(y) \frac{dk(y)}{dy} \right] + k(y)\Lambda(t). \tag{10.22}$$

In other words, Eqs. (10.18) and (10.20) express an identical population model which is *different* from the one expressed by Eqs. (10.21) and (10.22). However, the question remains of what stochastic equations are to be used to account for random perturbations acting on a population whose deterministic growth is expressed by a first order differential equation. Of course, the considerations just made on the different discrete schemes that can be associated with a logistic growth law already suggest the answer. However, here let us refer to an arbitrary population growing in a finite carrying capacity environment and let us denote by $x_{n\tau}$ be the number of individuals counted by an observer at the instants $t = n\tau$

$(\tau > 0; n = 0, 1, 2, ...)$. In the absence of any other information, one may model this growth process by the difference equations

$$\Delta x_{n\tau} = A\tau h(x_{n\tau}) \quad (n = 0, 1, ...) \tag{10.23}$$

or by the differential equation

$$\frac{dx}{dt} = Bh(x). \tag{10.24}$$

A and B are quantities (relating to the Wrightian fertility and to the Malthusian parameter, respectively) to be experimentally determined. Since (10.23) and (10.24) have been taken as describing the identical population, we require that for all n $x_{n\tau} \equiv x(n\tau)$. In other words, we impose that the solution to Eq. (10.24) at the discrete times $t = n\tau$ yields the values determined by the discrete model (10.23). What we now wish to obtain is a relationship between quantities A and B securing that the identity between the considered population sizes actually holds. To this purpose we integrate equation (10.24) over a time interval of duration τ to obtain:

$$B = \frac{1}{\tau} \int_{x_{n\tau}}^{x_{(n+1)\tau}} dz [h(z)]^{-1} \tag{10.25}$$

or, on account of (10.23):

$$B = \frac{Ah(x_{n\tau})}{\Delta x_{n\tau}} \int_{x_{n\tau}}^{x_{(n+1)\tau}} dz [h(z)]^{-1}, \tag{10.26}$$

where $\Delta x_{n\tau}$ is given by the right hand side of (10.23). A Taylor series expansion of the right hand side of (10.26) then easily yields:

$$B = A - \frac{1}{2} \left[\frac{dh(x)}{dx} \right]_{x = x_{n\tau}} A^2 \tau + o(\tau). \tag{10.27}$$

This relation shows that in general B is density dependent, the only exception being the Malthusian growth process for which $h(x)$ is a linear function.

Let us now suppose that the considered population grows in an intrinsically discontinuous fashion and include in (10.23) the random environment assumption by the substitution $A\tau \rightarrow A'_{n\tau}$, where $A'_{n\tau}$ $(n = 0, 1, ...)$ is a sequence of independent identically distributed random variables with

$$P\{A'_{n\tau} = \sigma\sqrt{\tau}\} = \frac{1}{2} + \frac{A\sqrt{\tau}}{2\sigma}$$

$$P\{A'_{n\tau} = -\sigma\sqrt{\tau}\} = \frac{1}{2} - \frac{A\sqrt{\tau}}{2\sigma} \tag{10.28}$$

and $\sigma > 0$ is an arbitrary constant. Then we have

$$E[A'_{n\tau}] = A\tau,$$
$$E[(A'_{n\tau})^2] = \sigma^2\tau, \tag{10.29}$$
$$E[(A'_{n\tau})^{2+p}] = o(\tau) \quad (p = 1, 2, ...).$$

Note that the outlined procedure amounts to substituting the deterministic population size $x_{n\tau}$, evolving according to Eq. (10.23), by the stochastic process $X_{n\tau}$ modeled by the equations

$$\Delta X_{n\tau} = A'_{n\tau} h(x_{n\tau}) \quad (n = 0, 1, ...). \tag{10.30}$$

Furthermore, from (10.29) it follows:

$$\lim_{\tau \downarrow 0} \{\tau^{-1} E[\Delta X_{n\tau} | X_{n\tau} = x]\} = Ah(x),$$
$$\lim_{\tau \downarrow 0} \{\tau^{-1} E[\Delta X_{n\tau})^2 | X_{n\tau} = x]\} = \sigma^2 h^2(x), \tag{10.31}$$
$$\lim_{\tau \downarrow 0} \{\tau^{-1} E[\Delta X_{n\tau})^{2+p} | X_{n\tau} = x]\} = 0 \quad (p = 1, 2, ...).$$

Hence, we conclude that in the limit as $\tau \to 0$ (and, consequently, $n \to \infty$) the *Markov* process $X_{n\tau}$ converges to the diffusion process $X(t)$ whose drift and infinitesimal variance are given by

$$A_1(x) = Ah(x)$$
$$A_2(x) = \sigma^2 h^2(x). \tag{10.32}$$

Recalling (10.21), (10.22), and (8.44) we conclude that the diffusion process $X(t)$ with infinitesimal moments (10.32) is modeled by either equation

$$dx(t) = Ah(x)dt + h(x)dW(t) \tag{10.33a}$$

$$\frac{dx(t)}{dt} = \left[A - \frac{\sigma^2}{2} h(x) \frac{dh(x)}{dx} \right] + h(x)\Lambda(t). \tag{10.33b}$$

Hence, we conclude that Ito's equation (10.33a) (and the related calculus) is the natural tool for modeling the continuous approximation to an intrinsically discontinuous growth process in random environment.

Let us now assume that the considered growth process takes place according to an intrinsically continuous scheme and thus let us take Eq. (10.24) as the deterministic model of it. The following system of difference equations then holds:

$$B\tau = \int_{x_{n\tau}}^{x_{(n+1)\tau}} dz[h(z)]^{-1} \quad (n = 0, 1, ...) \tag{10.34}$$

while the random environment assumption yields:

$$B'_{n\tau} = \int_{Y_{n\tau}}^{Y_{(n+1)\tau}} dz [f(z)]^{-1} \qquad (n=0, 1, \ldots) \tag{10.35}$$

with

$$P\{B'_{n\tau} = \sigma\sqrt{\tau}\} = \tfrac{1}{2} + \frac{B\sqrt{\tau}}{2\sigma}$$

$$P\{B'_{n\tau} = -\sigma\sqrt{\tau}\} = \tfrac{1}{2} - \frac{B\sqrt{\tau}}{2\sigma} . \tag{10.36}$$

In analogy with the case previously examined, it is now easy to prove that the following relations hold:

$$\lim_{\tau \downarrow 0} \{\tau^{-1} E[\Delta Y_{n\tau} | Y_{n\tau} = y]\} = h(y) \left[B + \frac{\sigma^2}{2} \frac{dh(y)}{dy} \right]$$

$$\lim_{\tau \downarrow 0} \{\tau^{-1} E[(\Delta Y_{n\tau})^2 | Y_{n\tau} = y]\} = \sigma^2 h^2(y) \tag{10.37}$$

and that all higher order infinitesimal moments vanish. This means that $Y_{n\tau}$ converges to the diffusion process $Y(t)$ defined by either equation

$$\frac{dy(t)}{dt} = [B + \Lambda(t)] h(y)$$

$$dy(t) = \left[B + \frac{\sigma^2}{2} \frac{dh(y)}{dy} \right] h(y) dt + h(y) dW(t) . \tag{10.38}$$

Hence, we are led to the conclusion that an intrinsically continuous growth process in random environment should be modeled by Stratonovich's equation (with the appropriate calculus).

What if one ignores whether the population grows according to an intrinsically continuous or an intrinsically discontinuous scheme? The foregoing discussion clearly indicates that either equation

$$dx(t) = A h(x) + h(x) dW(t) \tag{10.39a}$$

or

$$\frac{dy(t)}{dt} = [B + \Lambda(t)] h(y) \tag{10.39b}$$

may be used, provided A and B are *not* taken as independent quantities but as mutually related as specified by (10.27).

As a realistic straightforward example of the foregoing remarks, let us refer to a particular logistic growth process by setting:

$$h(z) \equiv z \left(1 - \frac{z}{K} \right),$$

$$B \equiv m, \tag{10.40}$$

$$A \equiv w.$$

Equations (10.23) and (10.26) then read:

$$\Delta x_{n\tau} = w\tau x_{n\tau} \left(1 - \frac{x_{n\tau}}{K} \right)$$

$$m = \frac{w x_{n\tau}}{\Delta x_{n\tau}} \left(1 - \frac{x_{n\tau}}{K} \right)^{x_{(n+1)\tau}} \int_{x_{n\tau}}^{x_{(n+1)\tau}} dz \left[z \left(1 - \frac{z}{K} \right) \right]^{-1}. \tag{10.41}$$

After some straightforward calculations one then finds:

$$w\tau = \frac{1 - e^{-m\tau}}{(1 - e^{-m\tau})K^{-1}x_{n\tau} + e^{-m\tau}}. \tag{10.42}$$

Let us now make the random environment assumption and change $w\tau$ into $W_{n\tau}$ and $m\tau$ into $M_{n\tau}$, where $W_{n\tau}$ and $M_{n\tau}$ are random variables of the type earlier considered with $E[W_{n\tau}] = w\tau$ and $E[M_{n\tau}] = m\tau$. From (10.27) it then follows:

$$E[M_{n\tau}] = w\tau - \frac{\sigma^2}{2} \left(1 - \frac{2x}{K} \right) \tau + o(\tau) \tag{10.43}$$

if the $W_{n\tau}$'s are looked upon as independent. If, instead, the $M_{n\tau}$'s are viewed as the independent variaibles, one finds:

$$E[W_{n\tau}] = m\tau + \frac{\sigma^2}{2} \left(1 - \frac{2x}{K} \right) \tau + o(\tau). \tag{10.44}$$

The relations existing between the Malthusian parameter m and the Wrightian fertility w that make the two models equivalent then easily follow:

$$m \equiv \lim_{\tau \downarrow 0} \{ \tau^{-1} E[M_{n\tau}] \} = w - \frac{\sigma^2}{2} \left(1 - \frac{2x}{K} \right) \tag{10.45}$$

and:

$$w \equiv \lim_{\tau \downarrow 0} \{ \tau^{-1} E[W_{n\tau}] \} = m + \frac{\sigma^2}{2} \left(1 - \frac{2x}{K} \right). \tag{10.46}$$

The above consideration sheds some light upon the question raised earlier: How does one construct stochastic diffusion analogues of population growth processes? It also provides a simple explanation of the incorrectly named "Ito-Stratonovich controversy". Further remarks can be found in Ricciardi (1977, 1980), and in Capocelli and Ricciardi (1979).

Acknowledgements

I wish to express my warmest thanks to Dr. Amelia G. Nobile for her assistance and valuable comments during the writing of this chapter.

References

Arnold, L. (1974). Stochastic Differential Equations. New York: Wiley

Capocelli, R.M., Ricciardi, L.M. (1974). A diffusion model for population growth in random environment, Theor. Pop. Biol. 5, 28–41

Capocelli, R.M., Ricciardi, L.M. (1974). Growth with regulation in random environment, Kybernetik 15, 147–157

Capocelli, R.M., Ricciardi, L.M. (1979). A cybernetic approach to population dynamics modeling, J. Cybernetics 9, 297–312

Feldman, M.W., Roughgarden, J. (1975). A population's stationary distribution and chance of extinction with remarks on the theory of species packing, Theor. Pop. Biol. 7, 197–207

Feller, W. (1951). Two singular diffusion processes, Ann. Math. 54, 173–182

Feller, W. (1952). Parabolic differential equations and associated semigroup transformations, Ann. Math. 55, 468–518

Feller, W. (1954). Diffusion processes in one dimension, Trans. Amer. Math. Soc. 77, 1–31

Giorno, V., Nobile, A.G., Ricciardi, L.M. (1985). On some diffusion approximations to queueing systems. J. Appl. Prob. (in press)

Giorno, V., Nobile, A.G., Ricciardi, L.M., Sacerdote, L. (1985). Some remarks on the Rayleigh process. J. Appl. Prob. (in press)

Gray, A.H., Caughey, T.K. (1965). A controversy in problems involving random parametric excitation, J. Math. and Phys. 44, 288–296

Karlin, S., Taylor, H.M. (1975). A First Course in Stochastic Processes, New York: Academic Press

Karlin, S., Taylor, H.M. (1981). A Second Course in Stochastic Processes, New York: Academic Press

Jazwinski, A.H. (1970). Stochastic Processes and Filtering Theory, New York: Academic Press

Levins, R. (1969). The effect of random variations of different types on population growth, Proc. Nat. Acad. Sci. 62, 1061–1065

Lewontin, R.C., Cohen, D. (1969). On population growth in randomly varying environment, Proc. Nat. Acad. Sci. 62, 1056–1060

May, R.M. (1972). Stability in random fluctuating versus determinsitic environments, Amer. Natur. 107, 621–650

May, R.M. (1973). Stability and Complexity in Model Ecosystems, Princeton: Princeton Univ. Press

Mortensen, R.E. (1969). Mathematical problems of modeling stochastic nonlinear dynamic systems. J. Stat. Phys. 1, 271–296

Nobile, A.G., Ricciardi, L.M., Growth and extinction in random environment, in Applications of Information and Control Systems (D. G. Loiniotis and N. S. Tzannes, eds.) pp. 455–465, Dordrecht: D. Riedel Publ. Co.

Nobile, A.G., Ricciardi, L.M. (1984a). Growth with regulation in fluctuating environments. I. Alternative logistic-like diffusion models, Biol. Cybernetics *69*, 179–188

Nobile, A.G., Ricciardi, L.M. (1984b). Growth with regulation in fluctuating environments. II. Intrinsic lower bounds to population size, Biol. Cybernetics *50*, 285–299

Pawula, R.F. (1967). Generalizations and extensions of the Fokker-Planck-Kolmogorov equations, IEEE Trans. Information Theor., IT *13*, 33–41

Ricciardi, L.M. (1977). Diffusion Processes and Related Topics in Biology, Heidelberg. Springer-Verlag

Ricciardi, L.M. (1979). On a conjecture concerning population growth in random environment, Biol. Cybernetics *32*, 95–99

Ricciardi, L.M. (1980). Stochastic equations in neurobiology and population biology, in Vito Volterra Symposium on Mathematical Models in Biology (C. Barigozzi, ed.), pp. 248–263, Heidelberg: Springer-Verlag

Soong, T.T. (1973). Random Differential Equations in Science and Engineering, New York: Academic Press

Stratonovich, R.L. (1963). Topics in the Theory of Random Noise, Vol. I, New York: Gordon and Breach

Stratonovich, R.L. (1968). Conditional Markov Processes and their Applications to the Theory of Optimal Control, New York: Elsevier

Wang, M.C., Uhlenbeck, G.E. (1945). On the theory of the Brownian motion, II, Rev. Mod. Phys. *17*, 323–342

Wong, E. (1971). Stochastic Processes in Information and Dynamical Systems, New York: McGraw-Hill

Wong, E., Zakai, M. (1965). On the convergence of ordinary integrals to stochastic integrals. Ann. Math. Stat. *36*, 1560–1564

Part IV. Communities and Ecosystems

Community Dynamics in a Homogeneous Environment

Thomas G. Hallam

Outline

Biomathematics, Vol. 17, Mathematical Ecology
Edited by T. G. Hallam and S. A. Levin
© Springer-Verlag Berlin Heidelberg 1986

1. Introduction to Communities

Populations do not exist as isolated entities in a physical environment. They interact with other biological populations on a regular long term basis and, because of these interactions, often coevolve as an ecological unit. An assemblage of two of more biotic populations is called a *community*. The simplest structure, one composed of two species, and the possible interactions between these two components will be discussed first. These would not be considered communities in the classical ecological literature, but I will be consistent in using this term whenever species interactions are involved.

There exist traditional classifications of two-population systems in terms of the nature of the interactions, and these will provide the point of departure for these notes. Such distinctions, however, are often difficult to ascertain since roles can depend upon life cycle stage, environment, and many other circumstances.

1.1 Predation

The resource-consumer interaction described previously is an example of a more general two population interaction called *predation*. One population, called the predator population, utilizes the other population, called the prey, as a resource. The association is traditionally viewed as "beneficial" from the prey to the predator and as "detrimental" from predator to the prey. Host-parasite interactions are often put into this category despite obvious differences in detail. From an energy flow viewpoint, a diagram of the predator-prey association is given in Fig. 1.1.

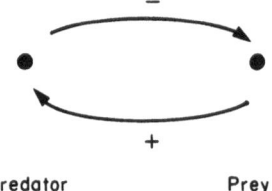

Predator   Prey

Fig. 1.1. The signs represent the energy flow in the predator-prey influence diagram

Predation can result in a negative per capita growth rate in the prey population; if it remains negative, then extinction of the population could result. Were the predator limited to that prey, the predator itself would go extinct. However, as prey become scarcer and harder to find, predators switch their diets to less expensive individuals. The same principle applies within a population: predators tend to take those that are the easiest to capture: the young, the old, and the weak. This has caused some to remark that stable predator-prey associations involve "prudent" predation.

Dyer (1980) has demonstrated that grazing can stimulate growth of the grazed plant species. Even more remarkable is the manner in which this stimulation occurs: the saliva of some herbivores can contain a hormone that initiates plant growth. There is a current debate on the beneficial aspects of grazing with the discussion focusing upon the idea that grasses and grazers have coevolved so that neither could coexist without the other. The references (Silvertown, 1982; Thompson and Uttley, 1982; Owen and Wiegert, 1983; Stenseth, 1983) make interesting reading.

The manner in which predation reacts to prey density is called the *predator functional response*. Examples of the typical types of functional responses that exist in the literature are those listed under resource-consumer interactions (Sect. 4 in the Population Ecology Chapter): linear sigmoid, and hyperbolic.

1.2 Competition

Section 2.1 describes the intraspecific competition between individuals of the same population for a set of resources. If two species must struggle for the same resources then *interspecific competition* results. Again, from an energy flow perspective, an influence diagram representing interspecific competition has each component exhibiting a negative influence upon the other (Fig. 1.2).

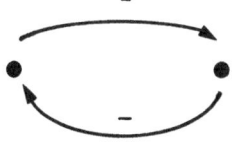

Competitor I   Competitor 2

Fig. 1.2. An influence diagram for a community composed of two competitors

It is often convenient for modelling purposes to distinguish between two aspects of competition: *exploitation* and *interference*. Interference competition refers to a mechanism, usually behavioral, that keeps a competitor from utilizing available resources. Exploitation competition occurs when a competitor actually utilizes the available resource.

There is much current argument about the role of competition in the determination of structure of an ecological community. Indeed, this discussion is rather spirited (Science, Vol. 221, 19 August 1983, p. 737), resulting in attacks on data, data interpretation, and even individuals.

1.3 Cooperation

Another type of community which we shall consider is represented by the influence diagram in Fig. 1.3.

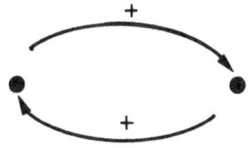

Cooperator I   Cooperator 2

Fig. 1.3. An influence diagram for energy flow in a cooperative community

These *cooperative* communities are based on the mutual benefaction of the cooperator species. Although such effects may be indirect, the traditional classifications of cooperation generally refer only to the situation where both species growth rates are increased directly by the presence of the cooperating species. The past decade has seen a widening recognition of such interactions as fundamental components of many ecosystems.

1.4 Classical Examples of Two-Species "Communities"

In Sect. 1.4 of the Population Dynamics Chapter, Figs. 1.16 representing the dynamics of the lynx and snowshoe hare and 1.17, the bean weevil and the wasp, represent special cases of predator-prey "communities", although the strength of the interaction between the lynx and the hare is open to debate (e.g. Hutchinson, 1978).

A classical laboratory predator-prey community involves the ciliates *Paramecium* and *Didinium*. *D. nasutum* is considerably larger than *P. aurelia* and will, in an unrestricted aquatic environment, consume all the *Paramecium* and then go to extinction (Luckinbill, 1973). Adding methyl cellulose to the medium (which increases the viscosity of the liquid) and decreasing the food supply of the *Paramecium* will allow persistence (Fig. 1.4: from Maynard Smith, 1974). Oscillation, as in these examples, in an important characteristic of many predator-prey communities (although see Sect. 2.4).

The dynamics of competitive communities have been studied for many systems, and examples are given in Figs. 1.5–1.7.

This system is discussed in more detail in Sect. 3.4.

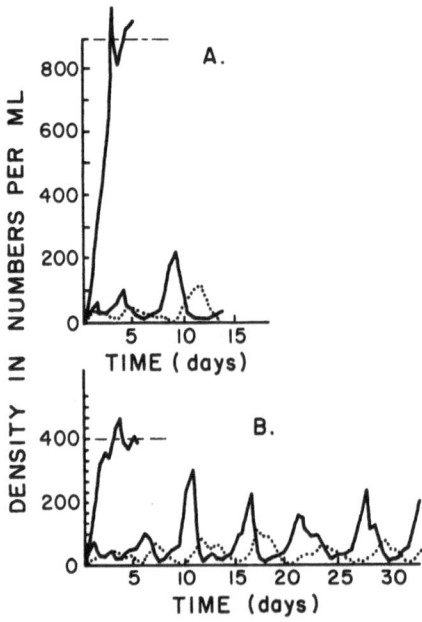

Fig. 1.4. Coexistence of *Paramecium aurelia* (solid curve) and *Didinium nasutum* (dashed curve), after Luckinbill (1973). A, medium with methyl cellulose; B, medium with methyl cellulose and reduced food for prey. In each graph, the upper solid curve is for *P. aurelia* grown by itself

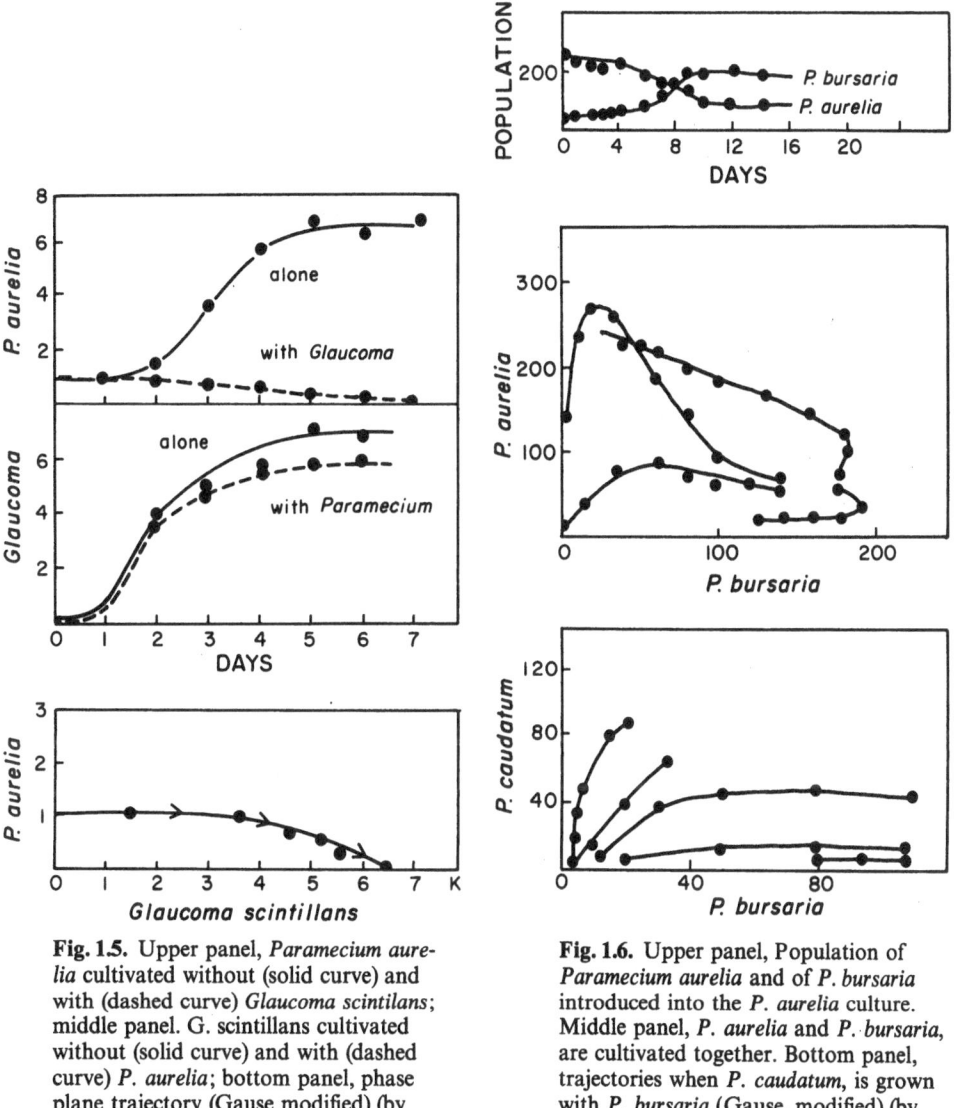

Fig. 1.5. Upper panel, *Paramecium aurelia* cultivated without (solid curve) and with (dashed curve) *Glaucoma scintilans*; middle panel. G. scintillans cultivated without (solid curve) and with (dashed curve) *P. aurelia*; bottom panel, phase plane trajectory (Gause modified) (by Hutchinson)

Fig. 1.6. Upper panel, Population of *Paramecium aurelia* and of *P. bursaria* introduced into the *P. aurelia* culture. Middle panel, *P. aurelia* and *P. bursaria*, are cultivated together. Bottom panel, trajectories when *P. caudatum*, is grown with *P. bursaria* (Gause, modified) (by Hutchinson)

When cooperative (mutualistic, symbiotic) communities are discussed, the most often cited (e.g., Boughey, 1973; De Angelis et al., 1986) examples are lichens (fungus-algae) and the interactions between the clown fish-sea anemone, cleaner wrasse-large fish, ant-acacia, nitrogen fixing bacteria-legumes, mycorrhizal fungi-plants, and plant-pollinators. However, the average ecology textbook devotes much less space to such interaction than to predator-prey or competitive interactions. A possible reason for this is that the basic principles of cooperative communities do not seem to have been as developed as those of other types of interactions. Some theoretical aspects of cooperation are explored later.

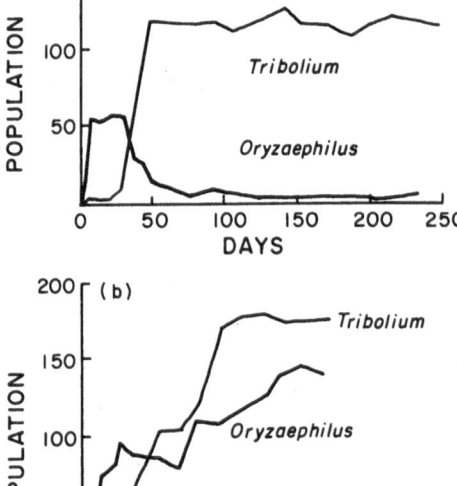

Fig. 1.7. (a) populations of the two grain beetles, *Tribolium confusum* and *Oryzaephilus surinamensis*, grown togeth-ether in flour; **(b)** the same when refuges (short lengths of capillary tubing) are mixed with the flour (Crombie)

1.5 Two Species Community Models

Let $x_1 = x_1(t)$ and $x_2 = x_2(t)$ be measurements of two populations at time t; they might, for example, denote population numbers, biomass, or densities. It is reasonable to assume that the birth rate, B_i, and the death rate, D_i, of population i, are functions not only of x_i but also of the other interacting population.

$$B_i = B_i(x_1, x_2), \quad D_i = D_i(x_1, x_2), \quad i = 1, 2.$$

The per capita growth rate is the difference between the birth rate and the death rate, and the resulting model is

$$\frac{1}{x_1}\frac{dx_1}{dt} = B_1(x_1, x_2) - D_1(x_1, x_2) \equiv f_1(x_1, x_2)$$

$$\frac{1}{x_2}\frac{dx_2}{dt} = B_2(x_1, x_2) - D_2(x_1, x_2) \equiv f_2(x_1, x_2)$$

or

$$\frac{dx_1}{dt} = x_1 f_1(x_1, x_2)$$

$$\frac{dx_2}{dt} = x_2 f_2(x_1, x_2).$$

(1.1)

Table 1.4. The signs of the partial derivatives determine the category of the two
species interaction

	$\dfrac{\partial f_1}{\partial x_2}$	$\dfrac{\partial f_2}{\partial x_1}$
Predation	−	+
Competition	−	−
Cooperation	+	+

For convenience, it is assumed that the functions f_i have continuous partial derivatives in the two dimensional nonnegative cone $\mathbb{R}_+ \times \mathbb{R}_+$. This particular form of a community model is called a model of *Kolmogorov type*.

The partial derivatives of f_i determine the classification of the community (see also Ginzburg, 1983). These are given in Table 1.4. The partial derivatives $\partial f_i/\partial x_i$ represent intraspecific competition effects or density dependent interactions.

The Kolmogorov model has numerous properties that are useful in analyzing behavior of the community. From the theory of ordinary differential equations, we know that through each point in $\mathbb{R}_+^2 = \mathbb{R}_+ \times \mathbb{R}_+$ there exists a unique trajectory of (1.1). The space $\mathbb{R}_+^2$ is invariant for (1.1); indeed, by the uniqueness of solutions to initial value problems, any trajectory emanating from the first quadrant in $x_1 - x_2$ space, $\{\mathbb{R}_+ - \{0\}\} \times \{\mathbb{R}_+ - \{0\}\}$ remains there for all time. This has implications when extinction is the objective of a study, as there can be no finite time extinction for populations modelled by (1.1).

1.6 Lotka-Volterra Systems

A simple form of f_i is a linear function: $f_i(x_1, x_2) = a_i + b_{1i}x_1 + b_{2i}x_2$. This traditional hypothesis results in logistic dynamics in the absence of the interacting species, and is built on a mass action interaction term. With a linear per capita growth rate, this model is often called a *Lotka-Volterra* model in reference to the men who first used it to study two dimensional communities. Lotka-Volterra models have been much studied and much criticized in the ecological literature for many reasons. They are difficult to apply in particular situations because the interaction coefficients cannot be computed. However, though these models may not mimic data sets exactly, they are useful in building hypotheses. While these types of models remain prevalent in the literature, current efforts to model communities are becoming much more sophisticated (see Turelli, these notes).

2. Predation

2.1 Volterra's Principle

V. Volterra, in analyzing a problem posed by his son-in-law U. D'Ancona, concluded that a moderate amount of harvesting of a prey population can, increase the average number of prey while, decreasing the average number of predators in the system. If the level of harvesting is reduced, the predator population will increase and the prey population will decrease.

This result was obtained by using the model

$$\frac{dx_1}{dt} = x_1(a_1 - b_1 x_2)$$

$$\frac{dx_2}{dt} = x_2(-a_2 + b_2 x_1).$$

(2.1)

This system has two equilibria: I: $(0, 0)$, II: $\left(\dfrac{a_2}{b_2}, \dfrac{a_1}{b_1}\right)$. Linearization shows that I is unstable, but that no conclusion can be drawn about II. Hence, we must work harder to analyze II. Fortunately, (2.1) can be written as a first order equation (by eliminating t) and then solved in a closed form by separating variables. This leads to the solutions

$$\frac{x_1^{a_2}}{e^{b_2 x_1}} - \frac{x_2^{a_1}}{e^{b_1 x_2}} = c,$$

where c is constant. It can be shown that for each c this relation defines a closed curve in $x_1 - x_2$ space and, as such, represents a periodic solution of (2.1) which contains II in its interior. The equilibrium II is the time average of each periodic trajectory: e.g. $a_2/b_2 = \dfrac{1}{p}\int_0^p x_1(s)\,ds$, where p is the period of the trajectory.

To obtain Volterra's Principle, suppose that harvesting is indiscriminate and results in a fixed proportion (hx_i) of both predator and prey being removed. The model (2.1) with harvesting is

$$\frac{dx_1}{dt} = x_1(a_1 - h - b_1 x_2)$$

$$\frac{dx_2}{dt} = x_2(-a_2 - h + b_2 x_1).$$

(2.2)

The interior equilibrium of (2.2) is $\left(\dfrac{a_2 + h}{b_2}, \dfrac{a_1 - h}{b_1}\right)$ so, on the average, indiscriminate harvesting results in an increase in prey and a decrease in predators.

If the prey species is a desired species, as it was in Volterra's situation of edible fish (prey) and selachians (predators), harvesting is desirable. On the other hand, if the prey species is undesirable, as occurred with the prey population of cottony cushion scale insects *(Icerya purchasi)* and the predator populations of the ladybird beetle *(Novius cardinalis)*, harvesting is not beneficial. This latter predator-prey system was causing only minor difficulty for the California citrus growers until indiscriminate "harvesting" by the insecticide (DDT) was initiated. As predicted, the prey population exploded and trouble ensued (Braun, 1975).

An excellent discussion of Volterra's Principle may be found in Braun (1975). Harvesting in more general predator-prey models is treated in the papers by Brauer and Soudak (1979a, b), and Brauer et al. (1976), and Brauer (1984).

2.2 Asymptotic Stability in Predator-Prey Models

Harrison (1979), expanding on a technique of Hsu (1978), has discussed the global asymptotic stability of an equilibrium of a general predator-prey model. The model is

$$\frac{dx_1}{dt} = a(x_1) - f(x_1) b(x_2)$$

$$\frac{dx_2}{dt} = n(x_1) g(x_2) + c(x_2),$$
(2.3)

where f and g are positive on $\mathbb{R}_+$; $a(x_1)$ represents the growth rate due to all factors except predation; $c(x_2)$ represents the rate of increase or decrease of the predator: $n(x_1)$ and $b(x_2)$ are assumed to be nondecreasing functions; $f x_1) b(x_2)$ is the functional response of the predator and $n(x_1) g(x_2)$ is the numerical response of the predator.

Let (x_1^*, x_2^*) be a positive equilibrium for the system (2.3) and assume that

$$[n(x_1) - n(x_1^*)] [x_1 - x_1^*] > 0, \quad x_1 \neq x_1^*;$$
$$[b(x_2) - b(x_2^*)] [x_2 - x_2^*] > 0, \quad x_2 \neq x_2^*.$$

Theorem 2.1. *If in a neighborhood of (x_1^*, x_2^*), $a(x_1)/f(x_1)$ and $c(x_2)/g(x_2)$ are both nonincreasing with at least one strictly decreasing, then the equilibrium (x_1^*, x_2^*) is asymptotically stable.*

If, in addition to all previous hypotheses,

$$\begin{cases} a(x_1) \geq b(x_2^*) f(x_1) & 0 < x_1 < x_1^* \\ a(x_1) \leq b(x_2^*) f(x_1) & x_1^* < x_1 < \infty, \end{cases}$$

$$\begin{cases} c(x_2) \geq -n(x_1^*) g(x_2) & 0 < x_2 < x_2^* \\ c(x_2) \leq -n(x_1^*) g(x_2) & x_2^* < x_2 < \infty, \end{cases}$$

with the inequalities strict according to whether $a(x_1)/f(x_1)$ or $c(x_2)/g(x_2)$ is strictly decreasing, then (x_1^*, x_2^*) is globally asymptotically stable.

Example. The Lotka-Volterra predator-prey system

$$\frac{dx_1}{dt} = x_1(a_1 - b_{11}x_1 - b_{12}x_2)$$

$$\frac{dx_2}{dt} = x_2(-a_2 + b_{21}x_1)$$

satisfies the conditions of the theorem, so the equilibrium $\left(\dfrac{a_2}{b_{21}}, \dfrac{a_1 b_{21} - b_{11} a_2}{b_{21} b_{12}}\right)$ is globally asymptotically stable. The difference between this model and Volterra's is the inclusion of a carrying capacity for the prey species. The global asymptotic stability should be contrasted with the neutrally stable (cycles of the) Volterra model.

Indication of Proof of Theorem.

The function

$$V(x_1, x_2) = \int_{x_1^*}^{x_1} \frac{n(s) - n(x_1^*)}{f(s)}\, ds + \int_{x_2^*}^{x_2} \frac{b(s) - b(x_2^*)}{g(s)}\, ds$$

is a Liapunov function for (2.3). Since $\dot{V}$ can only be zero at x_1^* or x_2^*, LaSalle's theorem (LaSalle and Lefschetz, 1961) on the extent of asymptotic stability implies that all solutions approach (x_1^*, x_2^*) as t approaches infinity.

2.3 Generation of Cycles in Predator-Prey Models

Cyclic variation in communities is a documented phenomenon (Figs. 1.14–1.18). Many of these variations do not correlate with known periodic exogenous forces such as diel or seasonal cycles. Some occur in predator-prey relationships; for example, see Figs. 1.16 (lynx-hare system) and 1.17 (wasp-bean weevil system). Many explanations of cyclic behavior have been suggested, ranging from poor data to the hypothesis that the predator cycle has nothing to do with the prey cycle. I will not belabor these points, it is my purpose to demonstrate the existence of a reasonable community model that exhibits cyclic behavior. To this end, I consider the Kolmogorov model (1.1). The Lotka-Volterra model, with its simple non-linearities, cannot have a limit cycle.

A subset of the following hypotheses can lead to cyclic system behavior (e.g., Coleman, 1976).

(H1) $\quad \dfrac{\partial f_1}{\partial x_2} < 0$.

This is a portion of the assumption that the system is of predator-prey type; (H1) implies that an increase in the predator population decreases the per capita growth rate of the prey.

(H2) $\quad \dfrac{\partial f_2}{\partial x_1} > 0$.

This completes classification as a predator-prey system and states that an increase in the prey population benefits the predator population.

(H3) $\quad \dfrac{\partial f_1}{\partial x_1} < 0$.

Density dependent effects are imposed independent of population densities.

(H3a) $\left.\dfrac{\partial f_1}{\partial x_1}\right|_{x_2=0} < 0.$

An increase in the prey population has an adverse effect upon the prey growth rate when there are no predators around.

(H4) $\dfrac{\partial f_2}{\partial x_2} < 0.$

The predator population also is limited by effects of crowding.

(H5) $f_1(0, x_2^T) = 0$ for some $x_2^T > 0.$

There is a size of the predator population, x_2^T, beyond which the prey population is decreasing even when the prey population is small.

(H6) $f_1(x_1^c, 0) = 0$ for some $x_1^c > 0.$

There exists a carrying capacity, x_1^c, for the prey population in the absence of the predator population. For $x_1 > x_1^c$, the growth rate of the prey is decreasing by (H3).

(H7) $f_2(x_1^T, 0) = 0$ for some $x_1^T > 0.$

There exists a threshold prey level necessary to support the predator population

(H8) $x_1^c > x_1^T.$

If this inequality is not satisfied, extinction of the prey population will occur

(H9) The equation $f_1(x_1, x_2) = 0$ can be solved uniquely, via the Implicit Function Theorem, for $x_2 = h(x_1)$ where $h \in C^1[0, x_1^c]$, $h' < 0$, $h(0) = x_2^T$, $h(x_1^c) = 0.$

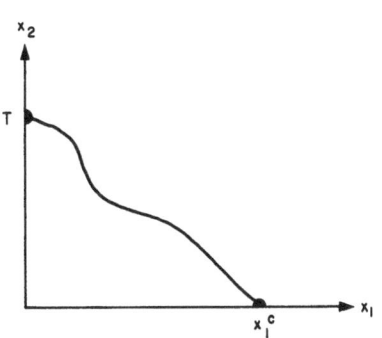 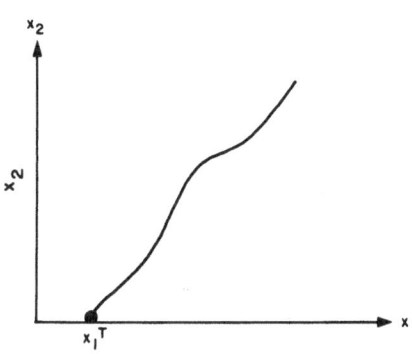

Fig. 2.1. The prey isocline defined by $x_2 = h(x_1)$ Fig. 2.2. The predator isocline defined by $x_1 = g(x_2)$

This hypothesis is, of course, related to (H3), but it is given to specify the prey isocline. The curve $x_2 = h(x_1)$ can be interpreted as the carrying capacity of the predator population at density x_1 of the prey population.

(H10) $f_2(x_1, x_2) = 0$ can be solved uniquely for $x_1 = g(x_2)$ where $g \in C^1[0, \infty)$, $g' > 0$, and $g(0) = x_1^T$.

(H11) $x_1 \dfrac{\partial f_1}{\partial x_1}(x_1, x_2) + x_2 \dfrac{\partial f_1}{\partial x_2}(x_1, x_2) < 0$.

Mathematically, this condition states that the change in f_1 along the outward normal vector emanating from the origin is negative.

(H12) $x_1 \dfrac{\partial f_2}{\partial x_1}(x_1, x_2) + x_2 \dfrac{\partial f_2}{\partial x_2}(x_1, x_2) > 0$.

(H13) The prey isocline has a hump, Fig. 2.3 (Rosenzweig, 1969).

This is an analogue of (H9) by replacing (H3) by (H3a).

(H14) $(x_1 - x_1^c) f_1(x_1, 0) < 0$,

(H15) $(x_2 - x_2^T) f_1(0, x_2) < 0$,

(H16) $(x_2 - x_2^c) f_2(x_1, 0) > 0$.

These last three conditions guarantee that equilibria on the axes are unique.

Theorem 2.2 (Limit Cycles). *Let f_1, f_2 satisfy* (H1), (H2), (H3a), (H4)–(H8), *and* (H13). *In addition, suppose that the prey-predator isoclines have the configuration in Fig. 2.4.*

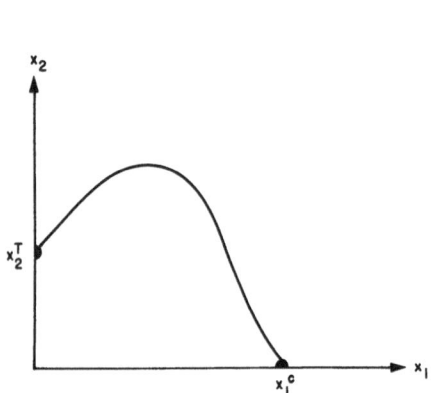

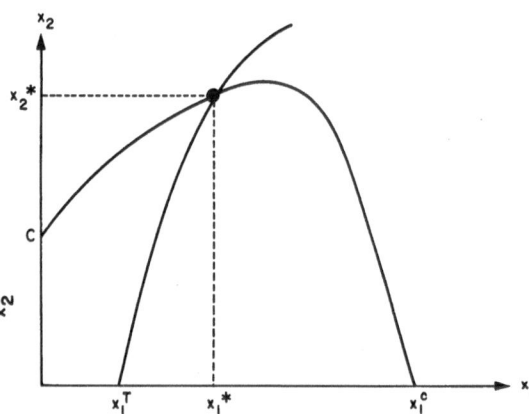

Fig. 2.3. The prey isocline has a hump

Fig. 2.4. An assumed structure for the predator prey isoclines

and that $a+d>0$, $ad-bc>0$ where

$$a=x_1^* \frac{\partial f_1}{\partial x_1}(x_1^*, x_2^*),$$

$$b=x_1^* \frac{\partial f_1}{\partial x_2}(x_1^*, x_2^*),$$

$$c=x_2^* \frac{\partial f_2}{\partial x_1}(x_1^*, x_2^*),$$

$$d=x_2^* \frac{\partial f_2}{\partial x_2}(x_1^*, x_2^*).$$

Then system (1.1) has a limit cycle in $\mathbb{R}_+^2$.

Indication of Proof. Consider the rectangle formed by the coordinate axes and the lines $x_1 = x_1^c$ and $x_2 = x_2^i$ where x_2^i is given by $f_2(x_1^c, x_2^i)=0$. The ω-limit set of any trajectory, $\omega(\Gamma^+)$, in this rectangle is a limit cycle. This may be demonstrated by showing that $\omega(\Gamma^+)$ contains no critical point. The equilibrium (x_1^*, x_2^*) is repellent; all extinction type equilibria are hyperbolic. To eliminate the possibility of a cycle graph, note that (x_1^*, x_2^*) cannot be in a cycle graph and if the extinction equilibria are in the cycle graph, $\omega(\Gamma^+)$ is unbounded. Hence, $\omega(\Gamma^+)$ is a limit cycle.

Another closely related result is

Theorem 2.3. *Let hypotheses* (H1), (H4), (H8), (H11), (H12), (H14), (H15), *and* (H16) *be satisfied. In addition, suppose* $f_1(0,0)>0$ *[that is, for small populations of predator and prey, the prey population increases]. Then, the predator-prey model* (1.1) *has a unique equilibrium with positive components. If this equilibrium is not asymptotically stable, there is a limit cycle in* $\mathbb{R}_+^2$ *which is asymptotically stable from the outside.*

Indication of Proof. For complete details see Albrecht et al. (1974). The hypotheses (H11), (H14), and (H15) imply the existence of a prey isocline, $x_2 = h(x_1)$, like that described in (H9) but $0 \le g' \le g(x_2)/x_2$; similarly, the existence of a predator isocline $x_1 = g(x_2)$ with properties similar to those indicated in (H10) but $h' \le h(x_1)/x_1$ follows. Since $x_1^c > x_1^T$, a single positive equilibrium exists. It can be shown that any trajectory must cycle about the equilibrium. The limit set could be the equilibrium or a limit cycle which is stable from the outside.

Remark. Another mathematical technique that is employed to generate cycles is a Hopf bifurcation. Waltman (1964) used this method to find periodic solutions to the Kolmogorov predator-prey system

$$\frac{dx_1}{dt} = \alpha x_1 f_1(x_1, x_2),$$

$$\frac{dx_2}{dt} = x_2 f_2(x_1, x_2).$$

There are many recent results that generate cyclic behavior by applying a bifurcation theorem (see Freedman, 1980).

Example. Another two dimensional model that uses logistic dynamics and mass action interactions for the prey population while the carrying capacity of the predator is a function of prey density is due to Leslie (1948). His model is

$$\frac{dx_1}{dt} = a_1 x_1 - b_1 x_1^2 - b_{12} x_1 x_2,$$

$$\frac{dx_2}{dt} = a_2 x_2 - b_2 x_2^2 / x_1.$$

This density dependence is probably best understood by considering the ratio of the number of prey per predator. If x_2/x_1 is small (so that there are many prey per predator), then the predators grow exponentially. If x_2/x_1 exceeds a_2/b_2 (that is, there are few prey per predator), then the predator population decreases.

The Liapunov function of Hsu-Harrison can be used to show that Leslie's model is globally asymptotically stable.

2.4 Do Predator-Prey Systems Approach Equilibria or Cycle?

Tanner (1975) studied numerous predator-prey communities and found evidence of stable equilibrium communities and cyclic behavior. While the majority of the systems he reviewed exhibited a dynamic behavior that approached an equilibrium, there was some evidence for factors that determine cycles. A propensity for a stable limit cycle seems to exist when the intrinsic growth rate of the prey population exceeds that of its predators. A prey population with a relatively high growth rate in an environment with a relatively large carrying capacity is needed for cyclic behavior (May, 1976).

Table 2.1. Life history data for 8 natural prey-predator systems (after Tanner, 1975)

Prey-predator	Geographical location	Apparent dynamical behavior
Sparrow – hawk	Europe	Equilibrium
Muskrat – mink	Central North America	Equilibrium
Hare – lynx	Boreal North America	Cyclic
Mule deer – mountain lion	Rocky Mountains	Equilibrium
White-tailed deer – wolf	Ontario	Equilibrium
Moose – wolf	Isle Royale	Equilibrium
Caribou – wolf	Alaska	Equilibrium
White sheep – wolf	Alaska	Equilibrium

2.5 Simple Food Chains

A simple food chain is a chain of predation where the dynamics of each population is determined by those species occupying the preceding and succeeding trophic levels. For example, the chain composed of a plant population, a herbivore population, and a carnivore population form a simple food chain.

A Lotka-Volterra model of a simple food chain of length n is

$$\frac{dx_1}{dt} = x_1(a_{10} - a_{11}x_1 - a_{12}x_2)$$

$$\frac{dx_2}{dt} = x_2(-a_{20} + a_{21}x_1 - a_{23}x_3)$$

$$\vdots \qquad\qquad \vdots \tag{5.1}$$

$$\frac{dx_{n-1}}{dt} = x_{n-1}(-a_{n-1,0} + a_{n-1,n-2}x_{n-2} - a_{n-1,n}x_n)$$

$$\frac{dx_n}{dt} = x_n(-a_{n0} + a_{n,n-1}x_{n-1}).$$

In the preceding model, all parameters are positive with the exception of a_{11}, which is nonnegative.

If the resource (lowest) level of the simple food chain has a carrying capacity, then solutions of (5.1) with positive initial conditions are bounded. If $a_{11} = 0$, then the unbounded growth of the resource is propagated throughout the system. First, the case $a_{11} > 0$ is developed; this model might be applicable in a situation where a resource is limited in supply and all other trophic levels are limited only by the available resource on the preceding trophic level.

Theorem 5.1. *All solutions of (5.1) with positive initial conditions are bounded provided $a_{11} > 0$.*

Indication of the Proof. The boundedness of the resource level component is readily established by using the comparison principle.

The function u is defined by

$$u(x) = \sum_{j=1}^{n} \left(\sum_{i=1}^{j-1} a_{i,i+1} \prod_{k=j}^{n-1} a_{k+1,k} \right) x_j,$$

and satisfies

$$\frac{du}{dt}(x(t)) \leq -mu(x(t)) + b,$$

where $m = \min_{1 \leq j \leq n} a_{j0}$, $b = \max_{x_1} \left(x, (2a_{10} - a_{11}x_1) \middle/ \prod_{k=1}^{n-1} a_{k+1,k} \right)$.

Solving this inequality leads to

$$u(t) \leq u(0) \exp(-mt) + b/m.$$

Since u is a linear function of x_i, each component x_i is bounded. The details of this argument as well as those of the next theorem can be found in Gard and Hallam (1979). $\square$

Persistence of the simple food chain is determined by a single system level parameter. Here persistence is defined in terms of the survival of the top predator: $\lim \sup_{t \to \infty} x_n(t) > 0$. The system level parameter is defined by

$$\mu = a_{10} - \frac{a_{11}}{a_{21}} \left[a_{20} + \sum_{j=2}^{k} \prod_{i=2}^{j} \frac{a_{2i-2, 2i-1}}{a_{2i, 2i-1}} a_{2j,0} \right]$$

$$- \sum_{j=1}^{l} \left(\prod_{i=1}^{j} \frac{a_{2i-1, 2i}}{a_{2i+1, 2i}} \right) a_{2j+1, 0},$$

where

$$k = \begin{cases} n/2 & \text{if } n \text{ is even} \\ \dfrac{n-1}{2} & \text{if } n \text{ is odd} \end{cases} \qquad l = \begin{cases} (n/2) - 1 & \text{if } n \text{ is even} \\ \dfrac{n-1}{2} & \text{if } n \text{ is odd}. \end{cases}$$

Theorem 5.2. *Let $a_{11} > 0$. The simple food chain modelled by (5.1) is persistent if $\mu > 0$; it is not persistent if $\mu < 0$.*

Indication of the Proof. Assume, for purpose of contradiction, that the food chain has a trajectory that satisfies $\lim_{t \to \infty} x_j(t) = 0$ for some j, $j = 1, 2, \ldots, n$. Then, again applying the comparison principle to (5.1) it follows that

$$\frac{dx_{j+1}}{dt} \leq -\frac{a_{j+1,0}}{2} x_{j+1}$$

and that $\lim_{t \to \infty} x_{j+1}(t) = 0$. In particular, if there is extinction, the top predator must go to extinction.

Now it will be shown by using a persistence function that the existence of a trajectory going to extinction leads to a contradiction.

Let $r_i > 0$, $i = 1, 2, \ldots, n$ and $x > 0$; define

$$\varrho(x) = \prod_{i=1}^{n} x_i^{r_i}$$

on the set $S = \{x \in R^n : 0 < x_n \leq \lambda\}$. By differentiating ϱ along trajectories on (5.1) and by proper choice of the r_i, some cancellations occur. This results in

$$\frac{d\varrho}{dt} = \varrho[r_1 \mu - r_{n-1} a_{n-1,n} x_n]. \tag{5.2}$$

On S, if λ is sufficiently small, the quantity in the brackets is positive; hence $d\varrho/dt > 0$. This implies that along trajectories $\varrho(t)$ is increasing; however, when $x_n \rightarrow 0$ so does ϱ. This contradiction shows that persistence is valid for (5.1). Conversely if $\mu < 0$, then ϱ satisfies (5.2) and

$$\frac{d\varrho}{dt} \leqq r_1 \mu \varrho .$$

Thus, $\varrho \rightarrow 0$ and an extinction must occur. $\quad\Box$

What is the situation when there is an apparently unlimited supply of a resource? (i.e., $a_{11} = 0$). There are lots of problems associated with dimensionality in Lotka-Volterra models and here we find some additional ones. The persistence-extinction parameter, μ_0, of a food chain of length $n = 2m + 1$ is

$$\mu_0 = a_{10} - \sum_{j=1}^{m} a_{2j+1,0} \prod_{i=1}^{j} \frac{a_{2i-1,2i}}{a_{2i+1,2i}} .$$

Theorem 5.3. Let $a_{11} = 0$. The food chain modelled by (5.1) is persistent provided $\mu_0 > 0$; it is not persistent if $\mu_0 < 0$.

Indication of the Proof. In the previous result with carrying capacity, the boundedness of solutions was required. As remarked above, it is not possible to demonstrate boundedness of solutions here; however, any solution that goes to extinction is bounded. This may be proved by using the classical Volterra auxiliary function:

$$V(x_1, x_2, \ldots, x_n) = \sum_{i=1}^{n} \alpha_i (x_i - \beta_i - \beta_i \log x_i/\beta_i) .$$

A proper choice of α_i and β_i (see Gard and Hallam, 1979) leads to $\dot{V}(t) < 0$ if t is sufficiently large. This shows the boundedness of solutions that go to extinction. The remainder of the argument is much like that of Theorem 5.2 and will be omitted. $\quad\Box$

As demonstrated in the next theorem, persistence in simple food chains with carrying capacity is related to the stability of an equilibrium. An interesting situation results for odd dimensional models without carrying capacity in that persistence can result even though there is no positive equilibrium. In this case, for dimension three it can be shown that the trajectories are unbounded.

Theorem 5.4. Let the system (5.1) with $a_{11} > 0$ have a positive equilibrium. Then, this equilibrium is asymptotically stable and the entire positive orthant is the domain of attraction. If the system (5.1) is persistent, it has a positive equilibrium which is globally asymptotically stable.

Indication of the Proof. The first part of the proof is due to Harrison, who uses La Salle's theorem on the extent of asymptotic stability. The proof allows for carrying capacities on each trophic level of the food chain. The system is assumed to be of the form

$$\frac{dx_i}{dt} = x_i(b_i + a_{i,i-1}x_{i-1} - a_{ii}x_i - a_{i,i+1}x_{i+1})$$

$$a_{ij} > 0, \ i \neq j, \ a_{11} > 0.$$

This is rewritten using the equilibrium x^* as

$$\frac{dx_i}{dt} = a_{i,i-1}[x_{i-1} - (x_{i-1}^*)]$$

$$- a_{ii}[x_i - (x_i^*)]$$

$$- a_{i,i+1}[x_{i+1} - x_{i+1}^*]x_i.$$

The Volterra Liapunov function

$$V(x) = \sum_{i=1}^{n} C_i\left[x_i - x_i^* - x_i^* \ln \frac{x_i}{x_i^*}\right]$$

with C_i chosen as $C_i a_i = C_{i+1} a_{i+1,i}$, has

$$\frac{dV}{dt} = - \sum_{i=1}^{n} C_i a_{ii}[x_i - x_i^*]^2 \leq 0.$$

Since $a_{11} > 0$, the set of points where $\dfrac{dV}{dt} = 0$ consists of only x^*. By La Salle's theorem, all solutions approach x^* as $t \to \infty$.

The last statement of the theorem may be proved by an inductive argument on n. □

2.6 Effects of Omnivory in Food Chains

To indicate some extensions of the classical models to which the persistence function techniques are applicable, Gard (1982) has considered the system

$$\frac{dx_1}{dt} = x_1\left[a_{10}(t, x) - \sum_{i=1}^{n} a_{1i}(t, x)x_i \right]$$

$$\frac{dx_2}{dt} = x_j\left[-a_{j0}(t, x) + \sum_{i=1}^{j-1} a_{ji}(t, x)x_j - \sum_{i=j+1}^{n} a_{ji}(t, x)x_i \right] \quad 1 \leq j \leq n-1 \qquad (5.2)$$

$$\frac{dx_n}{dt} = x_n\left[-a_{n0}(t, x) + \sum_{i=1}^{n-1} a_{ni}(t, x)x_i \right].$$

The a_{ij} are continuous functions of t and x that either vanish identically or satisfy, for some constants m_{ij} and M_{ij},

$$0 < m_{ij} \leq a_{ij}(t, x) \leq M_{ij} \qquad t \in R_+, x \in R_+^n. \qquad (5.3)$$

For $j \neq 0$, the symmetry condition means that $a_{ij} = 0$ if and only if $a_{ji} = 0$. Multiple level feeding can occur in these models.

Define the matrices A and b by

$$A = \begin{bmatrix} m_{21} & m_{31} & m_{41} & \cdots & m_{n1} \\ 0 & m_{32} & m_{42} & \cdots & m_{n2} \\ -M_{23} & 0 & m_{43} & \cdots & m_{n3} \\ \vdots & & & & \\ -M_{2,n-1} & \cdots & M_{n-2,n-1} & 0 & M_{n,n-1} \end{bmatrix}$$

$$b = \begin{pmatrix} M_{11} \\ M_{12} \\ \vdots \\ M_{1,n-1} \end{pmatrix}.$$

Theorem 6.1. *Assume that* (5.3) *holds. If there is an* $n-1$ *column vector* $r = (r_2, \ldots, r_n)^T$ *with* $r_i > 0, 2 \leq i \leq n$, *satisfying the vector matrix inequality* $Ar \geq b$ *and such that* $\mu(r) = m_{10} - \sum_{i=2}^{n} r_i M_{i0} > 0$, *then* $\limsup_{t \to \infty} x_n(t) > 0$ *for any solution* $x(t)$ *of* (5.2) *with* $x(0) > 0$; *i.e., the top predator persists.*

As an illustration of the criteria required for persistence in the case of omnivory and in the case of a simple food chain, an example is presented. In general, omnivory enhances top predator persistence from the perspective that the persistence criterion is more readily satisfied when omnivory is present.

Example. The Lotka-Volterra system

$$\frac{dx_1}{dt} = x_1(a_{10} - a_{11}x_1 - a_{12}x_2 - a_{13}x_3),$$

$$\frac{dx_2}{dt} = x_2(-a_{20} + a_{21}x_1 - a_{23}x_3),$$

$$\frac{dx_3}{dt} = x_3(-a_{30} + a_{31}x_1 + a_{32}x_2),$$

is a simple food chain if $a_{13} = a_{31} = 0$. It is a food chain with omnivory provided a_{13} and a_{31} are nonzero. The parameter that determines persistence for the simple food chain is

$$\mu_s = a_{10} - \frac{a_{11}}{a_{21}} a_{20} - \frac{a_{12}}{a_{32}} a_{30}.$$

To apply Theorem 6.1, r_2 and r_3 must be chosen so that

$$\begin{pmatrix} a_{21} & a_{31} \\ 0 & a_{32} \end{pmatrix} \begin{pmatrix} r_2 \\ r_3 \end{pmatrix} \geq \begin{pmatrix} a_{11} \\ a_{12} \end{pmatrix}.$$

A possible choice here is

$$r_3 = \frac{a_{12}}{a_{32}}, \quad r_2 = \frac{\left(a_{11} - \dfrac{a_{12}}{a_{32}} a_{31}\right)}{a_{21}}.$$

If r_2 is positive, the resulting persistence criterion is

$$\mu_0 = a_{10} - \left[\frac{a_{11} - \dfrac{a_{12}}{a_{32}} a_{31}}{a_{21}}\right] a_{20} - \frac{a_{12}}{a_{32}} a_{30} > 0.$$

It is possible for μ_0 to be positive and μ_s to be negative; hence, persistence of the top predator is enhanced by omnivory.

It is interesting that when omnivory is present, $\mu_0 > 0$ is not sufficient for the persistence of the food chain. In particular, if

$$v = a_{10} + \frac{a_{13}}{a_{23}} a_{20} - \frac{a_{11} + \dfrac{a_{13}}{a_{23}} a_{21}}{a_{31}} a_{30} > 0,$$

there are solutions close to the equilibrium

$$\left(\frac{a_{30}}{a_{31}}, \ 0, \ \frac{a_{10} - a_{11} \dfrac{a_{30}}{a_{31}}}{a_{13}}\right)$$

that approach this equilibrium. An argument similar to those above, using the function $\varrho = x_1^{-1} x_2^{r_2} x_3^{-r_3}$, may be employed to show that $v < 0$ is a persistence criterion for the intermediate level predator. The full food chain persistence criteria are $\mu_0 > 0$ and $v < 0$. An interpretation of these inequalities is that the intrinsic growth rate of the resource, a_{10}, must be large enough to support both predators.

For a general food web, Gard (1984) has employed the persistence function technique to arrive at a linear programming problem. He concludes that omnivory enhances trophic structure persistence.

2.7 Other Simple Food Chains

Freedman and Waltman (1977) have studied a general three dimensional model of a food chain:

$$\frac{dx_1}{dt} = x_1 g(x_1) - x_2 p(x_1),$$

$$\frac{dx_2}{dt} = x_2 [-r + cp(x_1)] - x_3 q(x_2), \tag{7.1}$$

$$\frac{dx_3}{dt} = x_3 [-s + dq(y)],$$

where r, s, c, and d are positive constants. They proved the persistence of (7.1) under fairly general conditions on the functions g and q. The interested reader can refer to the original paper or Freedman (1980). Freedman and Waltman (1984) have extended these food chain results to a Kolmogorov model and strengthened the results by establishing that components have a positive limit inferior as t approaches infinity.

The definition of persistence as $\lim_{t \to \infty} \sup x(t) > 0$ or $\lim_{t \to \infty} \inf x(t) > 0$ is for mathematical convenience rather than ecological reality. A more appropriate definition of persistence would involve a threshold for all components, but few results exist for such systems.

3. Competition

3.1 Lotka-Volterra-Gause Models

Gause (1934) developed a theory of competition based upon experimental work and theoretical studies grounded on the Lotka-Volterra type model,

$$\frac{dx_1}{dt} = x_1(a_1 - b_1 x_1 - b_{12} x_2),$$

$$\frac{dx_2}{dt} = x_2(a_2 - b_2 x_2 - b_{21} x_1). \tag{3.1}$$

It can be demonstrated that there are four ecologically feasible outcomes to the competition modelled by (3.1) (Fig. 3.1).

The two populations can coexist. In this case, the system has a unique positive equilibrium that is globally asymptotically stable. For later usage it is convenient to denote this coexistence by the symbol $x_1 \leftrightarrow x_2$ (Fig. 3.1a).

The positive equilibrium can also be a hyperbolic (saddle) point: The winner of the competition depends upon the initial population sizes. The function that governs the interaction is defined by the separatrices of the hyperbolic point. Notation for this outcome is $x_1 \oplus x_2$ (Fig. 3.1b).

The remaining outcomes are for the cases that one population dominates the other, so that independent of initial population size, the dominant population survives while the second goes to extinction. This is denoted by $x_1 \gg x_2$ or $x_2 \gg x_1$ according to whether x_1 or x_2 wins the competition (Figs. 3.1c and d). There is another type of system that is excluded from the above classification. This is the case where the parameters of one population, x_1, are a constant multiple, k, of the other population, x_2. This leads to an infinite number of equilibria and the relationship $x_1 = (\text{constant})(x_2)^k$ must hold between the two populations. This situation is related to the concept of competitive exclusion discussed later in Sect. 3.5.

The available resources can have an effect upon the competition between two species. While this topic is not developed here, the reader is referred to Leon and Tumpson (1975) and Hsu and Hubbell (1979).

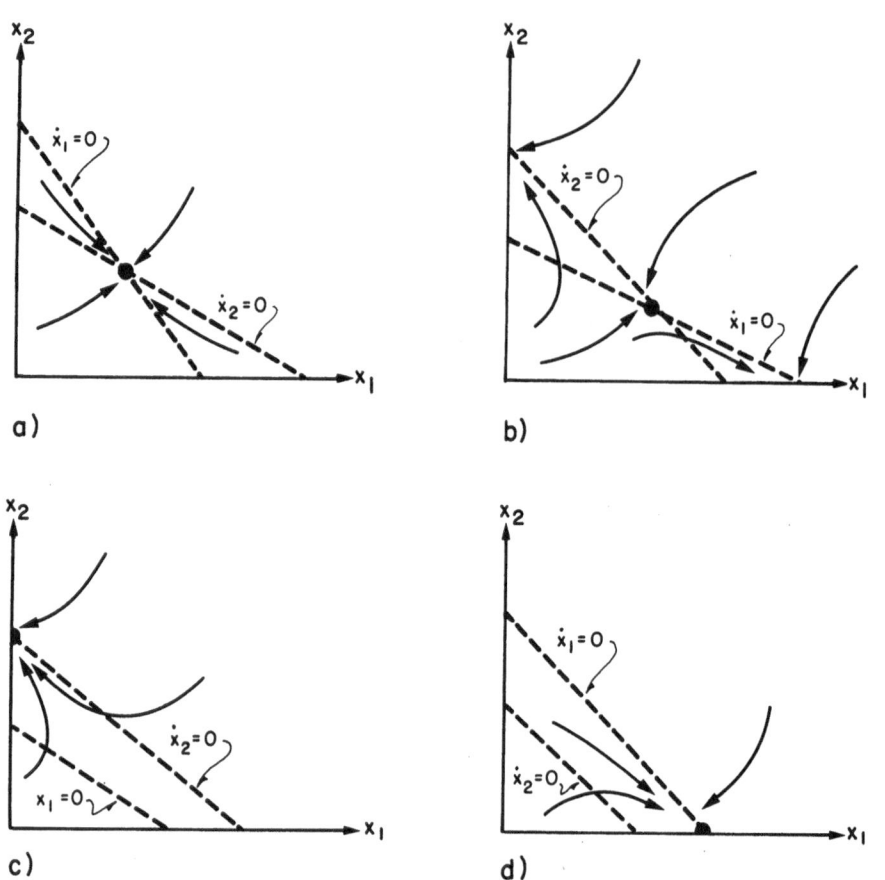

Fig. 3.1a–d. Outcomes of competition as determined by a Lotka-Volterra model. **a** $x_1 \leftrightarrow x_2$. **b** $x_1 \oplus x_2$. **c** $x_2 \gg x_1$. **d** $x_1 \gg x_2$. See text for details

3.2 Competition Models of Kolmogorov Type

The Kolmogorov model

$$\frac{dx_1}{dt} = x_1 f_1(x_1, x_2),$$

$$\frac{dx_2}{dt} = x_2 f_2(x_1, x_2),$$

$$(3.2)$$

with competitive interactions has a relatively restricted asymptotic behavior in that no limit cycles can arise. The hypotheses required to achieve this result include

(C1) $\dfrac{\partial f_1}{\partial x_2}(x_1, x_2) < 0; \quad \dfrac{\partial f_2}{\partial x_1}(x_1, x_2) < 0.$

Hence, if either population in the competition increases, the growth rate of the other species decreases.

(C2)   There exists a $K>0$ such that for $x_i \geqq K$, $i=1$ or 2, then both $f_i(x_1, x_2)$ are nonpositive.

When either population is sufficiently large, neither of them can grow.

(C3)   There exist carrying capacities x_1^c, x_2^c where $f_1(x_1, 0)>0$ for $x_1 < x_1^c$ and $f_1(x_1, 0)<0$ for $x_1 > x_1^c$; $f_2(0, x_2)>0$ for $x_2 < x_2^c$ and $f_2(0, x_2)<0$ for $x_2 > x_2^c$.

Theorem 3.1. *The limit of any solution of* (3.2) *exists and is an equilibrium, and hence populations tend to one of a finite number of limiting populations, provided* (C1), (C2), *and* (C3) *are satisfied.*

Indication of Proof. The complete details of the proof may be found in Hirsch and Smale (1974). The basic analysis employs the decomposition of the population quadrant into regions determined by flows across isoclines. Then each type of region is analyzed to determine the characteristics of the feasible ω-limit sets. As indicated in the theorem statement, all ω-limit points are equilibria (and in particular, there are no limit cycle behaviors associated with competitive systems of Kolmogorov type). The stability of these equilibria is relatively easy to determine and it is shown that there exists at least one asymptotically stable equilibrium.

3.3 Competition in Laboratory and Natural Communities:
Some Classical Examples

Gause, stimulated by the theoretical work of Volterra, undertook some laboratory experiments that led to outcomes much like the theoretical work predicted (Sect. 3.1). His work on two yeast populations was not definitive because of the production of ethyl alcohol (yeast are fine for making wine but not for interacting) which, in turn, shut down reproduction.

Apparently not discouraged, Gause continued his work in aquatic systems, this time using ciliates. Using *Glaucoma scintillans* and *Paramecium aurelia*, he found that the smaller organism, *Glaucoma*, was not inhibited by the *Paramecium* while the growth of the *Paramecium* population was hindered by the presence of *Glaucoma*.

Gause also employed three species of paramecium in some experiments: *P. aurelia*, *P. caudatum*, and *P. bursaria*. The outcomes of competition between these species were:

1. *P. caudatum* $\gg$ *P. aurelia* if metabolic products was completely removed.
2. *P. aurelia* $\gg$ *P. caudatum* in most other instances; hence the winner of the competition could be changed by a perturbation in environment.
3. *P. aurelia* $\leftrightarrow$ *P. bursaria*. (This might not be direct competition for a resource since *P. bursaria* tended to feed on the sediments.) The data indicates that multiple equilibria might result.

4. *P. caudatum* and *P. bursaria* mixtures led to inconclusive results. Stable equilibrium coexistence did occur in certain instances, and *P. caudatum* $\ll$ *P. bursaria*, occured if *P. bursaria* was initially present in sufficiently high densities.

Another classical competition experiment was that of Park (1954, 1962) using metazoa. *Tribolium confusum* and *T. castaneum* were used in a homogeneous environment, with different temperature and humidity ranges.

T. confusum, grown by itself in a hot, wet environment, reached peak densities. However, when it was grown with *T. castaneum* (which did moderately well in a hot wet environment), *T. castaneum* won the competition. In a cool dry environment, neither species did well. In this setting, *T. confusum* dominated the competition. Two aspects of classical ecological dogma are illustrated by these experiments. Environmental conditions are important factors in competition and extraction of information from population to community levels is not, in general, a feasible objective.

Many factors can provide a basis for changing the outcome of competition. These include refuge, predation, and genetic effects. Crombie (1945, 1946) utilized flour beetles and, by adding a refuge, changed the outcome of competition from one of competitive dominance to stable coexistence. Pimentel et al. (1965), using houseflies and blowflies, were able to change the dominance in this system through selection for superior competitors.

As indicated in Fig. 1.8, Connell (1961) studied the competition between two barnacle populations in an intertidal community. The barnacles of the genera *Balanus* and *Chthamalus* competed interspecifically for space on the rocks in the intertidal. The *Balanus* were vigorous and tended to dominate *Chthamalus* in the lower zones, while the situation was reversed in the upper regions.

3.4 Competition for a Single Nutrient in Continuous Cultures of Microorganisms

Hsu et al. (1977), (see also Novick and Szilard, 1950; Herbert et al., 1956) developed a theory of competition between microorganisms, such as phytoplankton, for a single limiting nutrient. Their modelling efforts were motivated by chemostat experiments in which the initial input, S^0, and dilution rate, D, of the nutrient were known constants and the environmental medium was homogeneous.

Let $x_i(t)$ denote the concentration of the ith population at time t; $S(t)$ denote the concentration of substrate at time t; m_i is the maximum growth rate of the ith population; y_i is the growth yield for the ith population and a_i is the Michaelis-Menten half saturation constant. The model is

$$\frac{dS(t)}{dt} = (S^0 - S(t))D - \sum_{i=1}^{n} \frac{m_i}{y_i} \frac{x_i(t) S(t)}{a_i + S(t)}$$

$$\frac{dx_i(t)}{dt} = \frac{m_i x_i(t) S(t)}{a_i + S(t)} - Dx_i(t),$$

$$S(0) = S^0,$$

$$x_i(0) = x_{i0} > 0.$$

Theorem 3.2 (Extinction). *Let* $b_i = m_i/D$. *If either* $b_i \leq 1$ *or* $\dfrac{a_i}{b_i - 1} > S^0$ *(when* $b_i > 1$) *then* $\lim\limits_{t \to \infty} x_i(t) = 0$.

Extinction results if the maximum growth rate m_i of the ith population is less than the dilution rate or if the metabolic needs of the population, $a_i/(b_i - 1)$ exceeds the initial amount of nutrient present in the system.

Theorem 3.3 (Persistence of one Population). *Let i be an integer, $1 \leq i \leq n$, and suppose $0 < a_i/(b_i - 1) < a_j/(b_j - 1)$ for all $j \neq i, j = 1, 2, \ldots, n$. Let $S^0 > a_j/(b_j - 1)$ and $b_j > 1$. Then*

$$\lim_{t \to \infty} S(t) = \frac{a_i}{b_i - 1},$$

$$\lim_{t \to \infty} x_i(t) = y_i \left(S^0 - \frac{a_i}{b_i - 1} \right),$$

$$\lim_{t \to \infty} x_j(t) = 0 \quad j \neq i.$$

The proof of this last theorem is long and involved, although not difficult to understand; the interested reader is referred to the original article for details.

This competition model has again led to a globally asymptotically stable equilibrium. Survival of a population is determined by the smallest of the ratios: $a_i/(b_i - 1)$. This indicates that when a single resource is limiting for a community, only one population can survive. The validity of this statement and the presence of cycles in competitive systems is explored in the next section.

3.5 The Proposition of Competitive Exclusion

Gause's experiments with *Paramecium caudatum* and *P. aurelia* resulted in *P. aurelia* dominating in the competition for a single limiting resource in most cases. From these experiments and from the mathematical theory developed by Volterra arose the proposition that an ecological community in which there are n species cannot persist on less than n limiting resources. (A resource is limiting if it is necessary for maintenance and development of the community and its supply is exhaustable by sufficient utilization.)

While Gause's research was in the laboratory, there also exist classical studies of competition with exclusion in natural ecosystems. R. MacArthur (1958) studied five species of warblers that appeared to be so similar in ecological preferences that competitive exclusion was violated. He found that they feed and occupy different levels in their forested environment and that competitive exclusion held for this community.

Theoretical aspects of competitive exclusion have been well developed in recent years. The work of Hsu et al. (1977) mentioned previously in Sect. 3.4 supported the concept of competitive exclusion if the ratio of the Michaelis-Menten parameters of each population was distinct from the others. They also demonstrated that whenever two species have equal Michaelis-Menten ratios, it is possible for both species to survive.

Related to competitive exclusion is the "paradox of the plankton" (Hutchinson, 1978). The coexistence of many species of phytoplankton in a well mixed body of water with only a few limiting nutrients (usually one) seems to violate competitive exclusion. The analysis of Hsu, Hubbell, and Waltman suggested that, for the species to survive, the Michaelis-Menten parameter ratios should be very similar. Theoretically, this allows exclusion to proceed very slowly.

Levin (1970) also provided a theoretical basis for a higher dimensional competition exclusion. He considered the model:

$$\frac{dx_1}{dt} = x_1 f_1(x_1, \ldots, x_n; y_1, \ldots, y_m)$$

$$\vdots \qquad\qquad \vdots \qquad\qquad\qquad (3.3)$$

$$\frac{dx_n}{dt} = x_n f_n(x_1, \ldots, x_n; y_1, \ldots, y_m),$$

where x_i are state variables representing species in the community and y_j represent environmental parameters. Any quantity that influences f_i is called a limiting factor. He also allowed combination of influences; for example, if a species requires and utilizes two resources R_1, R_2 with utilization efficiencies α_1, α_2 then $\alpha_1 R_1 + \alpha_2 R_2$ is a single limiting factor. Suppose that there exists a minimal independent set of limiting factors $\{z_1(x_1, \ldots, x_n; y_1, \ldots, y_m), \ldots, z_p(x_1, \ldots, x_n; y_1, \ldots, y_m)\}$ where $p \leq m+n$, and that the growth rates f_i are linear functions

(i) $f_i = \alpha_{i2} z_i + \alpha_{i2} z_2 + \ldots + \alpha_{ip} z_p + \gamma_i$.

Theorem 3.4. *No asymptotically stable equilibrium or periodic solution can be attained in a community modelled by (3.3) in which some r components are limited by less than r limiting factors.*

Indication of the Proof. If the first r components are limited by less than r factors, because of the linearity of the f_i, there exist β_i, δ not all zero such that

$$\beta_1 f_1 + \ldots + \beta_r f_r = \delta.$$

Employing the Eqs. (3.3), we obtain the expression

$$\beta_1 \frac{\dot{x}_1}{x_1} + \beta_2 \frac{\dot{x}_2}{x_2} + \ldots + \beta_r \frac{\dot{x}_r}{x_r} = \delta.$$

Integration leads to

$$x_1^{\beta_1}(t) \ldots x_r^{\beta_r}(t) = K e^{\delta t}.$$

Using the equilibrium condition, we obtain that δ must be zero, and that each solution lies on the surface

$$x_1^{\beta_1} \ldots x_r^{\beta_r} = K,$$

for some K. Any small perturbation will more the system to a different surface, indicating the impossibility of asymptotic stability. ☐

While the above mentioned works support the proposition of competitive exclusion, not all theoretical research does.

McGehee and Armstrong (1977) showed, for certain standard models where competitive exclusion occurs, that, topologically, the result is not robust. They, modify a model by several small nonlinear perturbations, and end up with a persistent system (in fact, one with a cyclic behavior).

Kaplan and Yorke (1977) demonstrate that Levin's work is not robust in that there exists a nonlinear n-dimensional system

$$\frac{dx_i}{dt} = x_i f_i(r_1(x), \ldots, r_k(x)), \quad k < n$$

which has an asymptotically stable periodic solution. Related competitive exclusion ideas will be discussed in the next section.

The concepts of competitive exclusion and niche theory have been interrelated in the literature. The reader is referred to Hutchinson (1978) and Whittaker and Levin (1975) for discussions on the theory of the niche.

3.6 Stability in Higher Dimensional Competitive Communities

The analytical theory of higher dimensional communities of competitors is only beginning to develop. To give an indication of some of the types of available results, I discuss persistence in a three dimensional Lotka-Volterra model. For other recent results the monograph of Waltman (1983) is a good reference.

Three Dimensional Lotka-Volterra Models of Competition

Persistence-Extinction Phenomena. While the case has been adequately made to not place too much biological faith in the interpretation and output of Lotka-Volterra type models, there are some mathematically interesting aspects that will be mentioned here.

The Lotka-Volterra type model for a competitive community consisting of three populations is

$$\frac{dx_i}{dt} = x_i \left(a_{i0} - \sum_{j=1}^{3} a_{ij} x_j \right) i = 1, 2, 3 . \tag{3.4}$$

The system (3.4) has a solution $x = (x_1, x_2, x_3)$ that goes to extinction if there exists a positive initial condition such that the solution through this initial value satisfies $\lim_{t \to \tau} x_i(t) = 0$ for some i, $i = 1, 2, 3$, and some $\tau \in (0, \infty]$. For (3.4) and, in fact, for all models of Kolmogorov-type with sufficient continuity requirements, finite time extinction is not possible provided initial value problems have unique solutions. The extinction planes $(x_i = 0)$ are invariant for such models and uniqueness of initial value problems guarantees that no trajectory that is ever in an extinction plane at a finite time can emanate from the population octant. If the system (3.4) has no solution that goes to extinction, it is called *persistent*.

To determine persistence of the community model (3.4), I first classify the types of extinction that can occur (Hallam et al., 1979). A comparison of components of solutions of (3.4) with appropriate logistic equations coupled with a differential inequality argument establishes all solutions of (3.4) are bounded. It is evident that complete extinction cannot result since for small x_i, all species have positive growth rates.

Two Population Extinction. Necessary conditions for the existence of a trajectory satisfying $\lim_{t \to \infty} x_k(t) > 0$ and $\lim_{t \to \infty} x_j(t) = 0, j \neq k$, are

$$a_{jk}a_{k0} \geq a_{j0}a_{kk} \quad j \neq k. \tag{3.5}$$

If the weak inequality in (3.5) is replaced with strict inequality then the inequality is also sufficient for the existence of a trajectory that has the two population extinction behavior described above.

Proof. Any extinction trajectory of this type must approach the equilibrium $x_k^* = a_{k0}/a_{kk}$; $x_j^* = 0, j \neq k$. This equilibrium can be attracting only if it is not situated in either of the regions where $dx_j/dt > 0, j \neq k$. This implies that inequality (3.5) must be satisfied.

Conversely, a linearization implies that the above equilibrium has a nontrivial stable manifold that intersects the positive population octant; and, hence, there exist solutions with the desired asymptotic behavior. □

Single Population Extinction. Define the system parameters

$$b_{ij} = a_{10}a_{jj} - a_{j0}a_{ij}$$

$$C_{ij} = a_{ii}a_{jj} - a_{ij}a_{ji}$$

$$d_k = a_{k0} - a_{ki}\left[\frac{b_{ij}}{C_{ij}}\right] - a_{kj}\left[\frac{b_{ji}}{C_{ij}}\right], \quad k \neq i, j, i \neq j.$$

A necessary condition for the existence of a trajectory satisfying $\lim_{t \to \infty} x_k(t) = 0$ and $\lim_{t \to \infty} x_j(t) > 0, j \neq k$, are

$$b_{ij}b_{ji} \geq 0 \quad i \neq j; i, j \neq k, \tag{3.6}$$

$$d_k \leq 0. \tag{3.7}$$

Indication of the Proof. The ω-limit set of any such extinction trajectory is the equilibrium $x_k^* = 0$, $x_i^* = \dfrac{b_{ji}}{C_{ij}}$, $x_j^* = \dfrac{b_{ji}}{C_{ij}}$. Thus, b_{ij} and b_{ji} must be of the same sign; hence, (3.6) is valid. The equilibrium cannot be located in the region where $dx_k/dt > 0$; geometrically, this requires that the equilibrium cannot be below the line which is the intersection of S_k and the $x_k = 0$ plane. This is inequality (3.7).

The converse of this extinction result is also valid. There are two cases. If both b_{ij} and b_{ji} are positive, there is a positive equilibrium in the $x_i x_j$-plane with a stable manifold that intersects the positive octant. When $b_{ij} < 0$, $b_{ji} < 0$, and $d_k < 0$, the equilibrium is a saddle point with only a single trajectory of single population extinction type. □

Persistence Results. Since the extinction results are almost necessary and sufficient, persistence can be determined when some of the extinction criteria are violated. The approach taken here is to assume that interactions between each of the two population community, as are known, then determine what conditions, are required for the system to be persistent.

In terms of the system parameters, the two species interactions can be described by

$$x_i \gg x_j \leftrightarrow b_{ij} > 0, \, b_{ji} < 0;$$

$$x_i \leftrightarrow x_j \leftrightarrow b_{ij} > 0, \, b_{ji} > 0;$$

$$x_i \Leftrightarrow x_j \leftrightarrow b_{ij} < 0, \, b_{ji} < 0.$$

Persistence can occur in the following cases:

A.1 $x_i \gg x_j$, $x_j \gg x_k$, $x_k \gg x_i$;

A.2 $x_i \leftrightarrow x_j$, $x_i \gg x_k$, $x_k \gg x_j$;

A.3 $x_i \leftrightarrow x_j$, $x_i \leftrightarrow x_k$, $x_j \gg x_k$;

A.4 $x_i \leftrightarrow x_j$, $x_j \leftrightarrow x_k$, $x_k \leftrightarrow x_i$.

Case A.1 has been studied extensively (e.g. May and Leonard, 1975; Gilpin, 1975; Grossberg, 1978). Grossberg demonstrated that the attractor is a cycle graph connecting the carrying capacities of the individual populations. This arrangement of two population interactions is rather unusual in that persistence is known automatically from the interactions.

For the arrangements A.2, A.3, and A.4, at least one additional relationship between the population parameters must be prescribed. These relationships require that a systems level parameter, d_k be positive. The biological interpretation of the d_k is that of an invasibility parameter indicating the ability of a population represented by x_k to invade a community at equilibrium: $x_i \leftrightarrow x_j$.

The persistence criteria are A.2, $d_k \geq 0$; A.3, $d_j \geq 0$ and $d_k \geq 0$; A.4, $d_i \geq 0, d_j \geq 0$, and $d_k \geq 0$. If, in any of the above arrangements, $d_k < 0$ then extinction of a population occurs. With the exception of A.1 just knowing all two population interactions is not sufficient to determine persistence. Any arrangement other than the forms above lead to extinction. In particular, any arrangement that contains an unstable competitive pair $(x_i \Leftrightarrow x_j)$ cannot be persistent. This is not a robust property and is a consequence of the Lotka-Volterra model.

Stability and Persistence. This model has some interesting dynamical features. The intransitive arrangement, A.1, has already been mentioned, with the ω-limit set being a cycle graph. It is interesting that the arrangement A.1 is persistent without auxiliary conditions satisfied, and has an asymptotic behavior where $\lim\sup_{t \to \infty} x_i(t) > 0$ for each $i = 1, 2, 3$, but it is not the case that $\lim\inf_{t \to \infty} x_i(t) > 0$ for an $i = 1, 2, 3$.

Goh (1977) has demonstrated that (3.4) with coefficients given by

$$a_{1j} = 2, 0.8, 0.5, 0.7;$$

$$a_{2j} = 1.5, 1, 0.2, 0.3; \, j = 0, 1, 2, 3,$$

$$a_{3j} = 2.1, 0.2, 1, 0.9$$

has an asymptotically stable equilibrium $(1, 1, 1)$ and a trajectory with components x_1 and x_3 that go to extinction. The underlying arrangement is of the extinction form $x_i \leftrightarrow x_k, x_j \gg x_k, x_i \oplus x_j$. Hence, it is possible for a competitive system to have an asymptotically stable interior equilibrium although the system is not persistent.

Strobeck (1973) gave necessary and sufficient conditions for local stability by employing the Routh-Hurwitz criteria. He also presented the two examples. The first has coefficients

$$a_{1j} = \tfrac{19}{3}, 1/3, 2/3, 4/3;$$

$$a_{2j} = 3/2, 1/18, 1/6, 1/3; \, j = 0, 1, 2, 3;$$

$$a_{3j} = 4, 1/3, 1/3, 1 \,.$$

This system has the two species arrangement $x_1 \leftrightarrow x_2, x_2 \leftrightarrow x_3, x_1 \leftrightarrow x_3$, and $d_1 > 0, d_3 > 0$. It is persistent and the equilibrium is asymptotically stable.

The second example multiplies the first and third components above by 3 to give coefficients $3a_{1j}; a_{2j};$ and $3a_{3j}$. This does not change the equilibrium or the species arrangement, but it does modify its stability to that of instability.

4. Models of Cooperation

Perhaps the most interesting and beneficial association between two species is the act of cooperation. This interaction has been suggested as an evolutionary objective of selection by Odum (1974) and others. The interaction can be classified as *obligatory* in the sense that survival of each population depends upon the presence of the other, or it can be *facultative* in that the association is not obligatory.

Classical examples of cooperation [which to various degrees have also been referred to as mutualism, symbiosis, commensalism, or amensalism (Odum, 1974)] include the algal and fungal components of lichens, the clown fish *(Amphirion percula)* and sea anemones, the ant-*Acacia* system (Janzen, 1966) and plant-pollinator systems.

4.1 Lotka-Volterra Models with Facultative Associations

In the absence of interspecific effects, the individual populations are assumed to be governed by logistic equations; hence the model with mass action interaction terms is

$$\frac{dx_1}{dt} = x_1(a_1 - b_1 x_1 + c_{12} x_2),$$

$$\frac{dx_2}{dt} = x_2(a_2 - b_2 x_2 + c_{21} x_1),$$

where

a_i, b_i, c_{ij} are positive constants.

This model has two possible types of asymptotic behavior. There can exist a positive equilibrium that is globally asymptotically stable (in the case when $b_1 b_2 - c_{12} c_{21} > 0$). The second type of behavior occurs if $b_1 b_2 - c_{12} c_{21} \leq 0$; the result is, as aptly described by May, "an orgy of mutual benefaction", specifically, there is unbounded growth for each component.

4.2 Obligatory Interactions as Modelled by Lotka-Volterra Kinetics

For obligatory interactions it is assumed that each population, in the absence of the interacting species, will decay exponentially and that interactions are represented by mass action formulations. Hence, the resulting model is

$$\frac{dx_1}{dt} = x_1(-a_1 + b_1 x_2),$$

$$\frac{dx_2}{dt} = x_2(-a_2 + b_2 x_1). \tag{4.1}$$

Models such as (4.1) can exhibit a stupendus orgy of mutuality since it can be demonstrated that they have solutions with a finite escape time [that is, there exists a $T < \infty$ such that $\lim_{t \to T} x_1(t) = \infty$ or $\lim_{t \to T} x_2(t) = \infty$]. For example, with $a_1 = a_2 = b_1 = b_2 = 1$ the substitution $v = x_1 - x_2$ leads to the temporal representation $V(t) = V(0)e^{-t}$. To demonstrate a finite escape, we can use the transformation $w = x_2^{-1}$ to show that

$$dw/dt + (-1 + V(0)e^{-t})w = -1.$$

The classification of those solutions w that vanish at finite time can be obtained, and these solutions correspond to those solutions of (4.1) with finite escape time. There is also a threshold below which initial populations of each component population tend to extinction.

A graphical solution of (4.1) is presented in Fig. 4.1. The equilibrium is a saddle point with regions of growth and extinction determined by the separatrices of the saddle point.

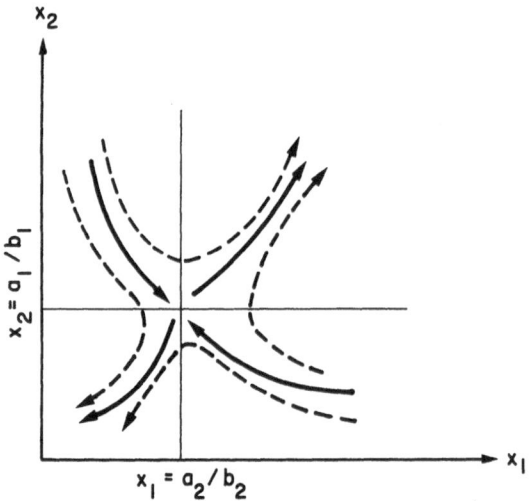

Fig. 4.1. Phase plane diagram of a Lotka-Volterra model of obligatory cooperation

4.3 Other Models of Cooperation

While the preceding model contains some desirable properties, such as an extinction threshold, the unbounded growth of solutions is certainly undesirable from a modeling perspective. Vandermeer and Boucher (1978) address the question "How should the isoclines be constructed for cooperative systems?" If interspecific interactions become weaker as population densities become large, then this might have the effect of curving the isoclines towards each other so that they again intersect. At this second intersection will be a stable equilibrium, and the unpleasant unboundedness of solutions present in the original model (4.1) does not occur here (Fig. 4.2).

The model might now have the form

$$\frac{dx_1}{dt} = x_1(-a_1 + b_1(x_1, x_2)x_2),$$

$$\frac{dx_2}{dt} = x_2(-a_2 + b_2(x_1, x_2)x_1),$$

where b_1, b_2 are decreasing functions of both x_1, x_2.

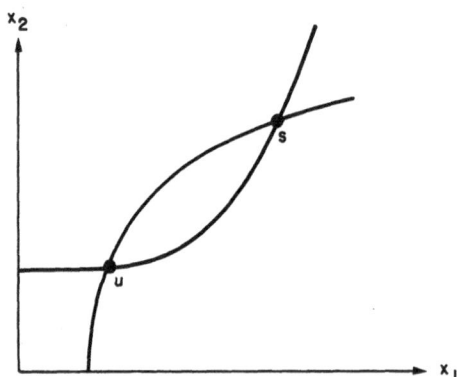

Fig. 4.2. Feasible isoclines for a cooperative system. u and s indicate an unstable and a stable equilibrium respectively

An example of a community where this situation might exist is the legume and bacteria *(Rhizobium)* system (Vandermeer and Boucher, 1978). Properties of this interaction include:

i) There is a minimal population of bacteria necessary for successful plant establishment. The few bacteria generally present are insufficient for crop growth and inoculation is often required to achieve an establishable community.

ii) Additional bacteria inoculum have little effect on nodulation (and thus, presumably, on plant growth and reproduction). This occurs above a certain threshold level; for example, on red clover seedlings grown in culture, additions above $10^4 (ml)^1$ of rhizosphere produce no discernible changes in growth.

iii) The number of bacteria present in the soil when symbiotic with legumes are usually substantially greater then the number needed for nodulation.

In another attempt to formulate a realistic model, May (1976) proposed modifying the carrying capacity of the logistic equation to reflect dependence upon the density of the complementary population. The system is written here in a slightly different form to be consistent with earlier discussions:

$$\frac{dx_1}{dt} = x_1 \left[r_1 - \frac{C_1 x_1}{B_1 + \alpha_1 x_2} \right],$$

$$\frac{dx_2}{dt} = x_2 \left[r_2 - \frac{C_2 x_2}{B_2 + \alpha_2 x_1} \right].$$

This formulation has the effect of increasing the equilibrium values of each of the components over the individual population carrying capacities.

Kolmogorov-Type Models

The general model of a cooperative two dimensional community of Kolmogorov-type is

$$\frac{dx_1}{dt} = x_1 f_1(x_1, x_2),$$

$$\frac{dx_2}{dt} = x_2 f_2(x_1, x_2),$$

where for $x_1 \geq 0$, $x_2 \geq 0$, the following requirements are imposed:

(M1) $\quad \frac{\partial f_1}{\partial x_2} > 0, \quad \frac{\partial f_2}{\partial x_1} > 0$ (the interaction is cooperative)

(M2) $\quad x_1 \frac{\partial f_i}{\partial x_1} + x_2 \frac{\partial f_i}{\partial x_2} \leq -\alpha < 0, \quad i = 1, 2$

$\quad$ (changes in f_i along outward vector from origin is negative)

(M3) $\quad f_i(0,0) > 0, \quad i = 1, 2$ (small populations grow)

(M4) $f_i(K_1,0)=f_2(0,K_2)=0$

(there is a carrying capacity for each population).

The reader is referred to Albrecht et al. (1974) who demonstrated that there is a feasible equilibrium which is globally asymptotically stable.

4.4 Stability in Higher Dimensional Cooperative Communities

An equilibrium, x^*, of the Kolmogorov system

$$\frac{dx_i}{dt}=x_ig_i(x),\quad i=1,2,\ldots,n,\ x=(x_1,x_2,\ldots,x_n)^T \tag{4.2}$$

is asymptotically stable if and only if the eigenvalues of the community matrix $S=(s_{ij})$,

$$s_{ij}=\frac{\partial}{\partial x_j}(x_ig_i(x))|_{x=x^*}$$

have negative real parts. Since $s_{ij}=x_i^*\frac{\partial g_i}{\partial x_j}(x^*)$, x^* is asymptotically stable if and only if all of the eigenvalues of DA gave negative real parts where $D=\mathrm{diag}(x_1^*,$ $x_2^*,\ldots,x_n^*)$ and A is the *interaction matrix*, $A=\frac{(\partial g_i(x^*))}{\partial x_j}$.

For competitive and predator-prey systems, equilibrium stability is independent of the stability of the interaction matrix. That is, there exist competitive and predator-prey communities for which the *community matrix DA* is unstable even though the interaction matrix A has eigenvalues with negative real parts; conversely, there exist communities for which the community matrix DA is asymptotically stable even though the interaction matrix is unstable (e.g. Strobeck, 1973).

A pleasant property of cooperative systems is that the above difficulties are simplified in that stability of an equilibrium is determined solely by the interaction matrix. Assuming that (4.2) is completely cooperative, that is,

$$\frac{\partial g_i}{\partial x_j}(x)\geqq0,\ x\in R_n^+,\quad i\neq j,$$

we obtain the following classification of stability.

Theorem 4.1. *A cooperative community modelled by (4.1) has an (asymptotically) stable, feasible equilibrium x^* ($x^*>0$) if and only if the interaction matrix A is asymptotically stable.*

Indication of the Proof. The concept of an M-matrix is useful in the subsequent arguments. The following criteria are equivalent
1. A is an M-matrix

2. All eigenvalues of A have positive real parts
3. A is nonsingular and $A^{-1} \geq 0$
4. There exists a $z > 0$ such that $Az > 0$
5. There exists a $y > 0$ such that $A^T y > 0$
6. The principal minors of A are positive.

The off diagonal elements of the matrix DA are nonnegative. The matrix DA is asymptotically stable if and only if $-DA$ is an M-Matrix. Property 5 yields that $-DA$ is an M-matrix is equivalent to the existence of a vector $x, x > 0$, such that $-(DA)^T x = -A^T Dx > 0$. Hence, this is equivalent to the existence of a $y > 0$ such that $-A^T y > 0$. Since $a_{ij} \geq 0$ for $i \neq j$, this is equivalent to the statement that $-A$ is an M-matrix. This results in the conclusion of the theorem. □

There are several interesting consequences of Theorem 4.1. Since $-A$ is an M-Matrix, the stability of an equilibrium, x^*, is equivalent to the existence of a vector $d > 0$ such that $Ad < 0$. Writing this statement in terms of the components, we obtain the inequality

$$d_i |a_{ii}| > \sum_{\substack{j=1 \\ j \neq 1}}^{n} d_j a_{ij}, \quad i = 1, 2, \dots, n. \tag{4.3}$$

When (4.3) holds, A is called quasi-diagonally dominant. An interpretation of (4.3) is that for stability of x^*, the intraspecific competition must dominate the interspecific interaction terms. In an analogous manner, a column diagonal dominance property can be found.

Employing property 6 of M-matrices, a simple algebraic relationship may be obtained for the stability of an equilibrium;

$$(-1)^K \begin{vmatrix} a_{11} & a_{12} & \cdots & a_{1K} \\ a_{21} & a_{22} & \cdots & a_{2K} \\ \vdots & & & \vdots \\ a_{K1} & a_{K2} & \cdots & a_{KK} \end{vmatrix} > 0, \quad K = 1, 2, \dots, n$$

(is equivalent to $-A$ has positive principal minors).

The above consideration has focused upon local stability properties. There are global stability results that can be obtained in a similar fashion.

Theorem 4.2. *For the Lotka-Volterra system of cooperation,*

$$\frac{dx_i}{dt} = x_i \left(r_{i0} + \sum_{j=1}^{n} a_{ij} x_j \right), \quad a_{ij} > 0 \tag{4.4}$$

a feasible equilibrium, x^*, is globally asymptotically stable if and only if all the principal minors of $-A$ are positive.

Indication of the Proof. It has been previously demonstrated that x^* is locally stable with this set of hypotheses. To establish global stability, a Liapunov function of Volterra type is useful. The function

$$V(x) = \sum_{i=1}^{n} C_i \left(x_i - x_i^* - x_i^* \ln \frac{x_i}{x_i^*} \right)$$

has derivatives along trajectories of (4.4) expressed in terms of a quadratic form with matrix $CA + A^T C$ where $C = \text{diag}(C_1, C_2, ..., C_n)$ (see Goh, 1977). If $CA + A^T C$ is negative definite, global asymptotic stability results. When the negative of an M-matrix is stable there exists a matrix $C = \text{diag}(C_1, C_2, ..., C_n)$, $C_i > 0$ such that $CA + A^T C$ is negative definite. Thus, global stability of x^* follows. $\square$

This material is related to that found in Siljak (1975), Goh (1979), and Travis and Post (1979).

5. Communities Composed of Populations with Different or Mixed Functional Roles

Models of communities of two and three dimension are explored in this section. First, the stability of a community in which the functional role of a population changes with the density is considered. Next, we turn to some three dimensional communities of Lotka-Volterra type with determinate roles for populations, but the coupling in the food web will be different than discussed previously. The community matrix role in three dimensional systems is investigated.

5.1 A Two Species Model with Density Dependent Functional Roles

Hastings (1978) has proved a general stability theorem for Kolmogorov type models.

Theorem 5.1. *Sufficient conditions for the global stability of an equilibrium* (x_1^*, x_2^*) *of*

$$\frac{dx_1}{dt} = x_1 f_1(x_1, x_2),$$

$$\frac{dx_2}{dt} = x_2 f_2(x_1, x_2), \qquad (5.1)$$

are

i) (x_1^*, x_2^*) *exists and is an unique equilibrium that is locally asymptotically stable.*

ii) *Both species sustain density dependent mortalities at all densities:*

$$\frac{\partial f_1}{\partial x_1} < 0, \quad \frac{\partial f_2}{\partial x_2} < 0.$$

iii) *There exist constants $A > 0$, $B > 0$ such that*
a) *for any $x_2 > B$, there is a $C > 0$ such that $f_1(C, x_2) < 0$.*
b) *For any $x_1 > A$, there is a $D > 0$ such that $f_2(x_1, D) < 0$.*

Indication of the Proof. Let (x_1^0, x_2^0) be any initial position. The rectangle bounded by the x_1, x_2 axes and the lines $x_1 = \tilde{x}_1 = \max(x_1^*, A, C, x_1^0)$, $x_2 = \tilde{x}_2 = \max(x_2^*, B, D, x_2^0)$ is invariant under the flow defined by (5.1). Transform the system by using the Volterra transformation $u_1 = \ln x_1$, $u_1 = \ln x_2$. This leads to

$$\frac{du_1}{dt} = f_1(e^{u_1}, e^{u_2}),$$

$$\frac{du_2}{dt} = f_2(e^{u_1}, e^{u_2}).$$

Since

$$\frac{\partial f_1}{\partial u_1} + \frac{\partial f_2}{\partial u_2} = \frac{\partial f_1}{\partial x_1} e^{u_1} + \frac{\partial f_2}{\partial x_2} e^{u_2} < 0,$$

the Bendixson nonexistence criterion implies that (5.1) does not have a limit cycle. Hence, global asymptotic stability results for (5.1). □

A Functional Role Determined by Density Dependence

It is not a trivial task to determine the functional role of a population in a community; indeed, it is often the case that a species will assume many different roles depending upon average age of the population, the density of the population, and other factors. The snail *(Thais)* is both a competitor and a prey for the starfish *(Pisaster)* (see Paine, 1966). Bluegill-bass interactions are also indeterminate in their interaction relationships: both predation and competition can occur between both species.

A system that models two populations where predation is the dominant interaction at high densities of the prey population and where competition dominates at low population levels is

$$\frac{dx_1}{dt} = x_1(1 - d_1 x_1 - d_2 x_2 - d_3 x_1 x_2),$$

$$\frac{dy_1}{dt} = x_2(1 - d_4 x_2 - d_5 x_1 + d_6 x_1^2).$$

Hastings (1978) has found sufficient conditions for a globally asymptotically stable equilibrium to be $4d_6 > d_5^2$, $d_4 > d_2$, and $d_2 d_6 > d_3 d_5$.

5.2 The Community Matrix

We have observed that the community matrix has an important role in discussions of stability. This role is now explored in more detail. The principal model with an essence of lower order nonlinearity is the classical Lotka-Volterra model

$$\frac{dx_i}{dt} = x_i\left(r_i - \sum_{j=1}^{n} b_{ij}x_j\right), \quad i = 1, 2, \dots, n.$$

These equations may be written in matrix form as

$$\frac{dx}{dt} = (\text{diag}(x_i))(r - Bx), \tag{5.2}$$

where $x = (x_1, x_2, \dots, x_n)^T$, $r = (r_1, r_2, \dots, r_n)^T$, and $B = (b_{ij})$. A nontrivial equilibrium x^* of (5.2) is (locally) asymptotically stable if and only if the eigenvalues of the matrix $-(\text{diag}(x^*))B$ all have negative real parts. This matrix, more commonly written as $-(\text{diag}(x_i^* b_{ij}))(I + A)$ where $A = (a_{ij})$, $a_{ij} = b_{ij}/b_{ii}$ if $i \neq j$ and $a_{ii} = 0$, will be called the *community matrix* of (5.2). There have been numerous attempts to derive methods of estimating a_{ij} from field and laboratory data, especially in the case of competitive communities (e.g., Gause, 1934; MacArthur and Levins, 1967; Vandermeer, 1969; Schoener, 1974; Hallett and Pimm, 1979). Because of the appeal of the community matrix and the fact that the parameters a_{ij} seem to have offered the best possibility for estimation in the past, it would be desirable to extract as much information as possible from the system (5.2) using only the matrix $I + A$. This approach can be developed without quantitative knowledge of r_i and b_{ii}.

As indicated in Sect. 4.4, the properties of $I + A$ are sufficient to determine the stability of n dimensional cooperative communities and also some communities of mixed mutualism and competition (Travis and Post, 1979). However, the examples of Strobeck (1973) in Sect. 3.6 show that the properties of $I + A$ are not sufficient to determine stability since both systems have the same community matrix.

The following result is valid only for dimension 3. Extensions to dimension 4 and higher are, at best, difficult (Clark and Hallam, 1982). Assumptions include that r is a 3-vector with positive entries and B is a 3×3 matrix with positive diagonal elements. To indicate parameter dependence, the system (2) will be denoted by $LV(r, B)$: the diagonal matrix $\text{diag}(x_i^* b_{ii})$, (when $B^{-1}r = x^* > 0$), by $D(r, B)$; and the community matrix $I + A = \text{diag}(b_{ii}^{-1})B$ by $CM(B)$. The second order principal minors of $I + A$ will be denoted by

$$M_1 = 1 - a_{23}a_{32}, \quad M_2 = 1 - a_{13}a_{31}, \quad M_3 = 1 - a_{12}a_{21}.$$

The next theorem gives conditions which are sufficient to ensure that the stability of a positive equilibrium of $LV(r, B)$ depends only on the community matrix. The conditions are also necessary in the sense that if they are not satisfied then either no positive equilibrium can exist for any choice of r and B or it is always possible to find examples, such as those of Strobeck, of distinct systems with the same community matrix but with different stability properties.

Theorem 5.2. Let A denote a 3×3 matrix whose diagonal elements are zero.
A. If $I + A$ satisfies either
(i) $\det(I + A) \leq 0$, or

(ii) $\det(I+A)>0$ and $M_i\leq 0$, $i=1,2,3$; then for any choice of r and B where $CM(B)=I+A$, $LV(r,B)$ cannot have a positive stable equilibrium.

B. If $I+A$ satisfies

(iii) $\det(I+A)>0$, $M_i\geq 0$, $i=1,2,3$, $\sum\limits_{i=1}^{3} M_i>0$, and

$$\sqrt{\det(I+A)}\leq\sqrt{M_1}+\sqrt{M_2}+\sqrt{M_3},$$

where equality can hold only if $M_1M_2M_3=0$, then for any choice of r and B where $CM(B)=I+A$, a positive equilibrium of $LV(r,B)$ is stable.

C. Suppose that $I+A$ satisfies none of the conditions, (i), (ii), or (iii). If (iv): there is no positive vector x such that $(I+A)x>0$, then for any choice of r and B where $CM(B)=I+A$, $LV(r,B)$ can have no positive equilibrium. If (v): there exists $x>0$ such that $(I+A)x>0$, then there exist matrices $r,B,\tilde{r},\tilde{B}$ with the properties $CM(B)$ $=CM(\tilde{B})=I+A$, $B^{-1}r=x^*>0$, $\tilde{B}^{-1}\tilde{r}=\tilde{x}^*>0$, x^* is a stable equilibrium of $LV(r,B)$, aNd $\tilde{x}^*$ is an unstable equilibrium of $LV(\tilde{r},\tilde{B})$.

Two observations are relevant to part C of Theorem 5.2. If $C(v)$ holds it is always possible to choose $x^*=\tilde{x}^*$. If $a_{ij}>0$ so that the system represents competition, then $C(iv)$ cannot hold and x^* and $\tilde{x}^*$ in $C(v)$ may be chosen arbitrarily.

Quasi Weak Diagonal Dominance.

A matrix $C=(c_{ij})_{n\times n}$ is weakly diagonally dominant if $|c_{ii}|>|c_{ij}|$ for $i=1,\dots,n$, and $j\neq i$. The matrix C is quasi weakly diagonally dominant if there exists a diagonal matrix D with positive diagonal elements such that $D^{-1}CD$ is weakly diagonally dominant. Quasi weak diagonal dominance might hold for many ecological systems as it relates interspecific and intraspecific interactions. The following theorem due to Fiedler and Ptak (1967) illustrates this and leads to some interesting observations.

Theorem 5.3. If $n\geq 2$ and C is any matrix, then C is quasi weakly diagonally dominant if and only if, for any set of distinct indices, $1\leq i_1,i_2,\dots,i_k\leq n$,

$$|c_{i_1i_2}c_{i_2i_3}\cdots c_{i_{k-1}i_k}c_{i_ki_1}|\leq|c_{i_1i_1}c_{i_2i_2}\cdots c_{i_ki_k}| \tag{5.3}$$

Applied to the matrix B of the system $LV(r,B)$, the condition (5.3) is a direct generalization of the well known condition which is necessary and sufficient for the stability of a positive equilibrium of a competitive system when $n=2$: $|b_{12}b_{21}|$ $\leq|b_{11}b_{22}|$. This condition is usually translated as "intraspecific interactions are stronger than interspecific interactions", and this interpretation also seems appropriate when $n>2$. Note that (5.3) holds for B if and only if it holds for $CM(B)$ $=I+A$, and, in this case, takes the form:

$$|a_{i_1i_2}a_{i_2i_3}\cdots a_{i_ki_1}|\leq 1. \tag{5.4}$$

It is well known that $\det(I+A)>0$ is a necessary condition for stability and also that it is not, in general, sufficient if $n>2$. However, in the three species case, condition (5.4) simplifies matters considerably and leads to a nice classification of stable equilibria.

Theorem 5.4. *Let $n=3$; $r>0$; and B be a matrix with positive diagonal elements such that B (or $I+A$) is quasi weakly diagonally dominant. Then, a positive equilibrium of $LV(r,B)$ is stable if and only if $\det(I+A)>0$.*

As remarked previously, quasi weak diagonal dominance might be valid for many community models. For a competitive community, it is a consequence of some of the formulations of the competition coefficients a_{ij}. As an illustration it is noted that one of the more familiar formulations first suggested by Gause for $n=2$ and generalized by Levins (1968) and MacArthur (1968), can be extended to include the case of a continuous resource spectrum as follows:

$$a_{ij}=\frac{\int\limits_S p_i(x)\,p_j(x)\,dx}{\int\limits_S p_i^2(x)\,dx}\,,\tag{5.5}$$

where $p_i(x)\,dx$ denotes the probability that species i will utilize the portion $(x, x+dx)$ of the resource spectrum in a unit of time, and S denotes the resource continuum. If "$\|\ \|$" denotes the inner product and norm in the appropriate inner product space, it follows, for distinct indices $i_1, i_2, \ldots, i_k$

$$a_{i_1 i_2} a_{i_2 i_3} \cdots a_{i_k i_1} = \frac{(p_{i_1}\cdot p_{i_2})(p_{i_2}\cdot p_{i_3})\cdots(p_{i_k}\cdot p_{i_1})}{\|p_{i_1}\|^2\,\|p_{i_2}\|^2\cdots\|p_{i_k}\|^2}$$

$$\leqq \frac{(\|p_{i_1}\|\,\|p_{i_2}\|)(\|p_{i_2}\|\,\|p_{i_3}\|)\cdots(\|p_{i_k}\|\,\|p_{i_1}\|)}{\|p_{i_1}\|^2\,\|p_{i_2}\|^2\cdots\|p_{i_k}\|^2}=1\,.$$

Therefore $I+A$ is quasi weakly diagonally dominant.

There are extensions of these results to the Kolmogorov-type system

$$\frac{dx_i}{dt}=x_i f_i(x),\qquad i=1,2,\ldots,n$$

where

$$\partial f_i/\partial x_i<0\,.$$

The proof of Theorems 5.2 and 5.3 are not given here; they may be found in Clark and Hallam (1982).

5.3 A Two Dimensional Competitive Subcommunity and Another Population

The effects of introducing a population into a competitive subcommunity is now explored. We have previously studied the case where the added population was a competitor with each of the other two populations. In this section some consequences of introducing a cooperative population or a predator population are described. The mathematical details are similar to those for the Lotka-Volterra competitive model.

The Third Population is a Cooperator (Hallam, 1980)

For a Lotka-Volterra model of a competitive subcommunity and an added cooperator, certain hypotheses about the coefficients are required to eliminate the "orgy" effect. With these imposed, extinction of the populations can be classified. Employing two-population interactions, there are some interesting outcomes. The introduction of cooperator can destroy the stable competitive subcommunity by driving one of the competitors to extinction. This can be accomplished by the symbiotic population helping one of the competitors too much.

The Third Population is a Predator

Models of a community containing a predator and two prey populations are numerous in the recent literature (e.g., Cramer and May, 1972; Vance, 1978; Gilpin, 1978; Freedman and Waltman, 1984). Most analyses have focused upon the processes of predation and competition as mechanisms that can generate diversity in communities. Two of these mechanisms that can be identified by model analysis are predator-mediated coexistence and competition-induced coexistence.

Predator-mediated coexistence is concerned with regulation of a dominant competitor by predation in order that the complete community might persist. Instances of predator mediated coexistence are well documented in the ecological literature. Classical experiments relating to predator mediated coexistence include those of Utida (1953) where the presence of a parasitic wasp, *Neocatolaccus mamezophagus*, could lead to coexistence of two bean weevil populations and Slobodkin (1961, 1964) where effects of predation on competing hydra populations were investigated. Paine (1966) performed a classic experiment where a top predator in a marine system was removed, and a collapse of the lower trophic system occurred. See Connell (1975) and Caswell (1978) for additional instances of occurrence or nonoccurrence of predator mediated coexistence.

Other references, related at least peripherally to the model studied here, in which a predator is introduced into a competitive subcommunity include Neill (1975) and Addicott (1974). Yodzis (1976) discussed effects of constant rate predation on competitive systems.

Employing a Lotka-Volterra model, an analysis of persistence and extinction (Hallam, 1981), shows that there can be two forms of predator mediated coexistence; these are given by the arrangements

$$v_1 \gg v_2; \quad v_1 \leftrightarrow p; \quad v_2 \leftrightarrow p;$$

and

$$v_1 \gg v_2; \quad v_1 \leftrightarrow p; \quad p \downarrow v_2.$$

In these arrangements, the competition notation is as in Sect. 3.6 for the prey populations v_1, v_2. The interactions between predator and prey indicated by $v_i \leftrightarrow p$ and $p \downarrow v_i$ represent an asymptotic stability coexistence for both predator and prey and the survival of the prey population only respectively. Both arrangements above require invasion capability of complementary species at equilibrium subcommunities if persistence is to occur.

Gilpin (1978), using a model employed by Vance (1978), numerically demonstrated that chaotic behavior can arise in a three dimensional system composed of two prey and a predator. While the parameter set utilized to find chaotic behavior is a limiting case of an arrangement whose persistence development is indicated in Hallam (1981), it can be shown there are parameter sets in the persistent arrangement that lead to chaotic motion. Not only is the phenomena of predator-mediated coexistence of ecological interest, the mathematical description of the dynamics can be very complicated as well.

The persistence analysis also leads to another possible mechanism of coexistence, namely, "competition induced coexistence". An analysis of the Lotka-Volterra model shows that this can occur in two ways:

$$v_1 \leftrightarrow v_2; \quad v_1 \leftrightarrow p; \quad p \downarrow v_2$$

and

$$v_1 \leftrightarrow v_2; \quad p \downarrow v_1; \quad p \downarrow v_2.$$

Two arrangements of subcommunities were classified in this category. The largest increase in diversity occurs in an arrangement where a predator cannot survive on either of the prey species but it can persist if it is able to invade the stable competitive subcommunity at its equilibrium density. Certain herbivore plant systems could theoretically fit into this category.

The terms "predator mediated coexistence" and "competition induced coexistence" refer to persistence in a community attained by effects of species upon subcommunities. Analysis indicates that determination of coexistence can depend up on all subcommunities as well as upon species interactive capabilities. In the case of predator mediated coexistence, a persistent community can be achieved theoretically by introduction of a prey competitor into a predator-prey subsystem in which the predator need not even be able to survive. The phenomenon of predator mediated coexistence might be masked in such a situation. Persistence, since it is dependent upon subsystems composition and interrelationships, is a community property and phrases as simple as "predator mediated coexistence" are probably not totally adequate descriptions.

As a concluding comment, I recommend the books of Cohen (1978) and Pimm (1982), which contain many documented food webs. It is clear from examining the diagrammed webs that structure can be very complex and that the analysis presented here is only an embarkation into a study of food webs.

References

Addicott, J.E. (1974). Predation and prey community structure: an experimental study of the effect of mosquito larvae on the protozoan communities of pitcher plants. Ecology 55: 475–492

Albrecht, F., Gatzke, H., Haddad, A., Wax, N. (1974). The dynamics of two interacting populations. J. Math. Anal. Appl. 46: 658–670

Boughey, A S. (1973). Ecology of Populations, 2nd ed., MacMillan, New York, 182pp.

Brauer, F. (1984). Constant yield harvesting of population systems. In Mathematical Ecology: Proceedings of the Conference, Trieste. (S.A. Levin and T.G. Hallam, eds.). Lecture Notes in Biomathematics 54: 234–242

Brauer, F., Soudak, A.C. (1979a). Stability regions and transition phenomena for harvested predator-prey systems. J. Math. Biol. 7: 319–337

Brauer, F., Soudak, A.C. (1979b). Stability regions in predator-prey systems with constant rate harvesting. J. Math. Biol. 8: 55–71

Brauer, F., Soudak, A.C., Jarosch, H.S. (1976). Stabilization and destabilization of predator-prey systems under harvesting and nutrient enrichment. Int. J. Control. 23: 553–573

Braun, M. (1975). Differential Equations and Their Applications, Springer-Verlag, New York, 718pp.

Caswell, H. (1978). Predator-mediated coexistence: a nonequilibrium model. Amer. Nat. 112: 127–154

Clark, C.E., Hallam, T.G. (1982). The community matrix in three species community models. J. Math. Biol. 16: 25–31

Cohen, J.E. (1978). Food Webs and Niche Space, Princeton University Press, Princeton, N.J., 189pp.

Coleman, C.S. (1976). Biological cycles and the five-fold way. MAA College Faculty Workshop. Cornell University

Connell, J.H. (1961). The influence of interspecific competition and other factors in the distribution of the barnacle Chthamalus stellus. Ecology 42: 710–723

Connell, J.H. (1975). Some mechanisms producing structure in natural communities. In M.C. Cody and J.M. Diamond (Eds.), Ecology and Evolution of Communities. Harvard University Press, Cambridge, Mass. 460–490

Cramer, N.F., May, R.M. (1972). Interspecific competition, predation, and species diversity: a comment. J. Theor. Biol. 34: 289–293

Crombie, A.C. (1945). The competition between different species of graminivorus insects. Proc. Roy. Soc. London 132 B: 362–395

Crombie, A.C. (1946). Further experiments on insect competition. Proc. Roy. Soc. Lond. 133 B: 76–109

Crombie, A.C. (1947). Interspecific competition. J. Anim. Ecol. 16: 44–73

DeAngelis, D.C., Post III, W.M., Travis, C.C. (1986). Positive Feedback Systems in Nature, to appear

Dyer, M.C. (1980). Mammalian epidermal growth factor promotes plant growth, Proc. Natl. Acad. Sci. USA 77: 4836–4837

Fiedler, M., Ptak, V. (1967). Diagonally dominant matrices. Czech. Math. J 17: 420–433

Freedman, H I., Waltman, P. (1977). Mathematical analysis of some three species food chain models. Math. Biosci. 33: 257–276

Freedman, H.I. (1980). Deterministic Mathematical Models in Population Biology. Dekker, New York, 254pp.

Freedman, H.I., Waltman, P. (1984). Persistence in models of three interacting predator-prey populations. Math. Biosci. 68: 213–231

Gard, T.C. (1982). Top predator persistence in differential equation models of food chains: the effects of omnivory and external forcing of lower trophic levels. J. Math. Biol. 14: 285–299

Gard, T.C. (1984). Persistence in food webs. In Mathematical Ecology: Proceedings of the Conference (S.A. Levin and T.G. Hallam eds.). Lectures Notes in Biomathematics 54: 208–219

Gard, T.C., Hallam, T.G. (1979). Persistence in food webs: I Lotka-Volterra food chains. Bull. Math. Biol. 41: 877–891

Gause, G.F. (1934). The Struggle for Existence. Hafner New York (reprinted 1964 by Williams and Wilkins, Baltimore, MD), 163pp.

Gilpin, M.E. (1975). Limit cycles in competition communities. Amer. Nat. 109: 51–60

Gilpin, M.E. (1978). Spiral chaos in a predator-prey model. Amer. Nat. 113: 306–308

Ginzburg, L.R. (1983). Theory of Natural Selection and Population Growth. Benjamin/Cummings, Menlo Park, CA, 160pp.

Goh, B.S. (1977). Global stability in many species systems. Amer. Nat. 111: 135–143

Goh, B.S. (1979). Stability in models of mutualism. Amer. Nat. 113: 261–275

Grossberg, S. (1978). Decisions, patterns, and oscillations in nonlinear competitive systems with applications to Lotka-Volterra systems. J. Theor. Biol. *73*: 101–130

Hallam, T.G., Svoboda, L., Gard, T.C. (1979). Persistence and extinction in three species Lotka-Volterra competitive systems. Math. Biosci. *46*: 117–124

Hallam, T.G. (1980). Effects of cooperation on competitive systems. J. Theor. Biol. *82*: 415–424

Hallam, T.G. (1981). Effects of competition and predation on diversity of communities. Applied Systems and Cybernetics. Vol. IV, G. Lasker (Ed.) 2054–2058

Hallett, J.G., Pimm, S.L. (1979). Direct estimation of competition. Amer. Nat. *113*: 593–600

Harrison, G.W. (1979). Global stability of predator-prey interactions. J. Math. Biol. *8*: 159–171

Hastings, A (1978). Global stability in two species systems. J. Math. Biol. *5*: 399–403

Herbert, D., Elsworth, R., Telling, R.C. (1956). The continuous culture of bacteria: A theoretical and experimental study. J. Gen. Microbiol. *14*: 601–622

Hirsch, M.W., Smale, S. (1974). Differential Equations, Dynamical Systems, and Linear Algebra. Academic Press, New York, 358 pp.

Hsu, S.B., Hubbell, S., Waltman, P. (1977). A mathematical theory for single-nutrient competition in continuous cultures of microorganisms. SIAM J. Appl. Math. *32*: 366–383

Hsu, S.B. (1978). On global stability of a predator-prey system. Math. Biosci. *39*: 1–10

Hsu, S.B., Hubbell, S.P. (1979). Two predators competing for two prey species: an analysis of MacArthur's model. Math. Biosci. *47*: 143–171

Hutchinson, G.E. (1978). An Introduction to Population Ecology, Yale University Press, New Haven, 260 pp.

Janzen, D.H. (1966). Coevolution of mutualism between ants and acacias in Central America. Evolution *20*: 24–75

Kaplan, J.L., Yorke, J.A. (1975). Competitive exclusion and nonequilibrium coexistence. Amer. Nat. *111*: 1030–1036

LaSalle, J.P., Lefschetz, S. (1961). Stability by Liapunov's Direct Method, Academic Press, New York, 134 pp.

Leon, J.A., Tumpson, D.B. (1975). Competition between two species for two complementary or substitutable resources. J. Theor. Biol. *50*: 185–201

Levin, S.A. (1970). Community equilibria and stability, and an extension of the competitive exclusion principle. Amer. Nat. *104*: 413–423

Levins, R. (1968). Evolution in Changing Environments. Princeton University Press, Princeton, N.J.

Luckinbill, L.S. (1973). Coexistence in laboratory populations of *Paramecium aurelia* and its predator *Didinium nasutum*. Ecology *54*: 1320–1327

MacArthur, R.H. (1958). Population ecology of some warblers of northeastern coniferous forests. Ecology *39*: 599–619

MacArthur, R.H. (1968). The theory of the niche. In: Lewontin, R.C. (ed.) Population Biology and Evolution. Syracuse University Press, Syracuse, N.Y. 159–176

MacArthur, R.H., Levins, R. (1967). The limiting similarity, convergence, and divergence of coexisting species. Amer. Nat. *101*: 377–385

May, R.M. (Ed.) (1976). Theoretical Ecology, Principles and Applications. Saunders, Philadelphia. 317 pp.

May, R.M., Leonard, W.J. (1975). Nonlinear aspects of competition between three species. Siam. J. Appl. Math. *29*: 243–253

Maynard Smith, J. (1974). Models in Ecology, Cambridge University Press, Cambridge, 146 pp.

McGehee, R., Armstrong, R.A. (1977). Some mathematical problems concerning the ecological principle of competitive exclusion. J. Diff. Equations *23*: 30–52

Neill, W.E. (1975). Experimental studies of microcrustacean competition, community composition and efficiency of resource utilization. Ecology *56*: 809–826

Novick, A., Szilard, L. (1950). Description of the Chemostat, Science *112*: 715–716

Odum, E.P. (1974). Ecology (2nd ed.). Holt, Rinehart, and Winston, New York, 244 pp.

Owen, D.F., Wiegert, R.G. (1983). Grasses and grazers: is there a mutualism? Oikos *38*: 258–260

Park, T. (1954). Experimental studies of interspecific competition. II. Temperature, humidity, and competition in two species of *Tribolium*. Physiol. Zool. *27*: 177–238

Park, T. (1962). Beetles, competition, and populations. Science *138*: 1369–1375

Paine, R.T. (1966). Food web complexity and species diversity. Amer. Nat. *100*: 65–76

Pimentel, D., Feinberg, E.H., Wood, P.W., Hayes, J.T. (1965). Selection, spatial distribution, and the coexistence of competing fly species. Amer. Nat. *99*: 97–109

Pimm, S.L. (1982). Food webs. Chapman and Hall, London

Rosenzweig, M.L. (1969). Why the prey curve has a hump. Amer. Nat. *103*: 81–87

Schoener, T.W. (1974). Some methods for calculating competition coefficients from resource utilization spectra. Amer. Nat. *108*: 332–340

Siljak, D.D. (1975). When is a complex ecosystem stable? Math. Biosci. *35*: 25–50

Silvertown, J.W. (1982). No evolved mutualism between grasses and grazers. Oikos *38*: 253–259

Slobodkin, L.B. (1961). Growth and Regulation of Animal Populations. Holt, Rinehart, and Winston, New York

Slobodkin, L.B. (1964). Experimental populations of Hydrida. J. Ecol. *52*: 131–148

Stenseth, N.C. (1983). Grasses, grazers, mutualism, and coevolution: a comment about handwaving in ecology. Oikos *41*: 152–153

Strobeck, C.N. (1973). *N* species competition. Ecology *54*: 650–654

Tanner, J.T. (1975). The stability and intrinsic growth rates of prey and predator populations. Ecology *56*: 855–867

Thompson, K., Uttley, M.G. (1982). Do grasses benefit from grazing? Oikos *39*: 113–116

Travis, C.C., Post III, W.M. (1979). Dynamics and comparative statics of mutualistic communities. J. Theor. Biol. *78*: 553–571

Utida, S. (1953). Interspecific competition between two species of bean weevil. Ecology *34*: 301–307

Vance, R.R. (1978). Predation and resource partitioning in one predator-two prey model communities. Amer. Nat. *112*: 797–813

Vandermeer, J.H. (1969). The competitive structure of communities: An experimental approach with protozoa. Ecology *50*: 362–371

Vandermeer, J.H., Boucher, D.H. (1978). Varieties of mutualistic interaction in population models. J. Theor. Biol. *74*: 549–558

Waltman, P. (1964). The equations of growth. Bull. Math. Biophys. *26*: 39–43

Waltman, P. (1983). Competition Models in Population Biology. CBMS 45 SIAM. 84pp.

Whittaker, R.H., Levin, S.A. (Eds.) (1975). Niche: Theory and Application. Dowden, Hutchinson, and Ross, New York

Yodzis, P. (1976). The effects of harvesting on competitive systems. Bull. Math. Biol. *38*: 97–109

Interacting Age Structured Populations

Alan Hastings

Both the study of interacting species without age structure and single species with age structure have had a long history in theoretical ecology, beginning with studies in the "golden age" of ecology (Scudo and Ziegler, 1978) by Lotka and Volterra. Only much more recently have there been studies on the dynamics of interacting, age structured models. Such models can quickly get extraordinarily complex, so in this review I will concentrate on the simplest models.

There are many biological examples where size or age clearly plays a dominant role in the dynamics of interacting species. One general area is those organisms with complex life cycles (see review in Wilbur, 1980). Here, of necessity, juveniles compete with juveniles and adults compete with adults.

Competition among species of different sizes may mean that size or age is important. Werner (1977) has discussed sunfish of different size, where the juveniles of the larger species compete with the adults of the smaller species for food items of similar size.

Age or size structure may be important in predator prey systems or host parasitoid systems. Lotka-Volterra models ignore juvenile periods in predators and assume a type III (exponential) survivorship curve. Any deviation from these assumptions may have important consequences for the stability and persistence of predator prey systems. Additionally, both predators and parasitoids do not eat all ages of prey indiscriminately. Well documented cases for predators include the wolves on Isle Royale (Mech, 1966), which preferentially hunt very young or very old moose. Many parasitoids use a particular stage in their host. If the generations of host and parasitoid overlap and are not synchronized, such age or stage specificity must be included in reasonable models.

Although this review will concentrate on ecological interactions between species, interactions of a similar nature occur among different ages or stages within a single species. Cannibalism is a very common and important natural occurrence. Competition between different stages of a single species which share resources is also common. Such interactions are modelled in a similar way to the situations covered more completely here.

Finally, this is not an exhaustive review of all the literature on interactions between age or size structured populations. I have chosen to emphasize studies that illustrate the various ways in which such interactions can be modelled.

Biomathematics, Vol. 17, Mathematical Ecology
Edited by T. G. Hallam and S. A. Levin
© Springer-Verlag Berlin Heidelberg 1986

Model Formulation

The simplest (from the viewpoint of model formulation) way to produce a model of age structured, interacting species is to start with the very detailed models already available for age or stage structured populations, the Leslie matrix approach and the von Foerster or McKendrick partial differential equation approach. (See the Chaps. III.2 and III.3 of these Notes.) One can use these equations, with suitable inclusions of interaction terms, to develop models for interacting species. This leads to very general models with large numbers of parameters.

For the Leslie matrix approach one simply makes the entries in the Leslie matrix functions of the density of (particular stages of) the other species. Examples of this approach are Pennycuick et al. (1968) and Travis et al. (1980).

Similarly, for the McKendrick-von Foerster model one just makes the coefficients depend on numbers of the other species. For this model a one species form is:

$$\partial u / \partial t + \partial u / \partial a = - \mu(a, t) u, \tag{1}$$

where $u(a, t)$ is a density function on age a at time t for population size, and $\mu(a, t)$ is the age and time dependent death rate. Information about births is contained in the boundary condition needed to complete the model:

$$u(0, t) = \int_0^\infty B(a, t) u(a, t) da, \tag{2}$$

where $B(a, t)$ is the age and time dependent fecundity. By simply making the functions B and μ depend on the other species one can create a model for interacting populations. (These functions may also depend on the species itself to introduce density dependence.) For this partial differential equation model, unlike the Leslie model, there may be questions of existence and uniqueness of solutions, which are covered in Prüb (1982). I will not consider these mathematical questions further.

Biological Results

Before exploring alternative approaches to model formulation, I will examine the kinds of results possible with models of this kind. A complete and detailed analysis of these general models is clearly impossible, so I will review results which show that the inclusion of age structure leads to new results (not possible without age structure) of biological importance.

In experiments with Drosophila sp. competing in the laboratory, Gilpin and Ayala (1973) showed that the isoclines (lines of zero population growth for given species in the phase plane) were nonlinear. They used this result to suggest that Lotka-Volterra competition models, which of necessity lead to linear isoclines, were inadequate. The question explored here is whether the results of nonlinear isoclines could arise from the effects of age structure alone. This would mean that the Lotka-Volterra competition assumptions would not be refuted.

I will now summarize results of Hassell and Comins (1976) which shed light on this question. An alternative way of viewing this problem will be discussed below. Assume here that there are two age classes in each species, juveniles, and adults. Assume that at any given time, both stages are present, and that a discrete time model is appropriate. Let X_t and Y_t be the number of adults in each of the species at time t, with x_t and y_t the corresponding number of juveniles. Then a discrete time version of the Lotka-Volterra competition model is the following:

$$
\begin{aligned}
X_{t+1} &= x_t \exp[-a(x_t + \alpha y_t)], \\
Y_{t+1} &= y_t \exp[-c(y_t + \beta x_t)], \\
x_{t+1} &= X_t \exp[r - a'(X_t + \alpha' Y_t)], \\
y_{t+1} &= Y_t \exp[s - c'(Y_t + \beta' X_t)].
\end{aligned}
\tag{3}
$$

Here a and c are parameters measuring intraspecific competition in juveniles, α and β are measures of interspecific competition in juveniles, with the primed parameters similar measures for adults. Since results are similar for different assumptions, I will concentrate here on a simple case. Assume that the only interspecific competition is between juveniles, so $\alpha' = \beta' = 0$, yielding:

$$
\begin{aligned}
X_{t+1} &= x_t \exp[-a(x_t + \alpha y_t)], \\
Y_{t+1} &= y_t \exp[-c(y_t + \beta x_t)], \\
x_{t+1} &= X_t \exp[r - a'(X_t)], \\
y_{t+1} &= Y_t \exp[s - c'(Y_t)].
\end{aligned}
\tag{3'}
$$

Now, the isoclines can be found by setting:

$$
\begin{aligned}
x_{t+1} &= x_t, \\
y_{t+1} &= y_t, \\
X_{t+1} &= X_t, \\
Y_{t+1} &= Y_t,
\end{aligned}
\tag{4}
$$

and then eliminating x_t and y_t from (4). This clearly leads to nonlinear isoclines, as can be seen from the resulting pair of equations for X_t and Y_t. Hassell and Comins go further and do a stability analysis of this model, and show that there can be multiple stable equilibria. A recent study using Leslie matrix (discrete time) models of interacting competitors is Travis et al. (1980).

One of the earliest papers to study interacting species using a von Foerster type approach is that by Auslander et al. (1974), where a parasite host interaction is modelled. This paper gives a good idea of the sorts of biological complications that can be found in models of this kind.

Consider a parasite-host interaction where adult parasites lay eggs in early instars of hosts and generations can potentially overlap. The formulation of this model is fairly straightforward. Let $p(a, t)$ be a density function on age a for the number of parasites at time t. Similarly, let $h(a, t)$ be a density function on age a for the number of hosts of age a at time t. Then the model consists of writing down the von Foerster type model for each species with appropriate changes to include the species interaction terms. For the parasite, the life cycle can be broken up into two parts, adults, immature stages, and the birth process. The dynamics can be represented as:

$$p_t + p_a = -\mu(t, a)p \quad \text{for} \quad 0 \leq a \leq \gamma, \tag{5}$$

where γ is the maximum age of adult parasites. The positive effect of hosts on parasites enters through the boundary condition describing the birth rate of new parasites, $p(0, t)$. In the most general case, the death rate of hosts would depend on the number of adult or mature parasites, the number of hosts (as measure of intraspecific competition) and on the number of hosts at the age where they are subject to predation. Thus the dynamics of the host population are described by:

$$h_t + h_a = -\mu_h(t, a, H, H_0, P_0)h, \quad 0 \leq a \leq \delta + \beta, \tag{6}$$

where

$$H = \int_0^\infty h(a, t)da \tag{7}$$

is the total number of hosts,

$$H_0 = \int_\beta^{\beta+\delta} h(a, t)da \tag{8}$$

is the number of hosts that can be parasitized, and

$$P_0 = \int_\alpha^{\alpha+\gamma} p(a, t)da \tag{9}$$

is the number of adult parasites. The boundary conditions $p(0, t)$ and $h(0, t)$ take the following form. The birth rate of parasites depends on the number of adult parasites, those between ages α and $\alpha + \gamma$, and the number of hosts available for parasitization, yielding:

$$p(0, t) = \int_\alpha^{\alpha+\gamma} m_p(a, t, H_0)p(a, t)da. \tag{10}$$

The birth rate of hosts is similarly defined as:

$$h(0, t) = \int_\beta^{\beta+\delta} m_h(a, t, H_0(t-\tau))h(a, t)da, \tag{11}$$

where the dependence on $H_0(t-\tau)$ arises from loss due to parasites.

The next step is to choose specific functional forms for the model. Even then, the model is far too complex to study analytically and must be examined through computer simulation. The simulations of Auslander et al. (1974) yield the following important result. For many realistic choices of parameters the system exhibits a periodic solution in which hosts and parasites appear to have almost discrete generations. Parasite adults are common at the end of times when host larvae are common and vice versa. Thus the commonly used discrete generation models arise here as a consequence of the dynamics rather than being an a priori modelling assumption. Auslander et al. conjecture that this periodic solution arises as a result of a Hopf bifurcation.

Can similar results be found in a simpler context? Consider the following model studied by Beddington and Free (1976). Let there be a predator with both juveniles and adults, and a prey species in which age structure is ignored. Then a discrete time model describing this situation is:

$$N_{t+1} = N_t(1 + r(1 - N_t) - c_1 P_t^1 c_2 P_t^2 ,$$

$$P_{t+1}^1 = \alpha c_2 N_t P_t^2 , \tag{12}$$

$$P_{t+1}^2 = s P_t^1 .$$

Here N_t is the population of the prey at time t, P_t^1 is the number of juvenile predators, and P_t^2 is the number of adult predators. The meaning of all the constants (which are all positive) is obvious from context. Determining the unique nontrivial equilibrium (when it exists) and its stability is a straightforward, albeit lengthy, algebraic exercise. There are two conditions for stability, a complicated one, and the condition

$$1 > c_2 s/c_1 . \tag{13}$$

As in the previous model, when c_2 is increased, the equilibrium solution becomes unstable. One can show analytically that there then exists a stable two point cycle. Thus the general behavior of the complex model studied by Auslander et al. (1974) can be mimicked here by this simple model.

There are numerous other ways in which age structure affects predator-prey systems which have been studied. The model of Auslander et al. (1974) includes many features that are related to age structure. Since all the features are included in one model, a detailed analysis is difficult. I will now consider three models which can be viewed as attempts to look at parts of the interaction embodied in the model of Auslander et al. (1974). By considering just one effect at a time, analytical studies can be used to determine general principles. The role of a juvenile period in the predator in altering stability has been considered by Cushing and Saleem (1982). In the model they formulate the predator obeys equations of the form:

$$p_a + p_t = -\mu p ,$$

$$p(0, t) = \int_0^\infty b\beta(a) h(R, P) da . \tag{14}$$

Here $p(a, t)$ is a density function for predators of age a, μ is the age independent death rate, and b is a parameter determining total fecundity. The function h depends on the level of prey (resources) R and the total predator population size P, and the function $\beta(a)$ gives the relative age specific fecundities. Cushing and Saleem specialize the function $\beta(a)$ and show that the model becomes unstable for shorter maturation periods, a result that is opposite to the lore that time delays are necessarily destabilizing. An earlier, less comprehensive, study along similar lines but using stage classifications as discussed below is that of McDonald (1976, 1977).

Deviations in survivorship of adult predators away from the Lotka-Volterra ideal have been considered by Hastings and Wollkind (1982). Let the total number of prey be H, the total number of predators be P and let $p(a, t)$ be a density function on age a for the number of predators. Let the prey obey the following equation which ignores age structures:

$$dH/dt = Hg(H) - Pf(H), \tag{15}$$

where $g(H)$ is the per capita growth rate of prey in the absence of predators and $f(H)$ is the functional response of predators to prey.

The predators obey the following dynamics, which assume that all predators live to a fixed age and then die. This is an approximation to the life history of some arthropods, leading to:

$$\partial p/\partial t + \partial p/\partial a = -\mu(a, t)p, \quad 0 < a < T, \tag{16}$$

where T is the maximum life span and

$$P(t) = \int_0^T p(a, t)da. \tag{17}$$

The fecundity of all adult predators is assumed the same giving the following boundary condition:

$$p(0, t) = cPf(H). \tag{18}$$

It is then possible to determine the equilibrium of this model, find its stability and compare the stability to that in the corresponding model with an exponential death term. The stability is markedly different, as the region of local stability of the equilibrium is greatly reduced in the age structured model as compared with the age independent model (Hastings and Wollkind, 1982; Wollkind et al., 1982).

Another problem that has received much attention in recent years has been the problem of age dependent predation. Models describing this situation can be formulated using the von Foerster type approach or Leslie matrix approach described above (e.g. Gurtin and Levine, 1979). I will describe another approach used by Gazis et al. (1973) and Smith and Mead (1974) and recently by Hastings (1983). Divide the population into classes and define transition or maturity rates between classes. A simple example with two stages in a prey species and one in a predator species is the following, where the predator eats only juvenile prey:

$$dH_0/dt = BH_1 - aH_0 - Pf(H_0),$$
$$dH_1/dt = aH_0 - kH_1,$$
$$dP/dt = Pf(H_0) - P.$$
(19)

Here, the number of juvenile prey is H_0, the number of adult prey is H_1, the number of predators is P, $f(H_0)$ determines the predation rate, B is the per capita birth rate of adult prey, a is the maturation rate, and k is the death rate of adult prey. This model and related models are analyzed in Hastings (1983), where conditions for local stability and numerical results are reported. For additional details on other ways of deriving or thinking about this kind of model also see Cushing (1977) and McDonald (1978).

Discussion

The modelling of interacting age structured population is still in its infancy, and virtually no general conclusions are yet possible. The central theme is that the interactions between the time delays created by age structure and by ecological interactions such as predator prey oscillations are the key to understanding dynamic behavior. On the one hand, many interesting and complex behaviors are possible in the models being developed. However, this diversity may prove to be an embarassment of riches, as the ability to explain all phenomena may really mean the inability to truly explain anything. In particular, models based on the von Foerster equation and especially on Leslie matrices can involve an extraordinarily large number of parameters. For this reason, models based on stage classification like (19) may prove more useful in the future when trying to determine general principles. The relative simplicity of such models allows them to be analyzed in greater detail. In many ways such models fit more closely with the kinds of observations that can be made of natural populations. Secondly, the stochastic features that these models introduce also may be realistic in many populations.

Acknowledgement

Conversations with Lou Bottsford and the participants at the conference in Trieste have improved my understanding of these problems. Supported by NSF Grant DEB 8002593.

References

Auslander, D.M., Oster, G.F., Huffaker, C.B. (1974). Dynamics of interacting populations. J. Franklin Institute 297: 345–376
Beddington, J.R., Free, C.A. (1976). Age structure effects in predator-prey interactions. Theo. Pop. Biol. 9: 15–24

Cushing, J. (1977). Integrodifferential Equations and Delay Models in Population Dynamics. Lecture Notes in Biomathematics 20, Springer-Verlag

Cushing, J., Saleem, M. (1982). A predator-prey model with age structure. J. Math. Biol *14*: 231–250

Gazis, D.C., Montroll, E.W., Ryniker, J.E. (1973). Age-specific, deterministic model of predator prey populations: application to Isle Royale. IBM J. Res. Devel. *17*: 47–53

Gilpin, M., Ayala, F. (1973). Global models of growth and competition. Proc. Nat. Acad. Sci. *70*: 3590–3593

Gurtin, M.E., Levine, D.S. (1979). On predator-prey interactions with predation dependent on age of prey. Math. Biosci. *47*: 207–219

Hassell, M.P., Comins, H.N. (1976). Discrete time models for two species competition. Theo. Pop. Biol. *9*: 202–221

Hastings, A. (1983). Age dependent predation is not a simple process. I. Continuous time models. Theo. Pop. Biol. *23*: 347–362

Hastings, A., Wollkind, D. (1982). Age structure in predator-prey systems. I. A general model and a specific example. Theo. Pop. Biol. *21*: 44–56

McDonald, N. (1976). Time delay in predator-prey models. Math. Biosci. *28*: 321–330

McDonald, N. (1977). Time delay in predator-prey models. II. Bifurcation theory. Math. Biosci. *33*: 227–234

McDonald, N. (1978). Time Lags in Biological Models. Springer-Verlag, New York

Mech, D. (1966). The wolves of Isle Royale. U.S. Govt. Printing Office, Washington

Pennycuick, C.J., Compton, R.M., Beckingham, L. (1968). A computer model for simulating the growth of a population or of two interacting populations. J. theor. Biol. *18*: 316–329

Prub, J. (1981). Equilibrium solutions of age-specific population dynamics of several species. J. Math. Biol. *11*: 65–84

Scudo, F., Ziegler, J. (1978). The golden age of theoretical ecology: 1923–1940, Springer-Verlag, New York

Smith, R.H., Mead, R. (1974). Age structure and stability in models of predator-prey systems. Theor. Pop. Biol. *6*: 308–322

Travis, C., Post, W., De Angelis, D., Perkowski, J. (1980). Analysis of compensatory Leslie matrix models for competing species

Werner, E. (1977). Species packing and niche complementarity in three sunfishes. Amer. Nat. *11*: 553–578

Wilbur, H. (1980). Complex life cycles. Annu. Rev. Ecol. Syst. *11*: 67–93

Wollkind, D., Hastings, A., Logan, J. (1982). Age structure in predator-prey systems. II. Functional response and stability and the paradox of enrichment. Theo. Pop. Biol. *21*: 57–68

Population Models and Community Structure in Heterogeneous Environments

Simon A. Levin *

Introduction

Until recently, the bulk of the extensive mathematical literature in ecology ignored the role of spatial heterogeneity. There were exceptions, for example the classic work of Skellam (1951); but despite much earlier developments in the related areas of epidemics (Brownlee, 1911) and genetics (Fisher, 1937; Haldane, 1948), it has only been in the last few years that theoretical studies in ecology have begun to take account of the fundamental importance of geographic distribution. Recent reviews of the subject may be found in Levin (1976a, b).

In part, this current attention to spatial factors may be traced to parallel developments in the ecological experimental literature, especially island coloni-zation studies (Schoener, 1975; Simberloff, 1976; Simberloff and Wilson, 1969, 1970; Sutherland, 1974) and the associated theory of island biogeography (MacArthur and Wilson, 1967). Even ostensibly non-insular environments, such as forests, farms, or rocky shores, may often be viewed as mosaics of islands of different characteristics (Levin and Paine, 1974, 1975), in essence archipelagoes of microhabitats (Cromartie, 1975; Janzen, 1968; Platt, 1975; Root and Chaplin, 1976).

Further, Hutchinson (1951) emphasized long ago the influence of heterogene-ity on adaptive strategies, and there has been growing interest in recent years in colonization strategies, especially dispersal (Den Boer, 1968, 1971; Gadgil, 1971; Reddingius, 1971; Smith, 1972; Strathman, 1974) and dormancy (D. Cohen, 1970; Marks, 1974), and their relation to the spatial and spatio-temporal characteristics of the environment. Simulation studies have been undertaken to shed some light on these evolutionary questions (Allen, 1975; Caswell, 1976; Reddingius and Den Boer, 1970; Roff, 1974a, b, 1975), but analytical results are nearly nonexistent.

The second major factor contributing to the development of the mathematical theory of spatially distributed populations is the increasing theoretical interest in the prototype mathematical description, the diffusion-reaction system (Fife, 1976a, b; Greenberg and Hastings, 1977; Kopell and Howard, 1974; Othmer, 1975, 1977; Othmer and Scriven, 1971, 1974; Torre, 1975), especially in biological contexts (Aronson and Weinberger, 1975, 1977; Dubois and Adam, 1976; Fife and Peletier, 1977; Hadeler, 1977; Keller and Segel, 1970; Kolmogoroff et al., 1937;

* Reprinted by permission of the Mathematical Association of America from: *Studies in Mathematical Biology. Part II: Populations and Communities* (S. A. Levin, ed.), Vol. 16, MAA Studies in Mathematics 1978

Segel and Jackson, 1972; Segel and Levin, 1976; Smale, 1974; Steele, 1974a, b). A more complete review may be found in Levin (1976b).

The consideration of spatial effects fundamentally changes our view of the organization of ecological communities. This paper is an attempt to survey the most important implications.

Basic Mathematical Approach

Although numerous alternatives exist and may sometimes be mandated (Levin, 1976b), the most and usual description of spatially distributed populations is through a discrete or continuous version of a transport-reaction system. Although movement is not generally of purely diffuse nature, such systems are often referred to generically, and somewhat sloppily, as *diffusion-reaction* systems even when non-diffusive terms are included. These systems provide the most convenient starting point for a discussion of the mathematical approach, and I shall first discuss them and some of the ways that they have found usage in ecology.

The discrete formulation is a bit more flexible with regard to the transport schemes which can be conveniently accommodated, and assumes an environment consisting of m distinct cells (numbered $\mu = 1, ..., m$) immersed in a much larger "bath." For each i, the dynamics of the density u_i^μ of species i in patch μ are assumed to be governed by an equation of the form

$$du_i^\mu/dt = f_i(\mathbf{u}^\mu, \mathbf{v}^\mu) + \{\text{net exchange with other patches}\}$$
$$+ \{\text{net exchange with bath}\}, \tag{1}$$

in which $\mathbf{u}^\mu$ is the vector $(u_1^\mu, ..., u_n^\mu)$ of densities in cell μ and $\mathbf{v}^\mu$ is the corresponding vector of environmental parameters (e.g., altitude, temperature, moisture) and may be time dependent. The functions f_i^μ are arbitrary except that they are assumed to be defined and continuously differentiable on an open set the projection of which onto "$\mathbf{u}^\mu$-space" contains the set $R_\mu = \{\mathbf{u}^\mu : u_i^\mu \geq 0 \, \forall i\}$; and further, for $u^\mu \varepsilon R_\mu$, $f_i^\mu \geq 0$ if $u_i^\mu = 0$ (Levin, 1976a). The function f_i^μ, the localized growth rate for species i in patch μ, depends only on local densities and parameters. The assumption made above simply says that this localized growth rate cannot be negative if species i is not present. Some special cases of (1) will be considered later.

Many and perhaps most environments are truly patchy, and the discrete formulation (1) provides a better description than does a continuous one. However, a continuous model is more amenable to analytical treatment, and there are many situations when the continuous approximation may be justified. Indeed, for some (such as aquatic systems) it is probably to be preferred on biological grounds.

In the continuous version, the cell mosaic is replaced by a continuum, and each point identified by its spatial coordinates $\mathbf{x}$. In the simplest versions, movement is assumed to be by passive diffusion, although the diffusion coefficients may vary in the different directions (as will be discussed later). The result is a system of nonlinearly coupled partial differential equations of parabolic type for the species' densities. When the habitat is linear and only one spatial dimension is to be considered, these equations take the form

$$\partial u_i/\partial t = f_i(\mathbf{u}, \mathbf{v}) + \partial(D_i \partial u_i/\partial x)/\partial x, \tag{2}$$

in which D_i is the diffusion coefficient for species i and might depend on x, u, and t. The extension to two dimensions is obvious, yielding

$$\partial u_i/\partial t = f_i(\mathbf{u}, \mathbf{v}) + \partial(D_i^1 \partial u_i/\partial x_1)/\partial x_1 + \partial(D_i^2 \partial u_i/\partial x_2)/\partial x_2, \tag{3}$$

where D_i^1 and D_i^2 are the diffusion coefficients in the two directions. If $D_i^1 = D_i^2 = D_i$ and D_i is constant, (3) degenerates to the familiar form

$$\partial u_i/\partial t = f_i(\mathbf{u}, \mathbf{v}) + D_i \nabla^2 u_i. \tag{4}$$

A slightly more general movement regime introduces advection into the system, for instance due to water current velocity $w(t)$. In the one-dimensional case, this replaces (2) by

$$\partial u_i/\partial t = f_i(\mathbf{u}, \mathbf{v}) + \partial(D_i \partial u_i/\partial x)/\partial x - w(t)\partial u_i/\partial x. \tag{5}$$

Platt and Denman (1975), in one of the most detailed efforts to provide a mathematical model for plankton distribution, utilize a three-space-dimensional version

$$\partial b/\partial t = f(b, \mathbf{v}) + D_H \nabla_H^2 b + D_{x_3} \partial^2 b/\partial x_3^2 - \mathbf{w} \cdot \nabla b, \tag{6}$$

in which b is phytoplankton density, $\mathbf{w}$ is the water velocity vector, D_H and D_{x_3} are horizontal and vertical diffusion coefficients, and $f(b, \mathbf{v})$ has the special form

$$f(b, \mathbf{v}) = k_m[i_{x_3}/(K_i + i_{x_3})]b - R_m[1 - e^{-\lambda(b - b_0)}]^+. \tag{7}$$

In this particular formulation, the first term represents growth, where k_m is the maximal cell division rate (at optimal light intensity), $i_{x_3} = i_0 \exp(-\alpha x_3)$ is the availability of light at depth x_3 (α is an optical extinction coefficient), and K_i measures the strength of the dependence of cell division on light. The second term is the loss due to herbivory and is of a type due to Ivlev; R_m is the maximal feeding rate, b_0 is a food threshold level, and λ is a constant. The notation $[\ ]^+$ indicates that the value within the bracket appears in (7) if it is positive, but is replaced by zero if it is negative.

The particular form of (6) reflects the fact that the diffusion coefficients in the two horizontal directions may be taken to be equal, but the coefficient in the vertical direction will in general be quite different.

Equation (6) is an equation for one species only (or more correctly, for the sum of all phytoplankton species). Of course, in reality, the feeding rate must depend on zooplankton density (since these are the principal herbivores); if the particular form (7) is accepted, this means that R_m must depend on zooplankton density.

Since zooplankton density itself will be related to phytoplankton density, principally through effects upon rate of change of its density, (6) must at the very least be replaced by a system of two equations of predator-prey type. I shall return to discuss this problem further in the next section.

Ecological Applications of the Basic Approach

There is considerable literature, much of recent vintage (see for example Aronson and Weinberger, 1975, 1977; Fife, 1976a, b; Fife and Peletier, 1977; Greenberg and Hastings, 1977; Hadeler, 1977; Kolmogoroff et al., 1937; Kopell and Howard, 1974; Othmer, 1975, 1977; Segel and Levin, 1976; Torre, 1975) on the properties of the solutions to diffusion-reaction systems of the form (4), and a somewhat smaller literature (see for example Levin, 1974, 1976a; Othmer and Scriven, 1971, 1974; Segel and Levin, 1976) on the discrete version (1). Their interest is generally in the mechanisms which lead to the formation of spatial or spatio-temporal patterns, including waves, wave fronts, and pulses (Aronson and Weinberger, 1975; Fife, 1976a, b; Greenberg and Hastings, 1977; Hadeler, 1977; Othmer, 1977; Segel and Levin, 1976). No attempt will be made to survey those mechanisms here; rather, I shall illustrate by examples how these influence thinking about ecological communities.

1. Spread of a Population Released at a Single Point

This is one of the simplest of applications and has appeared often in the ecological literature (see for example Skellam, 1951). Growth is considered negligible, and assuming the spread is isotropic and two-dimensional with constant rate of spread the population is modeled by an equation of the form

$$\partial u/\partial t = D(\partial^2 u/\partial x^2 + \partial^2 u/\partial y^2) = D\Delta^2 u. \tag{8}$$

The one- and three-dimensional versions show no significantly different behavior. Equation (8) is a two-dimensional version of (5), but with $f_i = w(t) = 0$. The solution, subject to the specified initial conditions that the release took place at the origin, takes the well-known form of a normal distribution

$$u(x, y, t) = (c/t) \exp[-(x^2 + y^2)/4Dt], \tag{9}$$

in which c is a constant. Note that the variance of this distribution is $\sigma^2 = 2Dt$ which increases linearly with time. This formula (9) changes slightly if diffusion is not isotropic or if the number of dimensions is changed, but the key property that the variance increases linearly with time remains true.

   If the release does not take place at a single point, or if one simply wishes to predict the future dispersion of the population given the distribution at time zero, then one obtains the initial value problem

$$\partial u/\partial t = D(\partial^2 u/\partial x^2 + \partial^2 u/\partial y^2),$$
$$u(x, y, 0) = f(x, y).$$

(10)

The solution of this is well known, being constructed by superposition of fundamental solutions of the form (9):

$$u(x, y, t) = (1/4\pi t) \int_{-\infty}^{\infty} \int_{-\infty}^{\infty} f(\xi, \eta)$$
$$\cdot \exp-([(x-\xi)^2 + (y-\eta)^2]/4Dt)d\xi d\eta.$$

(11)

Note that this will reduce to the form (9) if f is a multiple of a δ-function centered at the origin, and further that

(i) The total population size, $U = \iint u(x, y, t)dxdy$ is independent of time, and is thus equal to

$$\int_{-\infty}^{\infty} \int_{-\infty}^{\infty} f(\xi, \eta)d\xi d\eta.$$

This may be verified directly from (11), making appropriate changes in the order of integration and using the identity

$$1 = (1/4\pi t) \int_{-\infty}^{\infty} \int_{-\infty}^{\infty} \exp[-(x^2+y^2)/4Dt]dxdy.$$

However, it simply reflects the fact that the underlying equations describe only the redistribution of the population, and not growth or decay.

(ii) The center of the population,

$$\mathbf{M} = (\bar{x}, \bar{y}) = \int_{-\infty}^{\infty} \int_{-\infty}^{\infty} (x, y) \cdot u(x, y, t)dxdy/U,$$

also is independent of time; and in fact $\bar{x} \equiv 0$, $\bar{y} \equiv 0$. This simply describes the fact that, due to symmetry, there is no change in the population center.

(iii) The variance of (x, y),

$$V = \int_{-\infty}^{\infty} \int_{-\infty}^{\infty} (x^2+y^2)u(x, y, t)dxdy/U,$$

increases linearly with time according to the relation

$$V = V_0 + 2Dt.$$

Okubo (1970) has shown that the above model does not precisely conform to diffusion in the surface layers of oceans and lakes, but needs to be modified somewhat. Indeed, dye diffusion experiments show a quadratic or cubic dependence on t for the variance, rather than the linear dependence of Fick's diffusion (above). His proposed modification replaces (8) by

$$\partial u/\partial t = (c/r)\partial/\partial r(r^m \partial u/\partial r) \tag{12}$$

in which c and m are constants and $r^2 = x^2 + y^2$. Equation (8) is a special case of (12) with $m = 1$, but Okubo's data places the correct value of m at approximately 2.7.

2. Chemical Communication

Animals within a species signal danger, resource location, and sexual availability by emitting chemicals known as pheromones. Rubinow (1973) and Bossert and Wilson (1963) discuss mathematical aspects of the spread of the alarm pheromone in the harvester ant, *Pogonomyrmex badius*, using a three-dimensional version of the basic diffusion approach described above, with an additional factor of 2 introduced in the solution because of reflection at the surface of the earth. (This is equivalent to imposing the boundary condition $\partial u/\partial z = 0$ at $z = 0$.) The basic formula (9) is then replaced by

$$u(x, y, z, t) = 2N/(4\pi Dt)^{3/2} \exp[-(x^2 + y^2 + z^2)/4Dt], \tag{13}$$

in which the point of release is considered to be virtually at the surface of the earth. Here, u is the pheromone concentration and N is the amount released at time zero.

Okubo (1977) discusses further developments in the theory, considering continuous emission from a moving source and emission in a wind, thereby introducing advective effects and the complications of turbulent diffusion.

3. Planktonic Patchiness

One of the earliest applications of the basic diffusion approach to ecological problems was due to Kierstead and Slobodkin, who were motivated by a need to understand the causes of the initiation of red tides, large blooms of certain phytoplankton (particularly dinoflagellates) which can be extremely harmful to fish populations. Qualitatively, what appears to trigger these blooms is a local accumulation of favorable nutrients and other requisite organic growth 'substances. The quantitative theory began with Kierstead and Slobodkin, who initially consider one horizontal dimension and introduce a simple logarithmic growth term into the one-dimensional version of (8). This results in the equation

$$\partial u/\partial t = D\partial^2 u/\partial x^2 + ru, \tag{14}$$

in which u is the phytoplankton concentration and r the "intrinsic rate of natural increase." The form (14) is equivalent to assuming a local growth rule of the form

$$du/dt = ru,$$

which would have as solution $u = u_0 \exp(rt)$; hence the name logarithmic or exponential growth. Although it is an oversimplified description of population growth and not in general valid, it may be justified in this case since interest is upon the problem of initiation of tides. Hence logarithmic growth may be regarded as simply a linear approximation near $u = 0$.

In an infinite region, the solution to (14) is obtained simply:

$$u(x, t) = [e^{rt}/(2\sqrt{\pi t^{1/2}})] \int_{-\infty}^{\infty} f(\xi) \exp[-(x - \xi)^2/4Dt] d\xi, \tag{15}$$

where $f(\xi)$ describes the initial data. It is easy to show, most simply by using the methods of Sect. (1) for the case $r = 0$, that the total population size grows exponentially with exponent rt, the population center remains static, and the variance increases linearly with t.

However, the problem of interest to Kierstead and Slobodkin did not involve an infinite region. Rather, it is assumed that a (long narrow) favorable region of horizontal diameter L (and transverse cross-section area A) develops and is surrounded by a totally unfavorable region in which phytoplankton concentration cannot attain positive values; this introduces the boundary value problem

$$\partial u/\partial t = D\partial^2 u/\partial x^2 + ru, \quad 0 < x < L,$$
$$u(0, t) = u(L, t) = 0. \tag{16}$$

Roughly speaking, the population size will increase at a rate related to the "volume" of the region, which is proportional to L, minus the losses due to dissipation through the boundaries of the region, which are independent of L. If L is too small, the population cannot grow fast enough to compensate for those losses; but beyond a critical diameter L_c, there will be a net growth. This value L_c should be a lower bound to the size of blooms, and the diameter of the smallest favorable region which will permit blooms to arise.

The dependence of L_c on the parameters D and r can be roughly determined by dimensional arguments. The units of D are (length)^2/time and of r, (time)$^{-1}$; and hence it is to be expected that

$$L_{cr} \propto \sqrt{D/r}.$$

This may be rigorously verified and the constant of proportionality determined by direct consideration of the boundary value problem (16), as I now demonstrate.

One way to find particular solutions of (14) is by separation of variables; that is, by looking for solutions of the form $A(x)B(t)$. When this is done, one finds the solutions $\exp[ikx + (r - k^2D)t]$, where k is an arbitrary real number known as the wave number. Since we are only interested in real solutions, we are restricted to the real and complex parts of these separable solutions; that is, $\cos kx \exp(r - k^2 D)t$ and $\sin kx \exp(r - k^2 D)t$. Further, and more important, one can establish using Fourier transform arguments that any solution to (14) may be constructed by (an infinite) superposition of such separable solutions and further without loss of generality restrict attention to wave numbers $k \geq 0$.

For the boundary value problem (16), however, not every wave number is available but only those which result in the boundary conditions being satisfied. Since $\sin kx = 0$ for $kx = 0$ and in general for kx any multiple of π, this means that $kL = n\pi$ for some $n > 0$. Thus the wave numbers k are limited to the "discrete spectrum" $n\pi/L, n = 1, 2, 3, \ldots$, which is to say (since the cosine terms will not satisfy the boundary conditions) that the general solution to (14) is of the form of a convergent series

$$u(x, t) = \sum_{n=1}^{\infty} B_n \cdot \sin(n\pi x/L) \exp(r - n^2 \pi^2 D/L^2)t. \tag{17}$$

For each n for which $r < n^2 \pi^2 D/L^2$, the corresponding term in (17) will be damped to zero as $t \to \infty$. In particular, if $r < \pi^2 D/L^2$, every term and hence $u(x, t)$ itself will be damped to zero as $t \to \infty$. This means that the phytoplankton population will die out unless $L^2 \geq \pi^2 D/r$, and hence the critical diameter is given by

$$L_{cr} = \pi \sqrt{D/r}. \tag{18}$$

[The identical result, but in a somewhat different setting, was obtained by Skellam (1951).] Compare this result with the heuristic relationship determined by dimensional arguments alone.

This is, of course, only the simplest imaginable model, but the results turn out to be quite robust. Kierstead and Slobodkin extend it to two horizontal dimensions by assuming that the region of interest is a cylinder of diameter L and depth h, and assume cylindrical symmetry. Formula (18) is then replaced by

$$L_{cr} = 4.81 \sqrt{D/r}, \tag{19}$$

which is seen to be only a slight modification of (18). Again, this parallels a result of Skellam (1951). Okubo (ms.) retains the cylindrical (that is to say, horizontally circular) form, but considers three different diffusion laws. These give different forms for the critical size problem; but on the basis of empirically estimated constants, all turn out to be of the same order of magnitude (1–10 km).

Dubois (1975a, b), Platt and Denman (1975), and Wroblewski et al. (1975) discuss certain minor modifications of the results of Kierstead and Slobodkin when interacting populations are considered. The latter two papers reinterpret the relation (18) by introducing a grazing term as in (7) and find only a small percent correction to the critical length-scale.

As mentioned earlier, although Platt and Denman (1975) introduce a grazing effect, they do not consider directly the dynamics of the grazing population.

Various attempts to do this, however, have been undertaken by Dubois (1975a, b), Levin and Segel (1976), Okubo (1974), and Steele (1973, 1974a, b). Part of the motivation is that, as Platt and Denman (1975) write, "fluctuations in the spatial distribution of phytoplankton have been found on all length-scales so far examined in the sea." Steele (1975) argues that the larger scale patchiness cannot be related to variation in physical parameters such as salinity or temperature, and so an alternative explanation in terms of biological interactions is needed.

Steele (1974a), as do several other authors, begins from the classical Lotka-Volterra equations

$$dP/dt = aP - bPH,$$
$$dH/dt = cPH - dH,$$
(20)

in which P is the phytoplankton concentration and H the concentration of herbivorous copepods (zooplankton). These equations, as is well known, exhibit rather peculiar behavior. Their solutions are closed orbits in the $P - H$ plane, but are not limit cycles. Rather, (20) describes something much like a harmonic oscillator in that P and H will oscillate in a pattern determined by initial conditions. Slight perturbations from the closed orbits will not be corrected, but instead will transfer the solution to a new closed orbit. An unstable equilibrium exists at $P = H = 0$ and a second, more interesting, equilibrium at $P = c/d$, $H = a/b$. The latter one, which is enclosed by the other orbits, is really a degenerate cycle and is, like the others, neutrally stable. Thus the system is not structurally stable in that small modifications to (20) will result in qualitatively different behavior. Therefore, they provide a shaky starting point for investigation.

To consider the effects of diffusion, Steele adds appropriate terms to (20) to produce the system

$$\partial P/\partial t = aP - bPH + D_1 \partial^2 P/\partial x^2,$$
$$\partial H/\partial t = cPH - dH + D_2 \partial^2 H/\partial x^2.$$
(21)

He specifically assumes $D_1 = D_2$, but we need not. The system has a spatially uniform equilibrium at $P = d/c$, $H = a/b$; to study its stability with respect to small perturbations p_0, h_0, one sets $P = (d/c) + p$, $H = (a/b) + h$ and discards higher-order terms to obtain the linearized system

$$\partial p/\partial t = (-bd/c)h + D_1 \partial^2 p/\partial x^2,$$
$$\partial h/\partial t = (ac/b)p + D_2 \partial^2 h/\partial x^2.$$
(22)

In particular, if one seeks solutions to (22) of the form $p = A(t)e^{ikx}$, $h = B(t)e^{ikx}$, one finds that $A(t) = A_0 e^{\sigma t}$, $B(t) = B_0 e^{\sigma t}$ where

$$\begin{pmatrix} -D_1 k^2 & -bd/c \\ ac/b & -D_2 k^2 \end{pmatrix} \begin{pmatrix} A_0 \\ B_0 \end{pmatrix} = \sigma \begin{pmatrix} A_0 \\ B_0 \end{pmatrix}.$$
(23)

σ is thus an eigenvalue of the matrix

$$M = \begin{pmatrix} -D_1k^2 & -bd/c \\ ac/b & -D_2k^2 \end{pmatrix},$$

and hence has negative real part. Thus all finite wavelength perturbations will be damped, and the system with diffusion is seen to be more stable than the original. This conclusion is not restricted to the special solutions determined above, but is general since any solution may be Fourier synthesized from them.

    Steele, however, (for a particular boundary-value problem) argues that this linear approach is inappropriate because, by neglecting higher-order (quadratic) terms in (22), one is missing the fact that small wavelength perturbations can interact to produce ones of longer wavelength. This complaint is inappropriately applied to the system (21), since the basic underlying system (20) is structurally unstable; Murray (1975) (considering the case $D_1 = D_2$) demonstrates the error. Steele (1974b), however, extends his considerations to the more general system

$$\partial P/\partial t = aP - f_1(P)H + D_1\partial^2 P/\partial x^2,$$
$$\partial H/\partial t = f_2(P)H - f_3(H) + D_2\partial^2 H/\partial x^2, \tag{24}$$

in which $f_1(P)$ is the grazing rate of H on P, $f_2(P)$ is the growth rate for herbivores, and $f_3(H)$ is the predation rate on the herbivores. Again Steele assumes $D_1 = D_2$, but this is not necessary. He further assumes that (24) has a uniform equilibrium P_0, H_0, which will satisfy

$$aP_0 = f_1(P_0)H,$$
$$f_2(P_0)H_0 = f_3(H_0). \tag{25}$$

If one linearizes about that equilibrium, one obtains

$$\partial p/\partial t = \alpha p + \beta h + D_1\partial^2 p/\partial x^2,$$
$$\partial h/\partial t = \gamma p + \delta h + D_2\partial^2 h/\partial x^2, \tag{26}$$

where

$$\alpha = a - f_1'(P_0)H_0, \quad \beta = -f_1(P_0),$$
$$\gamma = f_2'(P_0)H_0, \quad \delta = f_2(P_0) - f_3'(H_0). \tag{27}$$

Note that in general $\beta < 0$ and, since prey are good for predators, $\gamma > 0$. Using the equilibrium relations (25), one may rewrite the formulas for α and δ as

$$\alpha = H_0[f_1(P_0)/P_0 - f_1'(P_0)], \quad \delta = [f_3(H_0)/H_0 - f_3'(H_0)]. \tag{28}$$

Either may be legitimately expected to be positive, thereby introducing autocatalytic effects, since such forms of the predator "functional response" are well-known

(Holling, 1965; Oaten and Murdoch, 1975). Indeed the Ivlev-type grazing function discussed earlier (7) is of the type which would result in such autocatalysis.

If α and δ are positive, or indeed if simply $\alpha + \delta > 0$, then the uniform equilibrium (25) is unstable in the absence of diffusion ($D_1 = D_2 = 0$). For general perturbations of the type considered previously, the stability matrix has the form

$$M = \begin{pmatrix} \alpha - D_1 k^2 & \beta \\ \gamma & \delta - D_2 k^2 \end{pmatrix}. \tag{29}$$

This can have eigenvalues with positive real part if and only if k is sufficiently small; in other words, there will be a critical value $k_c > 0$ such that perturbations with wave number $> k_c$ will be damped, while those with smaller wave number will grow. In this situation short-wavelength perturbations may interact nonlinearly to reinforce, in particular, ones of longer wavelength. This may cause equilibria which are stable to small perturbations to become unstable in the presence of sufficiently large ones. Thus the uniform distribution is destabilized and patchiness initiated.

Levin and Segel (1976) and Okubo (1974) advance another hypothesis; they suggest that patchiness may arise by a process known as "diffusive instability," in which a stable uniform state is destabilized by the differential dispersal (diffusive) abilities of the species. The basic idea derives from an earlier suggestion by Turing (1952) and first advanced in an ecological setting by Segel and Jackson (1972) and Levin (1974).

Levin and Segel's hypothesis is an extension of considerations introduced by Segel and Levin (1976), based on the system

$$\partial P/\partial t = P(a + eP - bH) + D_1 \nabla^2 P,$$
$$\partial H/\partial t = H(-d + cP - gH) + D_2 \nabla^2 H, \tag{30}$$

in which all parameters are non-negative. [Segel and Levin (1976) also treat the discrete analogue, for which the basic conclusions are virtually identical.] The system may be regarded as obtained (in the non-diffusive case) by retention of the linear terms in Taylor expansions for $(dP/dt)/P$ and $(dH/dt)/H$. Since computation is near equilibrium, this is for most purposes justifiable. (Technically, for some nonlinear calculations at least the quadratic terms should be retained; but in what is described below these make no difference.) The positivity of e may be justified in the same manner as mentioned earlier, based on a grazing function which has the same general "diminishing returns" character as (7). With $D_1 = D_2 = 0$, the system (28) has a positive equilibrium

$$\hat{P} = (bd + ag)/(bc - eg), \qquad \hat{H} = (ac + de)/(bc - eg), \tag{31}$$

provided $bc > eg$, and it is stable provided $(ac + de)g > (bd + ag)e$. To simplify the arithmetic, Levin and Segel (1976) take $d = 0$, so that the second stability condition reduces to $c > e$. However, this does not change the qualitative conclusions and we treat here the more general case.

The uniform solution $P = \hat{P}$, $H = \hat{H}$ is also an equilibrium when D_1 and D_2 differ from zero, but that equilibrium is not necessarily stable. Consideration as before of disturbances proportional to $\exp(i\mathbf{k}\cdot\mathbf{x})$ gives rise to the stability matrix

$$\begin{pmatrix} e\hat{P}-D_1q^2 & -b\hat{P} \\ c\hat{H} & -g\hat{H}-D_2q^2 \end{pmatrix}, \tag{32}$$

where $q=|\mathbf{k}|$. Provided

$$R=D_2/D_1 < R_{cr} = 1/[\sqrt{b/g}-\sqrt{(b/g)-(e/c)}]^2, \tag{33}$$

the matrix in (32) admits no eigenvalues with positive real part for any value of q. For $R>R_{cr}$, however, a real positive eigenvalue and hence destabilization will occur for certain values of q. Segel and Levin (1976) study nonlinearly the bifurcation from the uniform state (with $d=0$) and show that the destabilized equilibrium is replaced by a spatially non-uniform steady state. Thus a pattern of patchiness can arise due to the effects of differential diffusive abilities in destabilizing a uniform pattern of phytoplankton and zooplankton. This adds yet another mechanism to those which may contribute to planktonic patchiness.

This mechanism for patch formation also found another and earlier application (Keller and Segel, 1970) in the consideration of the initiation of aggregation in cellular slime molds.

4. Coexistence in Heterogeneous Environments

Levin (1974, 1976a, b) develops a general theory of ecological interaction in heterogeneous environments based upon systems of the forms (1)–(4). One of the most important applications is to systems in which the underlying dynamics in the absence of diffusion admit multiple stable states. In numerous ecological situations, the outcome of competition (for example) is dependent on initial conditions (such as colonization episodes), or events which occur early in successional development. The existence of multiple basins of attraction means that the local environment can support alternative stable communities, and leads in patchy environments to increased resource partitioning and species diversity (Levin, 1974, 1976a, b; Whittaker and Levin, 1977). In such environments, "overall system pattern is a mosaic of equilibrium patches, each slightly modified by some input from nearby patches" (Levin, 1976b).

The proof that such pattern may result from multiple stable states is actually dependent on the following theorem (Levin, 1974, 1976a):

Theorem 1. *Assume that $F_i^\mu(\mathbf{U},\mathbf{D})\geq 0$ when $u_i^\mu=0$ and $\mathbf{U}>0$, and that the system $du_i^\mu/dt=F_i^\mu(\mathbf{U},\mathbf{D}_0)$ has a stable equilibrium (in the sense that the eigenvalues of the Jacobian matrix have negative real parts) at $\mathbf{U}=\mathbf{U}_0>0$. Then for $\mathbf{D}$ sufficiently close to $\mathbf{D}_0$, the system $du_i^\mu/dt=F_i^\mu(\mathbf{U},\mathbf{D})$ has a stable equilibrium at some point $\mathbf{U}_\mathbf{D}>0$, where $\mathbf{U}_\mathbf{D}$ tends to $\mathbf{U}_0$ as $\mathbf{D}$ tends to $\mathbf{D}_0$. Here $\mathbf{U}=(\mathbf{u}^1,\mathbf{u}^2,...,\mathbf{u}^m)$, where $\mathbf{u}^\mu=(u_i^\mu,...,u_n^\mu)$, and $\mathbf{D}$ is a matrix of parameters. The notation $\mathbf{U}>0$ means that all components of $\mathbf{U}$ are non-negative. F_i^μ is assumed continuously differentiable.*

Note that the notion of stability used is stricter than asymptotic stability in that the borderline case when eigenvalues have zero real part is not considered.

The proof of this theorem is given in Levin (1976a) as a special case of the following more general theorem (Theorem 2). A discrete time version was derived by Karlin and McGregor (1972).

Theorem 2. *Let* $\mathbf{y} = (y_1, ..., y_p)$, $\mathbf{z} = (z_1, ..., z_q)$; *and for* $j = 1, ..., p$, *let* $G_j(\mathbf{y}, \mathbf{z})$ *be a continuously differentiable function of* $\mathbf{y}$ *and* $\mathbf{z}$ *and satisfy the inequality* $G_j(\mathbf{y}, \mathbf{z}) \geq 0$ *provided* $y_j = 0$ *and* $\mathbf{y} \geq 0$. *Assume further that the system of differential equations*

$$\frac{dy_i}{dt} = G_j(\mathbf{y}, \mathbf{z}^0)$$

has a stable equilibrium at $\mathbf{y}^0 \geq 0$. *Then for* $\mathbf{z}$ *sufficiently close to* $\mathbf{z}^0$, *the system*

$$\frac{dy_j}{dt} = G_j(\mathbf{y}, \mathbf{z})$$

also has a stable equilibrium $\mathbf{y}(\mathbf{z}) \geq 0$, *and* $\mathbf{y}(\mathbf{z})$ *tends to* $\mathbf{y}^0$ *as* $\mathbf{z}$ *tends to* $\mathbf{z}^0$.

The assumption of stability of the original equilibrium is an essential ingredient not only for the stability of the perturbed non-negative equilibrium, but also for the existence of a feasible equilibrium. In particular, if the original equilibrium lay on a boundary plane ($y_i = 0$) and were unstable, then in general a perturbed equilibrium would exist but might have negative components.

As an application of the above theorem, consider the system of equations

$$du_i^\mu/dt = F_i^\mu(\mathbf{U}, \mathbf{D}) = f_i^\mu(\mathbf{u}^\mu) + \sum_{r=1}^{m} D_i^{\nu\mu}(u_i^\nu - u_i^\mu). \tag{34}$$

In this formulation as before u_i^μ is the density of species i in patch μ; but now, for simplicity, movement is assumed to be by passive diffusion, with $D_i^{\nu\mu}$ the exchange coefficient from patch ν to patch μ. The parameters $D_i^{\nu\mu}$ are positive, and $D_i^{\nu\mu} = D_i^{\mu\nu}$.

Suppose now that the system in patch μ, were it sealed off from the rest of the mosaic, would have a (non-negative) stable equilibrium at $\mathbf{u}_0^\mu$. Then in the absence of diffusion ($\mathbf{D} = 0$), the whole system has a stable equilibrium at $(\mathbf{u}_0^1, ..., \mathbf{u}_0^m)$; and the theorems guarantee that for slight diffusion (e.g., for $\mathbf{D} \neq 0$ but sufficiently close to 0) the system will still have a (non-negative) stable equilibrium which will be close to the equilibrium $(\mathbf{u}_0^1, ..., \mathbf{u}_0^m)$. In particular, if the individual equilibria $\mathbf{u}_0^\mu$ show differences from one another, the overall system will exhibit spatial heterogeneity. This heterogeneity might reflect local parameter differences which alter the local equilibrium states, but it might also occur when the patches represent identical habitats (all f_i^μ identical) provided the local dynamics (as represented by f_i^μ) admit more than one stable equilibrium. The extension of this and Theorems 1 and 2 to stable attractors is intuitively clear but has not been rigorously presented in the literature.

To specialize further, consider a network of m identical patches, in each of which considered alone one or more of the n species would become extinct but in each of which every species has the potential to survive given a sufficiently good beginning. Then, for any specified collection of species, there exists a number n such that, if the number of patches m exceeds n and if there is only slight migration between patches, a stable configuration is possible with all of those species present (Levin, 1974). In particular, if the matrix $\mathbf{D}$ is irreducible, then species i will at equilibrium be represented in every patch.

These results are exemplified by consideration of the admittedly over-simplified Lotka-Volterra competition equations:

$$du_1/dt = u_1(r - au_1 - bu_2),$$
$$du_2/dt = u_2(s - cu_1 - du_2),$$
(35)

where all coefficients are positive.

In an environment consisting of two patches, two competitors whose local dynamics are governed by (35) will be described by the equations

$$du_1^\mu/dt = u_1^\mu(r - au_1^\mu - bu_2^\mu) + D_1(u_1^\nu - u_1^\mu),$$
$$du_2^\mu/dt = u_2^\mu(s - cu_1^\mu - du_2^\mu) + D_2(u_2^\nu - u_2^\mu),$$
(36)

where $\mu \neq \nu$ and $\mu, \nu = 1, 2$. Assume D_1 and D_2 are positive, but close to zero. Then if $r/s > \max(b/d, a/c)$, (36) has only one stable equilibrium, $u_1^1 = u_1^2 = r/a$, $u_2^1 = u_2^2 = 0$; further, every trajectory beginning in the positive orthant will tend to this equilibrium. If $r/s < \min(b/d, a/c)$, every such trajectory tends to the stable equilibrium $u_1^1 = u_1^2 = 0$, $u_2^1 = u_2^2 = s/d$. Hence, these cases always result in competitive exclusion of one species or the other.

Assume now r/s is intermediate between b/d and a/c. If $a/c > b/d$ (intraspecific competition outweighs interspecific), then (35) has a stable equilibrium with both species present, and this translates to the equilibrium [in (35)]

$$u_1^1 = u_1^2 = (rd - sb)/(ad - bc),$$
$$u_2^1 = u_2^2 = (sa - rc)/(ad - bc).$$
(37)

It is non-negative, and is the only stable equilibrium for (36). As in the previous two cases, it is spatially homogeneous, and every trajectory originating in the positive orthant tends to it. [For a more general treatment, see Hastings (1977c). For the system (36) one can easily write down a Liapunov function valid for the whole region.] Thus, in all of the three cases so far considered, the system tends asymptotically to a spatially homogeneous one.

One case remains, however, defined by the inequalities $b/d > r/s > a/c$ (inter-specific competition outweighs intraspecific). In this case, the system (35) admits two stable equilibria,

$$u_1 = r/a, \quad u_2 = 0$$
(38)

and

$$u_1 = 0, \quad u_2 = s/d.$$
(39)

In other words, the two species cannot coexist. The system (36), however, admits not only the two homogeneous equilibria corresponding to (38) and (39); but also two heterogeneous ones in which in one patch u_1 and u_2 are approximately given by

$$u_1 = r/a + O(D_1 + D_2), \quad u_2 = O(D_1 + D_2) > 0, \tag{40}$$

and in the other by

$$u_1 = O(D_1 + D_2) > 0, \quad u_2 = s/d + O(D_1 + D_2). \tag{41}$$

To illustrate this more clearly, suppose for simplicity that $a = d$, $b = c$, $r = s$, $D_1 = D_2 = D$, and $b > a$. Then an equilibrium exists at

$$\begin{aligned}
u_1^1 &= u_2^2 = (r - 2D)/2a \\
&\quad + \sqrt{(r - 2D)[r - 2D(b+a)/(b-a)]}/2a, \\
u_1^2 &= u_2^1 = (r - 2D)/2a \\
&\quad - \sqrt{(r - 2D)[r - 2D(b+a)/(b-a)]}/2a,
\end{aligned} \tag{42}$$

provided D is not too large $[D \leq r(b-a)/2(b+a)]$. The stability of this equilibrium is determined by the eigenvalues of the matrix $I_2 \otimes J + DK \otimes K$, where

$$I_2 = \begin{pmatrix} 1 & 0 \\ 0 & 1 \end{pmatrix}, \quad K = \begin{pmatrix} 0 & 1 \\ 1 & 0 \end{pmatrix},$$

and

$$J = \begin{pmatrix} r - 2au_1^1 - bu_2^1 - D & -bu_1^1 \\ -bu_2^1 & r - bu_1^1 - 2au_2^1 - D \end{pmatrix}$$

(Levin, 1974), or equivalently (Friedman, 1956) by the eigenvalues of the matrices $J \pm DK$. It is straightforward to show (Levin, 1974) that the equilibrium is locally stable provided $D \leq r(b-a)/2(2b+a)$; the global behavior of the system remains an unsettled question. It is not surprising that the heterogeneous equilibrium destabilizes for D large, since in this case mixing is so complete that no heterogeneous pattern can be maintained. This example shows that coexistence is possible in a patchy environment between species which would otherwise exclude one another; that is, species may avoid competitive exclusion by habitat partitioning. Why this occurs may be understood in another way by writing (following Peter Yodzis, unpublished) equations for the mean densities of u_1 and u_2,

$$\bar{u}_1 = (u_1^1 + u_1^2)/2, \quad \bar{u}_2 = (u_2^1 + u_2^2)/2.$$

Using (36) in the simple case just considered, one gets (after rearrangement)

$$\begin{aligned}
d\bar{u}_1/dt &= \bar{u}_1(r - a\bar{u}_1 - b\bar{u}_1) - a\delta_{u_1}^2 - b\,\mathrm{cov}(u_1 u_2), \\
d\bar{u}_2/dt &= \bar{u}_2(r - b\bar{u}_1 - a\bar{u}_2) - a\delta_{u2}^2 - b\,\mathrm{cov}(u_1 u_2),
\end{aligned} \tag{43}$$

in which $\delta^2_{u_1}$ and $\delta^2_{u_2}$ are respectively the variance of u_1 and u_2 about their means, and $\mathrm{cov}(u_1u_2)$ is the covariance. Coexistence has been made possible because the basic interaction, as described by (35), has been modified to account for the fact that the rates of growth of the populations are influenced by the dispersion patterns of the species over the patchwork environment. Were the populations uniformly distributed, then $\delta^2_{u_1}$, $\delta^2_{u_2}$, and $\mathrm{cov}(u_1u_2)$ would all be zero and (43) would reduce to (35); indeed, this does define the behavior of (36) on the invariant two-dimensional subspace $u_1^1 = u_1^2$, $u_2^1 = u_2^2$. Naturally, (43) by itself is not a complete description of the system; two additional equations are needed to describe the rates of change of the dispersion patterns. The above analysis, however, shows that this can operate to stabilize a heterogeneous pattern.

This phenomenon is ecologically very important, because it means that in a patchy environment, a complex and spatially heterogeneous pattern of species abundance can result either due to micro-variation in parameters or due to multiple steady states, and small amounts of migration will not destabilize such patterns. The determinants of the pattern are factors not included in the Eqs. (1) or (34), such as extinction and colonization episodes. Often these could be accommodated in the framework (1) by allowing the equations to be stochastic, particularly colonization either from within the system or from the bath. However, as I shall discuss later, this may not be the best method for treating colonization and extinction, which often occur on a different time scale.

Although these factors also carry analogous implications in continuous environments, the precise mathematical results are more difficult to develop, and the patterns will in general not be spatially fixed. The relevant mathematical results are extremely difficult to obtain, but some sense of the problems may be realized by consideration of the equation

$$\partial u/\partial t = F(u) + \partial^2 u/\partial x^2 \tag{44}$$

which describes the dynamics of a single variable distributed over a linear habitat (the units have been chosen so that the diffusion coefficient is 1). Here u could represent the density of a single species, or it might represent the relative frequency of one species in a collection of species. Fisher (1937), in considering the spread of an advantageous allele, introduced an equation of the form (44) for the relative frequency of the allele, and the more general equation has come under extensive study in a variety of applications. In particular, Kolmogorov et al. (1937), Aronson and Weinberger (1975, 1977), and Kanél (1962) have intensively studied (44) with special attention to the existence of wave fronts (see also Fife, 1976a; Hadeler, 1977; Rinzel, 1975, 1977). For the cases when $F(u)$ has either two or three zeroes, the existence of such fronts has been demonstrated, waves which can in some cases transform a particular region from one (possibly stable) equilibrium state to another. Further, Fife (1977) has shown that for infinite regions or for finite ones with no-flux boundary conditions, no genuinely stable non-uniform stationary solutions to (44) exist. When more species are considered, these general results do not apply; stable patterns may exist as the case considered by Segel and Levin (1976) demonstrates. The latter example, however, was of a predator-prey system, and to my knowledge the question of the existence of such patterns in the competitive case is still open.

Other Approaches

In the preceding section, we examined a number of examples of the application of the basic "diffusion-reaction" approach to ecological problems. The litany was by no means exhaustive; we did not discuss in detail competition along environmental gradients, chemotactic or other directed movement, Skellam's computation of the steady-state distributions under logarithmic and logistic growth in a number of different habitats, or the potential ecological importance of periodic wave-train solutions such as those discussed by Kopell and Howard (1973, 1974). Instead, it is worthwhile to recognize that in a number of important situations, the previous description is inappropriate either in temporal or spatial scale; in these cases, a different approach will be necessary.

5. Equilibrium Island Biogeography and Regional Patterns

The theory of island biogeography was developed in the context of oceanic islands (MacArthur and Wilson, 1967; Simberloff and Wilson, 1969, 1970). However, it is in the consideration of mainland situations where the island approach may find its greatest application, since a large number of communities can, due to their patchy composition, be viewed as aggregates of patches or islands (Cromartie, 1975; Janzen, 1968; Levin and Paine, 1974; Root, 1973). In some of these situations, there will be no mainland in the usual sense, but there will be a "bath" which serves as the source of new colonists. In the marine inertidal, for example, "islands" appear as gaps in mussel beds and new colonists come from planktonic larvae. In forests, canopy gaps are colonized predominantly from "seed baths." In other situations, there may not even be any distinguished source of colonists; rather, the system is closed in that new colonists come from other patches. Moreover, it is often the case that colonists travel widely, and the short-range movement characteristic of the diffusion-reaction approach becomes inappropriate (except possibly on a faster time scale). Many species, for example, may undergo long-range dispersal to find new habitats at one stage in the life cycle, but engage in short-range, basically diffusive-type movements for feeding in between major dispersal episodes.

Given that the diffusive-type movement must be replaced by an alternative scheme, the problem still exists to specify what that should be. An intermediate solution, which retains features of both long-range and short-range movement, is utilized in the stepping-stone models of genetic correlation (Kimura and Weiss, 1964; Weiss and Kimura, 1965). A more extreme but probably often justified approach (MacArthur and Wilson, 1967) assumes that all movement is long-range and that the composition of an island's immediate neighbors is inconsequential. This permits one to classify each island according to its intrinsic characteristics (e.g., size, exposure, distance from source, etc.), and to study colonization patterns in terms of these parameters.

As we saw in the previous section, those characteristics may not uniquely determine even the end state of the island's successional development; this

theoretical conclsuion has been borne out in numerous field studies of colonization (Root, 1973; Simberloff and Wilson, 1969, 1970; Sutherland, 1974). Random
events may play an important role in an island's colonization, and this could
conceivably even apply to the recovery following such a major disturbance as a
glaciation (Davis, 1976). In this section, we restrict our attention to the so-called
"equilibrium theory" of island biogeography (MacArthur and Wilson, 1967;
Simberloff and Wilson, 1969, 1970), which places major importance on the long-
term (asymptotic) species composition of an island following a disturbance. As we
shall see later, transient phenomena are often very important and may not be
ignored if disturbance is common. There are, however, situations where it is
applicable, and it provides a convenient starting point.

The equations of the general form (1), which describe species interactions
within an island and possibly some short-range movements, are assumed in the
equilibrium theory to pertain to a very rapid time scale, so rapid that on the
colonization-extinction time scale the events of (1) may be regarded as having run
their course and reached some generalized form of equilibrium. That equilibrium
will be upset only when some new disturbance is superimposed on the system.

To simplify the considerations, we assume first that the environment is
homogeneous in that each island (patch) is identical as regards habitat characteristics and accessibility to new colonists. This assumption can be later relaxed. Each
patch is assumed to have a finite number of steady-state compositions, and our
interest is on the overall distribution of these states. The basic approach is
exemplified by considering that each island has two possible stable states, A_1 and
A_2, and one unstable one (the vacant state). We assume, however, that the
perturbations which can destabilize the vacant state (i.e., colonization events) do
not occur on the fast time scale. A_1 and A_2 may be characterized by a single species,
for example sugar maple (A_1) and beech (A_2) (see, for example, Smith, 1972;
Whittaker and Levin, 1977), or by complexes of species. Following Levin
(1976a, b), let $P_1 =$ the number of patches in state A_1, $P_2 =$ number in state A_2, and
$K =$ total number of patches, so that $K - P_1 - P_2$ are vacant. The basic model
might then take the form

$$dP_1/dt = F_1(P_1, P_2, t), \quad dP_2/dt = F_2(P_1, P_2, t), \tag{45}$$

where the time dependence represents parametric variation (e.g., seasonal) in
growth characteristics, disturbance, or colonization.

As a particular example, assume that K is constant and that the rate of
replacement of vacant patches by state i patches is of the form $(m_i P_i + b_i) \cdot V$, where
b_i represents the rate of "colonization" from the bath and $m_i P_i$ from other patches
in the complex. Assume further that e_i is the rate that type i patches are recycled to
"vacant," i.e., the rate at which they go exstinct. Then, since $V = K - P_1 - P_2$, (45)
becomes

$$dP_1/dt = (K - P_1 - P_2)(m_1 P_1 + b_1) - e_1 P_1,$$
$$dP_2/dt = (K - P_1 - P_2)(m_2 P_2 + b_2) - e_2 P_2. \tag{46}$$

This system will always (given positive initial conditions) tend to a stable balance between the two species or states, which is not surprising given that there is a perpetual source of new colonists.

Suppose now that the community is sealed off from the bath, so that $b_1 = b_2 = 0$. Then (46) reduces to the simpler form

$$dP_1/dt = P_1[(m_1K - e_1) - m_1(P_1 + P_2)],$$
$$dP_2/dt = P_2[(m_2K - e_2) - m_2(P_1 + P_2)]. \tag{47}$$

Clearly, (47) can have no stable equilibrium; the "species" which wins is the one with the larger value of m/e (if both are identical, the balance is neutrally stable, and either A_1 or A_2 will random walk to extinction).

However, add now another level of complexity. Assume that A_1 and A_2 each has the potential to invade the territory of the other and instantaneously displace it (although this form of invasion competition may be allowed to be very one-sided). Suppose, indeed, that A_1 transforms to A_2 at the rate $c_2P_1P_2$, and A_2 to A_1 at the rate $c_1P_1P_2$. In each case this means that the rate of invasion is proportional to both the number of invasible patches and the potential number of invaders. Invasion is a form of growth, being at no cost to the host patch which is assumed to have colonists to spare. (47) then becomes

$$dP_1/dt = P_1[(m_1K - e_1) - m_1P_1 - (m_1 + c_2 - c_1)P_2],$$
$$dP_2/dt = P_2[(m_2K - e_2) - (m_2 + c_1 - c_2)P_1 - m_2P_2], \tag{48}$$

which is of the same general form as the Lotka-Volterra equations introduced earlier (35). Clearly, if $m_1K < e_1$, species A_1 disappears from the system; and if $m_2K < e_2$, species A_2 does likewise. Assume therefore that for both species the critical m/e ratio exceeds $1/K$. (Note that the importance of this inequality is reflected in the fact that either species in isolation would reach an equilibrium in which it occupied $K - e/m$ of the patches.) Then species A_1 will survive in the system provided

$$\frac{K - e_1/m_1}{K - e_2/m_2} > 1 + \frac{c_2 - c_1}{m_1}, \tag{49}$$

and species A_2 provided

$$\frac{K - e_2/m_2}{K - e_1/m_1} > 1 + \frac{c_1 - c_2}{m_2}. \tag{50}$$

If the species were equal competitors, so that $c_1 = c_2$, one of these conditions would have to be violated and at least one species go extinct. If, however, say species A_1 is the superior competitor, so that $c = c_1 - c_2 > 0$, then coexistence is possible if species A_2 has a larger m/e ratio than does species A_1, but not so much larger that species A_1 is outcompeted. If coexistence results, it is because a balance has been reached between one species which is a superior competitor and the other which is a superior colonizer, an idea which goes back at least to Hutchinson (1951) and Skellam (1951). The superior colonizer is sometimes referred to as a "fugitive" or "opportunistic" species.

Levins and Culver (1971) wished to extend this notion to permit consideration of an intermediate state in which both species are present, and Horn and MacArthur (1972) do the same for an environment made up of two types of patches. Although this is, of course, possible to do (see Cohen, 1970; Levin, 1974; Slatkin, 1974), the original models of Levins and Culver and of Horn and MacArthur are not correct (Levin, 1974). As mentioned, these ideas can obviously be extended to consideration of multiple patch types. This we shall not pursue here.

Hastings (1977a) analyzes a predator-prey system in a patchwork environment, basically implementing a system simulated by Caswell (1976). The approach is much like the one described in this section, except that invasion of a "prey" patch by a predator results in the local extinction of both species after a suitable delay time. The system he studies is thus a differential-delay system, and interesting results are obtained concerning the survival of the system. Extension to three species (one predator, two prey) was simulated by Caswell (1976), and Hastings (1977b) also provides some preliminary discussion of this problem.

6. Nonequilibrium Island Biogeography

In the last examples of the preceding subsection, we have technically deviated from a classical equilibrium view by allowing "mutations" from one equilibrium state to another, but this involves no fundamental change in our approach. In Levin (1976a, b), the notion of steady state is extended to include the *plateau*, a long-lived state which nonetheless will at some point be replaced by a successor state. In both Levin (1976b) and Whittaker and Levin (1977), models are considered for a variety of successional schemes, both obligate and non-obligate, in which the successional stages are considered as plateaus. Within this view, indeed, it is appropriate to view even the most advanced ("climax") stage as a plateau, since it will in time be recycled through the earlier stages of succession. Indeed, because the later successional species are often the most susceptible to major disturbance (such as fire) and the early successional are best adapted to colonize following disturbance, ecological communities in many cases tend to have a built-in successional microcycle with all stages being ephemeral.

The definition of a "successional stage" is a somewhat arbitrary convenience, since ecosystem succession, especially in terms of such parameters as diversity, biomass, and nutrient accumulation, is basically a continuous process. However, the general approach described above, a view of succession as a procession of plateaus, can obviously be extended to any number of such plateaus. In the limit, as one approaches the successional continuum, the models assume a form similar to that employed by Levin and Paine (1974, 1975). Their general approach is to assume that community organization is determined by the local properties and relative frequencies of the *patches* (or *islands* or *gaps*) which major disturbance creates, and by the biological development of the individual patch. These patches may be the canopy gaps which tree falls create, the phytoplankton blooms created by local bursts of nutrients, or the holes in mussel beds created by wave damage (numerous other examples are given in Levin, 1976b).

In the model of Levin and Paine, each patch is characterized at time t by its successional age a and by a vector ξ of other characteristics (such as size, species composition, humus or nutrient level). Within a patch, obviously, $da/dt = 1$; but further, the vector ξ changes according to the equation

$$d\xi/dt = g(t, a, \xi), \tag{51}$$

where g is the vector of mean growth rates of the components of ξ. The deceptively simple form of (51) may describe dynamics as straightforward as the rate of growth or shrinkage of the area of the patch, or as elusive as the interactive dynamics of component species. In the simplest versions (Levin and Paine, 1974, 1975), ξ is a scalar measuring area.

Primary interest focuses upon the distribution of patches with respect to a and ξ at time t, defined by the density function $n(t, a, \xi)$. To describe the rate of change of n over time, one may use the same basic equation of von Foerster (1959) that Oster and Rubinow utilize elsewhere in this volume. Specifically, let $\mu(t, a, \xi)$ be the instantaneous extinction rate for patches of age a and size ξ. Then the dynamics of n are governed by the first-order equation

$$\partial n/\partial t + \partial n/\partial a + \mathrm{div}\,(ng) = -\mu n. \tag{52}$$

[This can be extended to allow for the variance about the mean growth rate **g** by the introduction of a second-order term in (52). For a discussion, see the paper by Oster in this volume.]

Given the initial distribution $n(0, a, \xi)$ and the rate of new patch formation $n(t, 0, \xi)$, Eq. (52) may in general be solved in closed form by the method of characteristics. When ξ is a scalar, this solution is given (for $t \geq a$) in Levin and Paine (1974). As an example, when μ is constant and $g(\xi) = c\xi$, the complete solution is

$$n(t, a, \xi) = \begin{cases} n(t-a, 0, \xi e^{-ca}) \cdot \exp[-(c+\mu)a], & t \geq a, \\ n(0, a-t, \xi e^{-ct}) \cdot \exp[-(c+\mu)t], & t < a. \end{cases} \tag{53}$$

Although solutions of this form are technically correct, they may not suffice in that the "birth" function $n(t, 0, \xi)$, which influences the solution, is in many applications dependent upon the whole distribution $n(t, a, \xi)$ at time t. One technique is to obtain a solution in a form analogous to (53), and then substitute this back into the renewal equation relating $n(t, 0, \xi)$ and the rest of the distribution. This approach is discussed in some detail by Oster and Takahashi (1974) and Rubinow (1973).

By solving the Eq. (52) subject to the appropriate boundary and initial conditions, one obtains an expression for the distribution n in relation to indicators of disturbance. In this way, for example, one can determine the degree of patchiness and successional age distribution of a region in relation to such habitat parameters as degree of exposure to waves (in the case of the intertidal) or wind (in the case of forests, where blowdowns are important), etc. Once this distribution is known, species composition of the community may be inferred and such macroscopic parameters as diversity determined and related to disturbance. Thus, for example, if disturbance is low, one would expect the successional age distribution to be such that late successional species would be predominant and diversity low. If disturbance is high, early successional species would be in great abundance and diversity again low. For intermediate disturbances, however, the greatest mixture of species and the greatest diversity will be achieved.

This approach provides a powerful tool for dealing with spatial heterogeneity, quite distinct from the diffusion-reaction paradigm. It has been applied already to the intertidal and to forest dynamics, and offers great potential in a variety of situations both for the treatment of theoretical questions and as a tool for management, in which for example disturbance regimes (as represented say by lumbering) are to be manipulated. Further, it provides a framework which is well suited to the study of such adaptations as seed dormancy and dispersal (Levin et al., 1977), which are evolutionary responses to environments unpredictable in space and time.

Summary

Until recently, mathematical ecology has paid little attention to the importance of spatial heterogeneity, and thus most models have ignored species dispersion patterns. This, however, has rendered the conclusions of such models of limited interest in an understanding of community structure, since spatial heterogeneity and the potential for dispersal fundamentally alter species interactions and dynamics and increase by several orders of magnitude the potential for coexistence. The classical approach, localized in space and asymptotic in time, focuses attention only on the competitive dominants and their associated predators and parasites. However, consideration of spatial extent not only allows for the coexistence of incompatible co-dominants, it also allows for spatio-temporal patterns of occupancy and death or emigration which translate into survival for ephemerals which occupy early or intermediate stages on the successional gradient.

In this paper a variety of models are considered, from those of the diffusion-reaction type to the "patch" models of more recent vintage, and applications to a variety of ecological situations are discussed.

References

Allen, J.C. (1975). Mathematical models of species interactions in time and space, Am. Nat., *109*, 319–342

Aronson, D.G., Weinberger, H.F. (1975). Nonlinear diffusion in population genetics, combustion, and nerve propagation, Partial Differential Equations and Related Topics, Lecture Notes in Mathematics, *446*, J. A. Goldstein, ed., Springer-Verlag, Heidelberg

Aronson, D.G., Weinberger, H.F. (1977). Multidimensional nonlinear diffusion arising in population genetics, Advances in Math.

Bossert, W.H., Wilson, E.O. (1963). The analysis of olfactory communication among animals, J. Theoret. Biol., 5, 443–469

Brownlee, J. (1911). The mathematical theory of random migration and epidemic distribution, Proc. Roy. Soc. Edinburgh, 31, 262–289

Caswell, H. (1976). Predator mediated coexistence of prey, manuscript

Cohen, D. (1980). A theoretical model for the optimal timing of diapause, Am. Nat., 104, 389–400

Cohen, J.E. (1970). A Markov contingency table model for replicated Lotka-Volterra systems near equilibrium, Am. Nat., 104, 547–559

Cromartie, W.J., Jr. (1975). The effect of stand size and vegetational background on the colonization of cruciferous plants by herbivorous insects, J. Appl. Ecol., 12, 517–533

Davis, M.B. (1976). Pleistocene biogeography of temperate deciduous forests, Geosci. and Man, 13, 13–36

Den Boer, P.J. (1968). Spreading of risk and stabilization of animal numbers, Acta Biotheor. Leiden, 18, 165–194

Den Boer, P.J. (1971). Stabilization of animal numbers and the heterogeneity of the environment: The problem of persistence of sparse populations, Proc. Adv. Study Inst. Dyn. Numbers Popul. Oosterbeek, 1970, 77–97

Dubois, D.M. (1975a). Simulation of the spatial structuration of a patch of prey-predator plankton populations in the Southern Bight of the North Sea, Proc. Liège Colloq. Ocean Hydrodyn. 6th Mem. Soc. Roy. Sci. Liège, VII, 75–82

Dubois, D.M. (1975b). A model of patchiness for prey-predator plankton populations, Ecol. Model., 1, 67–80

Dubois, D.M., Adam, Y. (1976). Spatial structuration of diffusive prey-predator biological populations: Simulation of the horizontal distribution of plankton in the North Sea, System Simulation in Water Resources, G. C. Vansteenkiste, ed., North Holland Amsterdam, 343–356

Fife, P.C. (1976a). Singular perturbation and wave front techniques in reaction-diffusion problems, Proc. AMS-SIAM Symp., New York, Amer. Math. Soc., Providence

Fife, P.C. (1976b). Pattern formation in reacting and diffusing systems, J. Chem. Phys., 64, 554–564

Fife, P.C. (1977). Stationary patterns for reaction-diffusion equations, Nonlinear Diffusion, Pitman, London (to appear)

Fife, P.C., Peletier, L.A. (1977). Nonlinear diffusion in population genetics, Arch. Rational Mech. Anal. (to appear)

Fisher, R.A. (1937). The wave of advance of advantageous genes, Ann. Eugen. London 7, 355–369

Friedman, B. (1956). An abstract formulation of the method of separation of variables, Proc. of the Conference on Differential Equations, J. B. Diaz and L. E. Payne, eds., University of Maryland Bookstore, College Park, Maryland, 209–226

Gadgil, M. (1971). Dispersal: Population consequences and evolution, Ecology, 52, 253–260

Greenberg, J.M., Hastings, S.P. (1977). Spatial patterns for discrete models of diffusion in excitable media, SIAM J. Appl. Math. (in press)

Hadeler, K.P. (1976, 1977). Nonlinear diffusion equations in biology, Proc. of Conference on Partial Differential Equations, Dundee, Scotland, Lecture Notes in Mathematics, Springer-Verlag, Heidelberg

Haldane, J.B.S. (1948). The theory of a cline, J. Genet., 48, 277–284

Hastings, A. (1977a). Spatial heterogeneity and the stability of predator-prey systems, Theor. Pop. Biol., 12, 37–48

Hastings, A. (1977b). Spatial heterogeneity and the stability of predator-prey systems: Predator mediated coexistence of prey, manuscript

Hastings, A. (1977c). Models in population biology, Ph.D. Thesis, Cornell University

Holling, C.S. (1965). The functional response of predators to prey density and its role in mimicry and population regulation, Mem. Entom. Soc. Can., 45, 1–60

Horn, H.S., MacArthur, R.H. (1972). Competition among fugitive species in a harlequin environment, Ecology, 53, 749–752

Hutchinson, G.E. (1951). Copepodology for the ornithologist, Ecology, 32, 571–577

Janzen, D.H. (1968). Host plants as islands in evolutionary and contemporary time, Am. Nat., 102, 592–595

Kanél, I. (1962). Stabilization of solutions of the Cauchy problem for equations encountered in combustion theory, Math. Sb., *59*, 245–288 (Suppl.)

Karlin, S., McGregor, J. (1972). Polymorphisms for genetic and ecological systems with weak coupling, Theor. Pop. Biol., *3*, 210–238

Keller, E.F., Segel, L.A. (1970). Initiation of slime mold aggregation viewed as an instability, J. Theoret. Biol., *26*, 399–415

Kierstead, H., Slobodkin, L.B. (1953). The size of water masses containing plankton blooms, J. Mar. Res., *12*, 141–147

Kimura, M., Weiss, G.H. (1964). The stepping stone model of population structure and the decrease in genetic correlation with distance, Genetics, *49*, 561–576

Kolmogoroff, A., Petrovskij, I., Piskunov, N. (1937). Étude de l'équation de la diffusion avec croissance de la quantité de matière et son application à un problème biologique, Bull. Univ. Moscow Ser. Intl. Sect. AI, *6*, 1–25

Kopell, N., Howard, L.N. (1973). Plane wave solutions to reaction-diffusion equations, Stud. Appl. Math., *52*, 291–328

Kopell, N., Howard, L.N. (1974). Pattern formation in the Belousov reaction, Some Mathematical Questions in Biology, S. A. Levin, ed., *6*, 201–217. Lectures on Mathematics in the Life Sciences, vol. 7, Amer. Math. Soc., Providence

Levin, S.A. (1974). Dispersion and population interactions, Am. Nat., *108*, 207–228

Levin, S.A. (1976a). Spatial patterning and the structure of ecological communities, Some Mathematical Questions in Biology, S. A. Levin, ed., *7*, 1–36. Lectures on Mathematics in the Life Sciences, vol. 8, Amer. Math. Soc., Providence

Levin, S.A. (1976b). Population dynamic models in heterogeneous environments, Annu. Rev. Ecol. Syst., *7*, 287–310

Levin, S.A., Cohen, D., Hastings, A. (1977). Manuscript

Levin, S.A., Paine, R.T. (1974). Disturbance, patch formation, and community structure, Proc. Nat. Acad. Sci. USA, *71*, 2744–2747

Levin, S.A., Paine, R.T. (1975). The role of disturbance in models of community structure, Ecosystem Analysis and Prediction, S. A. Levin, ed., SIAM, 56–67

Levin, S.A., Segel, L.A. (1976). Hypothesis for origin of planktonic patchiness, Nature, *259*, (5545) 659

Levins, R., Culver, D. (1971). Regional coexistence of species and competition between rare species, Proc. Nat. Acad. Sci. USA, *68*, 1246–1248

MacArthur, R.H., Wilson, E.O. (1967). The Theory of Island Biogeography, Princeton University Press, Princeton, 203 pp.

Marks, P.L. (1974). The role of pin cherry (Prunus pensylvanica L.) in the maintenance of stability in northern hardwood ecosystems, Ecol. Monogr., *44*, 73–88

Murray, J.D. (1975). Non-existence of wave solutions for the class of reaction-diffusion equations given by the Volterra interacting-population equations with diffusion, J. Theor. Biol. *52*, 459–469

Oaten, A., Murdoch, W.W. (1975). Functional response and stability in predator-prey systems, Am. Nat., *109*, 289–298

Okubo, A. (1970). Oceanic mixing, Tech. Rep. 62, Chesapeake Bay Inst., Johns Hopkins University, Baltimore

Okubo, A. (1974). Diffusion-induced instability in model ecosystems: Another possible explanation of patchiness, Tech. Rep. 86, Chesapeake Bay Inst., Johns Hopkins University, Baltimore

Okubo, A. (1976). The critical size of water masses containing plankton blooms under scale-dependent diffusion, manuscript

Okubo, A. (1977). Ecology and Diffusion, Lecture Notes in Biomathematics, Springer-Verlag, Heidelberg (in press)

Oster, G., Takahashi, Y. (1974). Models for age-specific interactions in a periodic environment, Ecol. Monogr., *44*, 483–501

Othmer, H.G. (1975). Nonlinear wave propagation in reacting systems, J. Math. Biol., *2*, 133–164

Othmer, H.G. (1977). Current problems in pattern formation, Lectures on Mathematics in the Life Sciences, S. A. Levin, ed., vol. 9, Amer. Math. Soc., Providence (in press)

Othmer, H.G., Scriven, L.E. (1971). Instability and dynamic pattern in cellular networks, J. Theor. Biol. *32*, 507–537

Othmer, H.G., Scriven, L.E. (1974). Nonlinear aspects of dynamic pattern in cellular networks, J. Theor. Biol., *43*, 83–112

Platt, T., Denman, K.L. (1975). A general equation for the mesoscale distribution of phytoplankton in the sea, Mém. Soc. Roy. Sci. Liège, *7*, 31–42

Platt, W.J. (1975). The colonization and formation of equilibrium plant species associations on badger disturbances in a tall-grass prairie, Ecol. Monogr., *45*, 285–305

Reddingius, J. (1971). Gambling for existence: A discussion of some theoretical problems in animal population ecology, Acta Theor. Suppl. Climum, added to Acta Biotheor., Brill, Leiden, *20*, 208 pp.

Reddingius, J., Den Boer, P.J. (1970). Simulation experiments illustrating stabilization of animal numbers by spreading of risk, Oecologia, *5*, 240–248

Rinzel, J. (1975). Neutrally stable travelling wave solutions of nerve conduction equations, J. Math. Biol., *2*, 205–218

Rinzel, J. (1977). Integration and propagation of neuroelectrical signals, MAA Studies in Mathematical Biology, vol. 15: Cellular Behavior and the Development of Pattern, S. A. Levin, ed., Math. Assoc. of America, Washington

Roff, D.A. (1974a). Spatial heterogeneity and the persistence of populations, Oecologia, *15*, 245–258

Roff, D.A. (1974b). The analysis of a population model demonstrating the importance of dispersal in a heterogeneous environment, Oecologia, *15*, 259–275

Roff, D.A. (1975). Population dispersal and the evolution of dispersal in a heterogeneous environment, Oecologia, *19*, 217–237

Root, R.B. (1973). Organization of a plant-arthropod association in simple and diverse habitats: The fauna of collards *(Brassica oleracea)*, Ecol. Monogr., *43*, 95–124

Root, R.B., Chaplin, S.J. (1976). The life styles of tropical milkweed bugs, Oncopeltus (Hemiptera, Lygaeidae) utilizing the same host, Ecology, *57*, 132–140

Rubinow, S.I. (1973). Mathematical problems in the biological sciences, SIAM, 90 pp.

Schoener, A. (1974). Experimental zoogeography: Colonization of marine mini-islands, Am. Nat., *108*, 715–738

Segel, L.A., Jackson, J. (1972). Dissipative structure: An explanation and an ecological example, J. Theor. Biol. *37*, 545–559

Segel, L.A., Levin, S.A. (1976). Application of nonlinear stability theory to the study of the effects of diffusion on predator-prey interactions, Topics in Statistical Mechanics and Biophysics: A Memorial to Julius L. Jackson, R.A. Piccirelli, ed., Proc. AIP Conf., *27*, 123–152

Simberloff, D.S. (1976). Experimental zoogeography of islands: Effects of island size, Ecology, *57*, 629–648

Simberloff, D.S., Wilson, E.O. (1969). Experimental zoogeography of islands: The colonization of empty islands, Ecology, *50*, 278–296

Simberloff, D.S., Wilson, E.O. (1970). Experimental zoogeography of islands: A two-year record of colonization, Ecology, *51*, 934–937

Skellam, J.G. (1951). Random dispersal in theoretical population, Biometrika, *38*, 196–218

Slatkin, M. (1974). Competition and regional coexistence, Ecology, *55*, 128–134

Smale, S. (1974). A mathematical model of two cells via Turing's equation, Some Mathematical Questions in Biology, J.D. Cowan, ed., *5*, 15–26. Lectures on Mathematics in the Life Sciences, vol. 6, Amer. Math. Soc., Providence

Smith, F.E. (1972). Spatial heterogeneity, stability, and diversity in ecosystems, in Growth by Intussusception: Ecological Essays in Honor of G. Evelyn Hutchinson, Trans. Conn. Acad. Arts Sci., *44*, 309–335

Steele, J.H. (1973). Patchiness, Coastal Upwelling and Ecosystems Anal. Newsl., *2*, 3–7

Steele, J.H. (1974a). Spatial heterogeneity and population stability, Nature, *83*, 248

Steele, J.H. (1974b). Stability of plankton ecosystems, Ecological Stability, M.B. Usher and M.H. Williamson, eds., Chapman and Hall, London, 179–191

Steele, J.H. (1975). The Structure of Marine Ecosystems, Harvard University Press, Cambridge

Strathman, R. (1974). The spread of sibling larvae of sedentary marine invertebrates, Am. Nat., *108*, 29–44

Sutherland, J.P. (1974). Multiple stable points in natural communities, Am. Nat., *108*, 859–873

Torre, V. (1975). Synchronization of nonlinear biochemical oscillators coupled by diffusion, Cybernetics, *17*, 137–144

Turing, A. (1952). The chemical basis of morphogenesis, Philos. Trans. Roy. Soc. London Ser. B, *237*, 37–72

von Foerster, H. (1959). Some remarks on changing population, The Kinetics of Cellular Proliferation, F. Stohlman, Jr., ed., Stratton, New York, 382–407

Weiss, G.H., Kimura, M. (1965). A mathematical analysis of the stepping-stone model of genetic correlation, J. Appl. Probability, *2*, 129–149

Whittaker, R.H., Levin, S.A. (1977). The role of mosaic phenomena in natural communities, Theor. Pop. Biol., *12*, 117–139

Wroblewski, J.S., O'Brien, J.J., Platt, T. (1975). On the physical and biological scales of phytoplankton patchiness in the ocean, Mem. Soc. Roy. Sci. Liege, *7*, 43–57

Stochastic Community Theory:
A Partially Guided Tour

Michael Turelli

Contents

I. Introduction and Overview

This chapter is intended as an introduction to the biological questions, mathematical analyses, and biological conclusions of stochastic models for multispecies assemblages. Because of the large number of topics and models, little attention is devoted to detailed analysis of particular models. Instead, I attempt a survey of the key ideas and provide references for further study. The presentation is organized according to the mechanism(s) creating the probabilistic effects. Three classes are considered:

1. Demographic stochasticity (or "within-individual variability", Chesson, 1978) – Whether or not God plays dice, individuals who are apparently identical have different life lengths and produce different numbers of offspring. Integer-valued stochastic models are typically used to investigate the consequences of this variation.

2. Environmental stochasticity – Environments vary unpredictably through time in ways that affect all individuals equally. Most analyses of the consequences begin by introducing random variation into the parameters of a standard deterministic model. This produces stochastic difference and differential equations with continuous ranges.

3. Combined demographic and environmental stochasticity – In general, both sorts of variation are always present. However, there are relatively few attempts to analyze their joint consequences.

Biomathematics, Vol. 17, Mathematical Ecology
Edited by T. G. Hallam and S. A. Levin
© Springer-Verlag Berlin Heidelberg 1986

The excellent review by Chesson (1978) provides an alternative taxonomy that includes spatial as well as temporal variability.

Most stochastic multispecies models are derived by generalizing well-known deterministic models, such as the Lotka-Volterra equations, to include demographic or environmental stochasticity. The basic question underlying their analyses is whether there are biologically significant differences between the predictions of the stochastic models and the predictions of their simpler deterministic counterparts. If so, the stochastic fluctuations may play a central role in explaining the structure of natural communities. If not, the type of stochasticity analyzed can be regarded as a mathematical complication that can be safely ignored. In stochastic generalizations of deterministic models, a deterministic process such as competition or predation is assumed to be a central feature of community dynamics, and the stochastic fluctuations are regarded as an embellishment, albeit a possibly critical one. In contrast, there is a smaller class of stochastic models, specifically the so-called "null models" of community structure, that begin with the premise that observed patterns may be attributable almost solely to random processes. These models complement the more typical stochastic models by addressing the question of whether deterministic interactions need be invoked to explain observed patterns.

In addition to summarizing results, I would like to convey one key idea. The *qualitative* conclusions concerning broad questions like, "Does environmental variability limit the similarity of competitors?" can depend critically on both the model analyzed *and* the method of analysis used. Discrepancies must be reconciled by examining the biological and mathematical assumptions underlying both the model and its analysis. To avoid artifacts, the modeller must carefully suit the analysis to the biological question addressed and should determine the robustness of conclusions to the details of the model considered. All modellers should read Pielou's (1981) critical assessment of the field as well as Levin's (1981) constructive interpretation of theory's role in ecology.

The references given are a very limited and uneven sample of those available. Emphasis is placed on references that are biologically well motivated or mathematically elegant, provide extensive bibliographies, or were written by me. Because I am most familiar with the research on environmental stochasticity, it will be treated in more detail. References to parallel work in population genetics, which has a much more extensive stochastic theory, are provided at random.

II. Demographic Stochasticity

The first probabilistic analyses of ecological interactions to attract widespread attention were the attempts by Neyman et al. (1956), Leslie and Gower (1958), and Bartlett et al. (1960) to explain the indeterminate outcomes of some of Park's (1954) classic *Tribolium* competition experiments in terms of demographic stochasticity. For about fifteen years, this stochastic ecological explanation was eclipsed by a stochastic genetic explanation ("founder effect") involving genetic heterogeneity among the small initial populations used in the replicates (see Lerner and Ho, 1961; Lerner and Dempster, 1962; and Mertz et al., 1976). However, in one of the most convincing empirical applications of a stochastic ecological model, Mertz et al. (1976) revived the demographic stochasticity hypothesis with an elegant set of experiments. In addition to the ongoing tradition of generalizing deterministic models to include demographic stochasticity, this sort of individual-to-individual stochastic variation underlies several recent stochastic models of community structure, such as those of Caswell (1976), Horn (1975), and Hubbell (1979), that describe random replacement processes for individuals within a community.

Both continuous-time and discrete-time models have been analyzed. The random variation in population densities and community composition is essentially always Markovian, i.e. the future is independent of the past once present conditions are specified. No attempt will be made to describe the multispecies models or analyses in detail; most of the key mathematical ideas will reappear in the discussion of environmental stochasticity. The central conclusion is that demographic stochasticity produces little departure from deterministic dynamics if initial population sizes are large and the deterministic analog treats a collection of homogeneously interacting populations (see Chesson, 1978, 1981) whose joint deterministic dynamics possess a globally stable equilibrium at which all populations remain large. Conversely, if initial population sizes or equilibria are small, the populations are subdivided, or there are several or no deterministically stable equilibria, demographic stochasticity plays a central role in population dynamics and community structure. Slight elaborations of each of these themes follow.

To make the chat below slightly more concrete, it may be useful to have a simple class of models in mind that capture the flavor of typical models incorporating demographic stochasticity into the dynamics of homogeneously interacting populations. Begin with a discrete-time analog of the Lotka-Volterra competition equations such as

$$X_{i,t+1} = X_{i,t} g_i \left(\sum_{j=1}^{n} \alpha_{ij} X_{j,t}/K_i \right) \quad \text{for} \quad i=1, ..., n. \tag{1}$$

Here $X_{i,t}$ denotes the abundance of species i in generation t, and we assume that for each i, $g_i(x) > 0$, and $g_i'(x) < 0$ for $x > 0$ and $g_i(1) = 1$. A natural way to incorporate demographic stochasticity into (1) is to assume that the g_i describe the *mean* per capita reproductive rate for each species. An especially simple stochastic model is obtained by assuming that conditional on $\underline{X}_t = \underline{x}_t = (x_{1,t}, ..., x_{n,t})^T$, each individual of species i independently produces a Poisson distributed number of offspring, denoted $G_i(\underline{x}_t)$, with mean $g_i\left(\sum_{j=1}^{n} \alpha_{ij} x_{j,t}/K_i \right)$. This new non-negative, integer-valued stochastic model can be compactly described by

$$(X_{i,t+1}|\underline{X}_t = \underline{x}_t) \overset{D}{=} \sum_{k=1}^{x_{i,t}} G_{i,k}(\underline{x}_t), \tag{2}$$

with the $G_{i,k}(\underline{x}_t)$ all independent and identically distributed with

$$G_{i,k}(\underline{x}_t) \overset{D}{=} P\left[g_i\left(\sum_{j=1}^{n} \alpha_{ij} x_{j,t}/K_i \right) \right]. \tag{3}$$

[Here "$\overset{D}{=}$" denotes equality of probability distributions and "$P(\lambda)$" denotes a Poisson random variable with mean λ.] For our purposes, the central feature of the model follows from the fact that

$$(X_{i;t+1} \underline{X}_t = x_t) \overset{D}{=} P\left[x_{i,t} g_i \left(\sum_{j=1}^{n} \alpha_{ij} x_{j,t}/K_i \right) \right]. \tag{4}$$

Hence, the coefficient of variation (CV) of the demographic fluctuations satisfies

$$CV(X_{i,t+1}|\underline{X}_t = x_t) = \frac{\sqrt{\text{Var}(X_{i,t+1}|\underline{X}_t = x_t)}}{E(X_{i,t+1}|\underline{X}_t = x_t)} = \frac{1}{\sqrt{x_{i,t}}\sqrt{g_i\left(\sum_{j=1}^{n} \alpha_{ij} x_{j,t}/K_i\right)}} \tag{5}$$

The scaling of the relative magnitude of demographic fluctuations with the inverse of the square root of population size is a fairly general feature of models incorporating demographic stochasticity. It motivates most of the results reported below.

A. Large Deterministically Stable Equilibrium Population Sizes and Large Initial Values

Demographic stochasticity is mathematically interesting (i.e., nontrivial but relatively tractable) but biologically fairly boring if all populations are initially large, individuals interact homogeneously (i.e. each individual experiences the total population density, rather than only local density), and the deterministic analog has a globally stable equilibrium with all populations large. There is a large body of mathematically sophisticated analyses that demonstrate this (e.g., Barbour, 1976; Kurtz, 1981; McNeil and Schach, 1973; Wang, 1975). The basic idea is that the populations tend to settle into long-term "quasi-stationary" fluctuations about the deterministic equilibrium. These fluctuations can be nicely approximated by Ornstein-Uhlenbeck processes; and as indicated by (5), the coefficient of variation (standard deviation/mean) for each species is roughly proportional to the inverse of the square root of its equilibrium population size. The fluctuations are only "quasi-stationary" because *ultimate* extinction for each species is certain in the absence of recurrent immigration. However, the expected time until extinction results from demographic stochasticity will be of geological proportions for large populations, and thus ecologically irrelevant (see Bartlett, 1960; Goel and Richter-Dyn, 1974, Chap. 4; Ludwig, 1976; May, 1973, Chap. 2; Nisbet and Gurney, 1982).

B. Rare Invaders or Small Equilibrium Population Sizes

If initial population sizes are small or species have low equilibrium abundances (on the order of 10), relatively rapid extinction due to demographic stochasticity becomes likely. This represents a qualitative departure from deterministic results (see Becker, 1973). Most analyses concern the dynamics of colonizing species. The likelihood of successful invasion and long-term persistence is sensitive to the presence of competitors that slow growth rates and lower equilibrium population sizes. Typical multispecies analyses involve heuristic applications of single-species results obtained from birth and death processes or branching processes (e.g. MacArthur, 1972, Chap. 5, Appendix; Turelli, 1980). The basic theme is reminiscent of the classic observation by Fisher, Haldane, and Wright that selectively advantageous, rare mutations are often eliminated by chance (see Ewens, 1979, Chap. 1).

C. Multiple Stable Deterministic Equilibria

Chance fluctuations in population sizes can obviously play a decisive role in determining to which of several stable equilibria a set of populations is attracted, especially when the populations begin near a boundary for different domains of attraction. This observation led to the initial stochastic explanation of Park's (1954) two-species competition experiments in which apparently identical replicates ended with different winning species. The simplest models to capture this are stochastic analogs of the symmetric Lotka-Volterra competition equations

$$dX/dt = rX(K - X - \alpha Y)/K$$

and

$$dY/dt = rY(K - Y - \alpha X)/K \tag{6}$$

with $\alpha > 1$. According to this deterministic model, the species that is initially more abundant excludes the other, i.e. the deterministic separatrix is $X = Y$. When demographic stochasticity is introduced, the outcome no longer depends deterministically on the initial conditions, and the problem is to approximate the probabilities associated with the alternative outcomes. Clearly the probabilities will approach 0.5 as the initial conditions approach the separatrix. The original work on this problem relied on Monte Carlo simulations of birth-death processes modelling specific experiments (see Bartlett, 1960, Chap. 5; Bartlett et al., 1960). This approach was used by Mertz et al. (1976) to guide their *Tribolium* experiments. More recently, diffusion approximations of discrete-valued models have been used to obtain general analytical approximations (Barbour, 1976; Mangel and Ludwig, 1977).

D. Neutrally Stable Deterministic Analogs: "Null" Models of Community Structure

If deterministic mechanisms of species interaction are ignored or assumed to be relatively weak, demographic fluctuations (interpreted here as random replacement processes) become key determinants of relative abundances. Caswell (1976) exploited the elaborate machinery of the "neutral theory" of protein evolution (see Ewens, 1979) to analyze a "neutral" model of community structure that ignores species interactions. Horn (1975) and Hubbell (1979) analyzed mathematically related models for forest communities. The analyses of Caswell (1976) and Hubbell (1979) are philosophically closely allied to recent investigations of whether competition must be invoked to explain the relative sizes of coexisting species (see, for instance, Simberloff and Connor, 1981; Schoener, 1982). Unlike much of stochastic community theory, this area is dominated by analyses tailored to specific ecosystems (principally those governed by competition for space) and empirically motivated questions concerning the role of disturbance (Horn, 1975; Hubbell, 1979) and particular sorts of intraspecific interactions (Karlson and Jackson, 1981). Depending on whether immigration is allowed from a species pool of invariant composition, either transient or steady-state patterns of diversity are investigated. Harvey et al. (1983) review the rapidly expanding literature in this area.

E. Spatial Subdivision and Inhomogeneous Interactions: Topic A Revisited

Like much of mathematical ecology, the analyses referenced above assume that all individuals respond to the overall density of their population and the populations with which they interact. However, many populations exhibit spatial structure; and the *local* density experienced by an individual can differ appreciably from the *average* density of the entire population. This sort of local density variation would be generally expected in plant populations. Among animals it would exist for groups of primates, fish and birds that form social aggregates, and for species such as pitcher plant mosquitoes that spend a significant fraction of their life cycle within a single discrete resource. Chesson (1978) catalogs this as "within-patch variability". When local densities rather than average densities govern dynamics, the effects of demographic stochasticity can persist even in infinitely large populations if *local* densities remain small (see Chesson, 1981; Crowley, 1978). Within-patch variability induced by local demographic effects can qualitatively alter the outcome of interspecific interactions (e.g., Caswell, 1978; Crowley, 1981; Hastings, 1977; Slatkin, 1974) and natural selection (e.g., Wilson, 1980). At present, I know of no general quantitative criterion for determining when local population densities are sufficiently low for demographic fluctuations to be relevant. However, in the context of specific models, such as those involving local extinctions, one can compute the effect of population sizes on the expected persistence time. Clearly something like Wright's (1969, Chap. 8) definition of an effective population size for a large, continuously distributed population would be useful (cf. Crowley, 1978).

In many population subdivision models, the variation in local population densities can be attributed to environmental stochasticity that acts on a smaller spatial scale than that occupied by the total population. A fairly simple example of the effects of such variation is Gillespie's stochastic generalization of Levene's model for selection in subdivided populations (see Gillespie and Langley, 1974; Gillespie, 1978). For a very large number of patches undergoing independent environmental fluctuations, the dynamics of allele frequencies in the total population become deterministic. Nevertheless, the population subdivisions will be experiencing stochastic fluctuations in allele frequencies, and the environmental variation can critically influence the direction of selection.

III. Environmental Stochasticity

A. Sample Questions

A quantum leap in the mathematical sophistication of ecological modelling occurred when May (May and MacArthur, 1972; May, 1974) introduced stochastic differential equations (SDEs) to investigate limits to niche overlap in randomly fluctuating environments. [For mathematical background on SDE's and diffusion processes, see Arnold (1974) and Karlin and Taylor (1981, Chap. 15).] The relationship between environmental fluctuation and the coexistence of competitors has now been analyzed with a wide range of models and mathematical techniques. The answers range from May and MacArthur's (1972) claim that even small-scale environmental fluctuations significantly limit the similarity of competitors to Chesson and Warner's (1981) demonstration that large-scale fluctuations can make coexistence more likely. Interestingly, the supposedly widespread regularity of niche differentiation that the early papers sought to explain may not exist (Simberloff and Boecklen, 1981).

Environmental fluctuations ineluctably produce fluctuations in population levels. A general question is how species dynamics and interactions translate environmental fluctuations into temporal and spatial patterns of population abundances. For "low levels" of noise, stochastic linearization procedures suffice (see Roughgarden, 1979, Chap. 20; Nisbet and Gurney, 1982, Chap. 7).

B. Incorporating Noise: Discrete Versus Continuous Time

A central obstacle in stochastic community theory is the lack of mathematical machinery available to analyze nonlinear multidimensional stochastic processes. Thus, a critical problem is to incorporate noise into a deterministic model in a way that is biologically meaningful yet mathematically tractable. Because Markov processes are relatively "nice" and mathematical ecology was traditionally phrased in differential equations, SDEs, which represent multidimensional diffusion processes, appear as natural candidates for study. Unfortunately, they impose artificial biological constraints (e.g., noise can only be easily introduced into parameters that enter linearly) and they are mathematically ambiguous (see Feldman and Roughgarden, 1975; Turelli, 1977). As discussed in Turelli (1977), the mathematical ambiguity can be objectively resolved only by considering more realistic, but less tractable, models involving autocorrelated noise and/or discrete time. To make clear the biological assumptions underlying the final model, the optimal approach is to derive the diffusion approximation as a limit for a class of well posed models. [See Ewens (1979, Chap. 5) for a discussion of this procedure in population genetics and White (1977) for an ecological example.]

For many applications, for instance analyses of organisms with discrete breeding seasons, discrete-time models are natural. However, the global behavior of their stochastic analogs with continuous state spaces has only recently come under careful study, even in the Markovian case. Fortunately, this gap in applied probability theory is now being filled by workers directly interested in population biology (see Norman, 1975; Chesson, 1982; Ellner, 1984).

C. Methods of Analysis

A critical step in mathematically unravelling the consequences of environmental stochasticity is determining whether a local analysis will suffice (see Turelli, 1978a). If it will, you're in luck.

1. Local Analyses: Ornstein-Uhlenbeck Approximations

Local analysis of fluctuations about a deterministically stable equilibrium is straightforward, at least in principle, for both SDE and discrete-time stochastic models (see May, 1973, Chap. 5; Bartlett et al., 1960). SDE applications produce Ornstein-Uhlenbeck processes, i.e. diffusion processes with infinitesimal mean and variance-covariance functions of the form

$$M_i(x_1, ..., x_n) = \sum_{j=1}^{n} a_{ij} x_j \quad \text{for} \quad i = 1, ..., n, \tag{7a}$$

with negative real parts for all the eigenvalues of $A = (a_{ij})$, and

$$V_{ij}(x_1, ..., x_n) = b_{ij} \quad \text{for} \quad i, j = 1, ..., n \tag{7b}$$

with $B = (b_{ij})$ symmetric and positive definite. Here x_i measures the departure of the abundance of species i from its deterministic equilibrium $\hat{X}_i$, e.g. $x_i(t) = (X_i(t) - \hat{X}_i)/\hat{X}_i$. The stationary distribution for this process is multivariate normal with mean vector Q and variance-covariance matrix Σ satisfying

$$\Sigma A + A^T \Sigma = -2B. \tag{8}$$

Typically, the degree of stability of the stochastic model is quantified by Σ's dominant eigenvalue, $\lambda_{\max}$, which measures the variance of fluctuations about the deterministic equilibrium.

The assumptions underlying linear approximations and their dependence on the details of the nonlinear stochastic model are seen more easily in discrete time. This circumvents the problems associated with deriving the nonlinear SDE that (7a, b) approximates. Consider an n-species stochastic model of the form

$$X_{i,t+1} = X_{i,t} g_i(X_t, z_{i,t}) \quad i = 1, ..., n \tag{9}$$

with $\underline{X}_t = (X_{i,t}, \ldots, X_{n,t})^T$ and $z_{i,t}$ denoting a stochastic perturbation in the growth rate of species i satisfying $E(z_{i,t}) = 0$. Assume that the deterministic model

$$X_{i,t+1} = X_{i,t} g_i(\underline{X}_t, 0) \quad i = 1, \ldots, n \tag{10}$$

has a locally asymptotically stable equilibrium $\hat{\underline{X}}$. Expanding the right hand side of (9) about $(\underline{X}_t, z_{i,t}) = (\hat{\underline{X}}, 0)$ and introducing $x_{i,t} = (X_{i,t} - \hat{X}_i)/\hat{X}_i$ leads to the approximate recursion

$$\underline{x}_{t+1} = A \underline{x}_t + D \underline{z}_t, \tag{11}$$

in which $\underline{x}_t = (x_{1,t}, \ldots, x_{n,t})^T$, $\underline{z}_t = (z_{1,t}, \ldots, z_{n,t})^T$,

$$A = (a_{ij}) = (\delta_{ij} + \hat{X}_j \partial g_i(\hat{\underline{X}}, 0)/\partial X_j), \tag{12}$$

with $\delta_{ii} = 1$ and $\delta_{ij} = 0$ for $i \neq j$, and

$$D = \operatorname{diag}(\partial g_i(\hat{\underline{X}}, 0)/\partial z). \tag{13}$$

[Equations (11) and (12) correct Eq. (2.10) of Turelli (1981a); the modification affects none of the subsequent results in that paper.] Approximation (11) ignores all terms proportional to

$$x_i^a x_j^b z_k^c \quad \text{for} \quad a + b + c \geq 2 \tag{14}$$

with a, b, and c nonnegative integers and $1 \leq i, j, k \leq n$. Equation (11) is derived without assumptions concerning auto- or cross-correlation for the processes $\{z_{i,t}\}$.

A formal representation of the long-term behavior of the local approximation (11) can be obtained by iteration which yields

$$\underline{x}_t = \sum_{s=1}^{\infty} A^{s-1} D \underline{z}_{t-s}. \tag{15}$$

If the noise processes are assumed to have no autocorrelation, i.e., $E(\underline{z}_t \underline{z}_s^T) = \delta_{st} C$, with $\delta_{st} = 1$ if $s = t$ and 0 otherwise, then (15) implies that $V = E(\underline{x}_t \underline{x}_t^T)$, the variance-covariance matrix for the fluctuations, satisfies

$$V = \sum_{s=0}^{\infty} A^s D C D (A^T)^s. \tag{16}$$

[This series converges because local stability of $\hat{\underline{X}}$ implies that the eigenvalues of A satisfy $|\lambda_i| < 1$, see Sect. V.4 of Gantmacher (1959).] To make (16) more obviously informative, it is helpful to assume that the local stability matrix A is diagonalizable (i.e., has n linearly independent eigenvectors) so that

$$A = U \Lambda U^T$$

with $\Lambda = \mathrm{diag}(\lambda_1, \ldots, \lambda_n)$ and $U^T = U^{-1}$. Then (16) can be rewritten as

$$V = \sum_{s=0}^{\infty} U\Lambda^s U^T DCDU\Lambda^s U^T. \tag{17}$$

Although not particularly helpful for analyzing specific models, (17) displays two general features of stochastic linearization about interior deterministic equilibria. First, the stability of the stochastic model, as measured by the size of the elements of V, is proportional to the magnitude of the eigenvalues that govern the stability of the determinsitic equilibrium considered. In particular, $v_{ii} \to \infty$ as $\max_{1 \le i \le n} |\lambda_i| \to 1$. Second, (17) shows that V does not depend solely on A, the local stability matrix, and C, the variance-covariance matrix of the environmental noise, but also on $D = \mathrm{diag}(\partial g_i/\partial z)$, which reflects the way in which the stochastic perturbations are incorporated in the deterministic model. This point was ignored in May's (1973) original comparison of stability properties for stochastic models and their deterministic analogs (see Turelli, 1978a).

For $n = 1$, (17) simplifies to

$$V = \mathrm{Var}(x_t) = (\partial g(\hat{X}, 0)/\partial z)^2 \sigma^2/(1 - \lambda^2) \tag{18}$$

with $\sigma^2 = \mathrm{Var}(z_t)$ and λ the eigenvalue governing the local stability of $\hat{X}$. The dependence of V on the incorporation of noise can be seen by comparing

$$X_{t+1} = X_t \exp[r(1 - X_t/K_t)], \tag{19a}$$

with $K_t = K(1 + z_t)$, to

$$X_{t+1} = X_t \exp[r(1 - X_t/K)](1 + z_t). \tag{19b}$$

Stability of the deterministic equilibrium, $\hat{X} = K$, requires $0 < r < 2$. These models yield

$$V = r\sigma^2/(2 - r), \tag{20a}$$

and

$$V = \sigma^2/[r(2 - r)], \tag{20b}$$

respectively, and thus produce different predictions concerning the effects of the growth rate parameter r. Linearization-based predictions concerning the effects of intrinsic growth rates (and other model parameters) on stochastic population fluctuations in space and time will generally display this sort of model dependence. Thus generalizations based on specific models are not likely to be robust. Although this conclusion may be obvious in retrospect, several published accounts of stochastic models have ignored this source of model-dependent predictions (e.g., Roughgarden, 1979, Chap. 20). In a rare attempt at empirically distinguishing between models (19a) and (19b), Pimm (1984) has analyzed fluctuations in the abundance of various bird species. He finds that (20b) better accounts for the observations, and hence so does (19b).

In general, local analyses about internal deterministic equilibria cannot be relied on to provide conditions for species coexistence, which necessarily involves global dynamic behavior (Turelli, 1978a). An unavoidable consequence of interior (as opposed to boundary point) linearization is that environmental stochasticity is viewed as a destabilizing factor which jostles populations away from their deterministic equilibria (May, 1973, Chap. 5). For additional applications of linearization and refinements thereof see Bulmer (1976), White (1977) who rigorously treats both autocorrelation and time lags, Nisbet and Gurney (1982, Chap. 7), Poole (1978), and the discussion of approximate global analyses below.

2. Global Analyses

Global analyses are generally much more delicate. The key problem is that incorporating environmental stochasticity in a meaningful way usually eliminates all internal equilibrium *points* (or cycles). Hence stochastic convergence concepts that generalize the standard deterministic convergence-to-point-equilibria ideas are generally inapplicable. One is forced to apply persistence and coexistence criteria that allow populations to undergo undiminishing fluctuations without extinction. All criteria have some difficulties, either biological or mathematical.

An extremely appealing stochastic analog of deterministic global convergence is the existence of a unique nondegenerate stationary distribution that describes persistent long-term fluctuations. Unfortunately, very few results are available for discrete-time models. However, Chesson (1982) and Ellner (1984) have made considerable progress for one-dimensional models as well as some classes of two-species models (see below). The situation is only slightly better for SDEs. Turelli and Gillespie (1980) developed heuristic conditions for two-dimensional processes that look like exponential growth in a random environment near the extinction boundaries. Kushner (1972) has shown that Liapunov-like functions can yield sufficient conditions. Applications of this technique appear in Polansky (1979), Turelli and Gillespie (1980), and Turelli (1981a). In rare circumstances, one can guess the functional form of the multidimensional stationary density by generalizing results from one-dimensional or symmetric models (e.g., Turelli, 1981b).

In desperation, one can ignore mathematical rigor, apply a heuristic coexistence criterion, then do simulations to check its accuracy and hope for the best. One example is the invasion analysis applied in Turelli, (1978b, 1981a) and Turelli and Petry (1980). This criterion is based on the conjecture that a stationary distribution is likely to exist if each species can invade when rare and the remaining species are fluctuating at steady-state. Some weaknesses of this approach are described in Turelli (1981a) and a one-dimensional counterexample is provided by Chesson (1982). Nevertheless, the technique appears to be useful for analyzing stochastic models built by injecting small amounts of possibly autocorrelated noise (e.g. $CV \leq 0.1$) into difference equations. As applied in the papers cited above, this technique is built wholly on Taylor series approximations. However, unlike local linearization (which forms a part of the analysis) both linear and quadratic terms are taken into account. The necessity of complementing this sort of approximate analysis with computer simulations cannot be overemphasized. Chesson (1984) illustrates how the invasion criterion can be used to derive general qualitative conclusions for a class of multispecies models without explicitly approximating the dynamics of the resident species.

Chesson (1978) has proposed a stochastic persistence criterion that is weaker than the existence of a nondegenerate stationary distribution but retains its biological content by requiring that stochastic fluctuations lead to neither population extinction nor explosion. A population $X(t)$ is said to be *stochastically bounded* if for every $\varepsilon > 0$, there exist time independent constants $L_\varepsilon > 0$ and $U_\varepsilon < \infty$ such that

$$P(L_\varepsilon < X(t) < U_\varepsilon) > 1 - \varepsilon$$

for all t. Although it is generally quite difficult to establish necessary and sufficient conditions for stochastic boundedness, relatively simple examples of necessary conditions and sufficient conditions for one- and two-species models appear in Chesson (1982–1984), Ellner (1984) and Shmida and Ellner (1984). As shown by Chesson (1982), sufficient conditions can sometimes be obtained from straightforward invasion analyses. This connection between invasibility and stochastic boundedness has recently been extended to a class of discrete-time two-species competition models by Chesson and Ellner (pers. comm.). This is an area in which we can expect considerable mathematical progress that will be useful to population biologists.

Recognizing that ultimate extinction is certain in the presence of demographic fluctuations, Ludwig (1975) proposed calculating the mean extinction time (measured as the mean time to reach a low population threshold) as a measure of persistence. This general approach is also supported by Nisbet and Gurney (1982). Here again the mathematical difficulty of applying the criterion is formidable, even for multidimensional diffusions. Chesson (1982) and Turelli (1980, 1981a) discuss relationships among the persistence criteria and the difficulties associated with ecological applications.

D. Sample Applications: Stochastic Limiting Similarity

The dependence of biological conclusions on both the model analyzed and the analysis performed is nicely illustrated by two-species competition in a stochastic environment. Three approaches will be reviewed: 1) linearization of a discrete-time, stochastic generalization of the standard Lotka-Volterra equations, in the spirit of May and MacArthur's (1972) original analysis; 2) Turelli's (1981a) invasion analysis of the same model; and 3) Chesson and Warner's (1981) analysis of "lottery competition".

1. Linearization of a Two-Species Lotka-Volterra Model

Following Turelli (1981a) assume that

$$X_{i,t+1} = X_{i,t} g\left(\sum_{j=1}^{2} \alpha_{ij} X_{j,t} / K_i, z_{i,t} \right) \tag{21}$$

with $g(0,0)=1+r$, $g(1,0)=1$, $g(x,z)>0$ for $x>0$, $\partial g(x,0)/\partial x<0$ for $x>0$, $\alpha_{11}=\alpha_{22}=1$, $\text{Var}(z_{it})=\sigma^2$ for $i=1,2$, $\text{Cov}(z_{1,t},z_{2,t})=\theta\sigma^2$, and $\text{Cov}(z_{i,t},z_{j,s})=0$ for $t\neq s$. The stochastic linearization procedure is simplest to carry out with completely symmetric competition, i.e. $K_1=K_2$ and $\alpha_{12}=\alpha_{21}=\alpha$. Then, $\hat{X}_1=\hat{X}_2 = K/(1+\alpha)$ and (17) gives

$$v_{ij}=\text{Cov}(x_{i,t},x_{j,t})=(d\sigma)^2[(1+\theta)(1-\lambda_1^2)^{-1}$$
$$+(2\delta_{ij}-1)(1-\theta)(1-\lambda_2^2)^{-1}]/2 \tag{22}$$

with $d=\partial g(1,0)/\partial z$, $\delta_{11}=\delta_{22}=1$, $\delta_{12}=\delta_{21}=0$, $\lambda_1=1+\partial g(1,0)/\partial x$, and

$$\lambda_2=1+(1-\alpha)(1+\alpha)^{-1}\partial g(1,0)/\partial x.$$

Deterministic stability requires $0>\partial g(1,0)/\partial x>-2$ and $0<\alpha<1$. The stability criterion that underlies May and MacArthur's (1972) analysis of stochastic limiting similarity requires that the linearization-based estimates of species' fluctuations about their deterministic equilibria, i.e. $\text{Var}(x_{i,t})$, not exceed a chosen bound, denoted c. Thus, their coexistence criterion can be concisely written as

$$v_{ii}<c. \tag{23}$$

Substituting from (22) shows that the limit obtained on α from this criterion will depend on: i) the value of c chosen; ii) the way stochasticity enters the model, via d; iii) the underlying deterministic dynamics, via $\delta g(1,0)/\delta x$; and iv) the environmental noise, via σ^2 and θ. Clearly, the same dependence would hold for asymmetric models but would be more difficult to express. According to this criterion, the stochastic limit on α will always be smaller than the constant environment prediction. However, as discussed at length in Turelli (1978a), this criterion, which is based on near-equilibrium dynamics, is inappropriate for coexistence questions which are intrinsically related to the ability of rare species to avoid local extinction. This is the biological motivation for the invasion criterion for coexistence.

2. Invasion Analysis for the Two-Species Lotka-Volterra Model

The invasion criterion involves determining conditions under which the geometric mean growth rate of each species exceeds one when it is rare and its competitor is at its stochastic equilibrium. Thus, invasion analysis of (21) requires approximating

$$E\{ln[X_{i,t}g(\alpha_{ij}X_{j,t}/K_i,z_i)]\}>0 \quad \text{for} \quad i\neq j. \tag{24}$$

Here E denotes mathematical expectation with respect to the joint distribution of $(X_{j,t},z_{i,t})$ with $X_{j,t}$ the stationary process (assumed to exist) that is generated by the single-species model

$$X_{j,t+1}=X_{j,t}g(X_{j,t}/K_j,z_{j,t}). \tag{25}$$

As demonstrated in Turelli (1981a), (24) can be approximated by

$$K_1 - \alpha_{12}K_2 > 0 \qquad (26a)$$

and

$$K_2 - \alpha_{21}K_1 > 0, \qquad (26b)$$

which are precisely the deterministic coexistence criteria. [Analogous results based on SDE's appear in Turelli and Gillespie (1980).] The exact recovery of the constant environment results is an artifact of the symmetry assumptions. However, the existence of significant discrepancies between coexistence predictions based on linearization and invasion analyses is quite general. In sharp contrast to May and MacArthur's (1972) linearization-based conclusion that small amounts of environmental stochasticity can produce significant limits on the similarity of competition, the general conclusion from the biologically motivated invasion analysis is that small amounts of environmental stochasticity lead to only small departures from deterministic predictions (Turelli, 1978b, 1981a).

3. Chesson and Warner's Lottery Competition Model

From the Lotka-Volterra-based analyses above, environmental stochasticity emerges as a factor that makes coexistence more difficult (linearization) or does not affect it (invasion). A very different picture emerges from a model of competition for space analyzed by Chesson and Warner (1981). In its simplest form, the Chesson and Warner (1981) "lottery model" treats competition for an effectively infinite number of sites each of which can support only a single reproductive adult. The model assumes a discrete-time life cycle, with reproduction followed by possible adult death then settling of young into vacant sites. When applied to sessile organisms, the model assumes that each vacant site is colonized by a random sample of zygotes from the reproductive output of the entire system of patches. From these, a single individual is chosen at random as the ultimate occupant. (Biases in the outcome of this process due to competitive differences can be absorbed into the per capita birth rates.) Let $P_i(t)$ for $i=1,2$ be the fraction of patches occupied by adults of species i at time t, assume that $P_1(t)+P_2(t)=1$, and let $\beta_i(t)(\delta_i(t))$ be the per capita birth (death) rate of species i at time t. Environmental stochasticity enters the model through the birth and death rates. The assumptions above lead to the recursion

$$P_i(t+1) = [1 - \delta_i(t)]P_i(t) + \sum_{j=1}^{2} \delta_j(t)P_j(t)$$

$$\cdot \left[\beta_i(t)P_i(t) \Big/ \sum_{j=1}^{2} \beta_j(t)P_j(t) \right] \quad \text{for} \quad i=1,2. \qquad (27)$$

Because $P_2(t) = 1 - P_1(t)$, the two-species model is one-dimensional, i.e.

$$P_i(t+1) = P_i(t)f_i(P_i(t), \underline{\delta}(t), \underline{\beta}(t)) \quad \text{for} \quad i=1,2 \qquad (28)$$

with $\underline{\delta}(t) = (\delta_1(t), \delta_2(t))$ and $\underline{\beta}(t) = (\beta_1(t), \beta_2(t))$. Chesson (1982) presents a detailed analysis of stochastic boundedness for this model. However for a broad class of stochastic inputs, the same coexistence conditions can be obtained from an invasion analysis which simply requires

$$E\{\ln[f_i(0, \underline{\delta}(t), \underline{\beta}(t)]\} > 0 \quad \text{for} \quad i = 1, 2. \tag{29}$$

Chesson and Warner (1981) explore the resulting conditions in detail. Their principal qualitative conclusions are that coexistence is (essentially) impossible unless: i) generations overlap, i.e. $\delta_i(t) < 1$; and ii) the birth rates fluctuate. Moreover, coexistence occurs for a wider range of parameters, allowing larger mean advantage of one species over the other, as the level of birth rate fluctuations *increases*. In this setting, the coexistence-promoting effects of environmental fluctuations require overlapping generations which allow each species to "store" the effects of past environmental conditions that favored it. This "storage effect" is discussed in greater detail and generality in Chesson (1983, 1984).

IV. Joint Effects of Demographic and Environmental Stochasticity

In the real world both demographic and environmental stochasticity are unavoidable. Given the difficulties of analyzing each separately, it should come as no surprise that few multispecies analyses incorporate both. At least two types of analysis are relevant.

Because extinction of a population closed to recurrent immigration is certain under demographic stochasticity, it is of interest to compute the expected time (or more generally, the distribution of times) until one or more of a group of interacting species becomes extinct. No explicit calculations for multispecies models exist. However, in principle, they could be derived from asymptotic analyses of diffusion approximations (see Ludwig, 1975; Schuss, 1980; Leigh, 1981). Ludwig's (1976) analysis of mean persistence times for a single-species logistic model with both environmental and demographic stochasticity is the best available guide to the probable behavior of multispecies models. His approach is to obtain asymptotic approximations for the mean time until extinction (assuming a not-too-small initial size) by holding the level of environmental fluctuations fixed while letting stochasticity vanish by sending a parameter, denoted M, that reflects the deterministic equilibrium population size, off to infinity. Because the model without demographic stochasticity (i.e., $M = \infty$) does not allow extinction (see Chesson, 1978, for a discussion of this property), the mean persistence time approaches infinity as $M \to \infty$. Ludwig's (1976) key result is a demonstration that the rate of approach to infinity takes two distinct forms depending on the model's parameters. A central role is played by a positive parameter combination, denoted α, that governs the existence of a stationary density for the corresponding model without demographic stochasticity ($M = \infty$). For $\alpha < 2$, the $M = \infty$ model has a stationary density with no mass at zero, and it assigns less probability near zero as α decreases. For $\alpha > 2$, no stationary density exists and the population size converges to zero. As shown by Ludwig (1976), the asymptotic behavior of the mean persistence time as $M \to \infty$ is determined by

$$k = \frac{2}{\alpha}\left(1 + \frac{1}{\alpha M}\right). \tag{30}$$

For $k < 1$, the mean persistence time grows as $\log M$; for $k = 1$, it grows as $(\log M)^2$; and for $2 > k > 1$ (which requires $\alpha < 2$), it grows at the qualitatively faster rate M^{k-1}. [For algebraic reasons, Ludwig (1976) did not analyze $k \geq 2$, but the results for Model 2 of Tier and Hanson (1981) suggest that the algebraic increase of the mean persistence time with M is probably generally true for $k > 1$.]

The dependence of these results on α shows that the existence of a nonzero stationary distribution for the corresponding model without demographic stochasticity is generally necessary but not sufficient to insure long-term persistence in the face of demographic fluctuations. Clearly a random environment model whose stationary distribution assigns significant probability to very low populations [as the $M = \infty$ model of Ludwig (1976) does for α only slightly less than two] will lead to relatively short expected persistence times once demographic affects are taken into account. [See Chesson (1982) for a discussion of this same idea in the context of his lottery model.] Presumably, analogous results hold for multiple species models with a key role assigned to the smallest population mean in the corresponding model without demographic stochasticity. As discussed in Turelli (1980) and Chesson (1982), biological interpretation of extinction times depends critically on the relevant ecological time scale.

Assuming, or having established, that long-term coexistence is likely once each species is established, attention shifts to computing the probability that a rare species will successfully invade. By hybridizing the invasion analysis discussed above, with results for branching processes in random environments (see Keiding, 1975, 1976), fairly simple approximations can be derived (see Turelli, 1980). At least for simple stochastic generalizations of Lotka-Volterra equations, it appears that demographic stochasticity plays a far greater role than small to moderate amounts of environmental stochasticity in determining the fate of rare invading species. This follows because irrespective of a small population's intrinsic tendency to increase, it can be driven extinct by demographic fluctuations.

Acknowledgements

Despite its remaining inadequacies, for which I am unfortunately responsible, this review was significantly improved by suggestions from P. L. Chesson, S. Ellner, L. J. Gross, J. F. Quinn, T. W. Schoener, and an anonymous reviewer. This work was supported in part by N.I.H. grant GM22221.

References

Arnold, L. (1974). Stochastic Differential Equations: Theory and Applications. Wiley, New York

Barbour, A.D. (1976). Quasi-stationary distributions in Markov population processes. Adv. Appl. Probab. *8*: 296–314

Bartlett, M.S. (1960). Stochastic Population Models in Ecology and Epidemiology. Methuen, London

Bartlett, M.S., Gower, J.C., Leslie, P.H. (1960). A comparison of theoretical and empirical results for some stochastic population models. Biometrika *47*: 1–11

Becker, N.G. (1973). Interactions between species: some comparisons between deterministic and stochastic models. Rocky Mount. J. Math. *3*: 53–68

Bulmer, M.G. (1976). The theory of predator-prey oscillations. Theor. Pop. Biol. *8*: 137–150

Caswell, H. (1976). Community structure: A neutral model analysis. Ecol. Monographs *46*: 327–354

Caswell, H. (1978). Predator-mediated coexistence: A nonequilibrium model. Am. Nat. *112*: 127–154

Chesson, P. (1978). Predator-prey theory and variability. Ann. Rev. Ecol. Syst. *9*: 323–347

Chesson, P.L. (1981). Models for spatially distributed populations: The effect of within-patch variability. Theor. Pop. Biol. *18*: 288–325

Chesson, P.L. (1982). The stabilizing effect of a random environment. J. Math. Biol. *15*: 1–36

Chesson, P.L. (1983). Coexistence of competitors in a stochastic environment: The storage effect. Lecture Notes in Biomath. *25*: 188–198

Chesson, P.L. (1984). The storage effect in stochastic population models. Lecture Notes in Biomath. *54*: 76–89

Chesson, P.L., Warner, R.R. (1981). Environmental variability promotes coexistence in lottery competitive systems. Am. Nat. *117*: 923–943

Crowley, P.H. (1977). Spatially distributed stochasticity and the constancy of ecosystems. Bull. Math. Biol. *39*: 157–166

Crowley, P.H. (1978). Effective size and the persistence of ecosystems. Oecologia *35*: 185–195

Crowley, P.H. (1981). Dispersal and the stability of predator-prey interactions. Am. Nat. *118*: 673–701

Ellner, S. (1984). Asymptotic behavior of some stochastic difference equation population models. J. Math. Biol. *19*: 169–200

Ewens, W.J. (1979). Mathematical Population Genetics. Springer-Verlag, Berlin

Feldman, M.W., Roughgarden, J. (1975). A population's stationary distribution and chance of extinction in a stochastic environment with remarks on the theory of species packing. Theor. Pop. Biol. *7*: 197–207

Gantmacher, F.R. (1959). Matrix Theory, Vol. I. Chelsea, New York

Gillespie, J.H. (1978). A general model to account for enzyme variation in natural populations. V. The SAS-CFF model. Theor. Pop. Biol. *4*: 1–45

Gillespie, J.H., Langley, C. (1974). A general model to account for enzyme variation in natural populations. Genetics *76*: 837–848

Goel, N.S., Richter-Dyn, N. (1974). Stochastic Models in Biology. Academic Press, New York

Harvey, P.H., Colwell, R.K., Silvertown, J.W., May, R.M. (1983). Null models in ecology. Ann. Rev. Ecol. Syst. *14*: 189–211

Hastings, A. (1977). Spatial heterogeneity and the stability of predator-prey systems. Theor. Pop. Biol. *12*: 37–48

Horn, H. (1975). Markovian models of forest succession. In: Ecology and Evolution of Communities (Cody, M.L. and Diamond, J.G., Eds.), pp. 196–211

Hubbell, S.P. (1979). Tree dispersion, abundance, and diversity in a tropical dry forest. Science *203*: 1299–1309

Karlin, S., Taylor, H.M. (1981). A Second Course in Stochastic Processes. Academic Press, New York

Karlson, R.H., Jackson, J.B.C. (1981). Competitive networks and community structure: A simulation study. Ecology *63*: 670–678

Keiding, N. (1975). Extinction and exponential growth in random environments. Theor. Pop. Biol. *8*: 49–63

Keiding, N. (1976). Population Growth and Branching Processes in Random Environments. Preprint No. 9, Institute of Mathematical Statistics, University of Copenhagen

Kurtz, T.G. (1981). Approximation of Population Processes. SIAM, Philadelphia

Kushner, H.J. (1972). Stability and existence of diffusions with discontinuous or rapidly growing drift terms. J. Differential Equations *11*: 156–168

Leigh, E.G. (1981). The average lifetime of a population in a varying environment. J. Theor. Biol. *90*: 213–239

Lerner, I.M., Dempster, E.M. (1962). Indeterminism in interspecific competition. Proc. Natl. Acad. Sci. USA *48*: 821–826

Lerner, I.M., Ho, F.K. (1961). Genotype and competitive ability of *Tribolium* species. Am. Nat. *95*: 329–343

Leslie, P.H., Gower, J.C. (1958). The properties of a stochastic model for two competing species. Biometrika *45*: 316–330

Levin, S.A. (1981). The role of theoretical ecology in the description and understanding of populations in heterogeneous environments. Amer. Zool. *21*: 865–875

Ludwig, D. (1975). Persistence of dynamic systems under random perturbations. SIAM Rev. *17*: 605–640

Ludwig, D. (1976). A singular perturbation problem in the theory of population extinction. SIAM-AMS Proc. *10*: 87–104

MacArthur, R.H. (1972). Geographical Ecology. Harper & Row, New York

Mangel, M., Ludwig, D. (1977). Probability of extinction in a stochastic competition. SIAM J. Appl. Math. *33*: 257–266

May, R.M. (1973). Stability and Complexity in Model Ecosystems. Princeton University Press, Princeton

May, R.M. (1974). On the theory of niche overlap. Theor. Pop. Biol. *5*: 297–332

May, R.M., MacArthur, R.H. (1972). Niche overlap as a function of environmental variability. Proc. Natl. Acad. Sci. USA *69*: 1109–1113

McNeill, D.R., Schach, S. (1973). Central limit analogues for Markov population processes. J. R. Stat. Soc. B *35*: 1–23

Mertz, D.B., Cawthon, D.A., Park, T. (1976). An experimental analysis of competitive indeterminacy in *Tribolium*. Proc. Natl. Acad. Sci. USA *73*: 1368–1372

Neyman, J., Park, T., Scott, E.L. (1956). Struggle for existence. The *Tribolium* model. Proc. 3rd Berkeley Symp. on Math. Stat. and Prob. *3*: 41–79

Nisbet, R.M., Gurney, W.S.C. (1982). Modelling Fluctuating Populations. Wiley, Chichester

Norman, M.F. (1975). An ergodic theorem for evolution in a random environment. J. Appl. Prob. *12*: 661–672

Park, T. (1954). Experimental studies of interspecific competition. II. Temperature, humidity and competition in two species of *Tribolium*. Physiol. Zool. *27*: 177–238

Pielou, E.C. (1981). The usefulness of ecological models: A stock-taking. Quart Rev. Biol. *56*: 17–31

Pimm, S.L. (1984). Food chains and return times. In: Ecological Communities (D. L. Strong, Jr., D. Simberloff, L. G. Abele, and A. B. Thistle, eds.), pp. 397–412. Princeton University Press, Princeton

Polansky, P. (1979). Invariant distributions for multi-population models in random environments. Theor. Pop. Biol. *16*: 25–34

Poole, R.W. (1978). The statistical prediction of population fluctuations. Ann. Rev. Ecol. Syst. *9*: 427–448

Roughgarden, J. (1979). Theory of Population Genetics and Ecology: An Introduction. MacMillan, New York

Schoener, T.W. (1982). The controversy over interspecific competition. Amer. Sci. *70*: 586–595

Schuss, Z. (1980). Theory and Applications of Stochastic Differential Equations. Wiley, New York

Shmida, A., Ellner, S.P. (1984). Coexistence of plant species with similar niches. Vegetatio *58*: 29–55

Simberloff, D., Boecklen, W. (1981). Santa Rosalia reconsidered: Size ratios and competition. Evolution *35*: 1206–1228

Simberloff, D., Connor, E.F. (1981). Missing species combinations. Am. Nat. *118*: 215–239

Slatkin, M. (1974). Competition and regional coexistence. Ecology *55*: 128–134

Tier, C., Hanson, F.B. (1981). Persistence in density dependent stochastic populations. Math. Biosciences *53*: 89–117

Turelli, M. (1977). Random environments and stochastic calculus. Theor. Pop. Biol. *12*: 140–178

Turelli, M. (1978a). A reexamination of stability in randomly varying versus deterministic environments with comments on the stochastic theory of limiting similarity. Theor. Pop. Biol. *13*: 244–267

Turelli, M. (1978b). Does environmental variability limit the similarity of competing species? Proc. Natl. Acad. Sci. USA *75*: 5085–5089

Turelli, M. (1980). Niche overlap and invasion of competitors in random environments. II. The effects of demographic stochasticity. Lecture Notes in Biomath. *38*: 119–129

Turelli, M. (1981a). Niche overlap and invasion of competitors in random environments. I. Models without demographic stochasticity. Theor. Pop. Biol. *20*: 1–56

Turelli, M. (1981b). Temporally varying selection on multiple alleles: A diffusion analysis. J. Math. Biol. *13*: 115–129

Turelli, M., Gillespie, J.H. (1980). Conditions for the existence of stationary densities for some two-dimensional diffusion processes with applications in population biology. Theor. Pop. Biol. *17*: 167–189

Turelli, M., Petry, D. (1980). Density-dependent selection in a random environment: An evolutionary process that can maintain stable population dynamics. Proc. Natl. Acad. Sci. USA *77*: 7501–7505

Wang, F.J.S. (1975). Limit theorems for age and density dependent stochastic population models. J. Math. Biol. *2*: 373–400

White, B.S. (1977). Effects of a rapidly-fluctuating random environment on systems of interacting species. SIAM J. Appl. Math. *32*: 666–693

Wilson, D.S. (1980). The Natural Selection of Populations and Communities. Benjamin/Cummings, Menlo Park

Wright, S. (1969). Evolution and the Genetics of Populations. Vol. 2. The Theory of Gene Frequencies. University of Chicago Press, Chicago

A Theoretical Basis for Modeling Element Cycling

*Ray R. Lassiter**

Introduction

Elements that comprise the major portion of biomass (C, H, O, N, P, and S) exist in many different forms. Molecules of these forms are transient, appearing when converted from one form and disappearing when converted into another. Many of the transformations involve changes in oxidation-reduction (redox) state. For the most part these transformations proceed very slowly by abiotic chemical reactions. Organisms, however, possess enzyme systems that both catalyze the reactions and couple them to cellular biochemistry, producing energy for chemical and mechanical work (synthesis and movement). Element cycle transformation rates depend upon the concentration of organisms as well as upon concentrations of reactants, but generally, biotically-catalyzed rates are orders of magnitude faster than corresponding abiotic chemical reaction rates.

A biotically catalyzed redox reaction in any environment can be viewed both as an element cycle transformation and as a biotic process yielding energy for life processes. A cycle is completed when a series of reactions transform an element in one form through one or more others and back to the original form. Viewed as an energy yielding biotic process, complete cycles imply functionally complete ecosystems. One could say that the essential structure of an ecosystem is complete when species are present to catalyze all the transformations to complete the element cycles that are necessary to support life. This, however, does not conform to common usage of "ecosystem," because it equates completeness of ecosystems to completeness of life support functions, a constraint not usually imposed in selecting and delimiting systems for ecological study. Nevertheless, the idea of completeness of ecological function to support chemical cycles is perhaps useful, because of its parallel to completeness of element cycles.

Element cycle transformations are typically carried out by a multi-species complex. One visualizes that fluctuations in individual species of the complex are compensated by counter fluctuations in others, so that the process continues more smoothly than does the activity of any one species. If one were to model element cycling, including the response to outside perturbations, a multi-species representation of the processes might well be necessary in order to represent stability

* This paper has been reviewed in accordance with the U.S. Environmental Protection Agency's peer and administrative review policies and approved for presentation and publication
The work on which this report is based was partially supported through the Institute of Ecology, University of Georgia and was conducted while the author was in residence there during 1981–1982

through compensation. Taking such an approach would bring up the question of what level of redundancy is necessary as well as the practical consideration of what level can be handled.

The ideas and information summarized in this paper are not, I believe, restricted in their usefulness to the multispecies approach to modeling element cycles. The level of resolution selected for discussion of these processes is rather fine. This fineness of resolution is dictated by the nature of the controlling processes of element cycling as I now see them. I have chosen the level of resolution to be that of the species population, the chemical pool or finer. For example predator-prey relationships are developed explicitly in terms of multi-species interactions for two modes of feeding (filtering and pursuing). If, as has been reported (O'brien, 1979), the structure of whole ecosystems is heavily influenced by the activity of predators, it is to be expected that the structure of the consumer subcommunity is influenced by the action of predators on each other via both multispecies interactions and qualitatively different modes of feeding.

Element Cycles

Every process is some form of change from disequilibrium toward equilibrium conditions. In photosynthesis chloroplasts catalyze relaxation of the dis-equilibrium that exists when CO_2 and water occur together in the presence of light. The catalytic action of chloroplasts and sequentially coupled biochemical reactions produce both reduced (organic) carbon and molecular oxygen. Aerobic respiration is the converse process: molecular oxygen is reduced and organic carbon is oxidized. In this same relationship anaerobic respiration (in which oxidants other than oxygen are reduced) and chemosynthesis are converse processes. Energy for chemosynthesis is derived from energy released upon oxidation of reduced forms of N, S, and C. These processes proceed, predomi-nantly, in aerobic environments, because, predominantly, N, S, and C are oxidized by molecular oxygen. The oxidized products are used as oxidants (electron acceptors) by organisms in anaerobic environments. Thus the locations of formation of the forms of N, S, and C used for energy production are different from the locations of their use. Reduced forms are oxidized in aerobic environments and the oxidized forms must be transported to anaerobic environments if they are to be used in anaerobic respiration. Reciprocally, the products of anaerobic respiration, reduced forms of N, S, and C, must move to aerobic environments if chemosyn-thesis is to proceed. Transport can be an important and limiting factor in the control of these processes. In contrast, both light driven photosynthesis and its converse process, aerobic respiration, occur in aerobic environments.

Completeness of element cycles does not depend on transport between environments in every instance, however. Oxidized forms are reduced during biosynthesis regardless of the environment. Most organisms that live in aerobic environments can incorporate amino and thiol groups into their biomass by obtaining oxidized forms of N and S from the environment and reducing them internally. This is assimilatory reduction (to distinguish from dissimilatory reduction, the comparable term corresponding to respiration using electron acceptors other than oxygen). Upon death and decomposition in aerobic environments, these reduced forms are either assimilated directly or oxidized via chemosynthesis. By these mechanisms element cycles are complete in aerobic environments. The relative importance of assimilatory and dissimilatory pathways in maintaining local and global pool sizes of elemental forms is not clear, nor is the importance of the role of abiotic processes or boundary conditions imposed by geological and meteorological factors relative to biotic factors. If, in principle, these factors are included, however, it is fair to say that element cycling is responsible for pool sizes of geochemicals. These include the atmospheric gases, the composition of the oceans, and the earth's soils. The remarkable stability of these characteristics over geological time scales (Lovelock and Margulis, 1974) probably is indicative of the robustness of the component processes (perhaps due largely to functional compensation arising from species diversity). No comparative analysis of the role of biotic versus abiotic processes in element cycling has been made, but the number and importance of processes known to be carried out by organisms leaves no doubt that the characteristics of the earth depend on the continuation of biotic processes at about the level that they are presently occurring. Large scale environmental alterations, such as introduction of new xenobiotic chemicals into the biosphere, have been occurring for only a short time. Whether the robustness that apparently has been exhibited over geological time will continue in the presence of these alterations is not known. It is prudent, therefore, to attempt to gain a coherent view of element cycling so that objective analyses can ultimately be made of the likelihood of potential adverse effects.

Ecosystem Structure and Function

Structure and function are inseparable in ecology. Structures enable functions. It is my assumption that there is parsimony of structure and behavior in biotic systems (an assumption in keeping with the principle of Ockam's razor): structures do not exist without function. Generally, this is a reasonable assumption, because power is required to maintain structures, and the competitive advantage is held by organisms without the added burden of useless structures. In the theory of ecosystem function proposed here the power demand is calculated as the net cost of carrying out the functions. It is assumed that there are no overhead costs for maintaining structure, except structure that is used in obtaining energy, survival, or reproduction. These assumptions are in keeping with the overall assumption of parsimony. Certainly there is nearly an astronomical degree of variation in morphological of organisms. The parsimony assumption merely holds to the usual view that this structural variation is functional. One supposes that structural variation is highly correlated with variation in susceptibility to stresses.

The importance of the ecosystem concept is to account for phenomena that result from the interaction of smaller scale entities, such as physical factors, chemicals, species populations, and functional groups of populations. Examples of measures of entities that are hierarchically lower than the ecosystem are concentrations of chemical nutrients, population densities of component living species, various rate related measures such as productivities, and less easily quantitated concepts such as community structure. When these phenomena are viewed in the context of the whole ecosystem, *i.e.*, not in an experimentally controlled situation, they are often referred to as whole system phenomena.

Much theory exists in ecology, in the generic sense of theory. This theory is diffuse, and it has not been assimilated at the level of resolution that I have selected to provide the conceptual and theoretical basis for modeling element cycles. This theory incorporates concepts from chemical thermodynamics in much the same way as they are used in biochemistry. A biophysical approach is used to derive relationships so that the resulting equations are written primarily at the population level, with terms of the equations calculated using expressions reflecting underlying physiology, biochemistry, etc., all constrained by mass and energy accounting.

One of the more difficult problems to overcome in using concepts from several levels of resolution is that of achieving coherence and comprehensibility at the level of primary interest. This problem arises not simply from the need to use several disciplines in addressing a complex problem, but rather because the levels of resolution usually correspond to different levels in the hierarchical organization whose outer level is the ecosystem. This problem has been discussed widely (Allen and Starr, 1982). I have made no systematic attempt to develop or discover a set of rules by which to relate the hierarchical levels or to use results from one level in another. The approach taken (i.e., the selection of subject matter and the level of resolution) has been more intuitive. I view element cycling and most other macroscopic properties of ecosystems as deriving from interactions of organisms as they obtain useful energy via modes of metabolism suited to their energy sources. These modes of metabolism, unlike species, occur in diverse environments (see Ecology as the Environmental Pattern of Biochemical Energetics). I view a particular biotic community simply as a complement of species populations that are able to exploit the energy sources, grow and reproduce in a particular environment. These are not, I think, extreme views. They provide the rationale for selection of subject matter, level of resolution, hierarchical level, and the way that subjects are related to processes, populations, or chemical pools in the discussion.

Theoretical Basis

The theory presented here is not new in its entirety, but it does contain original elements. Perhaps more significantly, it departs from both holistic and reductionistic views. It departs from the traditional holistic view by attempting to synthesize system properties using a biophysical integration of finer scale components to the level required to represent the system. It departs from the reductionist view by concentrating on the ecosystem as the object to be represented and by virtue of interest in processes whose behavior cannot be considered to be a function of any single system component.

This theory like most theories of macroscopic phenomena, is based upon phenomenological models of its fundamental components. In an attempt to give the theory a greater predictive capability, I have selected as fundamental components, models representing energy transduction at the biochemical level. In these models organisms are considered as reaction systems that catalyze energy-rich, kinetically hindered redox reactions and, in the process, use part of the reaction energy for biosynthesis and mechanical work. Ultimately, means to represent rates of change of biotic populations or chemical pools are obtained from the biochemical models based upon redox chemistry. Ecological interactions result from dependencies among organisms for elements from which to form biomass and for compounds to serve as oxidation and reduction reactants to provide energy.

To maintain the identity of all organic compounds that could serve as chemical nutrients or as energy substrates in the working model would be an impossible computing problem. To overcome this problem, organic chemicals are aggregated into classes that are ecologically relevant. The underlying theory is worked out elsewhere (McCarty, 1971) and does not depend on this aggregation.

Ecology as the Environmental Pattern of Biochemical Energetics

Ecological Modes of Metabolism

Disequilibria that drive biotic processes are in two forms: light and chemical. Organisms that use light induced disequilibria are phototrophs, and those that utilize chemicals are chemotrophs (Doelle, 1975; Fenchel and Blackburn, 1979). Phototrophy and chemotrophy refer to the source of energy, i.e., whether the free energy is derived from photochemical or biochemical reactions. In either case the electron donor (reducing agent) can be either inorganic or organic (lithotrophy or organotrophy with the prefix photo- or chemo-) making a total of four major classes of energy metabolism. Figure 1 is a schematic of the processing of energy.

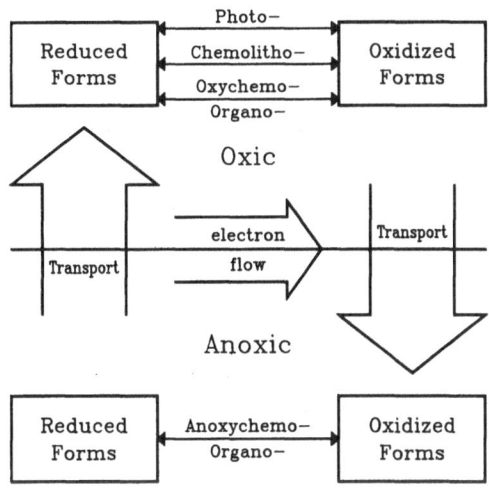

Fig. 1. Schematic for redox processes that yield energy useful for biotic processes. Elements cycle between reduced and oxidized forms as organisms catalyze redox reactions in their energy metabolism

For biosynthesis the source of carbon can be either CO_2 or an organic compound (autotrophy or heterotrophy). Combining the modes of energy metabolism with the modes of biosynthesis, eight categories are formed, all eight of which exist in nature along with variations on these major categories. For the initial theoretical development, it will suffice to consider only three of the major categories: photolithoautotrophy, chemolithoautotrophy, and chemoorganoheterotrophy. Referencing these categories by the shortened terms, photoautotrophy, chemoautotrophy, and heterotrophy will introduce no ambiguities in this limited context. Further subdivision of the latter category into macroheterotrophy and microheterotrophy, and microheterotrophy further into categories reflecting the electron acceptor used will also be convenient. I shall refer to the electron acceptor categories occasionally by the less specific terms, oxymicroheterotrophy and anoxymicroheterotrophy.

Figure 2 is a schematic of biosynthesis at the ecological scale. Figure 3 indicates the spatial and energetic relationships of microheterotrophy as a function of

Biosynthesis

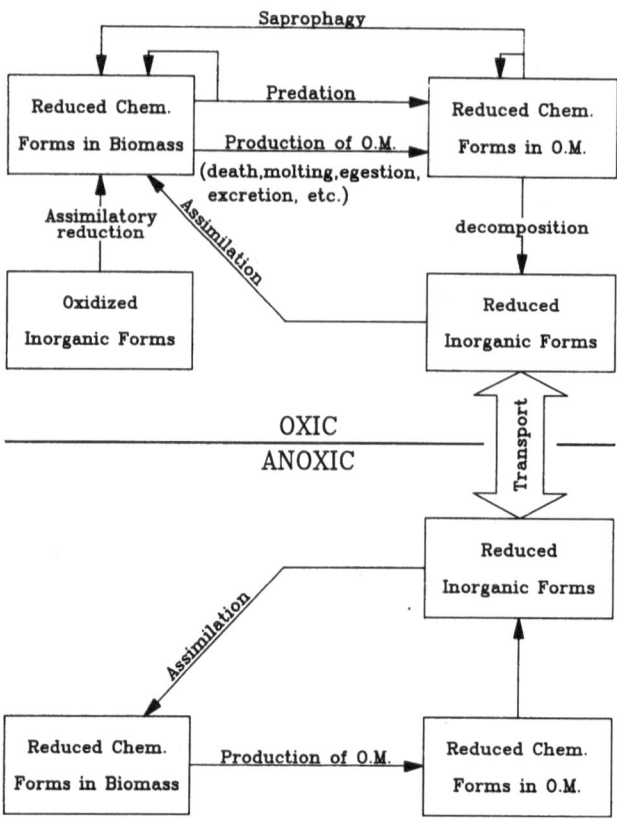

Fig. 2. Major pathways and processes important in biosynthesis

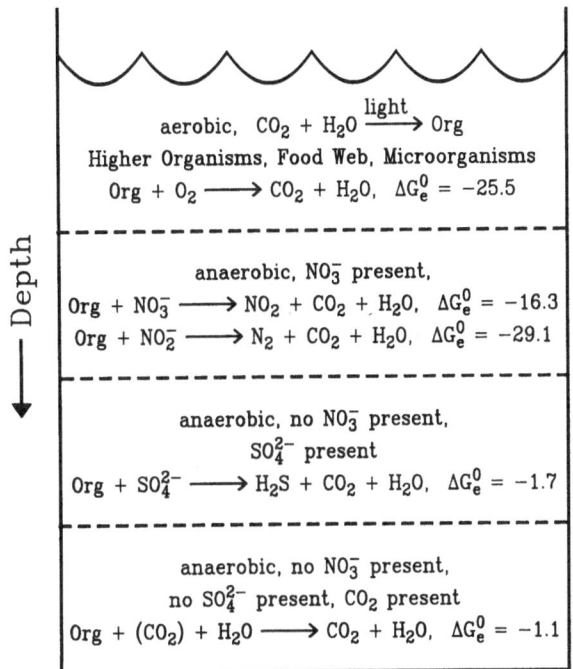

Depth ←

aerobic, $CO_2 + H_2O \xrightarrow{\text{light}} Org$

Higher Organisms, Food Web, Microorganisms

$Org + O_2 \longrightarrow CO_2 + H_2O, \quad \Delta G_e^0 = -25.5$

anaerobic, NO_3^- present,

$Org + NO_3^- \longrightarrow NO_2 + CO_2 + H_2O, \quad \Delta G_e^0 = -16.3$

$Org + NO_2^- \longrightarrow N_2 + CO_2 + H_2O, \quad \Delta G_e^0 = -29.1$

anaerobic, no NO_3^- present,
SO_4^{2-} present

$Org + SO_4^{2-} \longrightarrow H_2S + CO_2 + H_2O, \quad \Delta G_e^0 = -1.7$

anaerobic, no NO_3^- present,
no SO_4^{2-} present, CO_2 present

$Org + (CO_2) + H_2O \longrightarrow CO_2 + H_2O, \quad \Delta G_e^0 = -1.1$

Fig. 3. Differentiation in some of the ecological processes with depth; note especially the variation in terminal electron acceptor

presence of potential terminal electron acceptors. Although too complex to represent graphically, a composite of Figs. 1–3 would provide a rather complete, though abstract, scheme for element cycling.

Ecological Energetics

Energetics is simply a term for thermodynamics as applied to the analysis of biological processes. Biological processes, like all processes, proceed only when conditions of disequilibrium exist. More specifically, biochemical mechanisms proceed by transfer of electrons and associated energy to form, primarily, ATP and NADPH (or NADH). ATP is the universal mediator of biochemical reactions; NADPH and NADH are the major biochemical reductants. Processes in which these compounds participate are subcellular, yet processes by which they are formed and in which they are used are of utmost relevance at the ecological level of organization and of great utility in theoretical ecology. This is so because the macroscopic patterns in which they occur are interpretable, not at the cellular, individual, or population level – but only at the ecological level of organization. That is, the existence of a particular mode of metabolism, for example – chemoautotrophy, is a phenomenon that can be explained only with reference to ecological processes.

These processes are interactions of biotic and abiotic components that create and maintain environments suitable for particular modes of energy metabolism. Organisms can sustain a mode of energy metabolism in an environment only if the reaction on which this metabolism depends is thermodynamically favored (exergonic) in that environment. Specifically, organisms cannot maintain physiological conditions that permit a net derivation of energy from biochemical processes unless the net of those processes is exergonic in their environment. This fact can be used as a major organizing principle for theoretical ecology. It provides the basis for calculation of potential modes of energy metabolism for given environmental conditions without reference to organisms.

The Gibbs free energy function is the appropriate measure for such calculations because it is a measure of the useful work that can be extracted from the energy of a reaction, and it refers to conditions of constant temperature and pressure, but allows volume to vary (as, for example, occurs in a biotic process that results in the evolution of a gas). These are conditions most appropriate to biological systems. Such calculations are not of energy budgets as such. Calculations for energy budgets usually consider that energy is conserved (in keeping with the first law of thermodynamics) and accountable when all system gains and losses are considered. Conservation of energy, thus, refers to conservation of the sum of all forms of energy. Energy that can be used by biotic systems is the Gibbs free energy for the specific energy yielding reactions, such as the oxidation of organic compounds with sulfate as the terminal electron acceptor.

The useful energy that can be extracted from these processes is a function of the concentrations of the reactants and products. Hence, ecologically relevant calculations for the amount of useful energy that can be derived from chemical reactions are functions of the conditions of the environments in which the reactions occur, and, hence, not subject to simple application of conservation laws. In contrast, however, mass conservation is a principle that is useful for theoretical ecological computations, being essentially independent of environmental conditions.

Biophysical Models of Element Cycling Processes

An ecosystem model for element cycling is necessary for predicting effects of toxic chemicals on cycling processes. Morowitz (1968) has shown that flow of energy through a system from a source to a sink will necessarily result in cycling in steady state systems. Ecosystems approximate steady state systems that are far from equilibrium, a condition that is associated necessarily with structure (Nicolis and Prigogine, 1977). Ecosystems are usually considered to be open systems that are far from thermodynamic equilibrium. In the discussions to follow, the degree of approximation to steady state, however, is not explicitly assumed (except where noted for particular subsystems) but rather, the ecosystem is considered to be a dynamic system.

The organisms that carry out the ecological modes of metabolism comprise the structures that are necessary for element cycle transformations. For each mode of metabolism, stability is maintained in part by diversity among the populations in their response to external factors. As a consequence of diversity, compensatory changes occur, resisting fluctuations in the fluxes of materials and energy that result from element cycle transformations. There is a high degree of similarity at the biochemical level within each mode of metabolism. Diversity exists in morphological variation and is expressed in physiological and behavioral differences. The development that follows is based on the foregoing assumptions of biochemical similarity and morphological diversity within metabolic modes.

Every organism must cope with two problems, obtaining energy and obtaining elements from which to form biomass. Aquatic photoautotrophs potentially are limited in their energy supply rate by light availability because of light attenuation by various agents in water. For all microorganisms, obtaining chemicals from the environment is a process of molecular uptake of chemicals dissolved in water. Morphological and behavioral differences consist of various adaptations that function to reduce the diffusional limitation of moving chemicals to and across uptake surfaces.

Macroheterotrophs either move through the water or move water past themselves to obtain food. Mobile organisms use energy for evasion of predators. In these activities the power to overcome drag can account for a considerable fraction of the total power expenditure. For large aquatic animals the rate of food intake for energy and biosynthesis must be maintained in an environment in which potential food is varying and where pressure from predators is apt to be intense. In the following the supply rate of energy and materials is considered relative to the power expenditure for each of the metabolic modes. From this comparison, expressions for population rate of change are obtained.

A Model for Uptake and Feeding Rate with Potential Limitation by Simple Saturation Mechanisms

It is usually assumed that the uptake rate of dissolved chemicals by microorganisms is best described by a rectangular hyperbola as first proposed for microbial processes by Monod (1942). Such a description can be rationalized by any one of several specific mechanisms. All such processes appear to have a common characteristic, however, viz., a saturable component. In this respect many such processes exist, not all associated with microorganisms, and not all of which are biotic processes. Molecular sorption to particles is frequently described by the Langmuir isotherm, which, although not a rate description, is a rectangular hyperbola. In this instance the saturable component is the surface of the particle, or at least the capacity of the surface for the sorbate. The feeding of fishes was described in much the same way by Rashevsky (1959) and the feeding of insect predators by Holling (1959), the saturable component being the gut capacity in Rashevsky's analysis and available time in Holling's. Specifics of such processes will be discussed in the appropriate sections. The model is so generally applicable that it will be useful to indicate its derivation prior to specific uses.

The model consists of a description of each of two simultaneous processes: obtaining material (filling the saturable component) and removing material (restoring or emptying the saturable component). The rate of obtaining is proportional to the degree of unsaturation $(S - x)$ and to the concentration of the material being obtained (s):

$$\frac{dy}{dt} = k_0 s(S - x).$$ (1)

The rate of change of material associated with the saturable component is the difference between the rate of obtaining and the rate of emptying, the latter being proportional to the quantity of material associated with the saturable component:

$$\frac{dx}{dt} = \frac{dy}{dt} - k_e x.$$ (2)

In the above, y is the amount of material being obtained, x is the amount of the same material in the saturable component, S is the capacity of the saturable component, and k_0 and k_e are rate constants for obtaining and emptying, respectively.

For large populations over time intervals long with respect to characteristic times for the two processes, the degree of saturation will reach a stable value, implying that the rate of emptying is equal to the rate of obtaining. Thus

$$\frac{dy}{dt} = k_e x$$ (3)

from which

$$x = \frac{Ss}{k_e/k_0 + s}.$$ (4)

Substituting into Eq. (3) gives the expression for the rate of obtaining material when the processes are at a steady state:

$$\frac{dy}{dt} = \frac{k_e Ss}{k_e/k_0 + s}.$$ (5)

Transport Limited Chemical Uptake Rate

Even though microorganisms move, it is not clear what stimuli result in the movement, what terminates movement, whether movement ceases upon reception of other stimuli, etc. There is difficulty, therefore, in representing activity of microorganisms as a function of their energetic needs. It is more convenient to consider activity as part of the fundamental metabolism of microorganisms. As noted one aspect of microorganism physiology that is a result of size and the related absence of morphological features for feeding is the uptake of dissolved chemicals from the environment. For very small organisms viscosity of water is

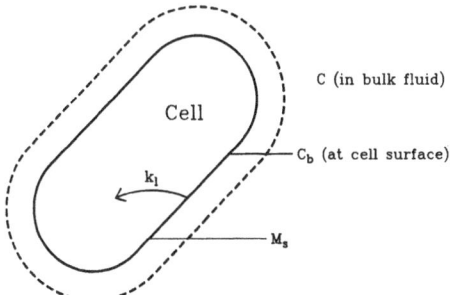

C (in bulk fluid)

C_b (at cell surface)

M_s

Fig. 4. Diagram of cell in aqueous medium illustrating components used in formulating model for uptake that incorporates diffusion limitation

high enough that movement results in little advantage in terms of increased food availability. Diffusion and other mixing processes, therefore, are important in bringing food to the organism (Purcell, 1977; Vogel, 1981).

The rectangular hyperbola is assumed to be the appropriate description of uptake of dissolved chemicals from the environment. Depending upon characteristics of the environment and the organisms, depletion of the chemical in the immediate vicinity of the cell surface can reduce the uptake rate below that expected when depletion is not considered. Pasciak and Gavis (1974, 1975) and Gavis and Ferguson (1975) analyzed this problem. Their model for the influence of local depletion of dissolved chemicals on uptake rate accounts for the effects of cell size and shape. If s is replaced by C_b, the concentration of the chemical at the cell boundary, and S by M, the capacity of the cell membrane for the chemical in moles cm^{-2}, then Eq. (5) represents a flux onto the cell membrane (Fig. 4). This flux can also be described in terms of Fick's first law:

$$J_b = -3.6 D (C - C_b) R^{-1}, \qquad (6)$$

where J_b is the flux at the cell membrane, D is the diffusivity of the chemical in water $(cm^2 \, s^{-1})$, C is the bulk concentration, R is the cell radius, and conversion to moles $cm^{-2} h^{-1}$ requires the conversion factor, 3.6. The flux to a spherical cell is Eq. (6) multiplied by $4\pi R^2$:

$$4\pi R^2 J_b = -14.4 \pi RD (C - C_b). \qquad (7)$$

When the two expressions for the flux into the cell [Eqs. (5) and (7)] are equated, a quadratic in C_b is obtained. Pasciak and Gavis (1974) write this equation in non dimensional form as

$$C_b^2 + (P^{-1} + 1 - C) C_b - C = 0, \qquad (8)$$

where $C_b = k_0 C_b / k_e$, $C = k_0 C / k_e$, and $P = -14.4\pi RD / k_0 M$. C_b is available via the quadratic formula and resubstitution of the above.

To account for cell geometries other than spheres, Pasciak and Gavis (1975) modified the above results to represent oblate and prolate spheroids (disks and spindles). This is accomplished by multiplying P by a shape factor that is a function of the eccentricity of the cells (eccentricity is a measure of the relationship of the major and minor axes). For a complete description and derivation see Pasciak and Gavis (1975) and their references.

This description of the uptake process is rather general, permitting uptake to be represented for any microorganism that uses dissolved substances. It does not provide, however, a description of uptake by aufwuchs or benthic communities. Solution to the latter problem in one dimension results in a form resembling equation (8). This problem is not discussed here.

Autotrophy

Production of biomass and O_2 by photoautotrophs is the fundamental biotic process. Production of other oxidants that are used by anaerobic heterotrophs to continue organic decomposition in the absence of molecular oxygen is carried out by chemoautotrophs. Chemoautotrophs obtain energy from oxidation of reduced inorganic compounds, usually with O_2 as the oxidizing agent (terminal electron acceptor, TEA).

Most chemoautotrophs are gram negative bacteria of pseudomonad related genera (Schlegel, 1975) and are, therefore, prokaryotes. Photoautotrophy is carried out predominantly by green algae and cyanobacteria. Green algae are eukaryotes and cyanobacteria, prokaryotes. It might be expected that extreme diversity of biochemical function exists among such diverse kinds of organisms. In regard to energy metabolism and biosynthesis (Fig. 5), however, there is little diversity. The Calvin cycle is the mechanism for fixation of CO_2 and production of hexoses (Schlegel, 1975). Lehninger (1975) gives the following reaction for biosynthesis in photoautotrophs:

$$CO_2 + 2\,H_2O + 2\,H^+ + 3\,ATP + 2\,NADPH \rightarrow (CH_2O) + 3\,P$$
$$+\,3\,ADP + 2\,NADP^+ . \qquad\qquad (9)$$

Autotroph Biosynthesis

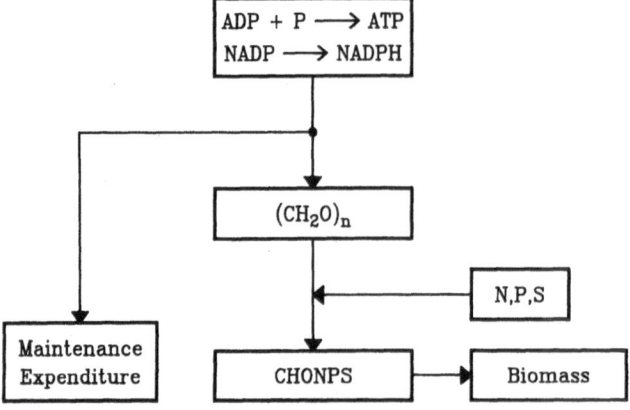

Fig. 5. Schematic of autotroph biosynthesis illustrating major components and relationships used as basis for general models of the process

Here (CH_2O) refers to 1/6 of a hexose. The degree to which the reaction is favored thermodynamically can be seen by decomposing it into its three component reactions.

$$ATP + H_2O \rightarrow ADP + P \qquad\qquad \Delta G^0(w) = -\quad 7.3$$
$$NADPH + 1/2\ O_2 \rightarrow NADP^+ + H_2O \quad \Delta G^0(w) = -\quad 52.5 \qquad (10)$$
$$CO_2 + H_2O \rightarrow (CH_2O) + O_2 \qquad\qquad \Delta G^0(w) =\quad 114.8\ .$$

When the standard free energies are used the net free energy for the reaction is -12.12. Under physiological conditions, however, the reaction is probably even more favorable. An indication of this can be obtained by introducing only the free energy of ATP hydrolysis ($-12.5\ kcal\ M^{-1}$) reported by Burton (1958). This increases the tendency of the reaction to proceed ($\Delta G = -27.72$). The analogous reaction for chemoautotrophic bacteria differs only in the use of NAD^+ and NADH rather than $NADP^+$ and NADPH (Schlegel, 1975). The energetics are approximately the same.

Diversity of autotrophy lies neither in energy metabolism nor biosynthesis but in the sources of energy; light for the photoautotrophs and a very large variety of redox reactions for the chemoautotrophs. Table 1 gives several of the reactions used by chemoautotrophs as energy sources.

Models of Autotrophic Energy Metabolism and Biosynthesis

The strategy adopted for representing autotroph growth is similar for both photoautotrophs and chemoautotrophs. Energy, whether light or chemical, and chemicals for biosynthesis have to be obtained separately. Therefore, energy or any element can limit the growth rate at a given time. First the growth rate is calculated that could be sustained on the supply rate of each required factor (element or energy), if no other factor were limiting. That is, given the concentration or density of a resource, the rate at which it can be obtained and used assuming that nothing else limits the use rate can be calculated. These calculations produce a set of potential resource limited growth rates. The actual growth rate is then taken to be the minimum of this set. Bannister's (1974) model for light limited growth of autotrophs is well worked out, whereas for chemoautotrophy I know of no comparable model. The model for energy limited growth of chemoautotrophs is based directly on the energetics and stoichiometry of carbon fixation and the redox reactions from which the energy is derived.

Light Limited Growth of Photoautotrophs in Aquatic Systems

Suspended substances reduce light by shading and dissolved substances by frequency selective sorption. The Beer-Lambert law expresses the reduction in intensity with depth (or mean optical path) as a function of a situation specific attenuation coefficient:

$$\frac{dI}{dz} = -\varepsilon I\ . \qquad (11)$$

Units of the rate are E absorbed m^{-2} of surface area m^{-1} depth day^{-1}, the units of I are $E\,m^{-2}\,$day^{-1}, and of the characteristic attenuation coefficient, ε, are m^{-1}. Steele (1962) proposed an equation for the rate of photosynthesis as a function of light intensity:

$$p = p_m \left(\frac{I}{I_m} \right) e^{1 - I/I_m}, \tag{12}$$

where p_m is the maximal rate of photosynthesis, reached when the light intensity is I_m. Equations (11) and (12) can be combined for the rate expected at any depth, etc., when the parameters are known.

Bannister (1974) developed production equations in terms of a parameter that is more closely related to the biochemical processes of energy gathering and utilization, ϕ_m, the maximal quantum yield. Bannister's analysis was carried out in terms of productivity, $M(C)\,m^{-3}\,$day^{-1}. It perhaps would be a purer analysis to express the quantum yield in terms of electron moles E^{-1}, and to work in terms of electron equivalents throughout the model. At this point, however, it would introduce an extra and apparently unnecessary step, and so I shall use Bannister's analysis including his units. He used three separate light curve equations. Because it includes the photoinhibition effect, I shall use Steele's equation as equation 12 above. The derivation is given here in abbreviated form.

Reduction of light itensity with depth can be partitioned into reduction by each of the causal components. Reduction of light intensity by suspended algae is caused primarily by the absorption of light by the cellular pigments. The molar absorption rate is

$$\frac{dI(A)}{dz} = k_a A I \tag{13}$$

with units of E absorbed $m^{-3}\,$day^{-1} (actually to be in keeping with the units of the derivative with respect to depth the units might better be expressed as $E\,m^{-2}$ of surface day$^{-1}\,m^{-1}$ of depth). Normally productivity is measured in $g(C)\,m^{-3}\,$day^{-1} or equivalent. Division by 12 converts to moles. The quantum yield is expressible as the ratio of the molar productivity to the molar adsorption rate, with units of $M(C)\,E^{-1}$, and an explicit expression for the quantum yield can be obtained by substituting from Eqs. (12) and (13)

$$\frac{p}{12} \bigg/ \frac{dI(A)}{dz}; \; \phi = \frac{p_m e^{1 - I/I_m}}{12\,I_m k_a A}. \tag{14}$$

As light intensity approaches zero, saturation effects on the photopigments disappear and the quantum yield approaches a maximum:

$$\phi_m = \frac{p_m e}{12\,I_m k_a A}. \tag{15}$$

This quantity theoretically is a constant, its value in natural waters being approximately 0.06 (Bannister, 1974) or 0.07 (Dubinsky and Berman, 1976) in units of $M(C) E^{-1}$ absorbed. Substituting for I in Eq. (12) and integrating with respect to depth gives the total photosynthesis rate per unit surface area. Further substitution for p_m [from Eq. (15)] and for ε in terms of the light attenuation factors provides the total photosynthesis rate in terms of parameters and the chlorophyll concentration. A. If this rate is integrated over time, production is obtained. To do so requires that I_0 be written as a time dependent light flux. This can be approximated in several ways (tabular data, function, etc.), but is not represented here.

A quantity that will often be useful is the depth-integrated photosynthesis rate of carbon assimilation, considering no limiting factors besides light. Two depths are of interest: that specified when a volume element of particular dimensions is considered and the depth at which the light compensation point is reached. In either instance the applicable equation for photosynthesis rate is

$$p_T = {}^{12} \phi_m I_m e^{-I/I_m} (1-L) A / \left(\frac{k_w}{k_a} + A \right) \tag{16}$$

in which L is the fraction of the incident light that is transmitted through the element, and is given by

$$L = I'/I_m = e^{-(k_w + k_a A) z'} . \tag{17}$$

If z' is a fixed depth, Eq. (17) can be used, but if z' represents the compensation depth, then the light level, I', at the compensation depth must be specified, and L is calculated directly as $L = I'/I_m$.

Equation (16) is the primary equation for calculating light-limited rate of growth of photoautotrophs. In calculations it will often be convenient to express the rate as a quantity per volume, rather than per surface area. This is a simple division of the quantity, p_T, by the applicable depth to which the total applies.

Energy Limited Growth of Chemoautotrophs in Auatic Systems

Energy is obtained for chemoautotrophy by uptake of reduced inorganic chemical species and of oxygen as an oxidizing agent, and by carrying out the redox reaction. Both the chemicals that are used for the energy reaction and those that are used in biosynthesis are obtained by uptake through a gradient. Therefore the concentration at the cell membrane can be the limiting factor [Eq. (8)]. The chemical that is used as the electron donor in the energy reaction can also be used as a substrate for biosynthesis. For example in nitrifying bacteria the following reaction (line 1, Table 1) is used for energy:

$$NH_4^+ + 3/2 O_2 \rightarrow NO_2^- + H_2O + 2 H^+ \quad \Delta G^0(w) = -64.92 . \tag{18}$$

Ammonia is also used for synthesis. It must be supplied at a rate that is commensurate with the other needs of growth, and therefore, the rates at which it is

Table 1. Some reactions used by chemoautotrophs for energy

Reaction	Free energy	
	kcal/M (substrate)	kcal/eq. wt. (substrate)
$NH_4^+ + 1.5\,O_2 = NO_2^- + 2\,H^+ + H_2O$	$- 64.92$	-10.82
$NO_2^- + 0.5\,O_2 = NO_3^-$	$- 18.50$	$- 9.25$
$HS^- + 2\,O_2 = SO_4^{2-} + H^+$	-190.30	-23.79
$S^0 + 5/2\,O_2 + H_2O = SO_4^{2-} + H^+$	-139.98	-23.33
$S_2O_3 + 2\,O_2 + H_2O = 2\,SO_4^{2-} + 2\,H^+$	-190.16	-23.77
$SO_3^{2-} + 1/2\,O_2 = SO_4^{2-}$	$- 29.54$	-29.27
$HS^- + 8/5\,NO_3^- + 3/5\,H^+ = SO_4^{2-} + 4/5\,N_2 + 4/5\,H_2O$	-177.92	-22.24
$S^0 + 6/5\,NO_3^- + 2/5\,H_2O = SO_4^{2-} + 3/5\,N_2 + 4/5\,H^+$	-130.74	-21.79
$H_2 + 2/5\,NO_3 + 2/5\,H^+ = 1/5\,N_2 + 6/5\,H_2O$	$- 53.60$	-26.80
$H_2 + 1/4\,SO_4^{2-} + 1/4\,H^+ = HS^- + H_2O$	$- 9.10$	$- 4.55$
$H_2 + 1/2\,CO_2 = CH_3COOH + 1/2\,H_2O$	$- 4.30$	$- 2.15$
$CH_4 + 2\,O_2 = CO_2 + 2\,H_2O$	-195.50	-24.44
$CH_4 + SO_4^{2-} + H^+ = CO_2 + HS^- + H_2O$	$- 5.20$	$- 0.65$

required for energy and for synthesis each need to be calculated and compared to the other needs. Equations (9) and (10) apply equally well to all autotrophs. From these equations it can be seen that three moles of ATP and two of $NADPH$ are required to synthesize one mole of carbon. Using standard free energy values, this requires that at least 126.9 kcal be supplied via the above reaction [Eq. (18)]. Assuming 38% efficiency [the same value often used for photoautotrophs (Clayton, 1980) and heterotrophs (Lehninger, 1975)], 334 kcal from about 5 moles of NH_2^+ must be released (a quantum yield of 0.2 M carbon synthesized $M^{-1} NH_4^+$ utilized for energy). A quantum yield of 0.2 was observed by Gunderson and Mountain (1976). It can be seen that for energy to be a non-limiting commodity, ammonia must be obtained at five times the rate of obtaining CO_2 and at $3^{1/3}$ the rate of obtaining O_2. The same kinds of calculations apply to each of the energy reactions indicated in Table 1.

The Achieved Growth Rate of Autotrophs

The above kinds of comparisons are valid if autotroph growth is resource limited, a situation that is expected in all but unusual transient conditions. Energy and all chemicals can be present transiently in abundant supply. Under these conditions growth is limited by some inherent intracellular property. Growth rate is then maximal, and the condition is transitory because the populations that are growing maximally increase rapidly until some resource again becomes limiting. In the absence of a better alternative, it can be assumed that the minimum of the maximum uptake rates, as limited by membrane transport, is the growth limiting factor for autotrophs. That is, the maximal growth rate that is achieved under these conditions is the minimum of the growth rate that would be achieved maximally for each of the required resources. Elemental composition of the organisms has to be considered when calculating these minima, however. One way to accomplish this is to reference each of the rates to the carbon assimilation rate. If the organism's elemental composition is $C_cN_nP_pS_s$, then the ratio of nitrogen to carbon is n/c, nitrogen is required at the rate $n/c \, dC/dt$, and thus the minimum of dC/dt and $c/n \, dN/dt$ would be the expected growth rate if either carbon or nitrogen were limiting (here and in the following equation dC/dt and dN/dt are used to indicate the uptake rates of C and N). In general the growth rate is given by

$$\min \left(E, \frac{dC}{dt}, \frac{c}{n}\frac{dN}{dt}, \frac{c}{p}\frac{dP}{dt}, \frac{c}{s}\frac{dS}{dt} \right). \tag{19}$$

Each of the rates of assimilation is described by an equation like Eq. (5).

Heterotrophy

Heterotrophs derive both energy and biomass from a complex mixture of chemicals. Some fraction of this mixture is essentially indigestible. The remainder is hydrolyzed to small compounds and further broken down to small monomers (7). Some of the monomers are used as reactants for energy production and some for biosynthesis. Essentially any of the monomers can be used for energy, but in biosynthesis discrimination among monomers occurs during synthesis of specific biomass for growth and repair of tissues. This general scheme characterizes the metabolism of both microheterotrophs and macroheterotrophs. Microheterotrophs, of course, deal with the complex organic mixture extracellularly first; then after it is broken down into soluble compounds, absorb it and use it for energy and synthesis. Excretion of products that cannot be used for energy or synthesis is an important process for microheterotrophs. Macroheterotrophs consume complex organic mixtures either as other organisms or as detritus, and the whole process of digestion and metabolism is carried out internally. Elimination of indigestible material and breakdown products that cannot be used for energy or synthesis are both important processes for macroheterotrophs.

A Model for Calculating Growth Rate of Heterotrophs

In keeping with the overall objective of representing ecological processes that are important in element cycling, the growth model is framed as a function of the supply rate of each of the important elements. As already discussed, processes that control these supply rates are driven by redox disequilibria that are created by photoautotrophy. These disequilibria provide continual input to the respiratory (redox) processes carried out by heterotrophs in obtaining energy. The model for calculating growth rate consists of three parts. One is the component for partitioning of food into power supply and biomass synthesis. The second is the calculation of power demand. The other is the calculation of food consumption rate in terms that permit calculation of power supply as well as the supply rate of food components whose composition is similar to that of the consumer.

Heterotrophy

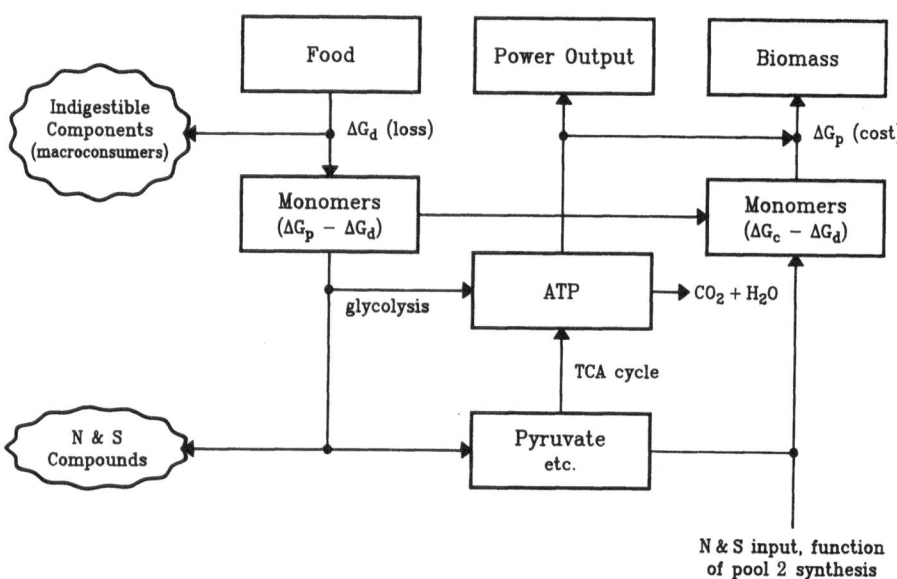

Fig. 6. Schematic of heterotrophy (energy metabolism and biosynthesis illustrating major components and relationships used as basis for general models of the process

Partitioning Food Into power Supply and Biosynthesis

Food consumption, power consumption, power production, and biosynthesis are rates that need to be maintained so that the power produced satisfies the heterotroph's power demand. This demand derives from activities of obtaining food and escaping predators, from the basic life processes, and from repair of tissues. Remaining power and materials are used to support growth. Figure 6 gives the scheme for accounting the energy and materials in partitioning the food intake rate into power output and biosynthesis. The scheme refers most directly to macroheterotrophs, but it is applicable to microheterotrophs if food is considered to consist of dissolved organic chemicals and elimination of the indigestible fraction is ignored. Figure 7 shows the assumed relationships of the categories of organic chemicals in the aquatic system. Processes and states inside the dashed line occur inside microbial cells.

The object of the model is to calculate growth rate of a heterotroph population. This is accomplished in several steps. Composition of the food is compared to composition of the heterotroph so that the rate of consumption of two food components can be calculated. The two components are biomasslike (referenced as pool 1 in the following) and non biomass-like (pool 2). (The term "pool" should be interpreted to indicate a dynamic quantity characterized by varying elemental composition as the composition of food varies.) The free energy content of total food at any time is assumed to be known or to be available via an accounting chain. The power supply is the free energy production rate from both of the pools.

Organic Matter

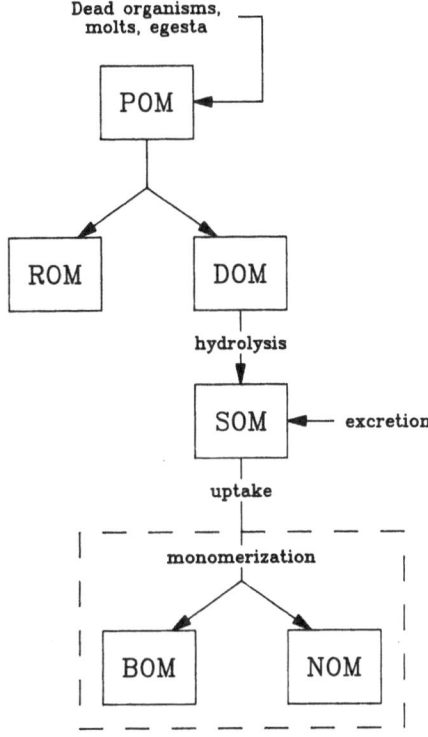

Dead organisms,
molts, egesta

POM

ROM   DOM

hydrolysis

SOM ◄— excretion

uptake

monomerization

BOM   NOM

Fig. 7. Categories of organic chemicals suggested as appropriate as a first approximation for modeling this very complex pool of chemicals. (P=particulate, R=refractory, D=digestible, S=soluble, B=biomass-like, N=non-biomass-like)

However, if the power demand is less than the power supply, the excess is not metabolized for energy but is synthesized. The specifics of the process are given below in the sequence of the descriptive overview presented in this paragraph.

The process of partitioning food into power supply and biosynthesis is developed in the following for macroheterotrophs. The comparable process for microheterotrophs is simpler in one aspect, because the organic chemicals on which they feed are represented as lumped categories. The development for macroheterotrophs can be applied to microheterotrophs by considering the categories of organics in place of the species of prey. But in another aspect it is more complex: macroheterotrophs use a single electron acceptor, oxygen, whereas microheterotrophs use a whole series of them.

Use of the term "energy content" is to be interpreted as the free energy per equivalent of the reaction between a chemical serving as an electron donor and another chemical serving as an electron acceptor. In much of the following the electron acceptor is assumed to be oxygen, and no single chemical will serve as electron donor, but rather electrons will derive from a mixture of chemicals, comprised of the biomass of higher organisms. In this situation also, I shall use the term "energy content." A discussion of energetics with electron acceptors other than oxygen will follow the section on food partitioning.

The food intake rate (Mass t^{-1} consumer^{-1}), is

$$F_{\cdot j} = \sum_i F_{ij}, \tag{20}$$

where i indicates a source of food (prey population or environmental pool of chemical), and j identifies the predator. The overall elemental composition of an organism (or of a chemical pool can be represented for the elements of concern as

$$C_c O_0 N_n P_p S_s, \tag{21}$$

where the upper case letters symbolize the elements as usual, and the lower case letters represent the mole numbers of the elements with which they are associated. Let n_{ik} represent the mole number of element k of species i, and w_k represent the gram atomic weight of element k. Then the fraction of food mass obtained from prey species i comprised of element m is

$$f_{im} = \frac{n_{im} w_m}{\sum_k n_{ik} w_k} \tag{22}$$

so that the rate of consumption of element m in the mixed food is

$$F_{\cdot jm} = \sum_i f_{im} F_{ij}. \tag{23}$$

The average molar rate of consumption of elements by consumer j is the set

$$\left\{ \frac{F_{\cdot jc}}{12}, \frac{F_{\cdot jo}}{16}, \frac{F_{\cdot jn}}{14}, \frac{F_{\cdot jp}}{31}, \frac{F_{\cdot js}}{32} \right\}. \tag{24}$$

Division of each of the members of this set by the mole number of the consumer for the corresponding element gives another set, the least value of which corresponds to the element of the food which is in least supply relative to the composition of the consumer. A set of mole numbers for the composition of the food normalized to the element in least supply can be obtained by dividing the members of the set of molar consumption rates [Eq. (24)] by the member corresponding to the element in least supply. A similar set can be obtained for the consumer by dividing the mole numbers of the consumer empirical formula by the mole number of the element in least supply. Then the food can be partitioned into the two pools. The normalized empirical formula of food can be represented as

$$C_{cF} O_{oF} N_{nF} P_{pF} S_{sF} \tag{25}$$

and, because pool 1 is like consumer biomass, it can be represented as

$$C_{cj} O_{oj} N_{nj} P_{pj} S_{sj}. \tag{26}$$

In the above formulae, the subindex F indicates that the mole numbers are for the food, and the subindex j indicates that those mole numbers are for pool 1 which is like consumer j. Pool 2 component constitutes the remainder of food. Its mole numbers are the difference between the mole numbers of the food and of pool 1:

$$C_{c_F - c_j} O_{o_F - o_j} N_{n_F - n_j} P_{p_F - p_j} S_{s_F - s_j} . \tag{27}$$

Now the rate of consumption of the two pools can be calculated. First let n_{kj} represent the mole number of element j in pool k. Conservation requires that

$$\sum_j n_{Fj} = \sum_j (n_{1j} + n_{2j}) . \tag{28}$$

The fraction of food that is pool k is

$$f_k' = \sum_j n_{kj} / \sum_u n_{Fu} , \tag{29}$$

and the rate of consumption of pool k $(k=1$ or $2)$ is

$$F'_{.jk} = f_k' F_{.j} . \tag{30}$$

If the free energy content of food is known, then it too can be partitioned. Morowitz (1968) discussed some of the problems of using free energies of biological materials and concluded that enthalpy (heat content) values are useful approximations. There are other approaches to this problem that could yield useful approximations, but rather than consider that practical problem, I shall merely assume that free energy values are available. The free energy content of food can be obtained if the free energy content of individual food items is known:

$$\Delta G_{Fj} = \frac{\sum_i \Delta G_{ci} F'_{ij}}{\sum_k F'_{kj}} . \tag{31}$$

ΔG_{ci} is the free energy content of species i.

During the process of digestion of the food to monomers, a small loss of free energy occurs, the polymer bond energy, ΔG_d. The free energy of pool 1 is the same as the free energy of the consumer less the free energy loss of depolymerization, $\Delta G_{m1} = \Delta G_{cj} - \Delta G_d$. Pool 2 free energy is the remainder less the energy loss of depolymerization: $\Delta G_{m2} = (\sum F_{ij} \Delta G_F - F_1 \Delta G_{cj}) / F_2 - \Delta G_d$. The power supply from the two pools can now be calculated. From pool 1 it is

$$P_{m1} = F_1 \Delta G_{m1} B_j^{-1} , \tag{32}$$

where B_j indicates the mass of an individual of species j, and from pool 2 it is

$$P_{m2} = F_2 \Delta G_{m2} B_j^{-1} . \tag{33}$$

Describing the way that this free energy is used to produce ATP and NADPH requires a model of cellular biochemistry. I assume that discrimination between the two pools occurs, so that pool 2 is used preferentially for energy and pool 1 as material for biosynthesis. Power demand, P_a is taken to be the sum of the energy expense rates for maintenance, repair, motility of hunting and escaping, and other activities associated with feeding. If the power demand is greater than power production from pool 2, then pool 1 is used as needed. If the power demand is less than power production from pool 2, then pool 2 remainder is converted into new biomass at a rate determined by the uptake of the necessary inorganics to achieve the consumer's elemental composition, or it is converted into storage compounds. The remainder, in excess of the rate of conversion by these processes, is eliminated. Four distinct cases arise as a result of the comparison of the power supply and demand:

Case 1: pool 2 power supply > power demand for activity plus synthesis of pool 1 into consumer biomass:

$$\phi_c P_{m2} > (P_a + F_{1j} \Delta G_p B_j^{-1}) \phi_s^{-1} . \tag{34}$$

Case 2: pool 2 power supply > power demand for activity but < power demand for activity plus synthesis of pool 1 into consumer biomass:

$$P_a \phi_s^{-1} < \phi_c P_{m2} < (P_a + F_{1j} \Delta G_p B_j^{-1}) \phi_s^{-1} . \tag{35}$$

Case 3: pool 2 power supply < power demand for activity, but pool 1 + pool 2 power supply > power demand for activity:

$$\phi_c P_{m2} < P_a \phi_s^{-1} < \phi_c (P_{m1} + P_{m2}) . \tag{36}$$

Case 4: pool 1 + pool 2 power supply < power demand for activity:

$$\phi_c (P_{m1} + P_{m2}) < P_a \phi_s^{-1} . \tag{37}$$

In the above ϕ_c represents the efficiency of obtaining free energy during the catabolic process, ϕ_s the efficiency of use of the free energy in activity or biosynthesis, and ΔG_p the free energy required for polymerization of the monomers into biomass.

Specific (per capita) growth rates can be calculated for each of the above cases. Case 1: all of pool 1 is synthesized, and the growth rate, therefore, is

$$\mu_j = \frac{dB_j}{B_j dt} = F_{1j} B_j^{-1} . \tag{38}$$

Some of pool 2 remains:

$$P_{m2r} = P_{m2} - (P_a + F_{1j} \Delta G_p B_j^{-1}) \phi_c^{-1} \phi_s^{-1} ,$$

$$F_{2r} = P_{m2r} B_j \Delta G_{m2}^{-1} . \tag{39}$$

Remaining pool 2 can be synthesized at a rate determined by uptake of necessary elements, or it can be turned into storage compounds. These processes are not discussed further here.

Case 2: pool 2 supplies the power demand of activity plus part of the power demand of synthesis. The growth rate on pool 1 monomers sustained by power derived from pool 2 is

$$\mu_{j2} = \phi_c \phi_s (P_{m2} - P_a) \, \Delta G_p^{-1} . \tag{40}$$

Power to synthesize the remainder of pool 1 must derive from pool 1, itself. The remainder of power in pool 1 is

$$P_{m1r} = P_{m1} - \mu_{j2} \Delta G_{m1} . \tag{41}$$

Thus, the growth rate sustained by the remainder of pool 1 is

$$\mu_{j1} = P_{m1r} (\Delta G_{m1} + \Delta G_p \phi_c^{-1} \phi_s^{-1})^{-1} . \tag{42}$$

The net growth rate is the sum of the two, which after simplification is

$$\mu_j = P_{m1} (\Delta G_p + \phi_c \phi_s \Delta G_{m1})^{-1} + \phi_c \phi_s (P_{m2} - P_a) \, \Delta G_p^{-1} . \tag{43}$$

Case 3: pool 2 is insufficient to supply the power demand of activity, so pool 1 is diverted at the necessary rate. The power available from pool 1 remainder is

$$P_{m1r} = (P_{m1} + P_{m2}) - P_a \phi_c^{-1} . \tag{44}$$

The growth rate that is sustained on pool 1 remainder is

$$\mu_j = \phi_c \phi_s (P_{m1} + P_{m2} - P_a \phi_s) \, (\Delta G_p + \phi_c \phi_s \Delta G_{m1})^{-1} . \tag{45}$$

Case 4: the power obtainable from the total food is less than the activity power demand. It is assumed that the consumer continues to attempt to carry out normal activities of feeding, escaping, etc., at the expense of body mass. The net effect to the whole population is a loss of biomass, expressible as a negative growth rate:

$$\mu^j = [\phi_c (P_{m1} + P_{m2}) - P_a] \, \Delta G_{m1}^{-1} . \tag{46}$$

The growth rates calculated above [Eqs. (38), (43), (45), and (46)] use values for power production rate and power utilization rate, in addition to the patterns of physiological processes and biomass composition. Power is used at the cellular level, but it can be accounted most easily at higher levels of organization. Power is required for biosynthesis, hence the accounting point is in the model for biosynthesis, i.e., in the above equations describing growth. There are two other major categories of power demand. One is the basic metabolic demand. The other is the demand due to motile activity. This latter demand is more readily accounted at a higher hierarchial level, the power required to move a body through a liquid medium. Power production rate is a function of the composition of the food, and of the rate at which it is obtained. Hence, a description of the feeding process is an important part of the model, because in its development the basis will be obtained for calculating power supply, supply of materials for biosynthesis, motile activity for obtaining food, and motile activity for escaping predators. In addition, loss rates caused by predation are obtained by simple rearrangement of the equations for feeding.

Models of Feeding by Macroheterotrophs

Macroheterotrophs capture discrete food particles by one of two general means. Either they feed upon all particles encountered that are within the size range that they can handle, or they select and capture individual food particles by explicit overt action. I shall refer to organisms characterized by the former behavior as filter feeders and to those by the latter as pursuit feeders.

The derivation for uptake and feeding given in Eqs. (1)–(5) follow the general pattern of Rashevsky's (10) derivation of stationary state feeding rate for fishes. His work was motivated by Ivlev's (1961) studies on the feeding of fishes. Ivlev's model for fish feeding is widely used in mathematical models of systems that include feeding by large consumers. It was derived as a formal mathematical expression that fitted his data. Rashevsky derived his model from assumptions about the way that fishes feed. Part of his purpose was to present a stationary state analysis based on Ivlev's model, which described the feeding of fishes as a function of the density of food. Ivlev's equation was of the form

$$R = R_m(1 - e^{-zB_w}),\tag{47}$$

where R is the food eaten per experiment (a fixed time), R_m is the maximum feeding rate for the fixed time, B_w is the biomass density of prey (mass per volume), and z is a fitted constant.

Rashevsky's derivation of an equation comparable to Ivlev's began with the statement that the feeding rate (F) equals the encounter rate (E) times the probability that the prey is eaten if encountered (P):

$$F = EP.\tag{48}$$

The encounter rate is derived by assuming that the predator sweeps out a right circular cylindrical volume as it swims in search of food. The encounter rate is then the product of this volume per time and the density of prey in the volume, $\pi r^2 vB$, where r is the radius of the cylinder, v is the swimming velocity, and B is the mass of the prey per volume of water. Rashevsky took the probability of consumption given an encounter to be proportional to the unfilled gut capacity, an assumption exactly equivalent to that made in Eq. (1).

The latter assumption is credible for situations in which the saturable component is filled via mechanisms that operate passively, such as the adsorption of dissolved chemicals onto a microbial cell membrane. Where the component does not have a quota of specific sites that can be occupied, but rather is filled in a non-specific way, such as the filling of a fish's stomach, it is not clear that the rate of filling is necessarily proportional to the unfilled capacity. Ivlev's fits of the model to data indicated no discrepancy from the model of direct proportionality, however, and therefore, the point is more cautionary than substantial. The point should simply be borne in mind that it is not necessary for a higher organism to reduce its feeding rate to the fraction of its maximum that corresponds exactly to the fraction of unfilled stomach capacity. In this particular regard the model of Rashevsky and the models that are developed here are not necessarily based upon an unassailable assumption.

The encounter rate is

$$E = \pi r^2 v B_w, \tag{49}$$

where r is the encounter radius, v is the swimming velocity, and B_w is the biomass density as in Ivlev's model. The probability of consumption of prey given an encounter is

$$P = c(M_s - M_g), \tag{50}$$

where M_s is the capacity of the stomach, M_g is the quantity of food in the stomach, and c is a constant of proportionality. If the probability of consumption is 1 when the gut is empty ($M_g = 0$) then the proportionality constant, c, is M_s^{-1}. Thus the feeding rate is

$$F = \pi r^2 v B_w \left(\frac{M_s - M_g}{M_s} \right). \tag{51}$$

Feeding rate, F, is the same as the rate of change of stomach content, dM_g/dt if the experiment is done over a short enough time interval that stomach emptying can be ignored (there is no loss term). Integrating and putting $M_g = 0$ when $t = 0$, the stomach content is

$$M_g = M_s \left(1 - e^{-\frac{\pi r^2 v B_w t}{M_s}} \right). \tag{52}$$

For Ivlev's experiment, $t = t_f$, the fixed experimental length, so that the constant, z, of Ivlev's model is equivalent to $\pi r^2 v t_f / M_s$ of Rashevsky's. As noted by Rashevsky (10), Ivlev's model applies only to the phenomenon of feeding as a function of concentration of food. That is, it does not take into account other factors that affect feeding, such as stomach emptying rate. It is not suitable, therefore, for use in a model in which time intervals are long enough that the other factors become important.

For a population considered over a time interval that is long compared to the characteristic times for stomach filling and emptying, the mean food intake rate equals the mean stomach emptying rate, i.e., steady state feeding is achieved:

$$\pi r^2 v B_w \left(\frac{M_s - M_g}{M_s} \right) = k M_g, \tag{53}$$

where k is the stomach emptying rate constant. The feeding rate is obtained by rearranging Eq. (53) so that kM_g is expressed in terms that do not include the stomach content, M_g, and recalling that at steady state feeding, feeding rate equals the stomach emptying rate, or

$$F = \frac{kM_s B_w}{\dfrac{kM_s}{\pi r^2 v} + B_w} \,. \tag{54}$$

Note that this equation is of the same form as Equ. 5, and that its equivalent components are interpretable similarly.

To the extent that this model represents the main features of predator-prey interactions in aquatic systems it has a property that is very useful, it permits complete specification of predator-prey interactions as a function of the characteristics of the individual predator and of the population densities of predator and prey. That is, no species specific, pairwise interaction coefficients are needed. Immediately, however, one rejects that predator-prey interactions involve characteristics of the individual predator but not the individual prey. In the following a more detailed derivation of predator-prey models is made for two major ecological modes of feeding by aquatic organisms, filter feeding and pursuit feeding.

Two modes of filter feeding are known for fishes (O'Brien, 1979). One is the continuous seiving of water through the gill rakers as the predator swims with its mouth open. The other is pump filtering in which a greater volume of water is filtered than would occur if the same organism were seiving. Pursuit feeders could also be termed visual feeders (O'Brien, 1979), visual detection of prey being the common feature of this form of predation. Organisms other than fishes can usually be described by one or the other of these types of feeding.

Filter Feeding

Consumers that feed indiscriminantly upon all organisms that they encounter that are within the size range possible for them to feed upon are grouped here under filter feeders. It is assumed that they move through the water creating a disturbance front detectable by prey with suitable sensory organs. The disturbance front can be a wave caused by the body of the oncoming predator or a directed flow field caused by the suction of pump feeders. Some of these prey are able to escape and some are captured. The scheme for this type of feeding is given in Fig. 8. If a prey organism swimming at velocity v_p swims normal to the path of the oncoming predator, reaching the edge of the encounter cross section, the distance $\varrho - r$, before the predator swimming at velocity v_c swims the encounter distance s_d, then the prey escapes; otherwise it is captured. All distances, τ, such that

$$(\varrho - \tau)/v_p = s_d/v_c \,, \tag{55}$$

are escape distances. All prey inside the radius, τ, at the point of encounter with the disturbance front preceding the predator, are captured. The feeding rate of an

Filter Feeding Schematic

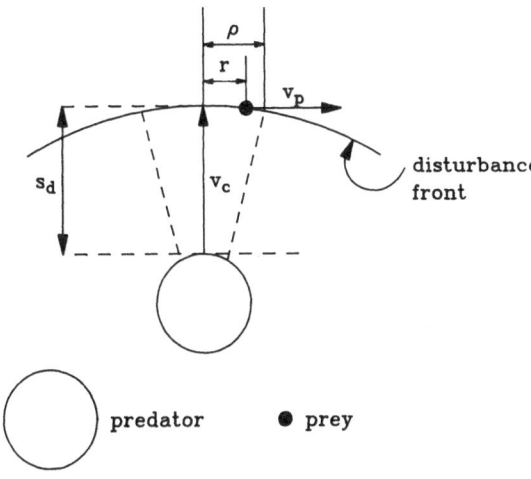

Fig. 8. Characteristics of a filter feeding predator and its prey that are used as basis for encounter model

individual predator feeding on a single prey of population density, M, is

$$F = \frac{kM_s \frac{\tau^2}{\varrho^2} B}{\frac{kM_s}{\pi \varrho^2 v_c} + \frac{\tau^2}{\varrho^2} B} = \frac{kM_s B}{\frac{kM_s}{\pi \tau^2 v_c} + B}. \tag{56}$$

The first of the two forms of F, although somewhat more cumbersome, is preferable, because it groups quantities together that are properties of the predator alone. Hence, in dealing with several prey populations and one predator these quantities are constants. They are the factor, kM_s, and the whole first term in the denominator. The second of the two forms indicates that the effective rate that the predator hunts is dependent upon both predator and prey. That is, in the first the volumetric search rate is $\pi \varrho^2 v_c$, but in the second the comparable term is $\pi \tau^2 v_c$, a smaller quantity that is a function of both predator and prey because τ is a function of the swimming velocities of both. The expression for feeding by predator j on n prey populations is

$$F_j = \frac{\frac{k_j M_{sj}}{\varrho_j^2} \sum_i \tau_{ij}^2 B_i}{\frac{k_j M_{sj}}{\pi \varrho_j^2 v_{cj}} + \frac{1}{\varrho_j^2} \sum_k \tau_{kj}^2 B_k}. \tag{57}$$

Pursuit Feeding

Consumers that swim at one velocity while searching for prey, then pursue the prey at another velocity are referenced here as pursuit predators. Figure 9 gives the

Pursuit Feeding Schematic

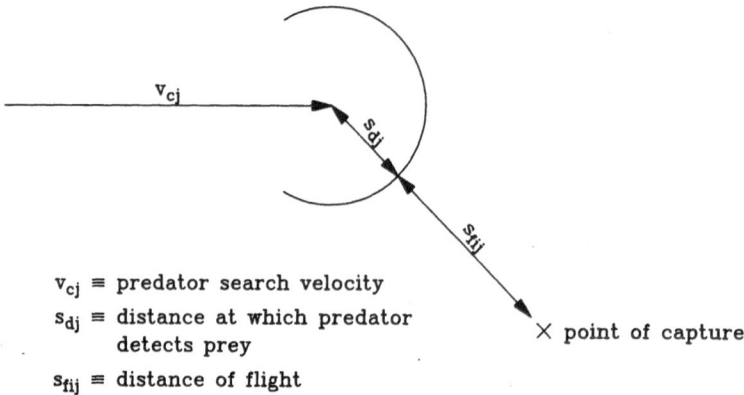

v_{cj} ≡ predator search velocity

s_{dj} ≡ distance at which predator detects prey

s_{fij} ≡ distance of flight

✗ point of capture

Fig. 9. Characteristics of a pursuit feeding predator and its prey that are used as basis for encounter model

schematic for this mode of food gathering. The searching velocity of the predator is v_{cj}. The distance at which the predator can detect prey is, for simplicity, here considered to be a constant that is characteristic of the predator, s_{dj}. In fact it is known to be a function of several variables including water clarity, light level (Confer et al., 1978) and prey size (Werner and Hall, 1974; Werner, 1974; Janssen, 1976). Upon detection both predator and prey begin swimming at the pursuit velocity, v_{pj}, and flight velocity, v_{fi}, respectively. The duration of the pursuit, t_{pij}, is given by

$$t_{pij} = s_{dj}/(v_{pj} - v_{fi}) = s_{dj}/\Delta v_{ij}. \tag{58}$$

The prey's distance of the flight, s_{fij}, is given by

$$s_{fij} = v_{fi} s_{dj}/\Delta v_{ij}. \tag{59}$$

It is arbitrarily assumed that prey that escape swim the same distance as prey that are captured, and similarly for predators. It is necessary to have information on other components of the process of feeding such as the conditional probabilities of pursuit given detection, of capture (or escape) given pursuit, and of ingestion given capture. The probability of escape is expected to be greater, the closer is v_{fi} to v_{cj}. That is if the prey can swim nearly as fast as can the predator, there is a high probability of escape. O'Brien (1979) discussed other factors. As with filter feeders the effective rate that the pursuit predator hunts is a function of both predator and prey. Gerritsen and Strickler (1977) developed a model for predation that depends on both predator and prey cruising velocities. This model is equivalent to the effective hunting rate, the rate of encounters while hunting:

$$a_{ij} = \pi s_{dj}^2 \left(\frac{u^2 + 3 w^2}{3 w} \right), \tag{60}$$

where $u = \min(v_{ci}, v_{cj})$, $w = \max(v_{ci}, v_{cj})$, and v_{ci} and v_{cj} are the cruising velocities of prey i and predator j, respectively. The rate of encounters while hunting is greater than the feeding rate, because pursuit feeders spend their time in two ways, searching and pursuing. (In general I shall aggregate all activities involved in feeding into pursuing, i.e., pursuit, capture, handling, ingestion. First, however, it is less cumbersome to develop the model in terms of overall rate of encounter, then later to introduce the probabilities of pursuit given encounter and of capture given pursuit.)

In parallel to the phenomenon of partitioning a saturable component into filled and unfilled portions, a pursuit predator partitions time into searching and capturing:

$$F_j^* = (T_{sj} + T_{pj})^{-1}, \tag{61}$$

where F_j^* is the encounter rate, T_s is the mean time spent searching per encounter, and T_p is the mean time spent pursuing per encounter. More specifically the mean search time per encounter and the mean pursuit time per encounter are

$$T_{sj} = \left(\sum_i a_{ij} N_i\right)^{-1}; \quad T_{pj} = \sum_i R_{ij} t_{pij}, \tag{62}$$

where R_{ij} is the relative frequency of the ij^{th} encounter:

$$R_{ij} = \frac{a_{ij} N_i}{\sum_k a_{kj} N_k}. \tag{63}$$

After rearrangement the encounter rate can be written as

$$F_j^* = \frac{s_{dj}^{-1} \sum_g \Delta v_{gj} \sum_i a_{ij} N_i}{s_{dj}^{-1} \sum_k \Delta v_{kj} + \sum_l a_{lj} N_l}. \tag{64}$$

Unlike the expression [Eq. (57)] for filter feeders, the components of Eq. (64) are not as simple dimensionally as is desirable for ease of interpretation. The terms of the denominator each represent an encounter rate, the first relating to pursuits and the second to detections of prey while the predator is searching. If the equation is rewritten by dividing both numerator and denominator by the sum of the velocity differences, the "half saturation" term becomes simply the detection distance for an encounter, a form that is perhaps more easily interpreted. Without exhausting the possible ways that the equation could be rewritten in search of a form that is most readily interpretable, it is apparent that the components of the expression are functions of characteristics of both predator and prey simultaneously to a greater degree than are the comparable components of Eq. (57). It is therefore comparably more difficult to separate the components of the equation in such a manner that it is as easily interpreted. In particular, the encounter rate, a_{ij}, is a function of both predator and prey. Note that in spite of its appearance in Eqs. (62)–(64), it is not fundamentally a pairwise coefficient. In Eq. (60) it is completely specified in terms of characteristics of individual predators and prey.

To complete the model for pursuit predation, the effect of stomach filling must be incorporated. In the general expression for the limitation of feeding rate by stomach filling [Eq. (51)], the probability of feeding was taken to be proportional to remaining stomach capacity. A similar assumption is made here. The encounter rate times the probability of encounter for pursuit feeders is the encounter rate, F^*, as developed above, times the probability of feeding given an encounter, $(M_{sj} - M_{gj})/M_{sj}$. At steady state feeding this rate equals the stomach emptying rate, kM_{gj}:

$$F_j = kM_{gj} = F_j^* \left(\frac{M_{sj} - M_{gj}}{M_{sj}} \right). \tag{65}$$

Rewriting to remove terms involving the variable, M_{gj}:

$$F_j = \frac{kM_{sj}F_j^*}{kM_{sj} + F_j^*}. \tag{66}$$

Equations (57) and (66) are expressions for feeding rate of the two major ecological modes of feeding by macroheterotrophs. By parameter variation they can be made to describe a wide variety of behavior by these types of organisms. For example, Equation (66) can be made to describe the feeding of an ambush predator by noting that the cruising velocity of such a predator is zero, which reduces the expression for the volumetric search rate to $a_{ij} = \pi s_{dj}^2 v_{ci}$. Predation on prey that do not actively swim to escape is described when $v_{fi} = 0$. An additional consideration is the feeding of macroheterotrophs in communities in which there is a wide disparity in sizes of the organisms present. Introduction of a factor in the numerator of both equations to account for the size window within which feeding occurs overcomes the potential problem for an operating model. Expressions can be obtained for mortality occurring in a population as a result of consumption by populations of filter and pursuit feeding predators, and for power used in moving about in the activities of feeding and escaping being fed upon.

Mortality Caused by Predation

The mortality rate experienced by a population is the sum of the mortalities caused by all the predators. The quantity, therefore, is obtained by summing the rates over all predators. One approach is to consider that the mortality caused by a predator population, j, on a prey population, i, is the total feeding rate of the predator multiplied by the fraction that the specific prey population comprises of the total prey encountered by the predator. The appropriate fraction, however, is not simply the prey population mass divided by total prey mass of all species. The appropriate expression for mortality by filter feeders is obtained by using the encounter radii, τ_{ij}, as weights, so that the fraction corresponds to the proportion of total encounters that occur with the prey population i:

$$f_{ij} = \frac{\tau_{ij}B_i}{\sum_k \tau_{kj}B_k} . \tag{67}$$

The mortality to i caused by j is

$$m_{ij} = f_{ij}F_j \tag{68}$$

and the mortality to i caused by all predators combined is

$$m_{if} = \sum_j f_{ij}F_j = B_i \sum_j \frac{k_j M_{sj}\tau_{ij}B_j}{\dfrac{k_j M_{sj}}{\pi v_{cj}} + \sum_k \tau_{kj}B_k} , \tag{69}$$

where m_{if} is the mortality rate to population i caused by filter feeding predators. This expression can also be arrived at by perhaps a more direct approach. Consider the expression for F_j, Eq. (57). This equation is a sum of n terms, any specific term of which refers to the portion of food that derives from a particular prey population. If an expression is formed that consists of the sum of specific terms for population i, one from each feeding rate expression (57), then Eq. (69) results directly.

For pursuit feeders essentially the same approach is used, except that encounter rate constants [Eq. (60)] are the appropriate weighting factors. The resulting expression is

$$m_{ip} = B_i \sum_j \frac{\dfrac{\dfrac{k_j M_{sj}}{s_{dj}}\sum_k \Delta v_{kj}}{s_{dj}^{-1}\sum_l \Delta v_{lj} + kM_{sj}} a_{ij}}{\dfrac{\dfrac{kM_{sj}}{s_{dj}}\sum_r \Delta v_{rj}}{s_{dj}^{-1}\sum_h \Delta v_{hj} + k_j M_{sj}} + \sum_u a_{uj}B_u} , \tag{70}$$

where m_{ip} is the mortality rate to population i caused by pursuit feeders.

Power Demand Resulting from Motile Activity

The power required to move around in a viscous fluid nominally is proportional to the third power of the velocity of movement:

$$P = 1/2 \, C_d \delta A v^3 , \tag{71}$$

where C_d is the drag coefficient, δ is the density of water, A is the appropriate area of the moving organism, and v is velocity. The drag coefficient, C_d, is not a constant, however, but is a complex function of velocity. In practice C_d has to be measured as a function of the velocity (or the Reynolds number) for each shape (13). This difficulty of operation will not be discussed further here. Instead, it is assumed that Eq. (71) is sufficient to develop the ideas for the model, thereby deferring this operational problem until the time when parameter values are sought for the model.

A heterotroph does not swim at constant speed, so its power demand is a variable. I assume that a heterotroph's time can be partitioned into four kinds of activities: searching, pursuing, escaping, and inactivity. (Filter feeders do not pursue, so for them this fraction is zero.) If every individual in the population behaves in this way, and the behavior is not synchronous, nor correlated time-wise, then an alternate view is that the fractions refer to the fractions of the population engaged in the four activities at any time. No account has been taken of the time fraction spent in reproductive activities as would be appropriate for higher organisms. There is, however, allowance for energy of reproduction in the cost of biosynthesis (see Partitioning of Food into Power Supply and Biosynthesis).

To obtain the fractions, I assume that an organism's first priority is self preservation. Therefore the fraction of time spent escaping predators can be calculated independently of all the other fractions, and the other fractions simply partition the remaining time. A given organism can be subject to predation by both filter feeders and pursuit feeders. The fraction of time spent escaping predators, in general, will be a function of both types of predation. The fraction of time remaining for feeding after escaping predators, in general, will alter feeding behavior, and Eqs. (57) and (66) must be correspondingly corrected to account for this effect.

Calculations of these fractions of time make use of the relationships already developed. When all ramifications are considered, however, while conceptually simple they are more tedious than is worthwhile for presentation in this report. Therefore they will not be presented in detail. Instead the general approach taken in obtaining the derivations will suffice.

The fraction of time spent in an activity is the time per unit (or event) of that activity times the rate of occurrence of that activity. Thus the fraction of time spent escaping predators is the time per escape times the encounter rate. (Naturally, the actual calculation is far more involved. For example, one complication is that the fraction of time spent fleeing is calculated as the difference between the fraction of time spent being pursued and the fraction of time spent being captured, because the power used by those that are captured is irrelevant to the surviving population.) After the fraction of time spent escaping, f_e, is accounted, the remaining time, f_r, is partitioned into feeding and inactivity. The expressions for feeding are corrected for reduction in available time by escaping predators by noting that stomach emptying occurs at the same rate regardless of the predation pressure, while feeding itself is limited to the remaining time. The power required for the activity is obtained by multiplying the velocity of swimming in each activity by the fraction of time spent in each activity and using Eq. (71) for organism j in the following form

$$P_j = 1/2 \ C_{dj} \delta A_j \sum_k (f_{kj} v_{kj})^3 , \tag{72}$$

where f_{kj} is the fraction of time spent in the kth activity by organism j, and v_{kj} is its velocity in that activity.

At this point power supply via feeding, power demand as a function of activity, and disposition of the elements comprising the food are all calculable for heterotrophs using the relationships developed. One additional aspect of metabolism needs to be considered for theoretical completeness, so that a closed system of equations can be developed to represent element cycling. That aspect is the use of different oxidizing agents as terminal electron acceptors in the energy metabolism carried out by all heterotrophs.

Energy Metabolism as a Function of Terminal Electron Acceptor

To represent organic chemicals in a general way as sources of materials for energy and synthesis, it is necessary to group them into categories. Criteria for establishing these categories include similarity of elemental composition, energy content, and the way in which organisms use them. These criteria do not specify the level of resolution, however, and in the final analysis the categories will be selected through experience with attempts to match several possible categorization schemes with the other model components to achieve the results with reasonable economy. An initial scheme is presented in Fig. 7. In this scheme the categories are particulate organic matter (POM), refractory organic matter (ROM), digestible organic matter (DOM), soluble organic matter (SOM), itself consisting of two components: biomass-like monomers (BOM) and non-biomass-like monomers (NOM). POM receives input from organisms deaths, molts, egesta, etc. DOM is readily hydrolyzed by exoenzymes of microorganisms into SOM, whereas ROM is only very slowly solubilized. SOM is absorbed by the organisms, and separation into BOM and NOM occurs. This categorization scheme reflects the mode of biological utilization more explicitly than similarity of composition or energy content. It is possible that for the latter criteria additional categories will be required. It might become necessary, for example, to represent CH_4, acetate, or other specific categories of chemicals that are important in microbial systems for the model to reflect certain aspects of their dynamics.

As organisms oxidize organic chemicals for energy, there is a sequence of utilization of electron acceptors that corresponds to the variation in redox potential for the reactions (Mc Carty, 1971; Stumm and Morgan, 1981). That is, a preferred electron acceptor is used until it is depleted, then the next preferred form is used, and so on. O_2 is used first, followed by NO_3^{-1} and NO_2^{-1}, SO_4^{-2}, and CO_2. Other oxidants are also used, such as other forms of sulfur. The interesting point of this sequence is that the preferred sequence is in the order of the energy released in the reactions. Oxidations of organic compounds using the more preferred electron acceptors result in greater energy yield than reactions using the less preferred. It is unlikely that this reflects any chemical necessity (Mc Carty, 1971), but more likely reflects the competitive advantage accruing to organisms that use the more highly productive energy sources. This sequence of reactions will be expressed wherever transport or regeneration of electron acceptors is slower than their use by organic decomposition. In natural systems, it is expressed temporally in highly eutrophic systems and spatially as vertical stratification in systems with highly organic sediments, such as wetlands and many water bodies.

The use of oxidants in sequence can be represented by the assumption that the strongest oxidant present is used. The problem that then remains is that of calculating the energy yield appropriate to the oxidant that is being used. One direct approach to this problem is that suggested as appropriate by McCarty (1971); the use of half reactions and equivalent weights as described in elementary chemistry texts, and as is commonly used in biochemical calculations (Lehninger, 1982; Chap. 17). If this approach is taken the standard free energies of the half reaction of their reduction (at pH 7, and unit activity for the other reactants), per equivalent weight of the oxidants is in the order corresponding to the sequence of their utilization. An exception to this is nitrite, whose value for $\Delta G^\circ(w)$ is higher than that of O_2. It is not clear whether this apparent reversal in tendency is a consequence of adaptation of the organisms that use nitrite as an electron acceptor or of the peculiar relationship of nitrite to nitrate. That is, nitrite is formed as a product of nitrate reduction and is available as a reactant, only after nitrate is reduced.

If the scheme for representing organic matter is to be used effectively, an accounting algorithm would necessarily be applied to ensure conservation of mass. Conservation of mass in an element cycling model, however, would be applied separately to each of the elements. This could be accomplished by accounting the concentration of the elements in each of the categories of organic matter. The categories can be treated in many respects as a single type of molecule with mole numbers proportional to the concentrations of the elements. (Except that for computations using the Gibbs free energy, it probably should not be assumed that the concentration of the molecule as a reactant is that of the whole pool. Further work is needed on this problem.)

To calculate the energy yield from reaction with the various oxidizing agents requires an additional algorithm. The oxidants are represented as half reactions in which an electron equivalent of the oxidant plus an electron (hence electron acceptor) yields a reduced form of the element. A complementary half reaction is required for the electron donor, a "molecule" from one of the categories of organic matter. The appropriate free energy for this half reaction is approximated by the heat of combustion with oxygen so that the energy of the pools can be accounted dynamically. To calculate the energy yield for reaction with another oxidizing agent, the reaction with oxygen is replaced by an equation for the appropriate reaction by first removing the half reaction for oxygen and then substituting the half reaction for the oxidizing agent of interest.

Conclusion

This work was carried out as an effort to provide a biophysical basis for modeling element cycling. Part of this effort was to find a simple organizational scheme that, on the one hand, would be of aid in finding the appropriate fundamental levels of resolution, and that, on the other, would transcend the primary level of description (population, chemical pool, etc.). In the first aspect of the scheme each of the chemical transformations necessary for complete element cycles is assumed to be catalyzed by organisms with the appropriate biochemical mode of energy metabolism. It is not apparent that macroheterotrophs are necessary to this scheme. Evidently the first aspect of the scheme alone is insufficient, because it does not require higher organisms. The first aspect of the scheme essentially consists of defining the organisms necessary to carry out element cycles. Considering element cycling from another viewpoint, however, one could ask what other kinds of organisms could be supported by complete element cycles. No other kinds of metabolism could be supported other than those that can derive energy via the electron transferring redox reactions without invoking some alien biochemical mode. Therefore the possibilities for other kinds of organisms are limited to variation of some of these modes (e.g., increased structural complexity) in which the variants are more efficient in some circumstances. In summary, the whole scheme consists of: element cycles, organisms necessary to completely catalyze all the component reactions, and higher organisms as structurally complex systems and as subsystems of more complex ecosystems, all to the degree of complexity supportable by the system power input and rate that elements are made available by the biotic system itself. At the simplest, elements cycle between oxidized and reduced forms. Organisms catalyze the redox reactions that change the reduced forms to oxidized forms, and vice versa, and in doing so derive the energy that drives life processes. These biotically catalyzed redox reactions are the fundamental transformations of element cycles. Life processes consist of mechanical and chemical work and are therefore power demanding. Mechanical work is associated more with higher organisms and consists of the activities of feeding, escaping predators, reproducing, etc. Chemical work is that of tissue repair, new biomass production, and carrying out other basic biochemical processes. Biomass production requires elements in rather tightly controlled stoichiometric ratio. In the limit organisms that persist are those that achieve a balance between their power supply and demand and that are able to obtain elements for biosynthesis at rates that match their rates of loss via deaths, excretion, emigration, etc. Comparison of power demand, power supply, and rates of obtaining elements as substrate for biosynthesis for the various modes of metabolism permit the calculation of population growth rates. Thus the foundation is laid for further consideration of element cycling rates via a system of coupled equations in which the equations represent populations of organisms carrying out the element cycle transformations.

Acknowledgements

I thank the following individuals and institutions without whose help and support I could not have made progress on this project: Dr. James Cooley, Director of the Institute of Ecology, University of Georgia for providing facilities and other support for the course of this research; Dr. Morris Levin, project officer for this project which was funded under the U.S. Environmental Protection Agency's Innovative Research Program; Dr. David Duttweiler, former director of the Athens Environmental Research Laboratory whose encouragement and support provided the initial impetus for the project; and Mr. Robert Hermann, graduate student in ecology whose willing assistance and participation in tiring discussions contributed significantly to whatever degree of continuity and coherence that there is in this chapter.

References

Allen, T.F.H., Starr, T.E. (1982). Hierarchy: Perspectives for Ecological Complexity. Univ. of Chicago Press, Chicago, Ill. xiv + 310 p

Bannister, T.T. (1974). Production equations in terms of chlorophyll concentration, quantum yield, and upper limit to production. Limnol. Oceanogr. *19*: 1–12

Burton, K. (1958). Energy of adenosine triphosphate. Nature *181*: 1594–1595

Clayton, R.K. (1980). Photosynthesis: Physical Mechanisms and Chemical Patterns. Cambridge Univ. Press, New York, 281 p.

Confer, J.L., Howick, G.L., Corzette, M.H., Kramer, S.L., Fitzgibbon, S., Landesberg, R. (1978). Visual predation by planktivores. Oikos *31*: 27–37

Doelle, H.W. (1975). Bacterial Metabolism. 2nd ed. Academic Press, New York, 738 p.

Dubinsky, Z., Berman, T. (1976). Light utilization efficiencies of phytoplankton in Lake Kinneret (Sea of Galilee). Limnol. Oceanogr. *21*: 226–230

Fenchel, T., Blackburn, T.H. (1979). Bacteria and Mineral Cycling. Academic Press, London, 225 p.

Gavis, J., Ferguson, J.T. (1975). Kinetics of carbon dioxide uptake by phytoplankton at high pH. Limnol. Oceanogr. *20*: 211–221

Gerritsen, J., Strickler, J.R. (1977). Encounter probabilities and community structure in zooplankton: a mathematical model. J. Fish. Res. Board Can. *34*: 73–82

Gunderson, K., Mountain, C.W. (1976). Oxygen utilization and pH change in the ocean resulting from biological nitrate formation. Deep Sea Res. *20*: 1083–1091

Holling, C.S. (1959). Some characteristics of simple types of predation and parasitism. Can. Entom. *41*: 385–398

Ivlev, V.S. (1961). Experimental Ecology of the Feeding of Fishes. English trans. by D. Scott. Yale Univ. Press, New Haven, 302 p.

Janssen, J. (1978). Feeding modes and prey size selection in the alewife (*Alosa pseudoharengus*). J. Fish. Res. Board Can. *33*: 1972–1975

Lehninger, A.L. (1975). Biochemistry. The Molecular Basis of Cell Structure and Function, 2nd. ed. Worth Publishers, Inc. New York, 1104 p.

Lehninger, A.L. (1982). Principles of Biochemistry. Worth Publishers, Inc. New York, 1011 p.

Lovelock, J.E., Margulis, L. (1974). Atmospheric homeostasis by and for the biosphere: the gaia hypothesis. Tellus *26*: 2–10

McCarty, P.L. (1971). Energetics and bacterial growth. In: Organic Compounds in Aquatic Environments. S.J. Faust and J.W. Hunter, eds., Marcel Dekker, Inc. New York, p. 495–529

Monod, J. (1942). Recherches sur la croissance des cultures bacteriennes. Hermann et Cie, Paris

Morowitz, H.J. (1968). Energy Flow in Biology, Biological Organization as a Problem in Thermal Physics. Academic Press, London, 179 p.

Nicolis, G., Prigogine, I. (1977). Self-Organization in Nonequilibrium Systems. From Dissipative Structure to Order through Fluctuations. John Wiley and Sons, New York, 491 p.

O'brien, J. (1979). The predator-prey interaction of planktivorous fish and zooplankton. Am. Sci. *67*: 572–581

Pasciak, W.J., Gavis, J. (1974). Transport limitiation of nutrient uptake in phytoplankton. Limnol. Oceanogr. *19*: 881–888

Pasciak, W.J., Gavis, J. (1975). Transport limited nutrient uptake rates in *Ditylum brightwelli*. Limnol. Oceanogr. *20*: 604–617

Purcell, B.M. (1977). Life at low Reynolds numbers. Am. J. Phys. *45*: 3–11

Rashevsky, N.F. (1959). Some remarks on the mathematical theory of nutrition of fishes. Bull. Math. Biophys. *21*: 161–183

Schlegel, H.G. (1975). Mechanisms of chemo-autotrophy. In: Marine Ecology. A Comprehensive, Integrated Treatise on Life in Oceans and Coastal Waters. O. Kinne, ed., John Wiley and Sons, New York, p. 9–60

Steele, J.H. (1962). Environmental control of photosynthesis in the sea. Limnol. Oceanogr. *7*: 137–150

Stumm, W., Morgan, J.J. (1981). Aquatic Chemistry. An Introduction Emphasizing Chemical Equilibria in Natural Waters. John Wiley and Sons, New York, 780 p.

Vogel, S. (1981). Life in Moving Fluids. The Physical Biology of Flow. Willard Grant Press, Boston, 352 p.

Werner, E.E. (1974). The fish size, prey size, handling time relation in several sunfishes and some implications. J. Fish. Res. Board Can. *31*: 1531–1536

Werner, E.E., Hall, D.J. (1974). Optimal foraging and the size selection of prey by the bluegill sunfish (*Lepomis macrochirus*). Ecology *55*: 1042–1052

Part V. Applied Mathematical Ecology

Bioeconomics and the Management
of Renewable Resources

Jon M. Conrad

1. Bioeconomics

Within the past decade many economists have become interested in natural
resource models which simultaneously consider economic flows (such as cost and
revenue) and population dynamics. Resource management is often cast as a
problem in dynamic optimization where the management objective may be to
maximize the present value of net benefits subject to the stock adjustments which
result from growth, natural mortality, and man's harvesting activities. When the
resource in question is a plant or animal, capable of regeneration, these resource
models are called *bioeconomic models.*

The basic bioeconomic model assumes that the renewable resource in question
can be adequately described by a single (state) variable measuring biomass; for
example, pounds or metric tons of fish. While such models have the advantages of
simplicity and mathematical tractability, they cannot take into account age or sex
related attributes, nor multispecies interactions. In spite of such shortcomings, the
basic model is a useful vehicle to introduce various biological and economic
concepts.

With the resource stock described by a single variable, we will characterize the
change in the resource by the difference equation

$$X_{t+1} - X_t = F(X_t), \tag{1}$$

where X_t denotes the resource stock (biomass) in period t. Equation (1) implies that
the change in the stock from period t to period $t+1$ is dependent on the stock in
period t. The function $F(X_t)$ will reflect factors affecting net growth of the resource
and environmental carrying capacity. A famous specification for $F(X_t)$ is the
logistic growth function which takes the form

$$X_{t+1} - X_t = rX_t\left(1 - \frac{X_t}{K}\right), \tag{2}$$

where r is a positive constant referred to as the intrinsic rate of growth and K, also a
positive constant, is the environmental carrying capacity. The logistic function is a
symmetric function with roots at $X_t = 0$ and $X_t = K$, and with a maximum
sustainable yield at $X_t = K/2$ (see Fig. 1).

In a "pristine" system, undisturbed by man's harvesting activities, a species
subject to the dynamics of the *continuous* version of the logistic growth function

Biomathematics, Vol. 17, Mathematical Ecology
Edited by T. G. Hallam and S. A. Levin
© Springer-Verlag Berlin Heidelberg 1986

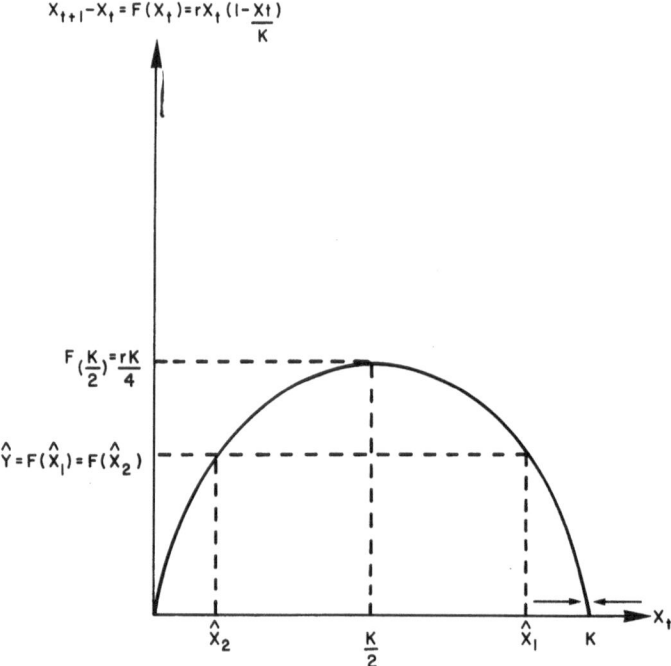

Fig. 1. The logistic growth function

would tend toward the stable equilibrium at K. In the discrete time (difference equation) given in (2), the dynamic behavior can be more complicated. If r, the intrinsic growth rate, is large, nonconvergent "overshoot" (or oscillatory) behavior can occur. This is shown in Figs. 2a–f. For Eq. (2), May (1975) has determined:

(a) If $0 < r \leq 1$, the population steadily approaches K without overshooting it (e.g., Fig. 2a and b),

(b) If $1 < r \leq 2$, the population overshoots but undergoes damped oscillations as it approaches K (e.g., Fig. 2c and d),

(c) If $2 < r \leq 2.449$, then the population settles down to a two-point cycle (e.g., Fig. 2e),

(d) If $2.449 < r \leq 2.570$, the population achieves a stable cycle with 2^n points, $n > 1$; the value of n depending on r, and

(e) If $r > 2.570$ – CHAOS – population size varies in a wholly unpredictable manner with different outcomes resulting from different values of X_0 (e.g., Fig. 2f).

There are many alternative specifications for the function $F(X_t)$. The logistic function belongs to a family of functions that is said to be "purely compensatory" and which generate a smooth and continuous yield response when the species is subject to commercial exploitation by man. The alternatives to purely compensatory models are models which exhibit depensation or critical depensation. These models will not yield smooth yield functions and admit the possibility of collapse and species extinction. Space precludes a discussion of these models and the interested reader is directed to Clark (1976).

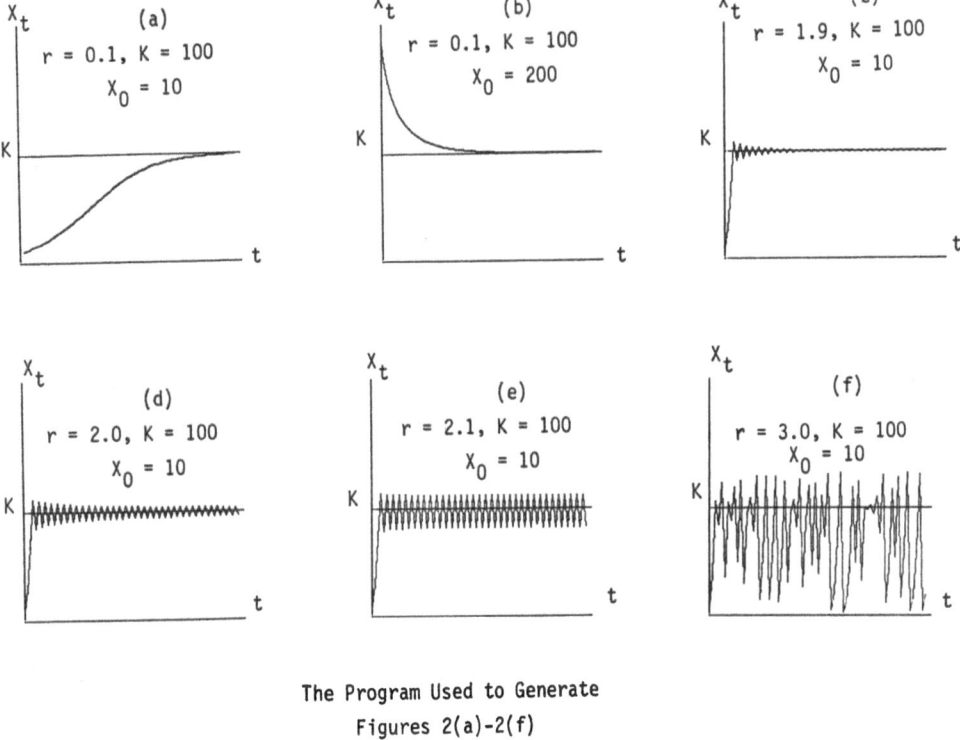

The Program Used to Generate
Figures 2(a)-2(f)

```
10:DIM X(71)
20:INPUT "r=";R
30:INPUT "K=";K
40:INPUT "X(0)=";
   X(0)
50:FOR T=0TO 69
60:X(T+1)=X(T)+R*
   X(T)*(1-X(T)/K
   )
70:NEXT T
80:GRAPH
90:COLOR 0
100:LINE (0,0)-(0,
    220)
110:LINE (0,0)-(22
    0,0)
120:LINE (0,K)-(22
    0,K)
130:COLOR 3
140:FOR T=0TO 69
150:LINE (3*T,X(T)
    )-(3*(T+1),X(T
    +1))
160:NEXT T
170:COLOR 0
180:END
```

Fig. 2a–f. Overshoot in the difference equation $X_{t+1}=X_t+rX_t(1-X_t/K)$

In summary, difference equation (1) describes the change in the resource stock for a species not commercially exploited by man. It is used to characterize the population dynamics of the species in its pristine, natural state. Commercial exploitation by man requires a modification of Eq. (1) to account for man's harvesting activities.

A production function defines the maximum level of output obtainable from a given bundle of inputs. It is the economist's way of characterizing the technology whereby inputs produce output. In a single species fishery the output from commercial fishing would be catch or yield denoted by Y_t. The bundle of manmade inputs utilized in catching fish are assumed capable of aggregation into a single input variable called "effort" and denoted by E_t. Fishing effort is directed at the fish stock X_t and results in a yield Y_t according to

$$Y_t = H(E_t, X_t),\tag{3}$$

where $H(E_t, X_t)$ is the production function for the fishery. The partial derivatives of $H(\cdot)$ are referred to as marginal products and assumed positive. If catch per unit effort is proportional to the fish stock one obtains the production function

$$Y_t = qE_tX_t,\tag{4}$$

where q is called the catchability coefficient. This production function underlies much of the early work in fisheries economics and has been studied extensively by Schaefer (1957).

When a fishery comes under commercial exploitation the equation describing population dynamics is modified to

$$X_{t+1} - X_t = F(X_t) - H(E_t, X_t).\tag{5}$$

In words: yield is deducted from the natural change in biomass to determine the net change in the fish stock. Attention is often focused on harvesting regimes that are sustainable in perpetuity. This will usually require attainment of a *steady state equilibrium* where $X_t = X$, $E_t = E$, and $Y_t = Y$ for all future periods. In such a equilibrium the left-hand side of Eq. (5) equals zero and

$$H(E, X) = F(X).\tag{6}$$

For the logistic function and the catch per unit effort production function (collectively referred to as the Gordon-Schaefer model), this implies:

$$qEX = rX\left(1 - \frac{X}{K}\right).\tag{7}$$

If one were to solve Eq. (7) for X as a function E, and multiply both sides by qE one would obtain a *yield-effort curve* expressing catch as a function of effort which for the Gordon-Schaefer model becomes

$$Y = Y(E) = qKE\left(1 - \frac{qE}{r}\right).\tag{8}$$

For our purpose we can arbitrarily set $q = 1$ and graph the resulting yield-effort curve in Fig. 3. It is also a symmetric curve with roots at $E = 0$ and $E = r$ and a maximum yield at $E = \frac{r}{2}$. As noted earlier, the compensatory nature of the logistic function results in a smooth continuous yield-effort curve where incremental changes in effort result in incremental changes in yield. This is in contrast to yield-

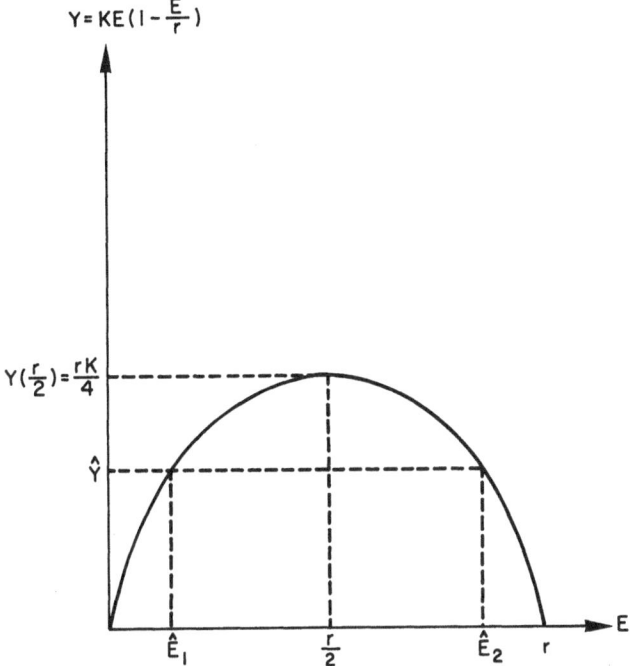

Fig. 3. The yield-effort curve for the Gordon-Schaefer model with $q=1$

effort curves where an incremental increase in effort may result in a fishery collapse with yield plummeting to zero.

To summarize, the yield-effort curve was derived for a fishery in steady-state equilibrium and was determined by the growth and production functions that underlie the fishery. The reader should note in Fig. 3 that any positive yield level less than $Y\left(\dfrac{r}{2}\right) = \dfrac{rK}{4}$ can be generated by two levels of effort. For example, $\hat{Y}$ can be obtained with effort $\hat{E}_1$, and stock $\hat{X}_1$ or with effort $\hat{E}_2$ and stock $\hat{X}_2$ where $\hat{E}_2 > \hat{E}_1$ and $\hat{X}_1 > \hat{X}_2$. (See Fig. 1 as well.) The level of effort, stock, and yield which emerge in an open access fishery and under an optimally managed fishery are discussed in the next two sections.

2. The Open Access Fishery

In an open access fishery, where the fish stock is treated as a common property resource, fisherman and vessels will enter the industry until profits are driven to zero. Suppose the cost of a unit of fishing effort is constant and denoted by the letter c. If E_t units of effort are directed at the fish stock in period t then the total cost in period t would be given by

$$C_t = cE_t. \tag{9}$$

Suppose further that the price per unit received by fishermen upon landing their catch is denoted by the letter p. Then total revenue would equal $pY(E_t)$ and for the Gordon-Schaefer model with $q=1$ we obtain:

$$R_t = pY(E_t) = pKE_t\left(1 - \frac{E_t}{r}\right).$$ (10)

With p a given positive constant the revenue curve would look identical to the yield-effort curve; only the scale of the vertical axis will have changed to measure revenues in dollars.

Profits or fishery rents are defined as

$$N_t = R_t - C_t = pY(E_t) - cE_t.$$ (11)

Under open access fishermen will enter until fishery rents are "dissipated" or driven to zero at which point $R_t = C_t$ or

$$\frac{Y(E_t)}{E_t} = \frac{c}{p}.$$ (12)

That is, yield per unit effort is equal to the cost/price ratio. Graphically this situation is portrayed in Fig. 4. The cost curve is shown as a ray from the origin with slope c while the revenue curve, similar in shape to the yield-effort curve, is shown as a quadratic $R = pY(E)$. Revenue equals cost, resulting in zero profits, at E_∞. The significance of the subscript ∞ will be explained shortly.

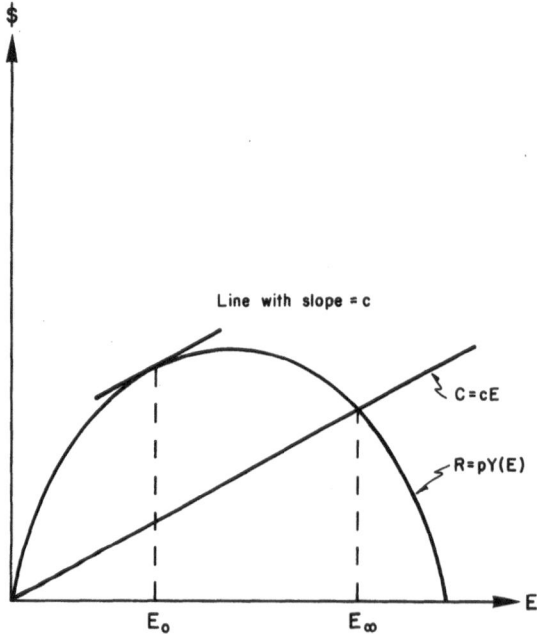

Fig. 4. Open access equilibrium and maximum sustainable economic rent

In an open access fishery E_∞ will often be associated with excessive capital, low fish stocks, and low yield. Open access stock denoted X_∞, may be determined for the Gordon-Schaefer model by noting that when $Y_t = qE_tX_t$ and $C_t = cE_t$ that

$$C_t = \frac{cY_t}{qX_t}. \tag{13}$$

Equation (13) is referred to as a cost function, which in this instance is *linear* in Y_t. The zero profit or rent conditions may be restated as

$$N_t = R_t - C_t = pY_t - \frac{cY_t}{qX_t} = 0 \tag{14}$$

and for $Y_t > 0$ this implies

$$X_\infty = \frac{c}{pq}. \tag{15}$$

After explaining how a competitively exploited open access fishery will tend toward overcapitalization, depleted stocks, and low yields, early fisheries economists suggested that the fishery should be managed so as to maximize *sustainable economic rent*; that is, set effort so as to

$$\max N_t = pY(E_t) - cE_t. \tag{16}$$

This situation is achieved when

$$pY'(E_t) = c \tag{17}$$

or in words, when marginal value product equals marginal cost [note: $Y'(E_t)$ is the derivative of the yield effort curve and $pY'(E_t)$ is marginal value product]. This situation is shown graphically in Fig. 4, where the level of effort which maximizes sustainable economic rent is denoted by E_0 and is determined by the tangency of a line with slope c and the revenue curve $R = pY(E)$.

As it turns out *neither* E_0 nor E_∞ will be optimal for a society with a positive but finite rate of time preference. The social rate of time preference, or society's discount rate, is the result of individuals' preferences for income now as compared to the same amount of income at a later date. To get a typical individual to give up (invest) current income today you usually need to repay that income plus a premium or interest payment at a later date. Society, comprised of many such individuals, will typically reveal a collective preference for income (or fish) today as compared to an equivalent amount of income (or fish) tomorrow.

The concept of social time preference or discounting raises numerous technical and ethical issues beyond the scope of these lectures. For our purpose it will be assumed that some positive but finite rate of discount is appropriate when society is considering how to allocate its resources over time. Within the simple fishery model developed thus far, the open access equilibrium level of effort will be optimal when society's rate of discount is infinite. The maximum sustainable rent level of effort will be socially optimal when society's rate of discount is zero. For a positive but finite rate of discount the optimal level of fishing effort will usually lie somewhere between E_0 and E_∞. Thus the open access and sustainable rent maximizing levels of effort, while in general *not* optimal, do serve to bracket the socially optimal level of effort which will depend on cost, price, the parameters of the growth and the production functions, *and* the discount rate. Determination of the optimal level for fish stock, yield, and effort will be examined next.

3. The Optimally Managed Fishery

A convenient persona for inquiring into the optimal management of a fishery is the sole owner. The sole owner has *exclusive* harvesting rights to a resource and it is possible to formulate a management problem for the sole owner which conforms to how the fishery should be managed from a social point of view.

Suppose δ represents the social rate of discount. Then $N(1+\delta)$ would be the amount of money one would have to pay an individual to obtain a loan of N dollars for one period (say, a year). By analogous reasoning, the value of a note promising to pay N dollars one year from now would be *discounted* to a value $\dfrac{N}{(1+\delta)}$. From another point of view δ may be thought of as the opportunity cost of investment or capital funds. It is the amount of money which could be earned on a dollar invested *elsewhere* in the economy. A single net cash payment N_t realized at period t in the future could be expressed as a *present value* equal to $\varrho^t N_t$, where $\varrho = \dfrac{1}{(1+\delta)}$ is referred to as a discount factor. The present value of a *stream* of net cash flows over the interval $[0, T]$ would be calculated according to

$$N = N_0 + \frac{N_1}{(1+\delta)} + \frac{N_2}{(1+\delta)^2} + \ldots + \frac{N_T}{(1+\delta)^T} = \sum_{t=0}^{T} \varrho^t N_t. \tag{18}$$

Suppose the sole owner is a price taker in that his level of fishing effort and yield do not affect per unit cost c or the per unit price p. In this case the cost function [Eq. (13)] may be written in the more general form

$$C_t = \frac{cY_t}{qX_t} = c(X_t)Y_t, \tag{19}$$

where $c(X_t)$ is a stock dependent average cost function. Since larger stocks can be expected to reduce average harvest costs the first derivative of $c(X_t)$, denoted as $c'(X_t)$, will be negative. The net revenues from harvest Y_t may be written as

$$N_t = pY_t - c(X_t)Y_t = [p - c(X_t)]Y_t \tag{20}$$

and the present value of all future net revenues becomes:

$$N = \sum_{t=0}^{\infty} \varrho^t[p-c(X_t)]Y_t. \tag{21}$$

A logical objective for the sole owner (and in this instance a desirable one for society as well) would be to maximize the present value of net revenues subject to the equation describing the change in the fish stock which may be rewritten as:

$$X_{t+1} = X_t + F(X_t) - Y_t. \tag{22}$$

In this form the sole owner's management problem is an example of a broader class of problems referred to as control problems. Specifically, the sole owner seeks to control the fish stock through harvesting so as to maximize the present value of net revenues. A solution to this problem may be achieved by an extension of the method of Lagrange multipliers. This method introduces a set of artificial variables (called multipliers) and adds the product of each constraint with its multiplier to the objective functional to form a Lagrangian expression. For the management problem confronting our sole owner, the Lagrangian expression takes the form:

$$L = \sum_{t=0}^{\infty} \varrho^t\{[p-c(X_t)]Y_t + \varrho\lambda_{t+1}[X_t + F(X_t) - Y_t - X_{t+1}]\}, \tag{23}$$

where λ_{t+1} is the Lagrangian multiplier associated with the constraint which defines X_{t+1}. Because the Lagrangian multiplier will indicate the value of an additional unit of the fish stock in period $t+1$, and given that we wish to maximize the *present* net value represented by L, λ_{t+1} is premultiplied by ϱ to yield the term $\varrho^{t+1}\lambda_{t+1}$ which can be interpreted as the present value of an additional unit of the fish stock in period $t+1$.

Necessary conditions for a maximum require that the partial derivatives of L be set equal to zero, demanding that

$$\frac{\partial L}{\partial Y_t} = \varrho^t\{[p-c(X_t)] - \varrho\lambda_{t+1}\} = 0, \tag{24}$$

$$\frac{\partial L}{\partial X_t} = \varrho^t\{-c'(X_t)Y_t + \varrho\lambda_{t+1}[1+F'(X_t)]\} - \varrho^t\lambda_t = 0, \tag{25}$$

$$\frac{\partial L}{\partial \lambda_{t+1}} = \varrho^{t+1}\{X_t + F(X_t) - Y_t - X_{t+1}\} = 0. \tag{26}$$

These conditions may be simplified somewhat and rewritten as:

$$p = c(X_t) + \varrho\lambda_{t+1}, \tag{27}$$

$$\lambda_t = -c'(X_t)Y_t + \varrho\lambda_{t+1}[1+F'(X_t)], \tag{28}$$

$$X_{t+1} = X_t + F(X_t) - Y_t. \tag{29}$$

Equation (27) requires that the marginal value of a fish harvested today, (p), equal the sum of marginal harvesting cost and *user* cost, $(c(X_t) + \varrho\lambda_{t+1})$. This latter term represents the discounted value of an additional unit of the resource *next period*. Thus, the marginal condition governing the harvest of the fish stock requires a balancing of current market value with harvest and user costs.

The Lagrange multiplier, (λ_t), may be interpreted as the marginal value of an additional unit of the fish stock in period t. It is sometimes referred to as a *shadow price*. Equation (28) requires that stock be maintained so that the shadow price in period t equals the sum of marginal stock induced reductions in average costs [note $-c'(X_t) > 0$] plus the discounted future value of a unit of the resource plus growth in period $t+1$. Thus, the current shadow price must equal current marginal stock induced cost savings plus the discounted value of an additional unit and associated growth in the next period.

In deriving the yield-effort curve we defined a steady-state as an equilibrium for a dynamic system where variables within the system are unchanging through time. A steady-state for the fishery under sole ownership would occur when $X_t = X$, $\lambda_t = \lambda$, and $Y_t = Y$ for all future t (note $E_t = E$ is also unchanging). While an unchanging physical or economic environment is unusual for any protracted period of time, the concept of a steady-state equilibrium is useful in identifying the long term effects of a change in important biological or economic parameters. In more complex, adaptive models of fishery management, short run decisions might be based on prevailing estimates of a steady-state equilibrium, with later decisions based on updated or revised estimates of steady-state. Thus the concept is fundamental to many stochastic model as well.

In steady-state, Eqs. (27)–(29) become a system of three equations in three unknowns; specifically:

$$p = c(X) + \varrho\lambda, \tag{30}$$

$$\lambda(1 - \varrho[1 + F'(X)]) = -c'(X)Y, \tag{31}$$

$$Y = F(X). \tag{32}$$

The left-hand side (LHS) of Eq. (31) can be manipulated to become:

$$\varrho\lambda(1 + \delta - 1 - F'(X)) = \varrho\lambda(\delta - F'(X)).$$

From Eq. (30) we note: $\varrho\lambda = p - c(X)$. Substituting into the LHS of (31) and solving for Y yields:

$$Y = \frac{[F'(X) - \delta][p - c(X)]}{c'(X)} = \phi(X). \tag{33}$$

Equation (33) is referred to as the catch locus (see Gould, 1972). In the positive orthant $(X > 0, Y > 0)$ the catch locus will typically have a positive slope $\left(\dfrac{dY}{dX} > 0\right)$ and may be graphed along with the growth function $Y = F(X)$ to identify the steady-state optimum. The precise position and shape of the catch locus will

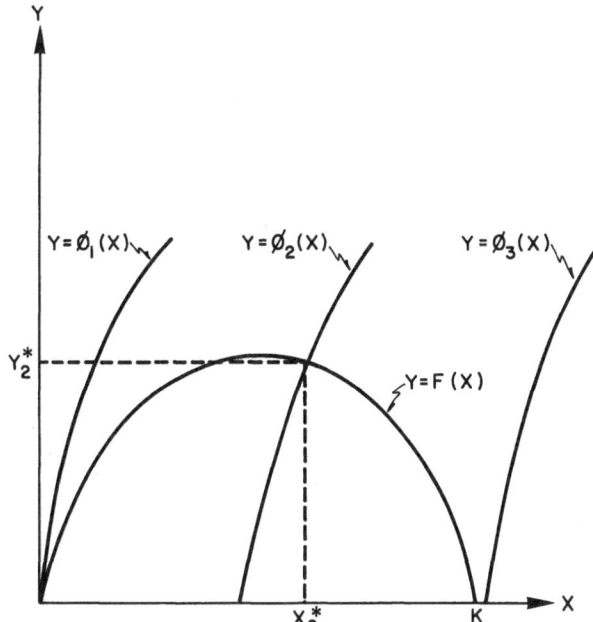

Fig. 5. Catch loci, the growth function and steady-state optimum

depend on the various bioeconomic parameters used to specify $F(X)$ and $c(X)$ as well as p and δ. Three situations are shown in Fig. 5.

The intersection of catch locus one ($\phi_1(X)$) and the growth function occurs at $X=0$, $Y=0$ implying that on purely economic grounds it is optimal to harvest the resource to extinction. Such a situation might arise if the intrinsic growth rate for the species is relatively low, the discount rate is high, and the cost of harvesting the last unit of the resource is finite and less than the market price ($c(0)<p$). Clark (1973) discusses in greater detail the situation where commercial harvesting can lead to extinction.

In the second case the intersection of $\phi_2(X)$ and $F(X)$ occurs at (X_2^*, Y_2^*). In this situation the optimal stock actually exceeds $X_{msy}=K/2$. Such a stock level might be justified on the basis of cost savings which result from higher stock levels.

In the third case the intersection of $\phi_3(X)$ and the X-axis occurs to the right of K (the species environmental maximum). In this instance the cost of harvesting is so high relative to the market price that it is *uneconomic* to commercially harvest the population, and the species goes unexploited by man. The majority of fish species would be described by this latter case.

The catch locus and growth function are simply a system of two equations in two unknowns. One could substitute $Y=F(X)$ into Eq. (33) to obtain a single equation for the optimal steady-state stock. This equation is commonly written as:

$$F'(X)-\frac{c'(X)F(X)}{[p-c(X)]}=\delta \tag{34}$$

[see, for instance, Clark (1976), p. 40] and requires that the steady-state stock equate the sum of the marginal growth rate and stock effect to the discount rate.

Thus, in steady-state the stock which is being maintained is providing returns in the fishery (in the form of growth and cost savings) which precisely equal the rate of return obtainable on other capital assets elsewhere in the economy (equal to δ). With the exception of extinction or no commercial exploitation, Eq. (34) can be solved for the optimal stock $X^* > 0$.

Up to this point, we have confined our discussion to steady-state equilibrium and have ignored the issue of *short run dynamics*. If the initial stock X_0 is not equal to the optimal stock X^*, what should be the optimal *approach path* for X_t? In general, the approach path will be asymptotic; that is, a more or less gradual approach to equilibrium with the $\lim_{t \to \infty} X_t \to X^*$. Under certain circumstances, the approach to equilibrium is optimal if it is *most rapid*. Spence (1974) has shown that a most rapid approach path (MRAP) is optimal if the objective function may be written as a quasi-concave and separable function of X_t and X_{t+1}. The objective of the sole owner was to maximize the present value of net revenues [see Eq. (21)]. The expression for net revenues in period t was

$$N_t = [p - c(X_t)] Y_t. \tag{35}$$

Solving the difference equation

$$X_{t+1} - X_t = F(X_t) - Y_t, \tag{36}$$

for Y_t and substituting into Eq. (35) we obtain what Spence and Starrett (1975) call the *derived utility function*:

$$N_t = W(X_t, X_{t+1}) = [p - c(X_t)] [F(X_t) + X_t - X_{t+1}]. \tag{37}$$

If $W(X_t, X_{t+1})$ is additively separable in X_t and X_{t+1}, it may be written as

$$W(X_t, X_{t+1}) = A(X_t) + B(X_{t+1}), \tag{38}$$

and the present value of net benefits may be written as:

$$N = A(X_0) + \sum_{t=1}^{\infty} \varrho^t V(X_t). \tag{39}$$

If $V(X_t)$ is quasi-concave MRAP is optimal.

Spence and Starrett (1975) discuss several problems where the above separability and quasi-concavity conditions may hold. For the fishery management problem, harvesting along a MRAP would be determined by the relationship of X_t to X^* such that

$$\begin{aligned} &\text{if } X_t > X^*, \quad Y_t = Y_{\max}, \\ &\text{if } X_t < X^*, \quad Y_t = 0, \\ &\text{if } X_t = X^*, \quad Y_t = Y^*. \end{aligned} \tag{40}$$

---

* Quasi-concavity requires

$$V(\alpha \underline{X}_t + (1 - \alpha) \bar{X}_t \geq \text{MIN}[V(\underline{X}_t), V(\bar{X}_t)]$$

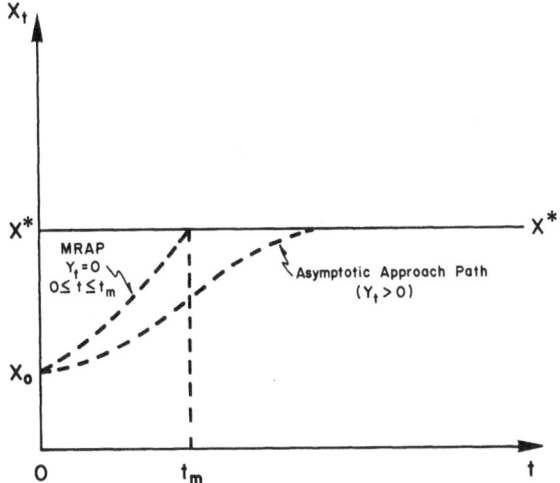

Fig. 6. Most rapid and asymptotic approach paths to the optimal stock X^*

Thus, if the initial stock where in excess of the optimal stock, one would wish to reduce it to X^* as rapidly as possible by harvesting fish at the maximum Y_{max}. If the initial stock were less than the optimal stock, one would impose a moratorium ($Y_t = 0$) until the stock reached X^*. If, by happenstance, the initial stock equalled the optimal stock, one would commence harvesting at $Y^* = F(X^*)$ and stay there forever.

Figure 6 shows the difference between the most rapid approach path (MRAP) and the asymptotic approach path for $X_0 < X^*$. Along the MRAP, $Y_t = 0$ for $0 < t < t_m$, until $X_t = X^*$ at $t = t_m$, at which time $Y_t = Y^* = F(X^*)$ for $t > t_m$. Along the asymptotic approach path, harvest is positive ($Y_t > 0$), but less than net natural growth, allowing the stock to slowly approach X^*: thus, as $t \to \infty$, $X_t \to X^*$, and $Y_t \to Y^*$.

Most rapid approach paths are relatively easy to calculate [one calculates $X_{t+1} = X_t + F(X_t) - Y_t$, where $Y_t = Y_{max}$ or $Y_t = 0$, until $X_{t+1} = X^*$]. Asymptotic approach paths are more difficult and would require the introduction of a transversality condition and solution of a two point boundary problem. A discussion of these concepts and procedures is beyond the scope of these lectures.

4. Management from a Bioeconomic Perspective

Up until the 1970s, the discussion of how commercial fisheries should be managed was almost exclusively dominated by biologists. With the development and application of dynamic optimization techniques, economists, while still in the minority, have begun to exert influence over the types of policies employed to manage marine fisheries. Let us examine some of the more traditional, often biologically based management policies, and then proceed to the policy implications of the basic bioeconomic model.

Traditional management policies have included closed seasons, gear restrictions, quotas, and limited entry.

Closed Seasons: This policy limits fishing to certain times during the year. It may be justified on a biological basis in terms of protecting the species during a critical period, such as spawning. A similar rationale might be used in prohibiting fishing in a particular *area* (for example, commercial salmon fishing in rivers or streams). Closed seasons can result in idle capital (vessels and processing equipment) and fishermen while the season is closed. A tremendous amount of fishing effort is often expended during the open season. In some commercial fisheries seasons have been limited to a few weeks. Excess capacity may be directed elsewhere; perhaps exacerbating the management of other depleted species.

Gear Restrictions: This policy attempts to reduce the effectiveness of harvesting units by prohibiting the use of certain technologies. This has the effect of raising (or maintaining) cost per unit effort and making the fishery less attractive financially. (For example, the State of Maryland restricts commercial harvest of oysters to vessels powered by sail.) Gear restrictions, while limiting the use of one input, might cause fishermen to use other inputs more intensively with little or no reduction in effective effort and yield.

Quotas: This policy establishes a maximum quantity that may be harvested per period. Quotas might be collective, setting a fleet limit, or assigned individually to each vessel. If the quota is collective, there is a strong incentive for an individual vessel to attempt to catch the largest share of the quota possible before the fishery is closed down (when the collective quota is reached). If assigned individually, each vessel has an upper limit to the amount of fish caught per period. Uniform quotas assigned individually, but based on vessel characteristics, are sometimes felt to penalize the most efficient fishermen, restricting their catch while reserving stock for harvest by higher cost (less efficiently run) vessels. (*Transferable quotas* will theoretically avoid this shortcoming and will be discussed in more detail.)

Limited Entry: This policy restricts harvesting to vessels which are licensed and controlled by the management agency. Licenses are often allocated to boats harvesting the resource at the time the limited entry program is introduced. The number of licensed boats may be reduced over time by attrition and subsequent licenses allocated by lottery. A yearly fee (sometimes substantial) is charged for a license. Current holders are often allowed to sell (transfer) their license to other potential fishermen. Allowing sale of licenses would theoretically result in only the most efficient boats gaining access to the fishery (they can pay the most for a license in a competitive situation). In addition to license fees, fleet or individual quotas may be imposed.

The traditional management policies are thus oriented at protecting the resource during critical periods in its life cycle and at reducing effort and catch. Such policies are usually of limited success. While they may reduce the degree of *biological overfishing* (the extent to which stock is less than X_{MSY}) they may exacerbate the problem of *economic overfishing* (zero fishery rents). Missing from these policies is an understanding of *user cost*.

In the analysis of an open access fishery we saw that effort would be expended to the point where fishery rents were driven to zero. The open access equilibrium may be denoted as $(E_\infty, X_\infty, Y_\infty)$. Open access was inefficient because there typically exists a bioeconomic optimum, denoted (E^*, X^*, Y^*) such that $E^* < E_\infty$, $X^* > X_\infty$, and $Y^* > Y_\infty$, which is capable of producing income (rents) that could make some fishermen better off without making anyone worse off. In the Gordon-Schaefer model, open access occurred when

$$N = \left[p - \frac{c}{qX_\infty} \right] Y_\infty = 0 \tag{41}$$

with $Y_\infty > 0$, we note $p = \frac{c}{qX_\infty}$ or $X_\infty = \frac{c}{qp}$.

The bioeconomic optimum was derived from the perspective of a sole owner who, for the Gordon-Schaefer model, would seek an equilibrium such that

$$p = \frac{c}{qX^*} + \varrho\lambda^*, \tag{42}$$

where $\varrho\lambda^*$ was defined as *user cost*, equal to the present value of an additional fish (in the water) toworrow. The user cost term plays a key role in defining fishery management policies that can correct for *economic overfishing*. Quite simply, the economist would like to establish policies which would create a real cost comparable to $\varrho\lambda^*$, and induce fishermen to take it into account.

Transferable Quotas and Landings Taxes

Suppose a team of biologists and economists could estimate (or assign) values to the parameters of the Gordon-Schaefer model ($p, c, \delta, r, K,$ and q) and determine the bioeconomic optimum (E^*, X^*, Y^*). Assume that the resource stock was moved from X_0 to X^*. The management team would now allow a total effort of E^*, resulting in a yield of Y^*. The total quota, Y^*, might be assigned to an arbitrary group of fishermen in, say, metric ton units. Any initial assignee could choose to harvest the permitted number of metric tons of fish or he could sell it to another fisherman. The fact that the individual quota is *transferable* to another creates an *opportunity cost* for its initial owner. He must decide whether the net revenues from harvesting his quota exceeds the amount he could get if he would sell it to another. Presumably, a market would develop for individual quotas, with fishermen who could harvest at least cost being able to outbid less efficient fishermen. What would the going price be for the right to harvest a metric ton of fish? If prices, costs, and other bioeconomic parameters were expected to remain unchanged, the economists would predict a per ton quota price of

$$P_{Q^*} = \varrho\lambda^*, \tag{43}$$

where P_{Q^*} is the per ton quota price and $\varrho\lambda^*$ is user cost. This relationship would result because at (Y^*, X^*)

$$p - \frac{c}{qX^*} = P_{Q^*} = \varrho\lambda^*. \tag{44}$$

That is, at the bioeconomic optimum, each metric ton harvested would yield a net revenue of $\varrho\lambda^*$. In deciding whether to "fish or sell," the rational fishermen would subtract the opportunity cost of holding the permit along with the per unit harvest costs to determine the "real" value of fishing. By defining exclusive but transferable rights of harvest, the management team would create incentives that (a) lead to the efficient (least cost) harvest of the total quota, Y^*, and (b) cause fishermen to individually consider the opportunity cost (equal to user cost) of harvesting another unit of the resource.

There is another way of establishing and maintaining the bioeconomic optimum (E^*, X^*, Y^*). The management team (or authority) could simply notify all fishermen that their catch had to be landed at certain locations and a per ton tax of

$$\tau^* = \varrho \lambda^* \tag{45}$$

would be levied on their catch. From a net revenue point of view, fishermen would harvest an additional metric ton until

$$p - \frac{c}{qX^*} - \tau^* = 0. \tag{46}$$

The per unit landings tax allows the introduction of an additional unit cost which in equilibrium would equal user cost. If initially $X_0 < X^*$, a tax set at $\tau^* = \varrho \lambda^*$ would actually choke off fishing effort until the stock increased (thereby reducing costs) to the optimum X^*.

The landings tax has the advantage (from the management authority's point of view) of generating tax revenues equal to

$$R_t = \tau^* Y^* \tag{47}$$

in equilibrium. These revenues might be earmarked for administration, enforcement, and research efforts by the management authority. Fishermen, of course, would much prefer freely assigned quotas to a landings tax. In theory, the management authority could employ *both*, and in equilibrium we would expect the following relationship:

$$P_{Q^*} = \varrho \lambda^* - \tau, \tag{48}$$

that is, the price emerging from the quota market would equal net revenue $\left(\varrho \lambda^* = p - \frac{c}{qX^*} \right)$ less the landings tax rate. Note that as $\tau \to \varrho \lambda^*$, $P_{Q^*} \to 0$; that is, as the landings tax is increased from zero to user cost ($\varrho \lambda^*$), the market price for the right to harvest a metric ton will decline toward zero. At some tax rate $\tau < \varrho \lambda^*$, the authority can generate some revenues to support their activities and still leave positive fishery rents to be captured by fishermen.

5. Two Applications: Yellowfin Tuna and Blue Whales

To illustrate some of the concepts presented in the preceding lectures we will examine two empirical studies. The first is a study of yellowfin tuna in the Eastern

Tropical Atlantic (ETA) and employs the Gordon-Schaefer model [see Eq. (7)]. The second is a study of the blue whale and will employ an alternative specification attributable to Spence (1974).

Yellowfin Tuna in the Eastern Tropical Atlantic (ETA)

In seeking to optimally manage a single species fishery we formulated a dynamic optimization problem that sought to maximize the present value of net revenues subject to a difference equation describing the change in the fish stock. A Lagrangian expression was formed and the first order (necessary) conditions derived. In steady-state these conditions lead to a system of three equations in three unknowns (X^*, Y^*, λ^*). We could eliminate λ^* from the system leaving a two equation system consisting of the catch locus and growth curve [see Eqs. (32) and (33)]. Further substitution led to a single equation in X^* [see Eq. (34)]. For the Gordon-Schaefer model with $c(X^*)=c/qX^*$ and $Y=rX^*(1-X^*/K)$ this single equation is a quadratic in X^* and the optimal stock level will equal the positive root according to:

$$X^* = \frac{K}{4}\left[\left(\frac{c}{qpK}+1-\frac{\delta}{r}\right)+\sqrt{\left(\frac{c}{qpK}+1-\frac{\delta}{r}\right)^2+\frac{8c\delta}{qpKr}}\right]. \tag{49}$$

If one had estimates of the bioeconomic parameters r, K, q, c, p, and δ, one could estimate the optimal stock X^* as well as yield and effort according to

$$Y^* = F(X^*), \tag{50}$$

and

$$E^* = \frac{Y^*}{qX^*}. \tag{51}$$

Estimates of these parameters for the yellowfin tuna fishery in the ETA were obtained by Adu-Asamoah and Conrad (1982) based on data from the International Commission for the Conservation of Atlantic Tuna (ICCAT), the National Marine Fisheries Service (NMFS) of the U.S. Department of Commerce, and earlier economic studies of purseseiners, baitboats, and longliners. The values of the parameters used to calculate maximum sustainable yield (MSY), open access, and the bioeconomic optimum were

$$r = 1.2883,$$

$$K = 351.2244 \times 10^3 \text{ MT},$$

$$q = 1.372 \times 10^{-2},$$

$$p = \$ 1,300/\text{MT},$$

$$c = \$ 2,000; \$ 2,500; \$ 3,000; \$ 3,500/\text{standard day at sea},$$

$$\delta = 0.00, 0.05, 0.10, 0.15, 0.20.$$

(52)

Table 1. MSY, bioeconomic, and open access equilibria for yellowfin tuna in the ETA*

Yellowfin parameters: $p = \$ 1300$, $q = 1.372 \times 10^{-2}$, $r = 1.2883$, $K = 351.2244$
Maximum sustainable: $X_{MSY} = 175.61$, $Y_{MSY} = 113.12$, $E_{MSY} = 46.9495$

	δ	$c = \$ 2,000$	$c = \$ 2,500$	$c = \$ 3,000$	$c = \$ 3,500$
Bio-	0.00	$X^* = 231.68$	$X^* = 245.70$	$X^* = 259.71$	$X^* = 273.73$
economic		$Y^* = 101.59$	$Y^* = 95.10$	$Y^* = 87.18$	$Y^* = 77.81$
equilibria		$E^* = 31.96$	$E^* = 28.21$	$E^* = 24.47$	$E^* = 20.72$
	0.05	$X^* = 228.21$	$X^* = 242.81$	$X^* = 257.35$	$X^* = 271.83$
		$Y^* = 102.97$	$Y^* = 96.56$	$Y^* = 88.61$	$Y^* = 79.16$
		$E^* = 32.89$	$E^* = 28.98$	$E^* = 25.10$	$E^* = 21.23$
	0.10	$X^* = 224.85$	$X^* = 240.02$	$X^* = 255.07$	$X^* = 270.00$
		$Y^* = 104.23$	$Y^* = 97.90$	$Y^* = 89.96$	$Y^* = 80.44$
		$E^* = 33.79$	$E^* = 29.73$	$E^* = 25.71$	$E^* = 21.71$
	0.15	$X^* = 221.58$	$X^* = 237.32$	$X^* = 252.87$	$X^* = 268.24$
		$Y^* = 105.37$	$Y^* = 99.15$	$Y^* = 91.23$	$Y^* = 81.65$
		$E^* = 34.66$	$E^* = 30.45$	$E^* = 26.30$	$E^* = 22.19$
	0.20	$X^* = 218.41$	$X^* = 234.71$	$X^* = 250.74$	$X^* = 266.54$
		$Y^* = 106.40$	$Y^* = 100.31$	$Y^* = 92.42$	$Y^* = 82.80$
		$E^* = 35.51$	$E^* = 31.15$	$E^* = 26.87$	$E^* = 22.64$
Open	$\delta \to \infty$	$X_\infty = 112.13$	$X_\infty = 140.17$	$X_\infty = 168.20$	$X_\infty = 196.23$
access		$Y_\infty = 98.34$	$Y_\infty = 108.51$	$Y_\infty = 112.92$	$Y_\infty = 111.56$
		$E_\infty = 63.92$	$E_\infty = 56.43$	$E_\infty = 48.93$	$E_\infty = 41.44$

* Stocks (Xs) and Yields (Ys) are measured in 10^3 metric tons. Effort is measured in 10^3 standard days at sea per year

The values for stock ($\times 10^3$ MT), yield ($\times 10^3$ MT), and effort ($\times 10^3$ standard day at sea) for the various equilibria are shown in Table 1. Maximum sustainable yield is $Y_{MSY} = rK/4 = 113,121$ MT occurring at $X_{MSY} = 175,612$ MT and $E_{MSY} = 46,950$ SDS. Open access stock, where net revenues are zero, occurs at $X_\infty = c/pq$. The four values of c used to test sensitivity to cost produced the estimates of open access equilibria at the bottom of Table 1. Bioeconomic equilibria for various combinations of c and δ are shown in the body of the table. Four aspects of these equilibria should be noted. First, for a given discount rate, increased costs result in increased stocks. Within the context of Fig. 5, increases in c cause the catch locus to shift to the right.

Second, for a given cost, increases in the discount rate result in lower optimal stocks. Increases in δ tend to shift the catch locus to the left.

Third, the optimal stock for all combinations of c and δ are in excess of X_{MSY}. The marginal stock effect [that is, the second term on the LHS of Eq. (34)] exceeds even the high value of $\delta = 0.20$ resulting in optimal stocks to the right of X_{MSY}. This situation is similar to the intersection of $\phi_2(X)$ and $F(X)$ in Fig. 5 where $X_2^* > X_{MSY}$, and the stock induced costsavings exceed the interest costs.

Finally, with the exception of the highest cost estimate, the open access stock is below X_{MSY}. For $c = \$ 3,500$/standard day the yellowfin stock would not be subject to *biological overfishing* (i.e., $X_\infty > X_{MSY}$) but would be subject to *economic overfishing* (i.e., $X^* > X_\infty$).

Blue Whales

One of the first attempts to apply control theory to a problem of renewable resource management was the bioeconomic analysis of the blue whale *(Balenoptera musculus)* by Spence (1974). While the techniques of analysis are similar to those encountered in previous lectures, Spence did *not* employ the Gordon-Schaefer specification but rather an alternative formulation for growth and production within the basic bioeconomic model. Let

X_t = the blue whale population in year t,

E_t = the number of fully equipped whale boats in year t.

Spence assumes that the growth function takes the following form:

$$X_{t+1} = F(X_t) - Y_t = AX_t^a - Y_t, \tag{53}$$

where $A > 1$ and $0 < a < 1$. This results in a concave (from below) growth function similar to the curve drawn in Fig. 7.

Fishing effort is directed at the stock of blue whales and yield is assumed to be some *fraction* of *next* year's stock, i.e.,

$$Y_t = F(X_t)[1 - e^{-bE_t}] = AX_t^a[1 - e^{-bE_t}], \tag{54}$$

where the term $[1 - e^{-bE_t}]$ determines the proportion of next year's stock harvested this year. Assuming $0 < b < 1$, we note

$$[1 - e^{-bE_t}] \begin{cases} =0 \\ >0 \\ =1 \end{cases} \quad \text{when} \quad E_t \begin{cases} =0 \\ >0 \\ E_t \to \infty \end{cases}. \tag{55}$$

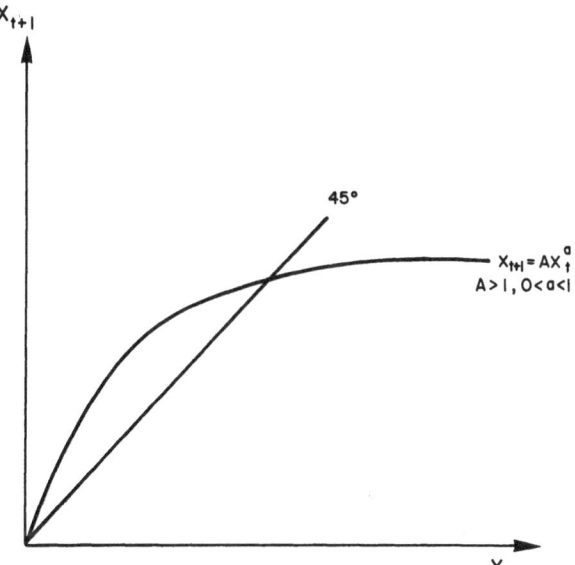

Fig. 7. Growth in Spence's model of the blue whale

If one substitutes Eq. (54) into Eq. (53), a simplified expression for X_{t+1} results:

$$X_{t+1}=AX_t^a e^{-bE_t}. \tag{56}$$

Denoting the average price of a blue whale as p and the yearly cost of operating a whaling vessel as c, the net revenues from the harvest of blue whales may be written as

$$N_t=pY_t-cE_t=pAX_t^a[1-e^{-bE_t}]-cE_t. \tag{57}$$

Maximization of the present value of net revenues subject to Eq. (56) can be accomplished by forming the Lagrangian

$$L=\sum_{t=0}^{\infty}\varrho^t\{pAX_t^a[1-e^{-bE_t}]-cE_t+\varrho\lambda_{t+1}[AX_t^a e^{-bE_t}-X_{t+1}]\}. \tag{58}$$

Recall $\varrho=\dfrac{1}{1+\delta}$ is our discrete time discount factor, and δ is the discount rate. In the appendix to this lecture, we derive the first order necessary conditions and ultimately a single equation defining the optimal stock of blue whales. The equation takes the form

$$\frac{abAX^a-a(c/p)}{[bX-c/p]}=(1+\delta), \tag{59}$$

where a, A, and b are parameters of the growth and production functions which Spence estimates using data from *International Whaling Statistics, c/p* is the ratio of yearly cost to price per average whale, and δ is the aforementioned discount rate. Spence obtained the following estimates: $a=0.8204$, $A=8.3560$, and $b=0.0019$. He set $\delta=0.05$ and then solved for the optimal stock of blue whales as a function of (c/p). His results are shown in Table 2. Spence admits to having limited data on the costs of operating a whaling vessel and thus wanted to examine the sensitivity of the optimal stock, X^*, to the cost/price ratio. Even for low c/p ratios, it is optimal to require *relatively large standing stocks*, much greater than the current estimate of 10,000 blue whales.

Given the functional forms employed by Spence, it is also possible to show that the approach path is *most rapid* (i.e., MRAP is optimal). In 1973, the estimated stock of blue whales was $X_0=1,639$. Since $X_0<X^*$ for any of the bioeconomic optima in Table 2, MRAP would call for a *moratorium* on commercial harvest of blue whales. How long? Spence calculates the length of time for the stock to grow from X_0 to X^* by iterating

$$X_{t+1}=AX_t^a \tag{60}$$

until $X_{t+1}>X^*$. The length of the moratorium will depend on this optimal stock, with larger stocks requiring a longer moratorium. The length of moratorium for

Table 2. Optimal stock of blue whales, boats, and catch for $\delta = 0.05$ and alternative values of (c/p) [a]

(c/p)	X^*	E^* (Boats)	Y^*	Y^*/E^* (Catch/boat)	N^*/c (Profits per year per $ of operating cost)
0	34,421	129	9,635	74.6	–
9	40,000	115	9,834	85.5	8.34
18	45,000	104	9,890	95.1	4.21
28	50,000	95	9,845	104	2.70
49	60,000	77	9,501	122	1.47
73	70,000	63	8,870	141	0.94
98	80,000	50	8,002	159	0.63
124	90,000	39	6,930	177	0.43
151	100,000	29	5.681	195	0.29
180	110,000	20	4,276	213	0.19
209	120,000	12	2,732	277	0.09
240	130,000	4.3	1,062	246	0.025
250	136,000	0	0	–	–

[a]   Table modified from Spence (1974)

Table 3. Length of moratorium on blue whales for $\delta = 0.05$, $X_0 = 1,639$, $p = 6,000$, and four alternative values for (c/p) [a]

(c/p)	X^*	E^*	Y^*	Years with no whaling	Profit per year at X^* (millions of $)
0	34,421	129	9,635	5	57.8
28	50,000	95	9,845	7	57.4
124	90,000	39	6,930	12	41.5
209	120,000	12	2,732	17	16.4

[a]   Table modified from Spence (1974)

$\delta = 0.05$ and for each of four values of (c/p) is shown in Table 3. Depending on the cost/price ratio, the moratorium would extend from 5 to 17 years.

As in the ETA tuna study discussed earlier, the estimates of a, A, and b should be regarded as preliminary. Spence rightfully suggests that the blue whale stock should be monitored during the moratorium to see if it is growing according to our estimates of a and A. If not, these parameters should be revised and new optima and moratoria calculated.

Spence concludes by noting that for the blue whale, extinction is *not* optimal on economic grounds and, given the large standing stocks at steady-state, there does not appear to be any conflict between economic and environmental objectives.

Bioeconomics of the Blue Whale: Mathematical Appendix

Maximization of the present value of net revenues for the growth and production functions used by Spence (1974) to characterize the blue whale led to the following

Lagrangian:

$$L = \sum_{t=0}^{\infty} \varrho^t \{ pAX_t^a [1 - e^{-bE_t}] - cE_t + \varrho \lambda_{t+1} [AX_t^a e^{-bE_t} - X_{t+1}] \}. \tag{A.1}$$

The first order conditions require:

$$\frac{\partial L}{\partial E_t} = \varrho^t \{ bpAX_t^a e^{-bE_t} - c - b\varrho \lambda_{t+1} AX_t^a e^{-bE_t} \} = 0, \tag{A.2}$$

$$\frac{\partial L}{\partial X_t} = \varrho^t \{ apAX_t^{a-1} [1 - e^{-bE_t}] + a\varrho \lambda_{t+1} AX_t^{a-1} e^{-bE_t} \} - \varrho^t \lambda_t = 0, \tag{A.3}$$

$$\frac{\partial L}{\partial \lambda_{t+1}} = \varrho^{t+1} \{ AX_t^a e^{-bE_t} - X_{t+1} \} = 0. \tag{A.4}$$

Equations (A.1) through (A.4) may be written as

$$bAX_t^a e^{-bE_t} [p - \varrho \lambda_{t+1}] = c, \tag{A.5}$$

$$apAX_t^{a-1} - aAX_t^{a-1} e^{-bE_t} [p - \varrho \lambda_{t+1}] - \lambda_t = 0, \tag{A.6}$$

$$AX_t^a e^{-bE_t} - X_{t+1} = 0. \tag{A.7}$$

In steady state:

$$[p - \varrho \lambda] = \frac{c}{bAX^a e^{-bE}}, \tag{A.8}$$

$$apAX^{a-1} - aAX^{a-1} e^{-bE} [p - \varrho \lambda] - \lambda = 0, \tag{A.9}$$

$$AX^a e^{-bE} = X. \tag{A.10}$$

Solving (A.8) for λ and using (A.10) yields:

$$\lambda = 1 + \delta \left\{ p - \frac{c}{bAX^a e^{-bE}} \right\} = (1 + \delta) \left\{ p - \frac{c}{bX} \right\}. \tag{A.11}$$

Substituting (A.8) and (A.11) into (A.9) yields

$$apAX^{a-1} - aAX^{a-1} e^{-bE} \left[\frac{c}{bAX^a e^{-bE}} \right] - (1 + \delta) \left[p - \frac{c}{bX} \right] = 0,$$

$$apAX^{a-1} - \frac{ac}{bX} - (1 + \delta) \left[\frac{pbX - c}{bX} \right] = 0,$$

$$\frac{apAX^{a-1}}{\left[\dfrac{pbX - c}{bX} \right]} - \frac{\dfrac{ac}{bX}}{\left[\dfrac{pbX - c}{bX} \right]} = (1 + \delta),$$

or finally,

$$\frac{abAX^a - a(c/p)}{[bX - c/p]} = (1 + \delta).$$ (A.12)

Given the bioeconomic parameters a, A, b, (c/p), and δ one can "iterate" (A.12) to solve for the optimal stock of blue whales. This was done by Spence for various values of (c/p) with $\delta = 0.05$ and the previously noted estimates of a, A, and b based on IWC data. This equation was used to generate the values of X^* shown in Tables 2 and 3 in the text of this lecture.

References

Adu-Asamoah, R., Conrad, J.M. (1982). Fishery Management: The Case of Tuna in the Eastern Tropical Atlantic. Staff Paper No. 82-15, Department of Agricultural Economics, Cornell University, Ithaca, New York, 14853

Clark, C.W. (1973). Profit Maximization and the Extinction of Animal Species. Journal of Political Economy *81*: 950–961

Clark, C.W. (1976). Mathematical Bioeconomics: The Optimal Management of Renewable Resources. New York: John Wiley & Sons

Gordon, H.S. (1954). Economic Theory of a Common-Property Resource: The Fishery. Journal of Political Economy *62*: 124–142

Gould, J.R. (1972). Extinction of a Fishery by Commercial Exploitation: A Note. Journal of Political Economy *80*: 1031–1038

May, R.M. (1975). Biological Populations Obeying Difference Equations: Stable Points, Stable Cycles, and Chaos. Journal of Theoretical Biology *51*: 511–524

Moloney, D.G., Pearse, P.H. (1979). Quantitative Rights as an Instrument for Regulating Commercial Fisheries. Journal of the Fisheries Research Board of Canada *36*: 859–866

Schaefer, M.B. (1957). Some Considerations of Population Dynamics and Economics in Relation to the Management of Marine Fisheries. Journal of the Fisheries Research Board of Canada *14*: 669–681

Scott, A.D. (1955). The Fishery: The Objectives of Sole Ownership. Journal of Political Economy *63*: 116–124

Spence, A.M. (1974). Blue Whales and Applied Control Theory. In: C. L. Zadeh, et al., eds., Systems Approaches for Solving Environmental Problems in the series: Mathematical Studies in the Social and Behavioral Sciences. Vandenhoeck and Ruprecht, Göttingen and Zürich

Spence, A.M., Starrett, D. (1975). Most Rapid Approach Paths in Accumulation Problems. International Economic Review *16* (June): 388–403

Wilen, J.E. (1979). Fisherman Behavior and the Design of Efficient Fisheries Regulation Programs. Journal of the Fisheries Research Board of Canada, *36*: 855–858

Population Biology of Microparasitic Infections

Robert M. May

Introduction

Much, though not all, of the material in my chapter has already been published in journals that are likely to be as accessible as this book. There is a constant temptation to repeat oneself in print; with the aim of avoiding this temptation, I have kept most of my presentation to the bare bones, adding flesh in those places where the work is not already published or where new avenues of investigation seem to me to be ready for study. The emphasis here is on the mathematical development of the subject; various kind of applications are discussed in the light of available data elsewhere (and references are given to these works, without repeating the presentation here).

In the study of host-parasite associations, a rough but useful distinction may be made between microparasites and macroparasites. Microparasites are broadly those having direct reproduction, usually at high rates, within the definitive host (as typified by most viral and bacterial, and many protozoan, infections); the duration of infection is usually short, relative to the expected life span of the host, and is therefore of a transient nature. Macroparasites are those having no direct reproduction within the host (as typified by most helminthic infections); infections are usually of a persistent nature, with hosts being continually reinfected. For microparasitic infections (typified, for example, by measles) it usually makes sense to divide the host population into a small number of discrete categories – susceptible, infected but not yet infectious (latent), infectious, recovered and immune – without distinguishing differing degrees of intensity of infection. Such *compartmental models* comprise the bulk of the literature on mathematical epidemiology. In contrast, for most macroparasites the pathogenic effects on a host, the egg output per parasite, the intensity of immune responses (if any), and a variety of other factors depend on the magnitude of the parasite burden borne by an individual host. Consequently, mathematical models for host-macroparasitic associations need to deal with the full distribution of parasites among the host population, rather than with a relatively small number of compartments. In particular, the simple microparasite models make no distinction between infection and disease (hosts either do or do not "have measles"), while for macroparasites there can be an important distinction between infection (having one or more parasites) and disease (having a parasite load large enough to produce illness).

This chapter deals exclusively with microparasites, effectively defined as those described by compartmental models. Discussion of mathematical aspects of host-

Biomathematics, Vol. 17, Mathematical Ecology
Edited by T. G. Hallam and S. A. Levin
© Springer-Verlag Berlin Heidelberg 1986

macroparasite associations, and of the implications for public health programs, is given by Anderson in the companion volume to these notes; see also May and Anderson (1979) and Anderson and May (1982a).

This chapter is organized as follows. Part I sets out the basic system of partial differential equations corresponding to a compartmental model, and indicates a variety of further refinements and complications that may be incorporated; some of these refinements and complications have been studied and some have not. Parts II and III deal with human populations and follow tradition in assuming the total magnitude of the host population is constant, unaffected by the microparasitic infection. In Part II, the statics of such host-microparasite associations are explored, with particular attention given to the effects of vaccination programs upon the transmission and maintenance of infection. Part III deals with the dynamics, outlining earlier work on the non-seasonal oscillations in incidence of many microparasitic infections, and presenting new studies of the short-term dynamic consequences of implementing specific vaccination programs. Part IV goes on to consider the possibility that many non-human animal populations may have their magnitude or geographical distribution regulated by parasitic infections; this section deals with the overall population biology of host-microparasite associations, with the total number of hosts now being itself a dynamical variable.

Part I. Compartmental Models

We begin quite generally by apportioning the total host population into three classes:

$X(a, t)$ = number susceptible, of age a, at time t,

$Y(a, t)$ = number infected, of age a, at time t,

$Z(a, t)$ = number immune, of age a, at time t.

The total number of hosts of age a, at time t, is, of course,

$$N(a, t) = X(a, t) + Y(a, t) + Z(a, t). \tag{1}$$

The basic set of partial differential equations for this system is:

$$\partial X/\partial t + \partial X/\partial a = -[\lambda(t) + \mu(a)]X(a, t), \tag{2}$$

$$\partial Y/\partial t + \partial Y/\partial a = \lambda X - [\alpha(a) + \mu(a) + v(a)]Y(a, t), \tag{3}$$

$$\partial Z/\partial t + \partial Z/\partial a = vY - \mu(a)Z(a, t). \tag{4}$$

These equations differ slightly from the set discussed by Bailey (1975), Dietz (1975, 1976), Hoppensteadt (1976), Waltman (1974), and others in that disease-induced mortality is explicitly included. The demographic and epidemiological parameters in these equations, which are discussed further below, are:

$\mu(a)$ = age-specific host mortality rate, per capita,

$v(a)$ = per capita recovery rate (which may be age-specific),

$\alpha(a)$ = per capita disease-induced mortality rate (possibly age-specific),

$\lambda(t)$ = "force of infection" at time t (discussed below).

An intuitive interpretation of this system is straightforward: susceptibles are lost by "natural" deaths (that is, deaths not associated with this infection) at a rate μ or are transferred to the infected class at a rate λ; infected individuals either die (naturally at rate μ or from infection at rate α) or recover into the immune class (at rate v).

To complete a description of the system, sets of initial or boundary conditions are needed. Usually these conditions are provided by specifying X, Y, Z for age zero at all times, and for time zero at all ages. That is, one condition is usually that at $t=0$ expressions for $X(a,0)$, $Y(a,0)$, $Z(a,0)$ are given, for all a. The second condition is typically provided by specifying that all hosts are born susceptible, so that at $a=0$ we have $Y(0,t)=Z(0,t)=0$ for all t, while $X(0,t)=B(t)$ where B is the net birth rate at time t.

A variety of comments, some conventional and some unconventional, may be made at this point.

(i) Latent and Other Classes

Other classes of hosts can obviously be incorporated in this general framework. The most usual such additional class is for hosts who are infected but not yet infectious, or "latent" (Bailey, 1975; Dietz, 1975, 1976; Anderson and May, 1983). This complicates the analysis without adding any substantially new features, and so will be omitted here.

(ii) Maternal Antibodies

Maternal antibodies may protect newly born infants for the first 3–9 months of life. This fact is important in the design of some immunization programs, because vaccination will not "take" during this protected period; as a result, immunization cannot usefully be implemented as a routine part of antenatal care, which otherwise would be administratively convenient. Thus for detailed analysis of some immunization programs it may be desirable to add an initial class of protected or immune hosts, into which class all infants are born; hosts typically will pass out of this class into the susceptible class in the first 3–9 months of life. Incorporation of such a class, and a test of the ensuing mathematical model against serological data for measles, has been studied by Anderson and May (1983).

(iii) Vertical Transmission

For some infections, there can be a degree of "vertical transmission," whereby the infection is passed directly to the newly born offspring of an infected parent (usually an infected mother, but also possibly an infected father). This pheno-menon has been discussed by Fine (1975), Anderson and May (1979, 1981) and others. Vertical transmission can be included in the system of Eqs. (2)–(4) by modifying the boundary condition at $a=0$ so that some fraction of the births are into the infected class: $X(0, t)=B_1(t)$, $Y(0, t)=B_2(t)$, $Z(0, t)=0$. These births, $B_2(t)$, in turn would represent some fraction of the births from infected people, and thus would depend in a complicated way on the number of infected people, $Y(a, t)$, convolved with some weighting function and integrated over the past 9 months.

(iv) Males and Females Treated Separately

If vertical transmission occurs differently from the father and from the mother, it may be necessary to move explicitly to a 2 sex model, with separate equations for X_i, Y_i, Z_i ($i=1, 2$) for males and females. Such a 2 sex model may be necessary for other reasons as well, when the epidemiological circumstances of males and females are not identical (as for many sexually transmitted infections).

(v) Loss of Immunity

Equations (2)–(4) assume that immunity, once acquired, is of lifelong duration. Loss of immunity can be incorporated in the model via terms [a loss term in Eq. (4), and an equal gain term in Eq. (2)] that describe the rate of transition from the immune class to the susceptible class.

(vi) Recovery

The recovery rate, $v(a)$, is usually treated as a constant (and, indeed, simply as an age-independent constant, v). While this is mathematically convenient, it is rarely realistic. More commonly, recovery may be after some defined period of time, T, has elapsed. These two extremes are illustrated in Fig. 1A and B: Fig. 1A corresponds to a constant recovery rate, v, and the probability for a given host to remain in the infected class declines exponentially with the passage of time, t; Fig. 1B corresponds to a defined interval of illness, and a given host remains infected for a time T and then recovers. More generally, most infections have etiology intermediate between these extremes (but closer to Fig. 1B), and a truly accurate model would employ some empirically-determined distribution of recovery times. As discussed by Hoppensteadt (1974) and others, such a model would have the simple vY terms of Eqs. (3) and (4) replaced by integrals, leading to integro-differential equations. Hoppensteadt (1974), Grossman (1980), and others have shown that – with important exceptions in a few situations – there is not much

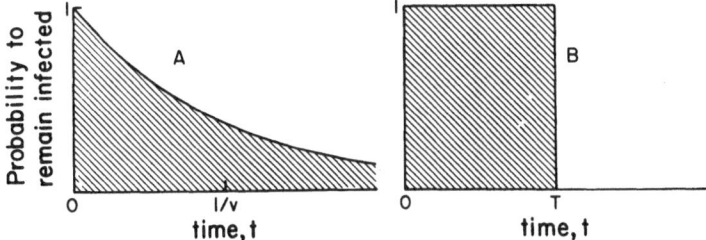

Fig. 1A, B. Type *A* recovery is illustrated in **A**; the probability to remain infected at time *t* after acquiring infection decays exponentially, corresponding to a constant recovery rate, v. Type *B* recovery is shown in **B**; here all individuals remain infected for a defined interval of time, T, and then recover. With the identification $T = 1/v$, the two recovery processes give results that are roughly equivalent in most contexts

difference between results obtained using what I shall call "type *A* recovery" (Fig. 1A) and those with "type *B*" (Fig. 1B). Henceforth, I consistently use type *A* recovery with age-independent v.

(vii) Natural Mortality

Likewise, essentially all the traditional literature of mathematical epidemiology (see, for example, Bailey, 1975) takes the natural mortality rate, $\mu(a)$, as an age-independent constant. This is to assume that a newly born individual's probability of surviving to age a declines exponentially with a (having the mathematical shape depicted in Fig. 1A), which is a grossly unrealistic assumption, especially in developed countries. A better approximation is that everyone survives to exactly age L, and then promptly dies (giving a survivorship relation with the shape depicted in Fig. 1B). Again, real survival curves are intermediate between these two extremes, although closer to the second. I shall henceforth label the first assumption as type *A* survival,

$$\text{TYPE } A: \quad \mu(a) = \mu = \text{const}\,(1/L), \tag{5}$$

and the second as type *B*,

$$\text{TYPE } B: \quad \mu(a) = 0, \quad \text{for} \quad a < L,$$
$$\mu(a) = \infty, \quad \text{for} \quad a > L. \tag{6}$$

For many purposes, predictions based on type *B* survival are effectively indistinguishable from those based on real survivorship data (Anderson and May, 1983).

(viii) Disease-Induced Mortality

The disease-induced mortality rate, $\alpha(a)$, has been omitted from most studies, and when it is included it is usually treated as a constant: $\alpha(a) = \alpha$ (Anderson and May,

1979). In reality, the mortalities associated with malaria, smallpox, measles, infant diarrhea, and many other infections typically are greater among infants, especially in developing countries where such mortality has been and is appreciable. We put $\alpha = 0$ in Parts II and III of this chapter, and take α to be an age-independent constant in Part IV; the complications introduced by realistic values of $\alpha(a)$ deserve attention, however, particularly in the planning of immunization programs against childhood infections in developing countries.

(ix) Transmission

The "force of infection", λ, is the per capita rate of acquisition of infection. That is, $\lambda(t)\Delta t$ represents the probability that a given susceptible host will become infected in the small time interval Δt. Sometimes, λ may be deduced directly or indirectly from epidemiological data. In other contexts, we may wish to "close" the system of equations by relating λ to the number of infected individuals. In this case, the almost-universal assumption (for example: Bailey, 1975; Dietz, 1975, 1976; Hoppensteadt, 1976; Waltman, 1974) is that λ is linearly proportional to the total number of infectious individuals:

$$\lambda(t) = \beta \int Y(a', t) da'. \tag{7}$$

Here β is a "transmission parameter," which combines a multitude of epidemiological, environmental, and social factors that affect transmission.

The conventional assumptions embodied in Eq. (7) imply that λ is not dependent on age (although it may vary in time as the total number of infectives varies). But for most microparasitic infections for which data are available, the indications are that the force of infection, λ, is age-dependent (for example, Anderson and May, 1982b). Such age-dependence could be incorporated relatively easily by assuming that the transmission parameter, β, is age-dependent, but that at all ages λ otherwise still depends simply on the total number of infectives:

$$\lambda(a, t) = \beta(a) \int Y(a', t) da'. \tag{8}$$

Although Eq. (8) represents a reasonable first step toward accounting for the observed fact that λ is usually age-dependent, a more realistic approach may be to admit that – to a degree – susceptibles of age a have relatively more contact with people of roughly their own age. This is particularly likely to be true for children. Formally, this would lead to $\lambda(a, t)$ being related to $Y(a, t)$ by an expression of the form

$$\lambda(a, t) = \int \beta(a, a') Y(a', t) da'. \tag{9}$$

Here $\beta(a, a')$ essentially represents the probability that an infective of age a' will infect a susceptible of age a. To a crude approximation [but probably better than Eq. (8)], one could take $\beta(a, a')$ to have two components: a smeared contribution, $\beta_1 = \beta = $ constant, when neither age of infective nor age of infectee matters; and a focused component, $\beta_2 = c\delta(a-a')$, reflecting contact within a given age class. This gives

$$\lambda(a,t) = \beta \int Y(a',t)da' + cY(a,t). \tag{10}$$

To my knowledge, none of these realistic complications – as described by Eqs. (8)–(10) or other alternatives – has been explored.

(x) Seasonality

A different kind of complication arises from the fact that many microparasitic infections exhibit marked seasonality in transmission. Such seasonality may derive from the effects of temperature or humidity on the survival of transmission stages of the viral, bacterial or other parasite, or indirectly from the effects of whether on social habits, or from such temporal patterns as the bringing of children together at the start of the school year. Whatever the cause, such seasonal variations in the effective magnitude of the transmission parameter β have been documented for measles, pertussis (whooping cough), chicken pox, and other microparasitic infections, and some of the dynamical consequences explored by Yorke and co-workers (Yorke and London, 1973; London and Yorke, 1973; Yorke et al., 1979) and Dietz (1975, 1976; see also Grossman et al., 1977; Grossman, 1980).

(xi) Homogeneous Mixing

Perhaps the most significant, and questionable, assumption embodied in Eqs. (2)–(4) and their friends and relatives is that of homogeneous mixing. All the local details – school, family, geography – are averaged out, and epidemiological and demographic processes are treated as occurring at rates that depend only on the average number or average density of susceptibles, infectives, and immunes. In particular, if λ is given by Eq. (7), the net rate of appearance of new infections is proportional simply to the product of the total number of susceptibles, X, and the total number of infectives, Y; infection is effectively described by binary collisions in an ideal gas. The underlying philosophy, well expressed by Bartlett (1960), is that "even if a multiplicity of detailed causes is operating to produce the observed broad classes of events, it is often an economy of thought ... to ignore these and appeal merely to the operations of chance and the laws of averages." Although this assumption of homogeneous mixing thus represents a sensible starting point for theoretical studies, we should always keep its potential inadequacies in mind, and be alert for chances to test its predictions against data.

Obvious breakdown of the assumption of homogeneous mixing occurs when – along with the more usual symptomatic carriers – there are asymptomatic carriers who can transmit infection; this complication can be included by distinguishing two different classes of infected individuals (Cooke, 1982). Similarly, Yorke and collaborators (Yorke et al., 1978; Hethcote et al., 1982; see also May, 1981) have found it necessary to distinguish two classes of transmitters (a sexually very active and promiscuous "core" class, and a separate less promiscuous class) to get a minimally accurate description of the epidemiology of gonorrhea. Becker and Angulo (1981) have found it necessary to distinguish between contacts within the family group and those outside in his discussion of smallpox in Brazil. Formally, it is possible to construct compartmental models with more and more compartments to describe regional or social subgroupings (for example: Kemper, 1980; Nold, 1980; Travis and Lenhart, 1983), and eventually such models will shade into those where there is a continuous distribution of hosts (corresponding to different genotypes, geographical location, social groupings, and so on). In what follows, I will restrict my attention to the basic insights that can be gained from the simplest models for homogeneously mixed populations.

(xii) Immunization

As discussed below, the overall statics and dynamics of the association between hosts and microparasites will be altered when a program of immunization is instituted. In Eqs. (2)–(4), any such program will be described formally by a loss term in Eq. (2) and a gain term in Eq. (4), representing transfer from the susceptible class to the immune class by immunization (rather than by natural infection and recovery).

Part II. Statics of Microparasitic Infections in Constant Host Populations

In this section, we focus attention on the steady state of human-host microparasite associations. That is, we assume no time dependence in the variables X, Y, Z of Eqs. (2)–(4), so that they become functions of the single variable a: $X(a)$, $Y(a)$, $Z(a)$.

Such analysis of the statics of host-microparasite interactions is accompanied by two assumptions that are so commonly made that they usually pass unremarked in the literature (for example, Bailey, 1975). The first assumption is that births and deaths are exactly balanced, to give a constant host population of magnitude N. The second assumption is to ignore mortality associated with the infection: $\alpha = 0$ in Eq. (3). With $\alpha = 0$, the first assumption corresponds to the birth rate B being given by

$$B = \int \mu(a)N(a)da, \tag{11}$$

where $N(a)$ is given by Eq. (1) (with the dependence on time t having dropped out). For a constant, type A mortality this equation reduces to $B = \mu N$ [with $N = \int N(a)da$].

Under these assumptions, the static versions of Eqs. (2)–(4) reduce to the system of ordinary differential equations (Bailey, 1975; Dietz, 1975, 1976):

$$dX/da = -[\lambda + \mu(a)]X(a), \tag{12}$$

$$dY/da = \lambda X - [v + \mu(a)]Y(a), \tag{13}$$

$$dZ/da = vY - \mu(a)Z(a). \tag{14}$$

Adding these three equations, we get a differential equation for the number of hosts of age a, $N(a) = X(a) + Y(a) + Z(a)$:

$$dN/da = -\mu(a)N(a). \tag{15}$$

The boundary conditions are $X(0) = N(0)$ and $Y(0) = Z(0) = 0$, where $N(0) = B$ is the net birth rate, or number of hosts of age zero.

As Dietz (1975, 1976) and others have shown, the integration of these equations is straightforward. Equation (15) leads immediately to the result

$$N(a) = N(0)\phi(a), \tag{16}$$

where $\phi(a)$ is defined for notational convenience as

$$\phi(a) = \exp\left[-\int_0^a \mu(s)ds \right]. \tag{17}$$

For $X(a)$ we obtain from Eq. (12) the result

$$X(a) = N(0)\phi(a)e^{-\lambda a}. \tag{18}$$

It follows that the *fraction* of hosts of age a who are susceptible is

$$x(a) = X(a)/N(a) = e^{-\lambda a}. \tag{19}$$

Clearly the expressions (18) and (19) could be generalized to the case when the force of infection is age-dependent, $\lambda(a)$. Finally, for $Y(a)$ and $Z(a)$ we may obtain the

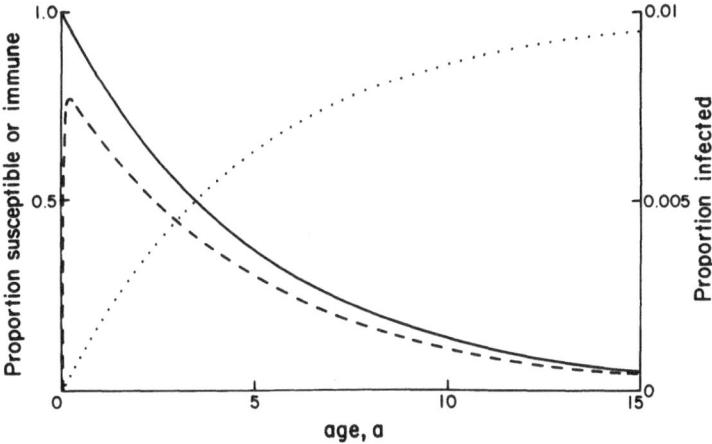

Fig. 2. This figure illustrates the proportions of a population who are susceptible (solid curve), infected (dashed curve), and recovered and immune (dotted curve), for an infection where the force of infection is $\lambda = 0.2 \, \mathrm{yr}^{-1}$ (corresponding to an average age at infection of 5 years) and where the recovery rate is $v = 25 \, \mathrm{yr}^{-1}$ (corresponding to an average duration of infection of about 2 weeks). Note that the proportion infected is always relatively small, being at most of the order of λ/v

results

$$Y(a) = \lambda N(0)\phi(a) [e^{-va} - e^{-\lambda a}]/(\lambda - v), \tag{20}$$

$$Z(a) = N(0)\phi(a) [1 - (\lambda e^{-va} - ve^{-\lambda a})/(\lambda - v)]. \tag{21}$$

These age-dependent patterns of susceptibility, infection and immunity are illustrated in Fig. 2, for a typical childhood infection.

Two quantities of epidemiological interest can usefully be discussed at this point.

Age at First Infection

It is intuitively plausible that the average age at which individuals acquire infection, A, is the reciprocal of the force of infection, λ. More rigorously, suppose we have serological studies of the host population, which determine empirically the fraction of hosts of age a who are susceptible, $x(a)$. Of these susceptibles, the number acquiring infection between age a and $a + d(a)$ is $\lambda x(a)da$, so that A is the first moment of the $\lambda x(a)$ distribution:

$$A = \frac{\int_0^\infty a\lambda x(a)da}{\int_0^\infty \lambda x(a)da}. \tag{22}$$

Substitution of the expression (19) for $x(a)$ leads to the identification

$$A = 1/\lambda. \tag{23}$$

More generally, the basic differential Eq. (12) in conjunction with the definition of $x(a)$, Eq. (19), can be used to perform a partial integration in Eq. (22), resulting in

$$A = \int_0^\infty x(a)da. \tag{24}$$

This result is true even when λ is a function of age, $\lambda(a)$.

The practical use of all this is that serological studies, or (less accurately) case notifications, can be used to estimate A for a particular infection, and thus to estimate λ from Eq. (23). Such empirical estimates of these epidemiological parameters are discussed more fully, with examples, by Anderson and May (1983).

Basic Reproductive Rate of Infection, R_0

Following the earlier work of Macdonald (1952), Dietz (1975, 1976), Smith (1970), Yorke et al. (1979), Anderson and May (1982b), and others, we define the basic reproductive rate of a microparasitic infection, R_0, as the number of secondary cases produced, on average, when one infected individual is introduced into a wholly susceptible population. Clearly, R_0 combines the biology (or etiology) of the infection with a variety of social, behavioral and environmental factors that influence contact rates. A direct assessment of R_0 by investigating all these various factors is usually quite impossible.

An indirect estimate of R_0 can, however, be obtained if the assumption of homogeneous mixing is regarded as reasonable. Under this assumption, if only a fraction X/N of the population is susceptible, then the effective reproductive rate, R, of the infection is reduced by this fraction:

$$R = R_0 X/N. \tag{25}$$

But at *equilibrium* we must have $R = 1$: by definition, at equilibrium each infected individual produces on average one and only one secondary infection. Thus we can estimate R_0 from knowledge of the fraction of the population that are susceptible at equilibrium:

$$R_0 = (N/X)_{equilibrium}. \tag{26}$$

Here X and N are the total number of susceptibles and of hosts, respectively:

$$N = \int_0^\infty N(a)da, \tag{27}$$

$$N = \int_0^\infty X(a)da. \tag{28}$$

To get explicit expressions for R_0 in terms of the epidemiological parameters A or λ, it is necessary to make some explicit assumption about the mortality curve, $\mu(a)$.

For type A mortality, Eq. (5), we have that the factor $\phi(a)$ of Eq. (17) is

$$\phi(a) = e^{-\mu a}. \tag{29}$$

Then, using Eqs. (16) and (27) to get N, and Eqs. (18) and (28) to get X, we have

$$N = N(0)/\mu, \tag{30}$$

$$X = N(0)/(\lambda + \mu). \tag{31}$$

Thence

$$R_0 = 1 + (\lambda/\mu). \tag{32}$$

Recalling that $\lambda = 1/A$, and that the average death rate μ may be related to average life expectancy L by $\mu = 1/L$, Eq. (32) can be re-expressed as

$$R_0 = 1 + (L/A). \tag{33}$$

This is a somewhat more direct derivation than that of Dietz (1975, 1976).

For the more realistic type B mortality, the analogous sequence of calculations gives

$$\phi(a)=1, \quad \text{for} \quad a<L,$$
$$\phi(a)=0, \quad \text{for} \quad a>L. \tag{34}$$

Then it follows that

$$N=N(0)L, \tag{35}$$

$$X=N(0)\,[1-e^{-\lambda L}]/\lambda, \tag{36}$$

$$R_0=\frac{\lambda L}{[1-\exp(-\lambda L)]}. \tag{37}$$

To a good approximation, this expression for R_0 reduces for $L\gg A$ (which is usually true) to

$$R_0\simeq L/A. \tag{38}$$

Immunization Programs and New Equilibria

Implementation of a vaccination or other immunization program has two main effects. The first, or *direct*, effect is that a fraction of the host population is removed into the immune class by immunization; such direct protection will, by itself, result in fewer infections. The second, or *indirect*, effect is that the smaller number of infections implies a weaker force of infection; an unvaccinated individual is indirectly protected, to a degree, by this diminution in the force of infection. One important consequence of these indirect effects is that one does not have to immunize everyone in order to eradicate an infection – once the vaccination coverage has reached some critical level, the infection will be unable to maintain its reproductive rate above unity.

To see how this works, under the assumption of homogeneous mixing, consider the circumstance where a fraction p of each cohort of the host population is successfully immunized at age b. Once the new equilibrium state is attained, let the new (smaller) force of infection be denoted by λ'. Then the appropriate modification of Eq. (12) gives

$$X'(a)=N(0)\phi(a)\exp(-\lambda'a); \quad a\leq b, \tag{39a}$$

$$X'(a)=(1-p)N(0)\phi(a)\exp(-\lambda'a); \quad a>b. \tag{39b}$$

The quantity $\phi(a)$ is, of course, still given by Eq. (17), and $N(a)$ still obeys Eq. (16). So long as the infection continues to persist at some new equilibrium state following the implementation of the vaccination program, the expression (26) relating R_0 to the equilibrium values of X and N continues to hold.

For type A survival, we have $\phi(a)$ given by Eq. (29). N is again given by Eq. (30), while X is given from Eqs. (28) and (39):

$$X = N(0)\,[1 - p\exp(-(\lambda' + \mu)b)]/(\lambda' + \mu). \tag{40}$$

Hence R_0 is related to p and λ' by

$$R_0 = \frac{1 + (\lambda'/\mu)}{[1 - p\exp(-(\lambda' + \mu)b)]}. \tag{41}$$

Similarly, for type B survival, we have N given by Eq. (35) and X by

$$X = N(0)\,[1 - p\exp(-\lambda'b) - (1 - p)\exp(-\lambda'L)]/\lambda'. \tag{42}$$

Hence

$$R_0 = \frac{\lambda'L}{[1 - p\exp(-\lambda'b) - (1 - p)\exp(-\lambda'L)]}. \tag{43}$$

In either case – the relatively realistic type B survival or the less realistic but more usually studied type A survival – the force of infection, λ', under a specified immunization program may be found by first estimating R_0 from pre-vaccination data [from Eqs. (37) or (33)], and then using Eqs. (43) or (41) to get λ' in terms of the parameters p and b that characterize the vaccination program. Other kinds of immunization programs (for example, immunization at some constant rate, or according to some complicated age-specific schedule) may be similarly assessed.

Criteria for Eradication

The infection will not be able to maintain itself at equilibrium in the population if the level of herd protection by immunization is such that the reproductive rate of the microparasite falls below unity. Formally, the critical level of vaccination coverage corresponds to the limit $\lambda' \to 0$ in Eqs. (41), (43) or other equivalent equations. Specifically, for type A survival, the critical immunization level, p_c, is obtained by putting $\lambda' \to 0$ in Eq. (41) to get

$$p_c = (1 - 1/R_0)\exp(b/L). \tag{44}$$

Alternatively, using Eq. (33) for R_0,

$$p_c = \left(\frac{L}{A + L}\right)\exp(b/L). \tag{45}$$

Likewise, for type B survival, the critical level of coverage, p_c, is found from Eq. (43) to be

$$p_c = \frac{1 - (1/R_0)}{1 - (b/L)}, \tag{46}$$

or

$$p_c = \frac{1-(A/L)(1-\exp(-L/A))}{1-(b/L)}. \tag{47}$$

In either case it is clear that eradication is easier if people are immunized at the earliest feasible age, b (possibly allowing for the complications of maternal antibodies, discussed earlier). Eradication is simply impossible once $b \sim A$. In the limit $b \rightarrow 0$, the above Eqs. (44) and (46) reduce to the familiar result (Smith, 1970):

$$p_c = 1 - 1/R_0. \tag{48}$$

The above analysis rests on the assumption of homogeneous mixing. One of the consequent predictions, namely that the total fraction of the population who remain susceptible at equilibrium is *independent* of the immunization program $[X/N = 1/R_0 = \text{constant, Eq. (26)}]$, can be tested against empirical data. One especially interesting such test has recently been made for measles by Fine and Clarkson (1982). Analyzing the data that are available for age-specific incidence and immunity levels for measles in England and Wales since 1950, they find the "total number of individuals susceptible to measles has remained relatively constant," at around 4 to 4.5 million. This corresponds to about 9% of the population being susceptible, implying R_0 is around 11–12; this accords with other independent estimates (Anderson and May, 1982b). As Fine and Clarkson emphasize, there are biases and deficiencies in the available data (having to do mainly with the notification of cases), and methods of correcting for such biases depend to a degree on the assumption of homogeneous mixing. Thus their test of this assumption is not altogether free of some circularity. I think, however, that Fine and Clarkson's analysis is important, showing that homogeneous mixing represents a useful working approximation for measles in England and Wales, and therefore possibly for broadly similar infections elsewhere. More such tests are greatly to be desired.

Does Immunization Always Reduce Disease?

As a result of the direct and indirect effects of an immunization program, fewer people will acquire infection. But the weaker force of infection in the post-immunization community means that those who do acquire infection will on average do so at an older age.

If fatalities or other complications are more likely at older ages, it is thus conceivable that certain vaccination programs could actually increase the incidence of serious cases. The likelihood of such a perverse outcome is not easily guessed, but rather requires a close analysis of the detailed interplay among several factors: the way serious illness depends on age, the details of the vaccination program, and the basic biology of the transmission process (summarized by R_0).

It is possible that the history of poliomyelitis represents a "natural experiment" of this general kind. The argument runs that, in earlier times, most people contracted this infection in the first few years of life, when it very rarely has serious consequences. Increasingly high standards of cleanliness in Western societies in the middle decades of this century led to lower values of R_0, and thence to a diminished force of infection and a consequent rise in the average age at which infection was acquired (albeit by fewer people, in total). Since paralysis and other complications are apparently much more likely for infections in teenage or older individuals, poliomyelitis paradoxically became more troublesome as its transmissibility declined. More recently, in a notable success story, the advent of safe, effective and cheap vaccines has effectively eradicated poliomyelitis.

Rubella is a particularly striking example of an infection whose seriousness depends on the age at infection. Usually a mild infection, rubella can cause the damaging congenital rubella syndrome (CRS) in the offspring of women who acquire the infection in the first trimester of pregnancy. Following a seminal study by Knox (1980), several people (Dietz, 1981; Hethcote, 1983; Anderson and May, 1983) have shown how the incidence of CRS may actually increase under some vaccination programs.

To explore these possibilities, we define $w(a_1, a_2)$ to be the number of people acquiring infection between the ages a_1 and a_2 at equilibrium after implementation of a specific immunization program, as a ratio to the corresponding number of infections at equilibrium in the pre-vaccination population:

$$w(a_1, a_2) = \int_{a_1}^{a_2} \lambda' X'(a) da \bigg/ \int_{a_1}^{a_2} \lambda X(a) da. \tag{49}$$

It is now a routine exercise to obtain explicit expressions for $w(a_1, a_2)$ for any specified immunization program: $X(a)$ and $X'(a)$ are calculated along the lines laid down above, and similarly λ' is computed from R_0 using an equation such as Eq. (41) or Eq. (43). In particular, if the policy consists of successful immunization of a proportion p of all children at some age $b < a_1$, it can be shown that for type A survival

$$w(a_1, a_2) = (1-p) \frac{\lambda'(\lambda+\mu)\left[\exp(-(\lambda'+\mu)a_1) - \exp(-(\lambda'+\mu)a_2)\right]}{\lambda(\lambda'+\mu)\left[\exp(-(\lambda+\mu)a_1) - \exp(-(\lambda+\mu)a_2)\right]}. \tag{50}$$

Here λ' is given by Eq. (41). For type B survival, the corresponding expression is

$$w(a_1, a_2) = (1-p) \frac{\left[\exp(-\lambda'a_1) - \exp(-\lambda'a_2)\right]}{\left[\exp(-\lambda a_1) - \exp(-\lambda a_2)\right]}, \tag{51}$$

with λ' given by Eq. (43). The detailed derivations are left as an exercise to the reader. The results can readily be extended to programs with vaccination according to any specified age schedule.

The solid lines in Fig. 3 show this ratio, $w(16, 40)$, in the incidence of infection in the age classes aged 16 to 40 after vaccination versus before, as a function of the proportion p who are successively immunized (at age 1 year). The curves are for a range of values of R_0 (or, equivalently, average age at infection, A), as indicated.

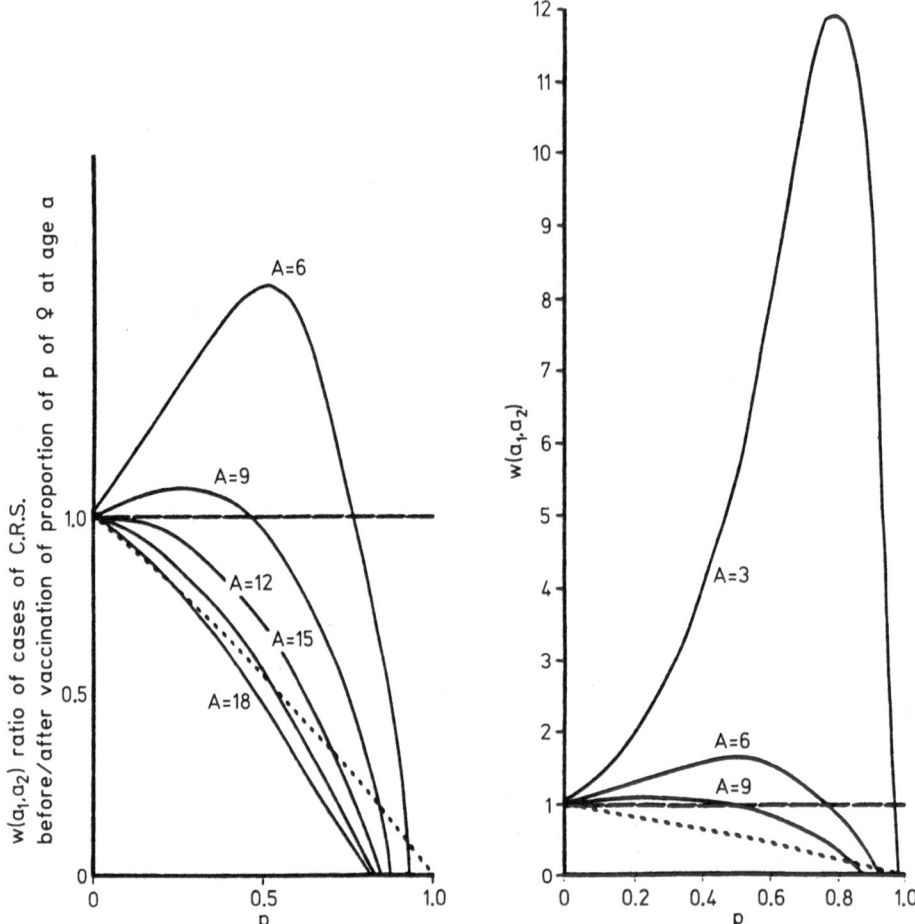

Fig. 3. The solid curves show the number of infections among females aged between 16 and 40 years when a proportion p of all children are successfully immunized at age 1 year, expressed as a ratio, $w(a_1, a_2)$, to the corresponding number of infections before any vaccination. This vaccination strategy roughly corresponds to that employed against rubella in the USA. The various curves represent different values of R_0, as equivalently expressed by different values of the average age at infection before widespread vaccination, A [see Eqs. (33) or (37)]; for rubella, $A \simeq 9$ years. The dashed curves show the same ratio, $w(a_1, a_2)$, between infection in the age range 16 to 40 years, at equilibrium after versus before vaccination, when a proportion p of girls, and only girls, are successfully immunized at age 12 years. This vaccination strategy roughly corresponds to that employed against rubella in the UK. Here the dashed curves are indistinguishable for the various values of A (or, equivalently, R_0) shown in the solid curves

The relatively realistic type B survival has been assumed, giving results indistinguishable from those with actual survival curves obtained from actuarial or demographic data. Rubella in the USA and UK has an A value around 9 years (corresponding to R_0 around 8 or 9). An accurate estimate of the incidence of CRS would involve integrating over the product of $\lambda X(a)$ times the pregnancy rate, but simply studying $w(16, 40)$ is a good first approximation (for a more full and accurate exploration of all these matters, see Anderson and May, 1983). From the figure, we see that protection levels of around 50% or less slightly increase the incidence of CRS. At high levels (around or above 90%), however, this vaccination strategy can eradicate rubella. In the USA, such coverage has been attained, and rubella seems on the way to eradication. But in Gambia, for instance, where the average age of rubella infection is substantially below 6 years, it would seem folly to institute vaccination of infants against rubella, as the result is likely to be marked increase in the incidence of CRS unless extraordinary levels of coverage of 95% or better can be achieved and maintained.

In the UK, where vaccination of children has never been required for entrance to school or kindergarten (as it is now in most states of the USA), p for rubella vaccination has been around 50% or so (although recent levels are around 80% or above). Consequently, the immunization strategy in the UK has been to vaccinate pre-pubescent girls, and only girls, at around age 12, thus taking advantage of the immunity acquired by natural infection in the first 12 years of life. Such a policy corresponds to vaccinating a fraction $p/2$ of the population (assuming a 50:50 sex ratio) at age $b = 12$: the consequent formula for determining λ' is the appropriate modification of Eq. (43),

$$R_0 = \frac{\lambda'L}{[1 - \frac{1}{2}p\exp(-\lambda'b) - (1 - \frac{1}{2}p)\exp(-\lambda'L)]}. \tag{52}$$

The formula (51), with the full factor $(1-p)$, still gives the relative incidence of rubella among females between the ages of a_1 and a_2. This policy of vaccinating at most half the population at age 12 has relatively little impact on the force of infection (that is, $\lambda' \simeq \lambda$), and so it is not capable of eradicating rubella even at high levels of coverage. But it does protect females directly, without the offsetting indirect effects that characterize "everyone-at-age-one" policy at modest values of p. The dashed line in Fig. 3, which is to be compared with the solid curves in Fig. 3, shows $w(16, 40)$ as a function of p, the proportion of 12-year-old girls successively immunized, for the UK strategy.

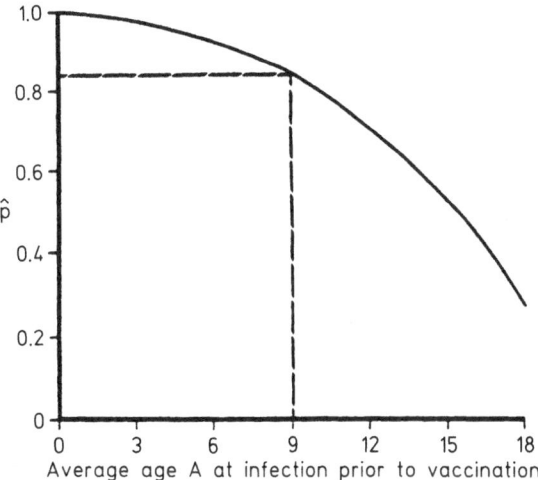

Fig. 4. This figure shows the critical level of successful immunization, p, above which the USA vaccination policy produces greater reduction in the incidence of CRS than the UK policy, as a function of the average age at infection, A, prior to vaccination. The dashed lines indicate this critical level of p for the actual value of A (around 9 years) that pertains in the UK and the USA

Figure 4 summarizes much of this work, by showing the value of p at which "UK" and "USA" policies are equally effective (leading to the same incidence of CRS), for various values of the average age of infection, A. For rubella, the cross-over point appears to be around 85% coverage: below this level (as in the UK), the UK policy affords better protection against CRS; above this level (as in the USA) the USA policy affords better protection.

A much more full account of these questions for rubella and for measles is given, along with comparison with serological and case notification data, in Anderson and May (1983).

Part III. Dynamics of Microparasitic Infections in Constant Host Populations

As mentioned in the introduction, most childhood infections exhibit non-seasonal oscillations, with periods ranging from around 2 years for measles in developed countries before vaccination to around 6–7 years for rubella before vaccination (and possibly around 7 years for smallpox in developing countries before the global eradication program). Evidence for these periodicities is summarized in Anderson and May (1982b); Figure 5 shows the data for measles and for pertussis in England and Wales before and after the advent of vaccination.

These periodic variations have excited attention since the turn of the century (for example, Soper, 1929). In this Part III of my chapter, I first outline a very simple

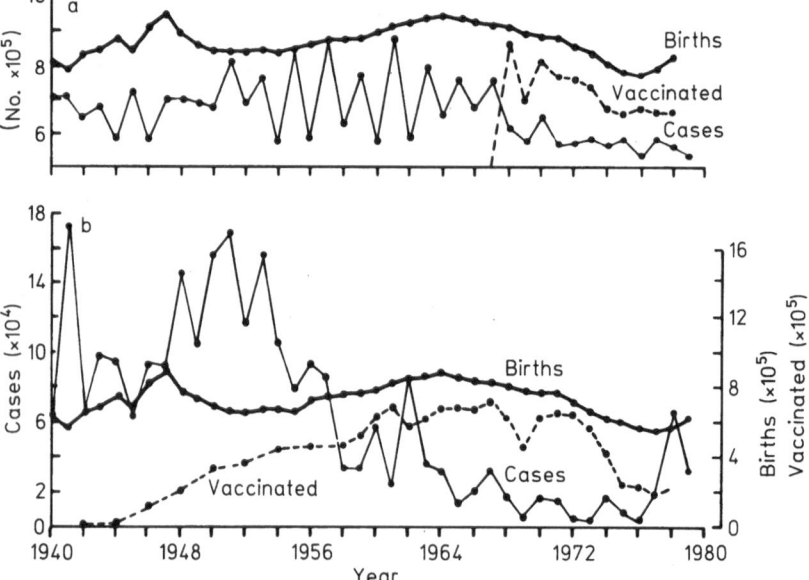

Fig. 5a, b. Reported cases of **a** measles and **b** whooping cough in England and Wales are shown, from 1940 to 1979. The figures show the total number of births (thick line), the total number of reported cases (thin line), and the total number of people vaccinated (dashed line) each year. From Anderson and May, 1982b

model for a microparasitic infection described in discrete time; this excessively simple model exhibits neutrally stable oscillations, which lay bare some of the essentials of the periodicities found in (or tortured out of) more complicated models. A series of realistic complications are then discussed in outline, beginning with the simplest model in continuous time, as originally studied by Soper. Finally, I turn to dynamic aspects of the implementation of vaccination programs, and show on the one hand how some basic features of such dynamics can be understood on the basis of the earlier discussion, while on the other hand some rather surprising practical implications emerge and can be tested against data from actual vaccination programs.

Microparasitic Infections in Discrete Time

Let us assume that the chain of infection and recovery is described as happening in discrete time steps, of duration τ: τ is the (average) interval between an individual acquiring infection and passing it on to the next infectee. For most childhood microparasitic infections, τ is around 6–20 days (latent plus infectious periods), although for some arthropod-borne viruses (for example, yellow fever), τ can be longer. We define:

$X_t =$ total number of susceptibles, at time t,

$Y_t =$ total number of infectious individuals, at time t,

$N =$ total population, assumed constant.

As always, $N = X_t + Y_t + Z_t$.

We now seek equations linking X_t and Y_t at time t with the corresponding quantities $X_{t+\tau}$ and $Y_{t+\tau}$ one time step later. Assuming homogeneous mixing, as discussed above, we get

$$Y_{t+\tau} = (R_0 X_t/N) Y_t. \tag{53}$$

Each case, of which there are Y_t, would give R_0 secondary infections if all hosts were susceptible, and each case gives $R_0(X_t/N)$ if a fraction X_t/N are susceptible. Susceptibles are lost by infection, Eq. (53), and gained by new births:

$$X_{t+\tau} = X_t - Y_{t+\tau} + B. \tag{54}$$

Here B represents net births in the interval τ. If the per capita annual birth rate is m, then

$$B = m\tau N. \tag{55}$$

Equilibrium in this simple model is found by putting $X_{t+\tau} = X_t = X^*$ and $Y_{t+\tau} = Y_t = Y^*$, to get

$$Y^* = B = m\tau N, \tag{56}$$
$$X^* = N/R_0. \tag{57}$$

Equation (57) is, of course, identical to Eq. (26).

The *local stability* of this equilibrium can be studied by the usual techniques. We put $X_t = X^* + x_t$ and $Y_t = Y^* + y_t$, and expand in Taylor series, ignoring terms of order x^2, y^2, xy, to get the pair of linear difference equations

$$y_{t+\tau} = y_t + (R_0 B/N)x_t, \tag{58}$$

$$x_{t+\tau} = x_t - y_{t+\tau}. \tag{59}$$

In these linearized equations, we can factor out the time dependence, via the eigenvalues Λ, essentially defined as

$$x_t \simeq x_0 \Lambda^n \quad \text{and} \quad y_t \simeq y_0 \Lambda^n. \tag{60}$$

Here $n = t/\tau$ is the number of time steps from 0 to t. Equations (58) and (59) thus reduce to

$$\Lambda y = y + \varepsilon x, \tag{61}$$

$$\Lambda x = x - \Lambda y. \tag{62}$$

For notational convenience, we have defined ε as

$$\varepsilon = R_0 B/N = (R_0 m)\tau. \tag{63}$$

We remember from Part II that R_0 is usually given to an excellent approximation by $R_0 \simeq L/A$ [where in a steady population the life expectancy L is the reciprocal of the birth rate, $L = 1/m$, and where A is the average age at infection: Eq. (38)]. Then Eq. (63) can equivalently be rewritten as

$$\varepsilon \simeq \tau/A. \tag{64}$$

Since τ is of order 6–20 days (or at most a month or so), while A is of order 2–10 years, we typically have $\varepsilon \ll 1$.

Returning to Eqs. (61) and (62), we obtain for the eigenvalues Λ the quadratic equation

$$\Lambda^2 - (2 - \varepsilon)\Lambda + 1 = 0. \tag{65}$$

The exact solution is

$$\Lambda = (1 - \tfrac{1}{2}\varepsilon) \pm i[1 - (1 - \tfrac{1}{2}\varepsilon)^2]^{1/2}. \tag{66}$$

That is, Λ is a complex number *with modulus unity*. Equation (66) for Λ can be rewritten as

$$\Lambda = \exp(\pm i\theta\tau), \tag{67}$$

with θ defined as

$$\theta = [\cos^{-1}(1 - \tfrac{1}{2}\varepsilon)]/\tau. \tag{68}$$

In other words, the time dependence of the perturbations to the equilibrium, x_t and y_t, are characterized by the behavior of the function $\Lambda^n = \Lambda^{t/\tau}$, which from Eq. (67) gives

$$x_t, y_t \sim \exp[i\theta t].\tag{69}$$

Since the magnitude of the eigenvalues is unity, a small disturbance to the equilibrium state of Eqs. (53) and (54) simply will oscillate indefinitely, exhibiting the neutral stability of the frictionless pendulum (like the similarly pathological neutrally stable oscillations in the Lotka-Volterra prey-predator model). The period, T, of these continuing oscillations is given by Eq. (69):

$$T = 2\pi/\theta.\tag{70}$$

For $\varepsilon \ll 1$, as discussed above, we have $\cos^{-1}(1 - \tfrac{1}{2}\varepsilon) = \sqrt{\varepsilon[1 + 0\varepsilon]}$, whence from Eqs. (68) and (70)

$$T \simeq 2\pi\tau\varepsilon^{-1/2} = 2\pi(A\tau)^{1/2}.\tag{71}$$

That is, the incidence of infection oscillates indefinitely – in a neutrally stable manner – with a period that is the geometric mean of the inter-infection interval (τ) and the average age at infection (A). The approximate expression (71) gives reasonably good agreement with the observed non-seasonal oscillations in many microparasitic infections (see Anderson and May, 1982b, Table 1). For example, for measles in developed countries we have $\tau \sim 0.04$ yr and $A \sim 4$–5 yr, leading to $T \sim 2$–3 yr. For rubella the corresponding figures before vaccination are $\tau \sim 0.06$ yr, and $A \sim 9$–10 yr, and thence $T \sim 5$ yr. Both estimates accord with epidemiological data from the USA and the UK before widespread vaccination.

More Complicated and More Realistic Models

Essentially any change in the structure of the above model will tip it off the razor's edge of neutral stability, either to damped oscillations, or to sustained fluctuations [which may be stable limit cycles, or even chaotic fluctuations: e.g., May (1976)]. In either case, however, the basic period will still be given approximately by Eq. (71). We now consider the continuous analog of Eqs. (53) and (54) in some detail, and indicate other possible refinements.

(i) The Soper (1929) Model

We define $X(t)$, $Y(t)$, and $Z(t)$ to be the number of susceptibles, infectives, and immunes, respectively, at time t; the total population, N, is taken as constant (that is, births exactly balance deaths). To simplify the analysis, we follow the (unrealistic) convention of taking the death rate (and therefore the birth rate) to be a constant, μ. As described by Eq. (7), the force of infection, λ, is related to the number of infectives by $\lambda = \beta Y$. We then have the system of differential equations.

$$dX/dt = \mu N - (\beta Y + \mu)X, \tag{72}$$

$$dY/dt = \beta YX - (v + \mu)Y, \tag{73}$$

$$dZ/dt = vY - \mu Z. \tag{74}$$

The term μN in Eq. (72) represents net births. Notice that adding all three equations gives $dN/dt = 0$, confirming $N = $ constant.

The *equilibrium* solutions of Eqs. (72)–(74) are obtained by putting the left-hand sides equal to zero, to get

$$X^* = (v + \mu)/\beta, \tag{75}$$

$$Y^* = (N - X^*)(\mu/(v + \mu)), \tag{76}$$

$$Z^* = (N - X^*)(v/(v + \mu)). \tag{77}$$

As usual, the *local stability* can be studied by expanding about this equilibrium $(X(t) = X^* + x(t), Y(t) = Y^* + y(t), Z(t) = Z^* + z(t))$, and disregarding second and higher order terms. Since $X + Y + Z = $ constant, for all t, there are effectively only two equations, and we therefore confine attention to the linearized versions of Eqs. (72) and (73):

$$dx/dt = -(\mu + \beta Y^*)x - (\beta X^*)y, \tag{78}$$

$$dy/dt = (\beta Y^*)x. \tag{79}$$

The time dependence can be factored out as $\exp(\Lambda t)$, where Λ represents the eigenvalues of the system. These eigenvalues will now obey the quadratic equation

$$\Lambda^2 + B\Lambda + C = 0, \tag{80}$$

where the coefficients B and C are defined as

$$B = \mu + \beta Y^* = \mu N/X^* = \mu R_0, \tag{81}$$

$$C = (\beta Y^*)(\beta X^*) = \mu(v + \mu)(N/X^* - 1)$$
$$= \mu(v + \mu)(R_0 - 1). \tag{82}$$

it follows that

$$\Lambda = -\tfrac{1}{2}B \pm i[C - (\tfrac{1}{2}B)^2]^{1/2}. \tag{83}$$

Moreover, we note that v (the recovery rate) is of order (1/days), while μ (the birth or death rate) is of order (1/decades), so that $v \gg \mu$ and consequently $C > B^2$. Thus Eq. (83) can be rewritten approximately as

$$\Lambda \simeq -\tfrac{1}{2}B \pm i[v\mu(R_0 - 1)]^{1/2}. \tag{84}$$

That is, the eigenvalues are complex numbers, describing damped oscillations. The damping time is of order $2/B \sim 2/\mu R_0 \sim 2A$ (with A the average age at infection); thus the damping time is characteristically of order 10 years or more. The period of the damped oscillations, T, is given by the imaginary term in Eq. (84),

$$T \simeq 2\pi(v\mu(R_0 - 1))^{-1/2}. \tag{85}$$

Using Eq. (33) to reexpress R_0 in terms of A, Eq. (85) gives

$$T \simeq 2\pi(A/v)^{1/2}. \tag{86}$$

This is essentially the same as Eq. (71) (allowing for the fact that we have omitted any account of the latent period in the above analysis). Note that the characteristic oscillation time is shorter than the damping time by the ratio $(\mu/v)^{1/2}$, which is typically of the general order of 10. In this sense, we note that the "Soper oscillations," although damped, are relatively *weakly* damped. The period of these weakly damped oscillations is as given by the simpler model, discussed above.

(ii) Demographic Stochasticity

The above analyses, both in discrete and in continuous time, are rigidly deterministic with birth, death, and infection all occurring at fixed rates. Although it may be mathematically inconvenient, humans in fact come quantized in integer units, so that ultimately it does not make sense to talk of a fraction of a birth or infection. But accounting for this fact inevitably introduces stochastic elements into the description of epidemiological processes. A variety of studies have shown that such "demographic stochasticity" [as distinct from "environmental stochasticity;" see May (1974)] can tip the kind of models discussed above from neutral stability or damped oscillations into *sustained* oscillations in the incidence of infection, with the period given essentially by the approximate equations (71) and (86).

A seminal such study is that by Bartlett (1957, 1960). He began with a discrete model along the lines of Eqs. (53) and (54) above, modified by replacing the "infection probability" $R_0 Y_t/N$ in Eq. (53) by the expression

$$[1 - (1 - R_0/N)^{Y_t}]. \tag{87}$$

The expression (87) gives a more accurate account of the binomial infection process, and clearly reduces to the previous expression when both R_0/N and $R_0 Y_t/N$ are small. This modification, by itself, serves to tip the neutrally stable model discussed above into (weakly) damped oscillations. Bartlett, however, showed that further incorporation of a term representing the variability introduced by demographic stochasticity had the effect of producing sustained oscillations (unless the population is too small to sustain the infection, which then exhibits "fade out"). Bartlett's work has been extended in a variety of elegant mathematical studies, many of which are surveyed by Bailey (1975). More recently, an extensive series of numerical simulations have shown that demographic stochasticity tends to produce sustained, if slightly irregular, oscillations at essentially the period T given by Eqs. (71) and (86) (Anderson and May, 1984).

(iii) Seasonality

As mentioned in Part I, a variety of biological and social effects can produce pronounced seasonal variations in the transmission rate. As shown by Yorke and London (1973), London and Yorke (1973), Yorke et al. (1979), Dietz (1976), Grossman (1980), Aron and Schwartz (1983), and others, such seasonality appears capable of "pumping" the system to produce longer-term oscillations, with peaks separated by intervals of years. The mechanism is most transparent when the basic period T of Eqs. (71) and (86) is around 2 years, but the mechanism may work more generally.

(iv) Time Lags, etc

Time delays in epidemiological processes, or defined intervals in the duration of latency or infectiousness (in contrast with constant rates: see Fig. 1), also in some extreme circumstances appear to be capable of "pumping" the system at the basic period T discussed above. These effects, which have been studied by Hethcote and Tudor (1980), Hethcote et al. (1981), Grossman (1980), Smith (1983), and others can be complicated, and there is room for further work here.

Overall, I regard this subject of non-seasonal oscillations in the incidence of many microparasitic infections as still a very open one. There is need for more study under the headings (ii)–(iv) above, and (even more important in my opinion) for synoptic and systematic study of the Fourier transforms of long runs of epidemiological data to determine exactly what the facts are.

Dynamics of Immunization Programs

The basic set of partial differential equations, Eqs. (2)–(4), may be used to study the dynamical response of the host population to the implementation of an immunization program. There have been relatively few such studies of the epidemiological dynamics of vaccination [Knox (1980) is an important exception]; most work has been confined to the kind of static analysis outlined in Part II above. As we shall now see, some things that are not intuitively obvious can happen en route to the final equilibrium. Some of these short-term dynamical phenomena seem likely to have practical implications. In what follows, I indicate the general nature of the analysis, and sketch some applications (comparing the simple theory with public health data) to vaccination against rubella in the USA and the UK. Both theoretical and empirical aspects of this work are amplified in much more detail in Anderson and May (1983).

If the age-specific rate of immunization is $c(a)$, Eq. (2) for $X(a, t)$ is modified to read

$$\partial X/\partial a + \partial X/\partial t = -[\lambda(t) + c(a) + \mu(a)]X(a, t).$$

(88)

Here the initial value $X(a, 0)$ at time $t=0$ is the pre-vaccination equilibrium distribution, described by Eq. (18) above. The other boundary condition is, as always, that $X(0, t) = N(0)$, the number of births, for all t. As emphasized at the beginning of Part II, all this rests on the assumption that the total host population is constant (with births balancing deaths).

Equation (88) can be integrated, using the method of characteristics (Hoppensteadt, 1974; Anderson and May, 1983), which is essentially a mathematical consequence of the biological fact that, as time t passes, each individual's age advances from a to $a+t$. Equivalently, Laplace transform techniques or Green's functions can be used in Eq. (88). In any event, the solution is

$$X(a, t) = N(0) \exp(-\psi(a, t)).$$ (89)

Here ψ is defined as

$$\psi(a, t) = \int_0^a \mu(s)ds + \int_{a-t}^a c(s)ds + \int_{t-a}^t \lambda(s)ds.$$ (90)

In the integrations in Eq. (90), it is to be understood that $c(s)=0$ for $s<0$, and that $\lambda(s)=\lambda_0$ (the force of infection in the initial, pre-vaccination population, as discussed in Part II) for $s<0$. In the frequently-met special case when a proportion p of hosts are successfully immunized at age b, Eqs. (89) and (90) take the following form: (1) For $a>t$, then if $t>a-b>0$

$$X(a, t) = (1-p)N(0)\phi(a) \exp\left[-\lambda_0(a-t) - \int_0^t \lambda(s)ds \right],$$ (91)

but if $b \geq a$, or $a > t+b$,

$$X(a, t) = N(0)\phi(a) \exp\left[-\lambda_0(a-t) - \int_0^t \lambda(s)ds \right];$$ (92)

(2) for $t>a$, we have for $a>b$

$$X(a, t) = (1-p)N(0)\phi(a) \exp\left[- \int_{t-a}^t \lambda(s)ds \right],$$ (93)

and for $b \geq a$,

$$X(a, t) = N(0)\phi(a) \exp\left[- \int_{t-a}^t \lambda(s)ds \right].$$ (94)

The mortality factor is always contained in the factor $\phi(a)$, defined by Eq. (17).

The above formulae give $X(a, t)$ in terms of known quantities and the time-dependent force of infection, $\lambda(t)$. We can, however, no longer determine the force of infection by the equilibrium methods employed in Part II, but instead must relate $\lambda(t)$ to the total number of infections, $\bar{Y}(t)$, by Eq. (17), $\lambda(t) = \beta \bar{Y}(t)$, or some equivalent relationship. Once this is done, we can integrate Eqs. (3) and (88) over all ages, to get a pair of coupled ordinary differential equations for the time-dependent quantities $\lambda(t)$ and $\bar{X}(t)$ [where $\bar{X}(t) = \int X(a,t)da$ is the total number of susceptibles]. As discussed more fully by Anderson and May (1983), the details of these ordinary differential equations for $\lambda(t)$ and $\bar{X}(t)$ will depend on the exact shape of the survival function, $\mu(a)$. For illustration, I now make the conventional assumption that survival is of type A ($\mu = \text{constant} = 1/L$), and outline the ensuing analysis; more realistic computations are carried out in Anderson and May (1983).

Further, for simplicity, let us assume that a proportion p of each cohort is successfully immunized at birth ($b = 0$), rather than at some finite age. Then, integrating Eq. (2) or (88) over all ages (and using the appropriate boundary conditions in evaluating the integral of the term $\partial X/\partial a$), we get the ordinary differential equation

$$d\bar{X}/dt = -(\lambda(t) + \mu)\bar{X}(t) + \mu(1-p)N .\tag{95}$$

Here N is the total population, related to the boundary condition $N(0)$ by $N(0) = \mu N$ (provided the birth rate equals the death rate). For $\lambda(t)$ an ordinary differential equation is obtained by integrating Eq. (3) over all ages, and putting $\lambda = \beta \bar{Y}$ (with β being constant):

$$d\lambda/dt = \beta \bar{X}(t)\lambda(t) - (v + \mu)\lambda(t) .\tag{96}$$

The notation in these equations can be simplified somewhat by defining $x(t) = \bar{X}(t)/N$ (so that x is the net *fraction* of the host population that is susceptible), and by using an earlier result, Eq. (75), to write $R_0 = N/X^*$ $= \beta N/(v + \mu)$. Then Eqs. (95) and (96) become

$$dx/dt = \mu(1-p) - (\mu + \lambda(t))x(t) ,\tag{97}$$

$$d\lambda/dt = (v + \mu)\lambda(t)[R_0 x(t) - 1] .\tag{98}$$

Equation (98) can indeed be integrated immediately:

$$\ln[\lambda(t)/\lambda(0)] = (v + \mu)\int_0^t [R_0 x(s) - 1]ds .\tag{99}$$

Equation (97), however, cannot be integrated in closed form, by virtue of the awkward nonlinear term λx.

Notice, incidentally, that these equations give the asymptotic ($t \to \infty$) values of $\lambda(\infty)$ and $x(\infty)$ that were found in Part II. By putting the left-hand side equal to zero in Eq. (98) we get $x(\infty) = 1/R_0$. The fact that the fraction susceptible is the same after the vaccination program has come to equilibrium as it was before vaccination (unless, of course, the vaccination coverage was high enough to achieve eradication) was previously noted and discussed. Looking at the integral on the right-hand side of Eq. (99), we see that the kernel thus vanishes both at $t = 0$

and as $t \to \infty$; the dip of $R_0 x$ below unity, on average, at intermediate times is associated with the change to a lower value of λ as $t \to \infty$. This asymptotic value of $\lambda(\infty)$ can be found explicitly by putting the left-hand side of Eq. (97) equal to zero, and is $\lambda(\infty) = \mu[R_0(1-p)-1]$; this result was found earlier, Eq. (41) with $b=0$.

Although Eqs. (97) and (98) can easily be solved numerically, it is instructive to look at the linearized approximation which is obtained by assuming p is small. To do this, we follow the standard procedure of putting $x(t) = x(0) + x'(t)$ and $\lambda(t) = \lambda(0) + y'(t)$, where $x(0) = 1/R_0$ and $\lambda(0) = \mu(R_0 - 1)$ are the initial equilibrium values. Terms of second or higher order in x' and y' are then discarded in a Taylor expansion, to arrive at the linearized differential equations

$$dx'/dt = -\mu p - \mu R_0 x' - y/R_0, \tag{100}$$

$$dy'/dt = \mu(v+\mu)R_0(R_0-1)x'. \tag{101}$$

This gives a standard second order differential equation for $x'(t)$:

$$d^2 x'/dt^2 + (\mu R_0)dx'/dt + \mu(v+\mu)(R_0-1)x' = 0. \tag{102}$$

The initial conditions are $x'(0) = 0$ and $dx'(0)/dt = -\mu p$. The solution is routine and is

$$x'(t) = -(\mu p/\omega)e^{-\alpha t}\sin(\omega t). \tag{103}$$

Here the damping rate α and the frequency ω are defined as

$$\alpha = \tfrac{1}{2}\mu R_0, \tag{104}$$

$$\omega = [\mu(v+\mu)(R_0-1)-\alpha^2]^{1/2}. \tag{105}$$

As discussed previously, the recovery rate v is typically of the order of (1/days), while μ is of order (1/decades), so that $v \gg \mu R_0$ even if $R_0 \sim 10$. This leads to the approximation

$$\omega \simeq [\mu v(R_0-1)]^{1/2}, \tag{106}$$

which is to say $\omega \simeq (v\lambda_0)^{1/2} = (v/A)^{1/2}$, as obtained earlier in Eq. (86). The frequency of the damped oscillations in the number of susceptibles, following implementation of the vaccination program, is exactly the fundamental period studied above. This result, of course, is not surprising: this period T or frequency ω is simply the characteristic period or frequency for the set of dynamical Eqs. (2)–(4). Using the approximation $v \gg \mu R_0$, we may substitute Eq. (103) into (101) and integrate to get

$$y'(t) = -\mu R_0 p[1 - e^{-\alpha t}\cos(\omega t)]. \tag{107}$$

The end result of this linearized study of Eqs. (97) and (98) is that $x(t)$ and $\lambda(t)$ are given (up to terms of order p^2) by

$$x(t) = (1/R_0)[1 - p(2\alpha/\omega)e^{-\alpha t}\sin(\omega t)], \tag{108}$$

$$\lambda(t) = \mu[R_0 - 1 - pR_0(1 - e^{-\alpha t}\cos(\omega t))]. \tag{109}$$

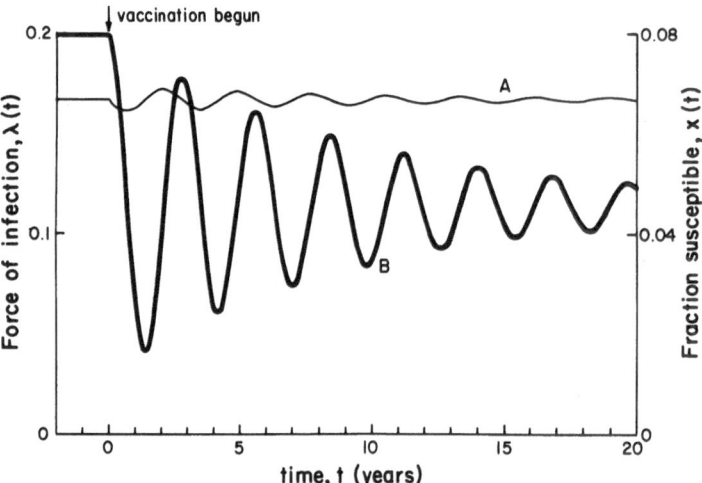

Fig. 6. This figure illustrates the changes in the net fraction of the host population who are susceptible, $x(t)$ (curve A), and in the force of infection, $\lambda(t)$ (curve B), following the institution of a program in which a proportion $p=0.4$ are successfully immunized at birth, beginning in year 0. The figure illustrates the behavior of the linearized Eqs. (108) and (109), with the epidemiological and demographic parameters having values $\lambda_0=0.2\,\text{yr}^{-1}$, $\mu=1/70\,\text{yr}^{-1}$ (whence $R_0=15$), and $v=25\,\text{yr}^{-1}$; it follows from Eqs. (104) and (105) that $\alpha=0.11\,\text{yr}^{-1}$ and $\omega=2.23\,\text{yr}^{-1}$. These numerical values are fairly representative of many childhood infections. The incidence of infection at time t is proportional to $\lambda(t)x(t)$, and so exhibits roughly the same dynamical behavior as $\lambda(t)$

Here the ω of Eq. (106) is $\omega=2\pi/T$, where T is the inter-epidemic period of Eqs. (37) and (86) that has already been discussed ad nauseam. These results are illustrated in Fig. 6 for $p=0.4$ and for the epidemiological parameters specified in the caption, which amount to $\alpha=0.11\,\text{yr}^{-1}$ and $\omega=2.23\,\text{yr}^{-1}$. Notice that perturbations to the fraction susceptible, $x(t)$, are very small even for this relatively large value of p; this is because the disturbance term scales relatively as $p\alpha/\omega$ in Eq. (108), and α/ω is of the order of $(\tau/A)^{1/2}$ which is small (τ is the duration of infection, $\tau=1/v$, and A is average age at infection, $A=1/\lambda$). The amplitude of the oscillations in the force of infection, $\lambda(t)$, as it moves from $\lambda(0)$ to $\lambda(\infty)$ are of significant magnitude, initially scaling simply with p relative to $\lambda(0)$ itself: this shows plainly in Fig. 6. A fully nonlinear and necessarily numerical study of Eqs. (97) and (98) gives results that are indistinguishable from the linearized approximation $x(t)$, and only slightly different from the linearized approximation for $\lambda(t)$ (unless p is large). For such detailed studies, see Anderson and May (1983, 1984).

Once the time dependence of $\lambda(t)$ has been obtained in the general manner just indicated, the full dynamical behavior of the age-specific number of susceptibles, $X(a,t)$, follows from the general Eqs. (89) and (90), or some appropriate particularization such as Eqs. (91)–(94). From this, it is then possible to estimate such quantities as the changing number of cases of CRS under a particular immunization program. An accurate such calculation would multiply the incidence of rubella at age a and time t $(\lambda(t)X(a,t))$ by the pregnancy rate at age a $[m(a)$, taken from tabulated demographic data], and integrate over the appropriate age range (say, 16 to 40 years).

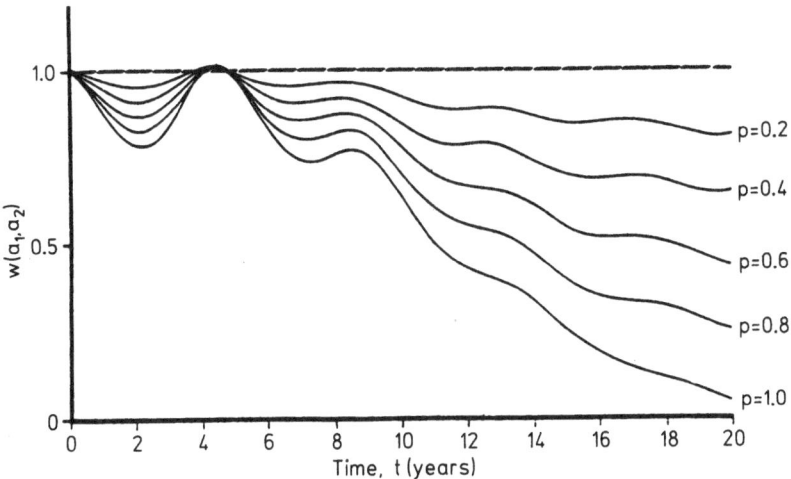

Fig. 7. This figure shows the temporal changes in the number of cases of Congenital Rubella Syndrome (CRS) when a program of successfully immunizing a proportion p of girls, and only girls, at age 12 is begun in year 0. This vaccination strategy roughly corresponds to that employed against rubella in the UK, where p has been around 0.5 until recently (rising to around 0.8 in the past year or so). The number of cases is expressed as a ratio, $w(a_1, a_2)$, to the number of cases at equilibrium in the pre-vaccination population; the ratio is calculated by multiplying the incidence of rubella among females in the age range a_1 to a_2 (specifically, 16 to 40 years) by the age-specific pregnancy rate. The calculations follow the lines sketched in the text (although they incorporate various additional details, such as latent periods), and assume the average age at infection is $A = 9$ years (roughly corresponding to rubella in the USA and UK) and the average life expectancy is $L = 75$ years. For details, see Anderson and May (1983)

The outcome of this procedure is depicted in Fig. 7, which shows numerical results for the incidence of CRS at time t after a "UK" vaccination program (proportion p of girls, and only girls, successfully immunized at age 12 years) is instituted, as a ratio to the incidence of CRS before vaccination is begun; as before, this ratio is labelled $w(a_1, a_2)$. These results, and those in Fig. 9 below, are based on numerical studies of the fully nonlinear set of equations. In the trajectories illustrated in Fig. 7, there is initially a pronounced oscillation, with a period of around 4–5 years (as estimated from the basic inter-epidemic period for rubella earlier in Part III). As time goes on, these oscillations damp out as the system settles to the post-vaccination equilibrium state discussed in Part II. Figure 8 shows data for the annual number of reported cases of CRS in the UK, from 1969 to 1980. The reality is less tidy than the theory, partly because the pre-vaccination state already shows oscillations (as discussed above) and the vaccination program was gradually implemented starting around a "trough." But there is indication of the existence of the predicted oscillation in incidence (with 4–5 year period), which is a consequential feature of the dynamics of the vaccination program that could not easily be guessed on purely intuitive grounds.

Similarly, Fig. 9 shows the relative number of cases of CRS, $w(a_1, a_2)$, as a function of time t after the institution of a "USA" vaccination program (proportion p of all children successfully immunized at age 1 year). The results are displayed in two parts, to avoid confusing the curves for various p-values. For small p there are

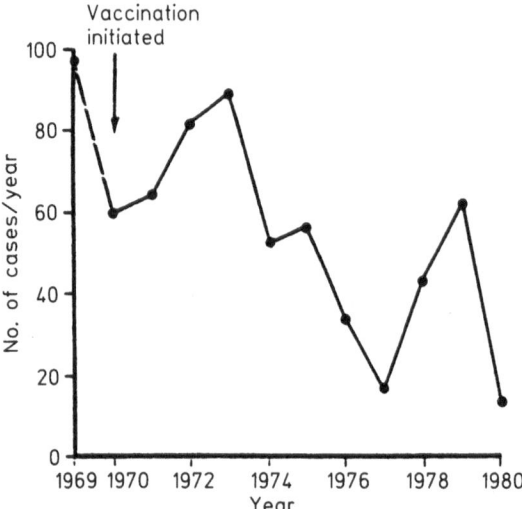

Fig. 8. The number of cases of CRS reported annually in the UK between 1969 and 1980. For each year, the total number of cases is based on diagnoses up to 4 years after the birth of the child (so that the figures for 1979 and 1980 are only provisional). From Anderson and May (1983)

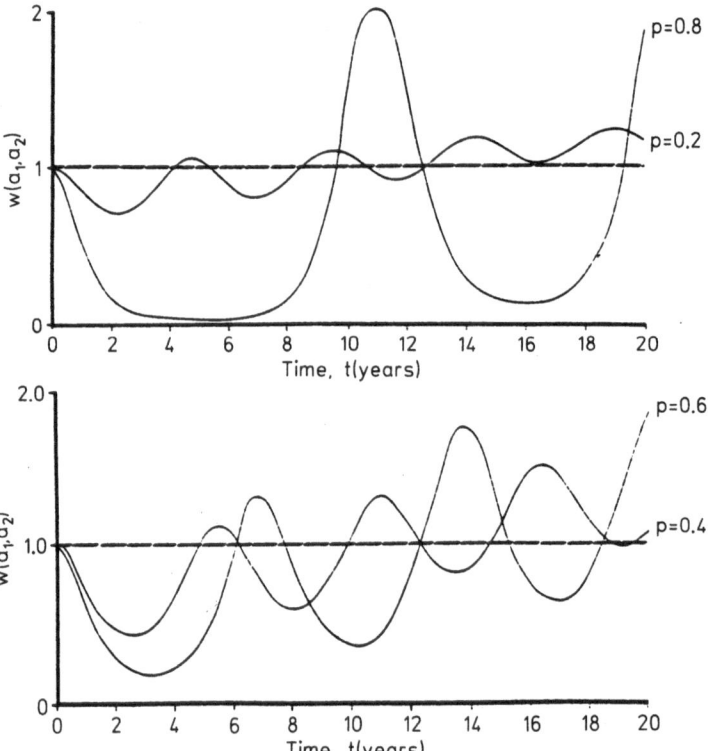

Fig. 9. This figure is similar to Fig. 7, except that it shows the dynamical response to the institution of a vaccination program like that employed against rubella in the USA, with a proportion p of all children being successfully immunized at age 1. The results for $w(a_1, a_2)$ as a function of time t are displayed separately for $p = 0.2$ and 0.8 and for $p = 0.4$ and 0.6, to avoid confusing the details. The epidemiological and demographic parameters are those for rubella in developed countries, as used in Fig. 7. The actual coverage, p, in the USA has been around 0.8 or better. For details and further discussion, see Anderson and May (1983)

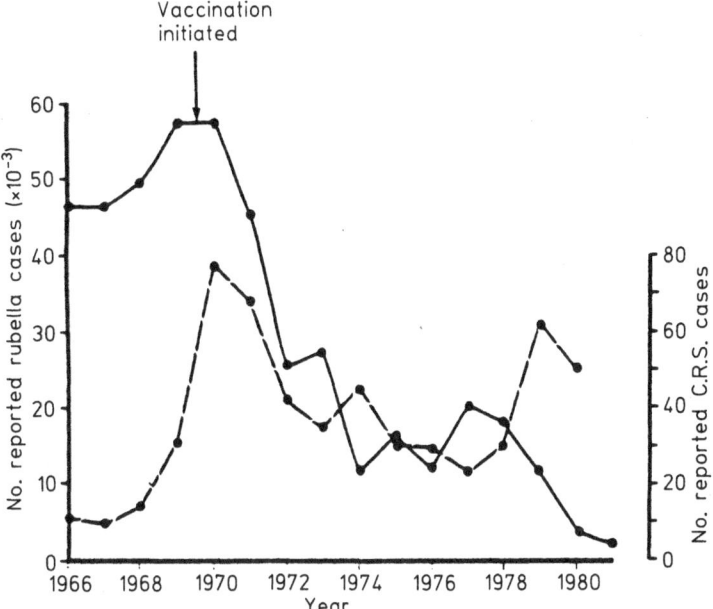

Fig. 10. The solid line shows the reported number of cases of rubella in the USA between 1966 and 1981, and the dashed line shows the reported number of cases of CRS. The features of these curves are discussed in the text; for a more full discussion of the comparison between Fig. 10 and Fig. 9, see Anderson and May (1983)

weakly damped oscillations with the by-now familiar period of around 4–5 years; at large p the nonlinear effects become pronounced, both lengthening the period [essentially by the factor $1/(1-p)^{1/2}$] and modifying the amplitude. Figure 10 shows data for the total number of cases of rubella and of CRS in the USA from 1966 to 1981. It can be seen that, following the institution of vaccination with a coverage of p about 0.8, beginning around 1969, the incidence of CRS has shown just the roughly 10 years oscillation predicted by Fig. 9; the amplitude, however, is not as large as predicted (probably because of the screening and pregnancy-termination policy that has been operating in conjunction with vaccination).

Figures 7–10 only hint at the kinds of theoretical analyses and comparisons with public health data that are possible. The agreement between theory and data is qualitative rather than quantitative, but is nevertheless encouraging in view of the extreme simplicity of the mathematical model. For a more full account of the above work, and suggestions for further studies of various kinds, see Anderson and May (1983).

Part IV. Regulation of Host Populations by Microparasitic Infections

In this final Part IV, I remove some of the constraints imposed at the start of Part II, to consider some of the complications that arise when there is disease-induced mortality ($\alpha \neq 0$) and when the disease-free death rate ($\mu(a)$) is not equal to the birth rate. In particular, it is interesting to examine the circumstances under which microparasitic infections may be capable of regulating the magnitude of the host population. What follows is a brief outline; much of this work has been set forth in greater detail elsewhere (Anderson and May, 1979, 1981, 1984; May and Anderson, 1979).

Regulation by Microparasitic Infections

Consider first a population where we abandon detailed description of age structure, and deal simply with the total number of susceptibles, infectives, and immunes, $X(t)$, $Y(t)$, and $Z(t)$, respectively. If the per capita birth rate v and disease-free death rate μ are both taken as density-independent constants, we may describe the dynamical behavior of the system by the set of differential equations (Anderson and May, 1979)

$$dX/dt = vN - \mu X - \beta XY + \gamma Z, \tag{110}$$

$$dY/dt = \beta XY - (v + \alpha + \mu)Y, \tag{111}$$

$$dZ/dt = vY - (\gamma + \mu)Z. \tag{112}$$

Here v is the recovery rate and α the disease-induced death rate (both as defined previously), and γ is the rate of loss of immunity ($\gamma = 0$ if immunity is lifelong; $\gamma \to \infty$ if there is no acquired immunity, as appears to be the case for most microparasitic infections of invertebrates). The force of infection has been represented as βY, as discussed earlier, Eq. (7). The total population is now itself a dynamical variable,

$$N(t) = X(t) + Y(t) + Z(t), \tag{113}$$

and obeys the equation

$$dN/dt = rN - \alpha Y. \tag{114}$$

Here $r = v - \mu$ is the per capita population growth rate, in the absence of the infection.

The *equilibrium* solutions are obtained by setting all time dependences equal to zero. Equation (111) gives

$$X^* = (v + \alpha + \mu)/\beta = N_T. \tag{115}$$

Here N_T corresponds to the threshold host population, below which the infection cannot be maintained (Bailey, 1975; Anderson and May, 1979). Equations (114) and (112) give

$$Y^* = (r/\alpha)N^*, \tag{116}$$

$$Z^*(v/(\gamma + \mu))Y^*, \tag{117}$$

respectively. Substituting Eqs. (115)–(117) into the equilibrium version of Eq. (110), we get the magnitude of the total host population at equilibrium:

$$N^* = \frac{\alpha N_T}{[\alpha - r\{1 + v/(\gamma + \mu)\}]}. \tag{118}$$

Clearly this equilibrium solution is only possible if the denominator in Eq. (118) is positive, which gives the criterion for the microparasite to be capable of regulating its host population:

$$\alpha > r[1 + (v/(\gamma + \mu))]. \tag{119}$$

In the special case where there is no acquired immunity, which pertains to most invertebrate hosts, the criterion (119) takes the simpler form

$$\alpha > r. \tag{120}$$

Provided the above equilibrium solutions exist, they can be shown to be stable, with perturbations decaying back to the equilibrium state.

If the microparasite is not sufficiently virulent to satisfy Eq. (119) – that is, if α is insufficiently large compared with the disease-free exponential growth rate r – the system of Eqs. (110)–(114) eventually settles to a state in which the total population grows exponentially at a rate ϱ given by

$$\varrho = [B^2 - (\mu + \gamma)(\alpha - r) + rv]^{1/2} - B, \tag{121}$$

with $B = \frac{1}{2}(\alpha + v + \gamma - v)$. This population growth rate is necessarily less than the disease-free one, $\varrho < r$. Asymptotically, the exponentially growing total population contains a roughly constant number of susceptibles, $X \simeq N_T$ with essentially all individuals being either infected or immune. The asymptotic prevalence of infection is

$$Y(t)/N(t) \rightarrow (r - \varrho)/\alpha. \tag{122}$$

All these results (the proofs of which are left in this chapter as an exercise to the reader) are derived and discussed more fully in Anderson and May (1979), and various refinements are explored in the invertebrate case when $\gamma \rightarrow \infty$ in Anderson and May (1981). Some of the biological implications are pursued in various chapters of the report of a recent Dahlem Conference (Anderson and May, 1982c).

In particular, it is informative to ask what fraction of all deaths must be caused by a disease, if that disease is indeed responsible for the density-dependent regulation of the population. The answer, within the framework discussed above, is that

$$\frac{\text{deaths from disease}}{\text{total deaths}} = \frac{v - \mu}{v}. \tag{123}$$

This is proved by observing that the ratio of disease-induced deaths (αY^*) to all deaths $(\mu N^* + \alpha Y^* = \nu N^*)$ at equilibrium is $\alpha Y^*/\nu N^*$; in conjunction with Eq. (116) this gives Eq. (123). Of course, rarely – if ever – will it happen that the birth rate and all mortality factors apart from the disease in question are density-independent (as assumed here), but the result is nevertheless interesting. For a population in which births and disease-free deaths are roughly equal, $\nu \simeq \mu$ (as for many large mammals), the population could have disease as its essential regulatory mechanism, even though few observed deaths are caused by the disease. Conversely, for invertebrates and other populations where usually $\nu \gg \mu$, essentially all deaths must be from the disease if it is indeed regulating the population.

Applications to Actual Populations

The above Eqs. (110)–(114) have – with appropriate modifications to account for particular features of specific systems – been applied to study several field or laboratory host-microparasite interactions for which data are available. Each such application has been discussed in detail in relatively accessible journals, and so all that is given here is a trail guide.

(i) Ectromelia or Pasteurella muris in Laboratory Populations of Mice

In the 1930s, Greenwood et al. conducted a classic series of long term experiments on the dynamics of laboratory populations of mice infected with the virus ectromelia or with the bacterium *P. muris*. The above models give an accurate account of the empirical findings (Anderson and May, 1979; May, 1983).

(ii) Invertebrate Host Populations

The interactions of invertebrate populations $(\gamma \rightarrow \infty)$ with microparasitic infections have been explored, taking account of many realistic complications, by Anderson and May (1981). Of special interest is the possibility that many forest insects may be regulated in 3–20 year cycles by baculoviruses or microsporidian protozoan infections (Anderson and May, 1981; May, 1983). This work has potential applications in setting preliminary criteria for the choosing of microparasites that may help control forest insect pests.

(iii) Rabies in Fox Populations in Europe

Modifying Eqs. (110)–(114) to take account of the kind of density-dependent effects associated with territoriality, Anderson et al. (1981) have shown that most of the main features associated with the outbreak of rabies in fox populations in Europe, beginning in Poland in 1939, can be captured. Specifically, such simple models – which concatenate a lot of behavioral and environmental detail in a few phenomenological parameters such as the transmission parameter β – appear to explain the observed prevalence levels of rabies in fox populations, and the 3–5 year cycles in previously steady fox populations following the establishment of endemic rabies.

(iv) Human History

More generally, it seems likely that the interplay between human population densities and the mortality associated with specific microparasitic infections (that can be maintained once threshold densities are exceeded) has played a significant role in human history, on both long and short time scales. These questions are discussed generally by, for example, McNeil (1976), McKeown (1976), and various authors in Anderson and May (1982c); a more specific discussion in the light of the above equations is by Anderson and May (1979; also May, 1983).

Microparasitic Infections in Growing Human Populations

In Parts II and III, I indicated how to go about estimating epidemiological parameters such as R_0, or assessing the dynamical consequences of a vaccination program, within a host population whose total size was unchanging. This is not a realistic assumption for many human populations, especially in developing countries. What follows is a terse account of some of the complications that ensue when birth rates and death rates are not assumed equal.

For simplicity, and at the expense of realism, we take the mortality rate (μ) to be constant, and likewise assume the per capita birth rate is an age-independent constant (v). Unlike the earlier analysis, however, we no longer take $v = \mu$, but rather assume $r = v - \mu > 0$. Disease-induced deaths are still ignored, $\alpha = 0$.

The partial differential equations (2)–(4) remain unaffected, as such; the key difference in the analysis is that the boundary conditions for the total number of hosts and of susceptibles, of age zero, respectively, are now $N(0, t) = X(0, t) = B(t)$. Here $B(t) = v\bar{N}(t)$, with $\bar{N}(t)$ the time dependent total host population, $\bar{N}(t) = \int N(a, t)da$. The resulting formula for the number of hosts of age a at time t can be found by methods set out in any ecology or demography text:

$$N(a, t) = N(0, 0)e^{rt - va}. \tag{124}$$

Here $N(0, 0)$ is the number of hosts of age zero at time $t = 0$. The total host population here increases exponentially at the rate r (remember, there is no mortality associated with the infection):

$$\bar{N}(t) = (N(0, 0)/v)e^{rt}. \tag{125}$$

The number of susceptibles at age a and time t can be found, along the lines sketched in Part III, and is

$$X(a, t) = N(0, 0) \exp\left[rt - va - \int_{t-a}^{t} \lambda(s)ds \right], \tag{126}$$

provided $t > a$. Integrating over all ages, we find the total number of susceptibles:

$$\bar{X}(t) = v\bar{N}(t) \int_0^\infty \exp\left[-va - \int_{t-a}^t \lambda(s)ds \right] da.$$ (127)

We could now solve to find $\bar{X}(t)$ and $\lambda(t) = \beta \bar{Y}(t)$ along the lines laid down at the end of Part III. An approximate analysis, however, provides some illumination. For human populations (and, indeed, for most other populations of large mammals), the per capita rate of population growth, $r = v - \mu$, is small relative to other rate constants in these host-microparasite systems. The integral, $\int \lambda(s)ds$, in the exponent in Eq. (127) may thus be approximated as $a\lambda(t)$, which leads to the result

$$\bar{X}(t) \simeq \bar{N}(t) \left[v/(v + \lambda(t)) \right].$$ (128)

The basic reproductive rate, R_0, is time dependent in this system (essentially because the host density continues to increase, thus making transmission easier and easier as time goes by). The value of R_0 can, however, still be calculated from Eq. (26), which now reads $R_0 = \bar{N}(t)/\bar{X}(t)$, or

$$R_0 \simeq 1 + \left[\lambda(t)/v \right].$$ (129)

As is intuitively understandable, it can also be shown, within this approximation, that the average age at infection is

$$A(t) = 1/\lambda(t).$$ (130)

Thus Eq. (129) can be reexpressed as

$$R_0(t) \simeq 1 + \left[L'/A(t) \right].$$ (131)

Equation (131) differs from the previous Eq. (33) in two respects. First, the average age at infection now varies over time (tending to decrease as the population grows). Second, the life expectance L of Eq. (33) is replaced by $L' = 1/v$, which is the reciprocal of the birth rate. Previously, when the host population was constant with $v = \mu$, it was of no consequence whether we wrote $L = 1/\mu$ or $L' = 1/v$, as the two expressions were identical; it is clear, however, that the appropriate quantity is the reciprocal of the birth rate (rather than the death rate) in a growing population. This second difference between Eq. (131) and Eq. (33) is of practical importance when one attempts to estimate values of R_0 for microparasitic infections in developing countries.

The above analysis of microparasitic infections in exponentially growing human populations, and its epidemiological implications, are pursued more fully elsewhere (Anderson and May, 1984).

Acknowledgements

This chapter is largely based on work done jointly with R. M. Anderson, and I am indebted to him and to many other people for helpful conversations and advice. The research was supported, in part, by NSF grant DEB81-02783.

References

Anderson, R.M., Jackson, H., May, R.M., Smith, T. (1981). The population dynamics of fox rabies in Europe. Nature *289*, 765–771

Anderson, R.M., May, R.M. (1979). Population biology of infectious diseases: Part I. Nature *280*, 361–367

Anderson, R.M., May, R.M. (1981). The population dynamics of microparasites and their invertebrate hosts. Phil. Trans. Roy. Soc. B *291*, 451–524

Anderson, R.M., May, R.M. (1982a). The population dynamics and control of human helminth infections. Nature *297*, 557–563

Anderson, R.M., May, R.M. (1982b). Directly transmitted infectious diseases: control by vaccination. Science *215*, 1053–1060

Anderson, R.M., May, R.M. (eds.) (1982c). Population Biology of Infectious Diseases. Springer-Verlag: Berlin and New York

Anderson, R.M. May, R.M. (1983). Vaccination against rubella and measles: quantitative investigations of different policies. J. Hyg. *90*, 259–325

Anderson, R.M., May, R.M. (1986). The Dynamics of Human Host-Parasite Systems. Princeton University Press: Princeton

Aron, J.L., Schwartz, I.B. (1984). Seasonality and period doubling bifurcations in an epidemic model. J. Theor. Biol. *110*, 665–679

Bailey, N.J.T. (1975). The Mathematical Theory of Infectious Diseases (2nd edn.). Macmillan: New York

Bartlett, M.S. (1957). Measles periodicity and community size. J. Roy. Stat. Soc. Ser. A *120*, 48–70

Bartlett, M.S. (1960). Stochastic Population Models. Methuen and Co.: London

Becker, N., Angulo, J. (1981). On estimating the contagiousness of a disease transmitted from person to person. Math. Biosci. *54*, 137–154

Cooke, K.L. (1982). Models for epidemic infections with asymptotic cases, I: one group. Math. Modelling *3*, 1–15

Dietz, K. (1975). Transmission and control of arbovirus diseases. In: Epidemiology (eds. D. Ludwig and K. L. Cooke), pp. 104–121. Society for Industrial and Applied Mathematics; Philadelphia

Dietz, K. (1976). The incidence of infectious diseases under the influence of seasonal fluctuations. In: Mathematical Models in Medicine; Lecture Notes in Biomathematics, II (eds. J. Berger, W. Buhlen, R. Regges, and P. Tautu), pp. 1–15. Springer-Verlag: Berlin

Dietz, K. (1981). The evaluation of rubella vaccination strategies. In: The Mathematical Theory of the Dynamics of Biological Populations, II (eds. R. W. Hiorns and D. Cooke), pp. 81–97. Academic Press: London

Fine, P.E.M. (1975). Vectors and vertical transmission: an epidemiological perspective. Ann. N.Y. Acad. Sci. *266*, 173–195

Fine, P.E.M., Clarkson, J.A. (1982). Measles in England and Wales, II. The impact of the measles vaccination programme on the distribution of immunity in the population. Int. J. Epidemiol. *11*, 15–25

Grossman, Z. (1980). Oscillatory phenomena in a model of infectious diseases. Theor. Pop. Biol. *18*, 204–243

Grossman, Z. Gumowski, I., Dietz, K. (1977). The incidence of infectious diseases under the influence of seasonal fluctuations – analytic approach. In: Nonlinear Systems and Applications to Life Sciences, pp. 525–546. Academic Press: New York

Hethcote, H.W. (1982). Measles and rubella in the United States. Am. J. Epidemiol. *117*, 2–13
Hethcote, H.W., Tudor, D.W. (1980). Integral equation models for endemic infectious diseases. J. Math. Biol. *9*, 37–47
Hethcote, H.W., Stech, H.W., Van den Driessche, P. (1981). Nonlinear oscillations in epidemic models. SIAM J. Appl. Math. *40*, 1–9
Hethcote, H.W., Yorke, J.A., Nold, A. (1982). Gonorrhea modeling: a comparison of control methods. Math. Biosci. *58*, 93–109
Hoppensteadt, F.C. (1974). An age dependent epidemic model. J. Franklin Inst. *297*, 325–333
Hoppensteadt, F.C. (1975). Mathematical Theories of Populations: Demographics, Genetics, and Epidemics. SIAM (Regional Conference Series and Applied Mathematics *20*): Philadelphia
Kemper, J.T. (1980). On the identification of superspreaders for infectious disease. Math. Biosci. *48*, 111–128
Knox, E.G. (1980). Strategy for rubella vaccination. Int. J. Epidemiology *9*, 13–23
London, W.P., Yorke, J.A. (1973). Recurrent outbreaks of measles, chickenpox, and mumps, I. Seasonal variation in contact rates. Amer. J. Epidemiol. *98*, 453–468
Macdonald, G. (1952). The analysis of equilibrium in malaria. Trop. Dis. Bull. *49*, 813–829
McKeown, T. (1976). The Modern Rise of Population. Edward Arnold: London
McNeill, W.H. (1976). Plagues and Peoples. Doubleday: New York
May, R.M. (1974). Stability and Complexity in Model Ecosystems (second edition). Princeton University Press: Princeton
May, R.M. (1976). Simple mathematical models with very complicated dynamics. Nature *261*, 459–467
May, R.M. (1981). The transmission and control of gonorrhea. Nature *291*, 376–377
May, R.M. (1983). Parasitic infections as regulators of animal populations. Amer. Sci. *71*, 36–45
May, R.M., Anderson, R.M. (1979). Population biology of infectious diseases: II. Nature *280*, 455–461
Nold, A. (1980). Heterogeneity in disease-transmission modeling. Math. Biosci. *52*, 227–240
Smith, C.E.G. (1970). Prospects for the control of infectious disease. Proc. Roy. Soc. Med. *63*, 1181–1190
Smith, H.L. (1983). Multiple stable subharmonics for a periodic epidemic model. Preprint
Soper, H.E. (1929). Interpretation of periodicity in disease prevalence. J. Roy. Stat. Soc. *92*, 34–73
Travis, C.C., Lenhart, S.M. (1985). Smallpox eradication: why was it successful? Submitted to Math. Bio. Science
Waltman, P. (1974). Deterministic Threshold Models in the Theory of Epidemics. (Lecture Notes in Biomathematics.) Springer-Verlag: New York
Yorke, J.A., London, W.P. (1973). Recurrent outbreaks of measles, chickenpox, and mumps, II. Systematic differences in contact rates and stochastic effects. Amer. J. Epidemiol. *98*, 469–482
Yorke, J.A., Hethcote, H.W., Nold, A. (1978). Dynamics and control of the transmission of gonorrhea. J. Sex. Trans. Dis. *5*, 51–56
Yorke, J.A., Nathanson, N., Pianigiani, G., Martin, J. (1979). Seasonality and the requirements for perpetuation and eradication of viruses in populations. Am. J. Epidem. *109*, 103–123

Author Index

Subject Index

Biomathematics

Managing Editor: S.A.Levin

Editorial Board: M.Arbib, H.J.Bremermann, J.Cowan, W.M.Hirsch, S.Karlin, J.Keller, K.Krickeberg, R.C.Lewontin, R.M.May, J.D.Murray, A.Perelson, T.Poggio, L.A.Segel

Springer-Verlag
Berlin Heidelberg New York Tokyo

Springer-Verlag
Berlin Heidelberg New York Tokyo